Improving soil health

It is widely recognised that agriculture is a significant contributor to global warming and climate change. Agriculture needs to reduce its environmental impact and adapt to current climate change whilst still feeding a growing population, i.e. become more 'climate-smart'. Burleigh Dodds Science Publishing is playing its part in achieving this by bringing together key research on making the production of the world's most important crops and livestock products more sustainable.

Based on extensive research, our publications specifically target the challenge of climate-smart agriculture. In this way we are using 'smart publishing' to help achieve climate-smart agriculture.

Burleigh Dodds Science Publishing is an independent and innovative publisher delivering high quality customer-focused agricultural science content in both print and online formats for the academic and research communities. Our aim is to build a foundation of knowledge on which researchers can build to meet the challenge of climate-smart agriculture.

For more information about Burleigh Dodds Science Publishing simply call us on +44 (0) 1223 839365, email info@bdspublishing.com or alternatively please visit our website at www.bdspublishing.com.

Related titles:

Managing soil health for sustainable agriculture Volume 1: Fundamentals
Print (ISBN 978-1-78676-188-0) Online (ISBN 978-1-78676-191-0; 978-1-78676-190-3)

Managing soil health for sustainable agriculture Volume 2: Monitoring and management
Print (ISBN 978-1-78676-192-7) Online (ISBN 978-1-78676-195-8; 978-1-78676-194-1)

Advances in Conservation Agriculture Volume 1: Systems and Science
Print (ISBN 978-1-78676-264-1) Online (ISBN 978-1-78676-266-5; 978-1-78676-267-2)

Advances in Conservation Agriculture Volume 2: Practice and Benefits
Print (ISBN 978-1-78676-268-9) Online (ISBN 978-1-78676-270-2; 978-1-78676-271-9)

Achieving sustainable crop nutrition
Print (ISBN 978-1-78676-312-9) Online (ISBN 978-1-78676-314-3; 978-1-78676-315-0)

Chapters are available individually from our online bookshop:
https://shop.bdspublishing.com

climate•SMART•publishing
IN AGRICULTURAL SCIENCE

BURLEIGH DODDS SERIES IN AGRICULTURAL SCIENCE
NUMBER 109

Improving soil health

Edited by Professor William R. Horwath, University of California-Davis, USA

burleigh dodds
SCIENCE PUBLISHING

Published by Burleigh Dodds Science Publishing Limited
82 High Street, Sawston, Cambridge CB22 3HJ, UK
www.bdspublishing.com

Burleigh Dodds Science Publishing, 1518 Walnut Street, Suite 900, Philadelphia, PA 19102-3406, USA

First published 2023 by Burleigh Dodds Science Publishing Limited

Library of Congress Control Number: 2022934578

British Library Cataloguing in Publication Data
A catalogue record for this book is available from the British Library

ISBN 978-1-78676-670-0 (Print)
ISBN 978-1-78676-673-1 (PDF)
ISBN 978-1-78676-672-4 (ePub)
ISSN 2059-6936 (print)
ISSN 2059-6944 (online)

DOI 10.19103/AS.2021.0094

Typeset by Deanta Global Publishing Services, Dublin, Ireland

Contents

Series list

Title	Series number
Achieving sustainable cultivation of maize - Vol 1 From improved varieties to local applications Edited by: Dr Dave Watson, CGIAR Maize Research Program Manager, CIMMYT, Mexico	001
Achieving sustainable cultivation of maize - Vol 2 Cultivation techniques, pest and disease control Edited by: Dr Dave Watson, CGIAR Maize Research Program Manager, CIMMYT, Mexico	002
Achieving sustainable cultivation of rice - Vol 1 Breeding for higher yield and quality Edited by: Prof. Takuji Sasaki, Tokyo University of Agriculture, Japan	003
Achieving sustainable cultivation of rice - Vol 2 Cultivation, pest and disease management Edited by: Prof. Takuji Sasaki, Tokyo University of Agriculture, Japan	004
Achieving sustainable cultivation of wheat - Vol 1 Breeding, quality traits, pests and diseases Edited by: Prof. Peter Langridge, The University of Adelaide, Australia	005
Achieving sustainable cultivation of wheat - Vol 2 Cultivation techniques Edited by: Prof. Peter Langridge, The University of Adelaide, Australia	006
Achieving sustainable cultivation of tomatoes Edited by: Dr Autar Mattoo, USDA-ARS, USA and Prof. Avtar Handa, Purdue University, USA	007
Achieving sustainable production of milk - Vol 1 Milk composition, genetics and breeding Edited by: Dr Nico van Belzen, International Dairy Federation (IDF), Belgium	008
Achieving sustainable production of milk - Vol 2 Safety, quality and sustainability Edited by: Dr Nico van Belzen, International Dairy Federation (IDF), Belgium	009
Achieving sustainable production of milk - Vol 3 Dairy herd management and welfare Edited by: Prof. John Webster, University of Bristol, UK	010
Ensuring safety and quality in the production of beef - Vol 1 Safety Edited by: Prof. Gary Acuff, Texas A&M University, USA and Prof. James Dickson, Iowa State University, USA	011
Ensuring safety and quality in the production of beef - Vol 2 Quality Edited by: Prof. Michael Dikeman, Kansas State University, USA	012
Achieving sustainable production of poultry meat - Vol 1 Safety, quality and sustainability Edited by: Prof. Steven C. Ricke, University of Arkansas, USA	013
Achieving sustainable production of poultry meat - Vol 2 Breeding and nutrition Edited by: Prof. Todd Applegate, University of Georgia, USA	014

Acknowledgements

We wish to acknowledge the following for their help in reviewing particular chapters:

- Chapter 5: Dr Diogenes Antille, Senior Research Scientist (Soil Physics)/ Team Leader (Resilient Soils) - CSIRO Agriculture and Food, Australia
- Chapter 6: Dr Randal Reeder, formerly Ohio State University, USA
- Chapter 8: Professor Stephen H. Anderson, University of Missouri, USA
- Chapter 9: Dr Mark Pawlett, Cranfield University, UK
- Chapter 10: Professor Ulrich Köpke, University of Bonn, Germany and Dr Melissa Wilson, University of Minnesota, USA

Chapter 1

Soil health: towards a sustainable world

William R. Horwath, University of California-Davis, USA

1 Introduction

The words soil and health were first used separately in publications in the early 1900s and predate the use of other terms including soil fertility and soil quality according to the Web of Science (Fig. 1). The terms soil and health were used separately to describe factors related to soil impacts on human and animal health up until about 1950. Many of the publications referred to pathogens, their effects on dental health, and their relationship to human and animal nutrition. Wrench (1939), a physician, referred to the health of soil and its relationship to human health. The relationship between soil fertility and human health from a dental perspective was described by Albrecht (1945). In a book entitled *Soil and Health: A Study of Organic Agriculture*, Howard (1947) described how cropping system management affects soil and human health. Many articles up until this time related soil to human and animal health, particularly soils as harbors for pathogens.

Soil quality as a single term was first used by Schultz (1925), describing vegetation effects on the quality of soil. All terms including soil health, quality, and fertility began to be used more extensively following the green revolution in the 1960s. Prior to the green revolution, soil fertility was the most published term used. By the 1930s, aspects of soil biology began to be recognized as an important link between soil and plant and animal health. Both soil quality and health became to be used frequently over time and rivaled the use of the term

http://dx.doi.org/10.19103/AS.2021.0094.21

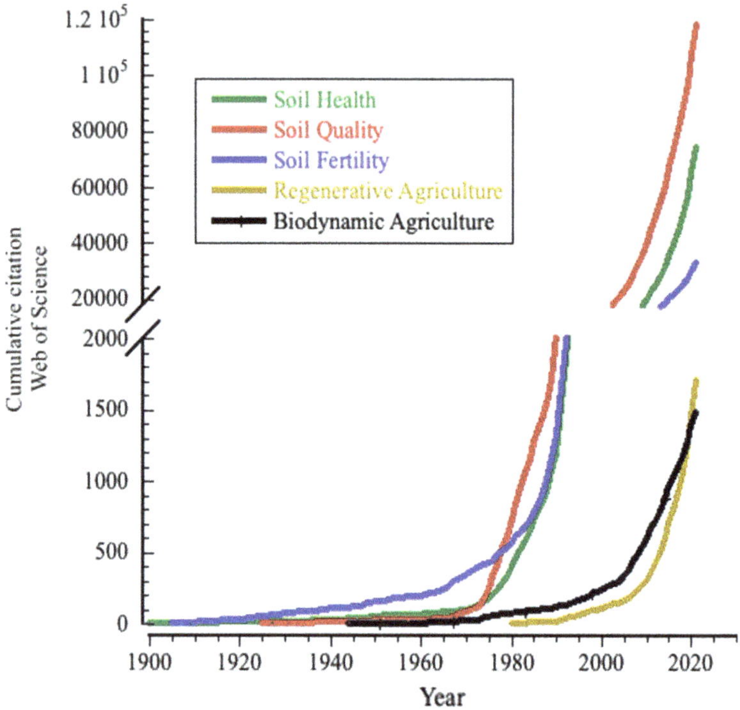

Figure 1 Citations from the Web of Science that cite the terms soil health, soil quality, soil fertility, regenerative agriculture and biodynamic agriculture from 1900 to 2021.

soil fertility. Within the last decade, the terms soil health and soil quality are the most frequently used terms with a dramatic drop in the use of the term soil fertility (Fig. 1).

One of the main reasons that soil quality and soil health began to be used extensively in the 1990s is the publishing of a report by the Board on Agriculture entitled 'Alternative Agriculture' by the National Research Council (1989) in the United States. Ever since 1990, these terms have dominated scientific literature and have become much more popular than the term soil fertility. The debate as to whether soil health and soil quality are synonymous terms continues today. More recently, especially during the last decade, the term soil health is becoming the most widely used term by scientists, practitioners, government officials, and nongovernment organizations. Soil health will likely become the dominant term within a few years according to the citation trends in Fig. 1. Soil health has been described as the capacity of the soil to function as a vital living system, within an ecosystem and land use boundaries, sustain plant and animal productivity, maintain or enhance water and air quality, and promote plant and animal health (Doran and Zeiss, 2000).

In this book, the authors discuss and present many aspects of soil health that include the influence of soil biology, soil amendments such as compost, biosolids, and animal manure, tillage, cover crops, and the effects of tillage and soil disturbance. The book starts with a discussion of definitions and underlying concepts by Wander (Chapter 2). The influence of soil biological processes on the soil ecosystem and its fundamental relationship to soil health is discussed by O'Kelly and Myrold in Chapter 3. The role of mycorrhiza as a component of soil health is presented by Adnan et al. in Chapter 4 in relation to improving soil health and ensuring an appropriate supply of soil nutrients to produce food, fiber, and biofuels for the growing population across the whole world.

Other chapters review the impact of particular cultivation practices on soil health. The influence of agricultural traffic and its impact on soil compaction outcomes in relation to soil health is presented by Misiewicz et al. in Chapter 5. The role of tillage and soil disturbance on soil health outcomes and cropping system productivity is addressed by Hamm and Manter in Chapter 6. In Chapter 7, Snapp et al. discuss the potential for positive soil health benefits through cover cropping, and the relationship to the complexity of real-world agricultural systems, and variable soil responses. The role of crop resides and management on the retention or removal of carbon in soil carbon sequestration and its influence on soil health is discussed by Gollany et al. in Chapter 8.

The final part of the book reviews the effects of organic amendments on soil health. The effect of compost application as a technology to stabilize organic waste and provide a source of nutrients and other soil improvements is discussed by Lazcano et al. in Chapter 9. The influence of the application of livestock and poultry manure on the ability of soil to deliver a range of ecosystem services that sustain plants, animals, people, and the environment is discussed by Tubeileh and Goss in Chapter 10. Nicholson reviews the evidence for both the beneficial and potentially harmful effects of applying biosolids on soil health and highlights areas where further research is still required in Chapter 11. Chapter 12 (Salomon et al.) discusses the role of arbuscular mycorrhizal fungi whilst Chapter 13 (Zampieri et al.) reviews the use of plant growth-promoting bacteria or rhizobacteria. Finally, in Chapter 14, Hayes et al. discuss the role of lime as an important soil pH controlling amendment in intensively managed cropping systems.

2 Early research on soil health

Promotion of soil health is increasingly popular in the research, practitioner, and regulatory community. Soils perform essential functions of regulating water, sustaining life, attenuating pollutants, cycling nutrients, and providing physical stability for crops. The leap from a hunter gatherer existence to organized societies was made possible through agricultural innovation and intensification. The intrinsic link of successful agricultural enterprises to soils was no doubt a key consideration for ancient peoples, but they sometimes lacked the knowledge to sustain the soil resource from uninformed mismanagement and other factors such as the lack of understanding of soil productivity and climate variability and change. The notion that soils provide water and nutrients to sustain crop production was mentioned in early writings, particularly by the Greeks and Romans. However, these writings were more observational and it's not clear whether a science behind nutrient management was being considered. The science of soil management began to flourish following the Middle Ages. However, the understanding that soil was more than a platform to support crops was only fully realized within the last 250 years.

Prior to 1900, important contributions were made describing the link of soil to plant performance. The identification of soil fertility indicators was discussed as a quantitative index to evaluate sustainability of agriculture. Thaer (1856), at the beginning of the nineteenth century, described the contributions of Palissy (1580) and Liebig (1840) and used them in his studies. Thaer developed a complex soil fertility scale based on soil properties, plant nutrient requirements, and components of cropping systems such as crop rotation. Thaer included economic outcomes and related them to intrinsic factors of soil fertility. Thaer can be recognized for developing the precursor of the soil quality index, an assessment tool widely used for soil quality evaluation today (Karlen et al., 2008). Thaer's legacy was tarnished when his 'humus theory' (plants derived all nutrients including carbon, hydrogen, oxygen, and nutrients directly from humus) was disproven by Sprengel (1826) and Liebig (1840) who both promoted the concept of soil mineral nutrition of plants. Liebig's law of the minimum (when an essential plant nutrient is deficient, plant growth will be poor even when all other essential nutrients are abundant) remains one of the most fundamental principles of modern-day nutrient management.

The evolution of the concepts of soil quality and soil health as a means for assessment is a recent occurrence. Alexander (1971) suggested that criteria for developing soil quality should be developed, particularly for agriculture's role in environmental improvement. Soil quality as a concept for soil resource assessment and land use planning was introduced by Warkentin and Fletcher (1977). The broadening of soil management principles from concentrating on preserving soil through erosion control to sustainable agriculture and environmental stewardship began to pick up momentum in the late 1970s to 1980s (Karlen et al., 2003). Larsen and Pierce (1991) presented a framework for the quantitative assessment of soil quality as affected by management. With the publication of methods for assessment of soil quality (Doran and Jones, 1996), the science of soil health became a mainstream research agenda for funders and researchers. The term soil health is now widely used when discussing soil quality and or fertility to understand factors related to soil productivity.

The concept of soil health involves ensuring soils perform the above essential functions to their maximum capacity. The National Resource Conservation Service (U.S. Department of Agriculture) defines the main principles of soil health management as follows:

1 maximize plant roots;
2 minimize disturbance;
3 maximize soil cover; and
4 maximize biodiversity.

Figure 2 shows how these management principles affect soil properties to promote soil health. By definition, a healthy soil will have a positive influence on human and animal health through nutritious food and feed, clean water and air through filtering and buffering of pollutants, benefiting pollinators and habitat in general and, finally, mitigating climate change by reducing GHG emissions and through soil carbon sequestration (SOC). Whilst some of these aspects of soil health have been considered for millennia, it has only been recently understood that soil health is optimized if all these management principles are consistently practiced together.

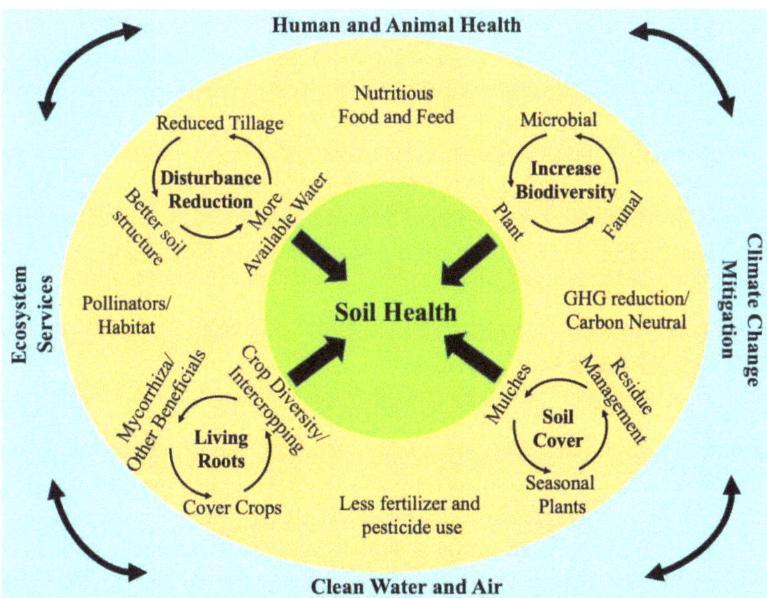

Figure 2 The main principles to manage soil for health according to the National Resource Conservation Service, United States Department of Agriculture.

Many publications, particularly recently, focus on soil carbon as an essential factor relating to both soil health and soil quality. In an unpublished thesis, Wallace (1910) wrote about the importance of humus, an operationally defined fraction of soil carbon. He mentioned the use of animal manure in maintaining soil health and humus. However, up until the 1990s, the concept of soil health mainly referred to the physical and chemical properties of soil. In what now seems to be a very forward-looking report, the U.S. Department of Agricultural Adjustment Administration published 'Soil Health and National Wealth' (1936) which suggested that animals, bacteria, molds, lichens, moss, algae, worms, etc. were important components of soil in supporting plant growth. One of the first scientific publications referring to soil biology was by Madaus (1938) who discussed the effects of soil biology on plants. Published articles on soil health and human and animal health continued to increase slowly until the 1980s. These early research studies were remarkably forward looking in making links between soil and human and animal health. Today, researchers are obsessed with developing soil health metrics and assessments, but often disassociate the link to crop production and nutrition and human and animal health.

Following the 1980s, there was a logarithmic increase in the number of publications on soil health and soil quality. The emphasis on soil health often excluded reference to crop production and economic outcomes. This resulted in somewhat of a disconnect between researchers' views and real outcomes experienced by farmers and land managers. Acton and Gregorich (1995) discussed soil health as a term related to soil productivity or fertility in relation to crop productivity, arguing that soil health should be related to soil fitness to support crop growth without resulting in soil degradation or other harm to the environment. This has resulted in numerous exercises in creating soil health and or quality indices such as those presented in Table 1. They include a range of chemical, physical, and biological indicators. Numerous definitions of soil health have been promoted but their relationship to plant productivity is not always sufficiently emphasized.

3 Soil health and soil carbon

Many (if not all) soil health indicators are influenced by soil carbon (C) and its management. Improved crop yield from increased soil C has primarily been attributed to its role in influencing the supply of water or nitrogen (Gregorich et al., 1994; Arshad and Martin, 2002; Lal, 2020). Both the amount and quality of soil carbon inputs influence soil carbon levels, a key indicator of soil health. As mentioned, increasing soil C affects the physical, chemical, and biological status of soil (Lal, 2020). Increasing soil carbon through additional crops in rotation such as cover crops is a common practice often leading to increasing

Table 1 Methods to assess soil health

Chemical indicators	Physical indicators	Biological indicators
pH	Texture	Microbial biomass
Organic matter	Bulk density	Earthworm populations
Total carbon	Penetration resistance	Nematode populations
Total nitrogen	Aggregate stability	Arthropod populations
Cation exchange capacity	Water holding capacity	Mycorrhizal fungi
Major and minor nutrients	Infiltration rate	Respiration rate
Heavy metals/plant toxins	Depth to hardpan	Soil enzyme activities
Electrical conductivity	Depth to water table	Pollutant detoxification
POX[a]	Porosity	Decomposition rate
Soil nitrate	Erosive potential	Potentially mineralizable N
Pollutants	Aeration	GHG emissions
	Particulate organic matter	
	Pore size	
	Texture	

[a]Permanganate oxidizable carbon.

soil C. Adding animal manure, compost, and other organic wastes such as food processing waste can also lead to increases in soil C.

Changes in crop rotation can alter the amount of carbon and its quality that can lead to overall changes in the amount of soil C. Conversely, some crop rotations can decrease C inputs, such as corn-soybean rotations versus corn–corn rotations (Poffenbarger et al., 2017). Though soil quality and health indicators are generally higher in rotations containing legumes such as soybeans, reduced nitrogen application can lead to lower soil C compared to conventional fertilization rates for corn. Legume-containing rotations also affect soil health outcomes positively through their effect on microbial activity and other functional soil properties affecting nutrient cycling and pest incidence. The results show that positive changes in social health are not entirely dependent on C inputs, but a shift in soil physical, chemical and biological outcomes is often dependent on the amount and quality of C inputs.

4 Tillage and soil health

Soil disturbance, particularly from tillage, can have negative consequences on soil health through erosional losses and impacts on microbial activity that can prime existing soil C. Soil health practices, such as conservation and zero (or no) tillage that reduce soil disturbance, do not add additional C inputs. Rather they change its distribution to increase soil C in the soil surface layer and can decrease it below the plow layer in comparison to fully tilled soils (Pittelkow et al., 2015). Reduced tillage practices can have a positive influence on soil water availability, aeration, and porosity.

In a meta-analysis, no till was found to have no observed effect on crop yields for oil seed, cotton, and legumes compared to tilled management (Pittelkow et al., 2015). They also found in general that no till performs best in rain-fed dry climates with crop yields being equal to or higher than conventional till management. Other outcomes of soil health assessments on conservation and no till management are often not fully considered, such as the effects on nitrogen losses. Mei et al. (2018), in a meta-analysis of nitrous oxide emissions under conservation tillage management, found an average 17.8% increase compared to conventional tillage management. Overall, the conservation tillage induced nitrous oxide emission factor increased by 0.40%. The increased nitrous oxide emissions under conservation tillage management potentially negate the positive effects on soil health and attenuate climate change benefits from any increase in soil carbon under reduced tillage.

5 Soil health indicators in the field

Recent studies on the role of soil health indicators have produced mixed results (Miner et al., 2020). The premise of linking crop productivity to soil health relies on the assumption that soil health practices alleviate yield limiting factors such as nutrient availability, water availability, disease risk, etc. Modern intensive cropping system management is designed to attenuate limitations on crop growth and yield by ensuring adequate resources and pest suppression. Crop yields are also influenced by other abiotic factors such as precipitation and temperature. Soil health management clearly does not affect abiotic factors directly. However, they can have a buffer effect in protecting against weather/climate variability through changes in water holding capacity for example. These effects are more prominent in rain-fed systems versus irrigated systems. In irrigation systems, soil health management may have less of an effect since water limitation is eliminated.

A recent study in rain-fed systems in North Carolina by Roper et al. (2017) could not differentiate among different agronomic management systems (that included organic and conventional chemical management with and without tillage) in soil health metrics using the Haney soil health test (HSHT) (Haney et al., 2018) or the Cornell comprehensive assessment of soil health (Moebius-Clune et al., 2016). In addition, no correlation between crop yield and soil health status could be established by applying any of the soil health assessment test outcomes. In another study in the Southeastern United States, evaluations using the HSHT also found no consistent response of soil health metrics to management-induced changes in soil health from changes in crop rotation, tillage reduction, or cover crop use (Singh et al., 2020). These results likely reflect that conventional nutrient management practices can offset the effects of soil health-promoting practices in terms of crop yields.

As mentioned earlier, many soil health promoting practices such as reduced or no tillage, organic amendments including manure, cover crops, and crop rotation have the potential to increase soil C, and therefore impact soil health outcomes. Cai et al. (2022) found that no-till management led to soil C losses within the first decade of adoption. It took an average of 14 years to recover soil carbon compared to conventional management. A global meta-analysis found that the effect of cover crops is generally to increase soil C in the surface soil by 15.5% (Jian et al., 2020). The effect of crop rotation is similar to the effects of cover crops in enhancing diversity in the above and below ground plant and soil system. In meta-analyses, McDaniel et al. (2014) and Reiss and Drinkwater (2018) found that adding one or more crops in rotation (compared to a monoculture) increased carbon by 2.2–3.6%.

Hijbeek et al. (2017) found no relationship between soil C and crop yields in a European study of arable crops. Their study showed that when sufficient nutrients are available, the effects of soil C on crop yields are insignificant. In a global meta-analysis, Oldfield et al. (2019) showed that crop yields only responded to soil C management when soil C was below 2%. These results indicate that management of soil health is dependent on the initial soil carbon state or concentration. Therefore, soil C as a soil health metric should consider the initial soil C condition when evaluating social health outcomes in relationship to crop yields.

6 The soil health debate continues

The debate around the definition of both soil health and quality is seemingly settled. However, many shortcomings of the soil health concept have been repeatedly raised (Sojka and Upchurch, 1999; Singer and Sojka, 2002; Sojka et al., 2003; Bünemann et al., 2018; Stewart et al., 2018; Powlson, 2020; Janzen et al., 2021). Norman Borlaug, considered the father of the green revolution, criticized the concept of soil quality as a misleading fad lacking scientific rigor, biased by societal values and conceptually incompatible with air and water quality (Sojka et al., 2003). Powlson (2020) makes the point that whilst almost everyone who works with soil (from gardeners to farmers) has some concept of a 'healthy' soil, and that the term 'soil health' has a value in promoting debate about soil management, it risks becoming over complex or meaningless if it is not clearly related back to the measurement of specific soil properties.

Can soil health or quality be quantified? Karlen et al. (1997) provided a numerical framework or index to quantify soil quality. Sojka and Upchurch

(1999) expressed concerns that endorsing the concept of soil quality risked a potentially subjective combination of individual soil health metrics which failed to account for the fact that soil properties are integrated and multifunctional. Janzen et al. (2021) argue that the term 'soil health' remains ambiguous, largely because the current use of the term 'health' denotes a metaphor, not a literal scientific concept. The discussion of soil health and assigning it a sliding scale or assessment of health status can ignore the integration of soil properties and functions that are diverse and often conflicting. Others, e.g. Lehmann et al. (2020) maintain that soil health is an overarching principle that contributes to sustainability goals rather than only a property to measure.

In soil health assessments, soil biological indicators such as microbial biomass size and activity, C and N mineralization, labile C, protein and enzyme status, and microbial community composition and structure are often promoted as important soil health indicators. This assumes a causal relationship exists between the microbial community and environmental factors such as disease or ecosystem function (Coyne et al., 2022). Fierera et al. (2021) concluded that many soil health metrics based on microbial measurements are not easy to interpret and may not relate to soil health status. Baveye (2021) describes this situation as 'soil health at a crossroads' and quotes Bouma (2020) who argues that, to embrace the soil health concept, we need to provide routine measurements such as those used for water and air quality. In reviewing current soil health research, Bümeman et al. (2018) concluded that it has not yet resulted in an operational measurement framework to relate soil health to crop production or environmental services.

7 The soil health research agenda

Historically, soil assessments, particularly fertility evaluation, focused on crop production. Today, numerous concepts relating to soil health often involve only soil processes or characteristics with little connection to crop productivity or nutrition. For example, microbial communities and processes play a role both in crop nutrition and in pathogenesis, providing a spectrum of desirable to unwanted outcomes (Fierera et al., 2021). Many studies make a link to ecosystem services but rarely include them in their experimental design showing mechanistic links or quantifiable benefits in terms of improved environmental outcomes. Many other studies cited in this chapter also found unclear relationships to how soil health evaluation metrics and indices relate to crop production and health. In a literature review, Miner et al. (2020) argue that 'soil health practices can also have neutral or negative yield impacts–only in select systems have consistent yield increases been realized, highlighting the need for language clarification and improvements in mechanistic understanding of regional-scale yield impacts'.

Regardless of how we evaluate soil health, it is clear that efforts to define it are necessary to produce healthy and nutritious food and maintain a functioning environment to support human and animal health. In addition, clear soil health assessments are needed in general to promote good water and air quality and global ecosystem health. It is also clear from the interactions with farmers that their assessments also align with efforts to both define and implement practices that improve soil for both food and fiber production and environmental stewardship (O'Neill et al., 2020). Therefore, the future soil health research agenda should include the following:

1 Establish a robust link between net primary productivity and soil health evaluation.
2 Provide causal relationships between soil health assessment and ecosystem services.
3 More research on the role of cover crops, soil organic amendments (manure, composts, other organic waste, etc.), and their effects on soil health.
4 Research to explore the link between soil health management and nutritious food; this requires a concerted research effort by agronomic and human and animal physiologists.
5 More research on the role of microbial communities, particularly those of symbionts such as mycorrhizae, to exploit the competitive abilities of crops to acquire nutrients efficiently.
6 All soil metrics are not equal and require a location or region-specific approach to determine best soil health assessments and outcomes.
7 Less emphasis should be on methods standardization to avoid a one size fits all approach, and rather adapt a 'minimum dataset' set approach of indicators to tailor soil health assessments for different soil types and regional cropping systems (Wander et al., 2019).

Soils have sustained humanity; it requires our most attentive efforts to define soil health to ensure planetary health that will benefit society and ensure astute environmental leadership and outcomes.

8 Where to look for further information

The concept of soil health continues to evolve from the ancients to modern day science concepts. Though the term 'health' is difficult to quantify as discussed in this chapter, it provides a framework to assess the soil's ability to continue to provide essential functions and services to benefit humanity and global ecosystem health. The references to soil health continue to increase, as seen in Fig. 1 citation, in count over the last century, both from a science and practitioner perspective.

The following are text resources on the topic of soil health for additional information.

- Soil Health Series: Volume 1 Approaches to Soil Health Analysis. Eds., Douglas L. Karlen, Diane E. Stott, Maysoon M. Mikha. DOI:10.1002/9780891189817. 2021 Soil Science Society of America, Inc.
- Soil Health. Eds., Bhoopander Giri, Ajit Varma. DOI: 10.1007/978-3-030-44364-1. 2020 Springer.

Other sources of information on soil health include organizations whose mission is to promote the science and its implementation.

- Soil Health Institute whose mission is to safeguard and enhance the vitality and productivity of soils through scientific research and advancement. https://soilhealthinstitute.org/about-us/. Morrisville, NC.
- Foundation for Food & Agriculture Research whose mission is to build collaborative partnerships to support audacious science addressing today's food and agriculture challenges. https://foundationfar.org/what-we-do/challenge-areas/soil-health/. Washington, DC.

9 References

Acton, D. F. and Gregorich, L. J. (Eds.) (1995). *The Health of Our Soils - Towards Sustainable Agriculture in Canada*. Centre for Land and Biological Resources Research, Research Branch, Agriculture and Agri-Food Canada, Ottawa, ON. pp. xiv + 138.

Agricultural Adjustment Administration (1936). *Soil Health and National Wealth*. United States Department of Agriculture, Washington, DC.

Albrecht, W. A. (1945). Soil Fertility and its health applications. *American Journal of Orthodontics and Oral Surgery* 31(5), 279–286.

Alexander, M. (1971). Agriculture's responsibility in establishing soil quality criteria. In: *Environmental Improvement-Agriculture's Challenge in the Seventies*. pp. 66–71. National Academy of Sciences, Washington DC.

Arshad, M. A. and Martin, S. (2002). Identifying critical limits for soil quality indicators in agro-ecosystems. *Agriculture, Ecosystems and Environment* 88(2), 153–160.

Baveye, P. C. (2021). Soil health at a crossroad. *Soil Use and Management* 37(2), 215–219.

Bouma, J. (2020). Soil challenges beyond publication issues. European Journal of Soil Science 72, 29–30.

Bünemann, E. K., Bongiorno, G., Bai, Z., Creamer, R. E., De Deyn, G., de Goede, R., Fleskens, L., Geissen, V., Kuyper, T. W., Mäder, P., Pulleman, M., Sukkel, W., van Groenigen, J. W. and Brussaard, L. (2018). Soil quality – a critical review. *Soil Biology and Biochemistry* 120, 105–125.

Cai, A., Han, T., Tianjing Ren, T., Sanderman, J., Rui, Y., Wang, B., Smith, P., Xu, M., and Li, Y. (2022) Declines in soil carbon storage under no tillage can be alleviated in the long run. *Geoderma* 425, 116028.

Coyne, M. S., Pena-Yewtukhiw, E. M., Grove, J. H., Sant'Anna, A. C. and Mata-Padrino, D. (2022). Soil health - it's not all biology. *Soil Security* 6, 100051.

Doran, J. W. and Jones, A. J. (1996). Methods for assessing soil quality. *Soil Science Society of America Special Publication Number 49*. Soil Science Society of America, Inc., Madison, WI. p. 410.

Doran, J. W. and Zeiss, M. R. (2000). Soil health and sustainability: managing the biotic component of soil quality. Applied Soil Ecology 15, 3-11.

Fierera,N., Wood, S. A. and Bueno de Mesquita, C. P. (2021). How microbes can, and cannot, be used to assess soil health. *Soil Biology and Biochemistry* 153, 108111.

Gregorich, E. G., Carter, M. R., Angers, D. A., Monreal, C. M. and Ellert, B. H. (1994). Towards a minimum data set to assess soil organic matter quality in agricultural soils. *Canadian Journal of Soil Science* 74(4), 367-385.

Haney, R. L., Haney, E. B., Smith, D. R., Harmel, R. D. and White, M. J. (2018). The soil health tool–theory and initial broad-scale application. *Applied Soil Ecology* 125, 162-168,https://doi.org/10.1016/j.apsoil.2017.07.035.

Hijbeek, R., van Ittersum, M. K., ten Berge, H. F. M., Gort, G., Spiegel, H. and Whitmore, A. P. (2017). Do organic inputs matter - a meta-analysis of additional yield effects for arable crops in Europe. *Plant and Soil* 411(1-2), 293-303.

Howard, Sir A. (1947). *Soil and Health, A Study of Organic Agriculture*. The University of Kentucky Press, University of Louisville, Lexington KY.

Janzen, H. H., Janzen, D. W. and Gregorich, E. G. (2021). The 'soil health' metaphor, Illuminating or illusory? *Soil Biology and Biochemistry* 159, 108167.

Jian, J., Du, X., Reiter, M. S. and Stewart, R. D. (2020). A meta-analysis of global cropland soil carbon changes due to cover cropping. *Soil Biology and Biochemistry* 143, 107735.

Karlen, D. L., Mausbach, M. J., Doran, J. W., Cline, R. G., Harris, R. F. and Schuman, G. E. (1997). Soil quality: a concept, definition, and framework for evaluation (A Guest Editorial). *Soil Science Society of America Journal* 61(1), 4-10.

Karlen, D. L., Andrews, S. S., Weinhold, B. J. and Doran, J. W. (2003). Soil quality: humankind's foundation for survival a research editorial by conservation professionals. *Journal of Soil and Water Conservation* July 58, 171-179.

Karlen, D. L., Andrews, S. S., Wienhold, B. J. and Zobeck, T. M. (2008). Soil quality assessment: past, present and future. *Journal of Integrative Biosciences* 6, 3-14.

Lal, R. (2020). Soil organic matter content and crop yield. *Journal of Soil and Water Conservation* 75(2), 27A-32A.

Larson, W. E. and Pierce, F. J. (1991). Conservation and enhancement of soil quality. In: J. Dumanski, E. Pushparajah, M. Latham, and R. Myers, eds. *Evaluation for Sustainable Land Management in the Developing World. Vol. 2: Technical Papers*. Proc. Int. Workshop., Chiang Rai, Thailand. 15-21 Sept. 1991. Int. Board for Soil Res. and Management, Bangkok, Thailand.

Lehmann, J., Bossio, D. A., Kögel-Knabner, I. and Rillig, M. C. (2020). The concept and future prospects of soil health. *Nature Reviews: Earth and Environment* 1(10), 544-553.

Liebig, J. (1840). *Die Chemie in ihrer Andwendung auf Agrikultur und Physiologie*. Vieweg und Söhne, Braunschweig.

Madaus, G. (1938). Plant studiea II: the issue of the maintenance of the health of soil with plants. *DMW - Deutsche Medizinische Wochenschrift* 64(28), 998–1000.
McDaniel, M. D., Tiemann, L. K. and Grandy, A. S. (2014). Does agricultural crop diversity enhance soil microbial biomass and organic matter dynamics? *Ecological Applications* 24(3), 560–570.
Mei, K., Wanga, Z., Huang, H., Zhang, C., Shang, X., Dahlgren, R. A., Zhang, M. and Xia, F. (2018). Stimulation of N2O emission by conservation tillage management in agricultural lands: a meta-analysis. *Soil and Tillage Research* 182, 86–93.
Miner, G. L., Delgado, J. A., Ippolito, J. A. and Stewart, C. E. (2020). Soil health management practices and crop productivity. *Agricultural and Environmental Letters* 5(1), e20023. https://doi.org/10.1002/ael2.20023.
Moebius-Clune, B. N., Moebius-Clune, D. J., Gugino, B. K., Idowu, O. J., Schindelbeck, R. R., Ristow, A. J., van Es, H. M., Thies, J. E., Shayler, H. A., McBride, M. B., Wolfe, D. W. and Abawi, G. S. (2016). *Comprehensive Assessment of Soil Health - the Cornell Framework Manual* (edn. 3.0). Cornell University, Geneva, NY.
National Research Council (1989). *Alternative Agriculture*. The National Academies Press, Washington, DC. https://doi.org/10.17226/1208.
Oldfield, E. E., Bradford, M. A. and Wood, S. A. (2019). Global meta-analysis of the relationship between soil organic matter and crop yields. *SOIL* 5(1), 15–32, https://doi.org/10.5194/soil-5-15-2019.
O'Neill, B., Christine, D., Sprunger, G. and Robertson, G. P. (2020). Do soil health tests match farmer experience? assessing biological, physical, and chemical indicators in the Upper Midwest United States. *Soil Science Society of America Journal* 85, 903–918.
Palissy, B. (1580, edition of 1880). *Oeuvres Complètes*. Charavay, P. (Ed.). Paris. p. 499.
Pittelkow, C. M., Linquist, B. A., Lundy, M. E., Liang, X., van Groenigen, K. J., Lee, J., van Gestel, N., Six, J., Venterea, R. T. and van Kessel, C. (2015). When does no-till yield more? a global meta-analysis. *Field Crops Research* 183, 156–168.
Poffenbarger, H. J., Barker, D. W., Helmers, M. J., Miguez, F. E., Olk, D. C. and Sawyer, J. E., Six, J. and Castellano, M. J. (2017). Maximum soil organic carbon storage in Midwest U.S. cropping systems when crops are optimally nitrogen-fertilized. *PLoS ONE* 12(3), e0172293. https://doi.org/10.1371/journal.pone.0172293.
Powlson, D. S. (2020). Soil health—useful terminology for communication or meaningless concept? Or both? *Frontiers of Agricultural Science and Engineering* 7(3). https://doi.org/10.15302/J-FASE-2020326.
Reiss, E. R. and Drinkwater, L. E. (2018). Cultivar mixtures: a meta-analysis of the effect of intraspecific diversity on crop yield. *Ecological Applications* 28(1), 62–77.
Roper, W. R., Osmond, D. L., Heitman, J. L., Wagger, M. G. and Reberg-Horton, S. C. (2017). Soil health indicators do not differentiate among agronomic management systems in North Carolina soils. *Soil Science Society of America Journal* 81(4), 828–843. https://doi.org/10.2136/sssaj2016.12.0400.
Schultz, A. (1925). Studies on the Vegetation in western Nyland (South Finland) and its Relationship with the Quality of the Soil. *Petermanns Mitteilungen* 43, 223–223.
Singer, M. J. and Sojka, R. E. (2002). Soil quality. In: E. Geller et al., eds. *McGraw-Hill Yearbook of Science & Technology*, pp. 312–314. McGraw-Hill Inc., New York.
Singh, S., Jagadamma, S., Yoder, D., Yin, X. and Walker, F. (2020). Agroecosystem management responses to Haney soil health test in the southeastern United States. *Soil Science Society of America Journal* 84(5), 1705–1721.

Soil Health, United States Department of Agriculture National Resource Conservation Service. Available at: https://www.nrcs.usda.gov/wps/portal/nrcs/main/soils/health/.

Sojka, R. E. and Upchurch, D. R. (1999). Reservations regarding the soil quality concept. *Soil Science Society of America Journal* 63(5), 1039-1054.

Sojka, R. E., Upchurch, D. R. and Borluag, N. E. (2003). Quality soil or soil quality management: Performance versus semantics. *Advances in Agronomy* 79, 1-68.

Sprengel, C. (1826). Ueber Pflanzenhumus, Humussäure undhumussaure Salze. *Archiv für die Gesamte Naturlehre* 8, 145-220.

Stewart, R. D., Jian, J., Gyawali, A. J. Thomason, W. E., Badgley, B. D., Reiter, M. S. and Strickland, M. S. (2018). What we talk about when we talk about soil health. *Agricultural and Environmental Letters* 3(1), 180033. https://doi.org/10.2134/ael2018.06.0033.

Thaer, A. D. (1856). *The Principles of Practical Agriculture*, Shaw, W. and Johnson, C. W. (Trans.). Saxton and Company Publishing, New York, p. 551.

Wallace, H. A. (1910). Relationship between livestock farming and the fertility of the land. Unpublished thesis. Iowa St. Univ. Ames, IA.

Wander, M. M., Cihacek, L. J., Coyne, M., Drijber, R. A., Grossman, J. M., Gutknecht, J. L. M., Horwath, W. R., Jagadamma, S., Olk, D. C., Ruark, M., Snapp, S. S., Tiemann, L. K., Weil, R. and Turco, R. F. (2019). Developments in agricultural soil quality and health: reflections by the research committee on soil organic matter management. *Frontiers in Environmental Science* 7, 109. https://doi.org/10.3389/fenvs.2019.00109.

Warkentin, B. P. and Fletcher, H. E. (1977). Soil quality for intensive agriculture. *Proceedings of the International Seminar on Soil Environment and Fertilizer Management in Intensive Agriculture*. Society for Science of Soil and Manure, National Institute of Agricultural Science, Tokyo, Japan. pp. 594-598.

Wrench, G. T. (1939). Health and the soil. *British Medical Journal* 1(4075), 276-277.

Part 1

Dimensions of soil health

Chapter 2

Soil health: definitions, history, key concepts and hurdles

Michelle M. Wander, University of Illinois at Urbana-Champaign, USA

1 Introduction

1.1 Challenges of normative science

Contributors to this text were invited to review topics related to soil health defined as 'the continued capacity of soil to function as a living ecosystem able to deliver a range of ecosystem services that sustain plants, animals, people and the environment in which they live'. As such, soil health is not distinguished from soil quality, with definitions of both emphasizing the living and dynamic nature of soil and associating interactions between its physical, chemical and biological properties with soil functions that underpin services (Doran and Parkin, 1994; Karlen et al., 1997). Both terms are normative and utilitarian and have been criticized for their diffuse nature associated with reliance on values-based judgment of non-subjective measurements that are made based on land use, land type, prioritized soil functions and context for application (Davis and Miller, 2000; Letey et al., 2003). This means quality and health will vary with a perspective that depends upon whether they are applied to cultural, aesthetic, industrial and ecosystem services that extend beyond the scope of agriculture (FAO, 2015). Even though this text will prioritize the productive function and provisioning and sustaining services needed for food and fiber production to narrow criteria for attribute evaluation, this alone cannot resolve the subjective and at times contradictory aspects of these concepts' application.

Patzel et al. (2000) argued for greater clarity and consistency in the use of the two terms, *quality* and *fertility*, by both ley and scientific communities

http://dx.doi.org/10.19103/AS.2021.0094.01

and concluded that making distinctions between these terms is helpful to both audiences by articulating concepts and acknowledging frames of reference. They also noted that while the concept of soil fertility is quite definite, it is dispositional or 'concealed' because judgments of what is, or is not, fertile are site specific. Additional weaknesses in the concept of soil quality were associated with undefined objectives and reliance on 'extensional definitions' that assert associations or linkages between properties and functions without providing sufficient evidence or specifying context. Schjønning et al. (2004) suggested that the emergence of the soil quality paradigm changed society's expectations of the scientific community by asking researchers to be conscious of societal priorities and their own values and goals. They noted soil quality and health are examples of reflective science, wherein the observer's efforts influence the object under consideration because research is undertaken within a cognitive space consisting of observational components that apply accepted methods within a civic setting for which the research is relevant using intentional processes developed to specify and then ideally satisfy goals tied to values.

Known challenges associated with the public's ability to place value judgements on soil indicator status and balance tradeoffs to determine how to use soil resources to appropriately meet societal needs (Patzel et al., 2000) must be addressed by structured steps that are not only reproducible and transparent but also acknowledge subjective, context-specific aspects of the work. While participatory research methods have been widely used in soil health research to satisfy these expectations (Andrews et al., 2003; Wander et al., 2002), it may be helpful to make a clearer distinction between soil quality and soil health to add clarity and transparency to a process that includes stakeholders in the process of assessment (Wander et al., 2019). Doran and Zeiss (2000) made this distinction noting soil health, defined as the capacity of soils to function, can be assessed using indicators of soil quality. By clearly associating soil quality with indicators used to assess soil condition and soil health with interpretive frameworks used to make judgments we can separate the scientific and socio-technical components of subjective health assessment. Distinctions between terms used by research communities that are delineated by expertise, location, and focus are made to define and shape the scientific discourse (Foucault, 1970). For example Grimm and Wissel (1997) found it beneficial to distinguish between terms used to describe ecological stability to efficiently communicate needed technical detail but that was harmful when distinctions reduce shared understandings. Of course, academic arguments like this are examples of privileged speech that are made moot when terms and related discourse are taken up by society (Landa and Angel, 2014). It is clear that the soil health framing is intuitive in nature and resonates with the public and this makes it well suited for social marketing purposes (Bouma and McBratney,

2013; Brevick et al., 2019). Social movements that equate soil health with human well-being (Wander, 2009) may be vulnerable to bad-faith actors intending to exploit public sympathies. Whether or not this occurs will depend upon whether or not the sociotechnical systems we devise to achieve related goals can avoid creation of false narratives and unrealistic expectations (Giampietro and Funtowicz, 2020). Frameworks must provide the technical specificity and transparency necessary for unbiased assessment of soil capability and condition and then, honestly communicate that information through evaluative frameworks designed to set policy, back-stop product claims or guide management.

1.2 History and context

The soil health discourse results from concern for human well-being that has driven efforts to understand and characterize soil in terms of soil capability or fitness for use since the dawn of civilization (Dale and Carter, 1955; Hillel, 1991). Even though 'soil quality' and 'soil health' are often used interchangeably, 'soil quality' appeared earlier and has been used with greater frequency in the scientific literature (Fig. 1). Liu et al., 2020 suggest that efforts have been focused on the USA and that there has been limited international collaboration. Their findings are at odds with this review which suggests there has been global interest in convergent, synergistic scientific histories. Efforts to evaluate the soil resource and encourage its stewardship pique in response to historical events that have caused economic insecurity resulting in resource exploitation. This was the case in the USA in the 1930s when economic depression, drought and westward expansion produced threats to soil health that were addressed

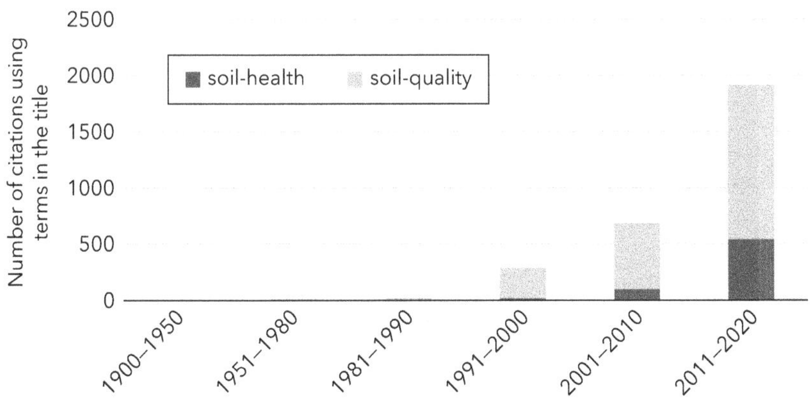

Figure 1 The increase in the number of citations using the terms 'soil health' in black or 'soil quality' in the article's title. These were recovered in the literature covered in the Web of Science.

Published by Burleigh Dodds Science Publishing Limited, 2022.

by governmental action (Worster, 2004). During the first half of the twentieth century, references to both quality and health were being made globally to broadly consider plant-soil interactions and land capability classification. For example, Schultz (1925) used soil quality to understand the distribution of vegetation in South Finland and, Hoon and Dhawan (1950) used patterns in vegetation to delineate soil quality in India. The concept of soil quality was used to classify Esthonian (Zimmerman, 1939) and German soils (Wolff, 1939). Forbes (1928) delineated soil quality based on suitability for food or timber production in the USA in an article that challenged the notion that all arable land should be used for food production. Unfortunately, the notion that the most heavily timbered soils were the most fertile was imported from Europe and it took decades for this to give way to the recognition that US grasslands and oak openings were more productive and thus had greater soil quality from an agricultural perspective (Peters, 1973). Even though the fact that soil quality is multivariate, site and context specific has been emphasized repeatedly, this crucial point is constantly being overlooked. Whether this is an example of intentional 'bypass' by the soil science community that was cited by Baveye (2020) is not clear. What is certain is that without specificity useful applications of soil quality for soil health assessment are impossible.

Beginning in the 1980s soil quality was invoked globally in association with environmental concern with resource inventories shifting in focus to consider management and remedy environmental problems (Breeuwsma et al., 1986), assess risk from erosion (Yassaglou and Kollias, 1989; Urzelai et al., 2000) and realize sustainable development goals (Wilson, 2000). An uptick in interest coincided with soil degradation rates that increased in the USA in the 1970s and 80s with the opening of global agricultural markets and industrialization of agriculture (Debailleul, 1990; Lekkerkerk et al., 1990). This was followed in the 1990s by disruptions in the agricultural sector in Eastern Europe resulting from the breakup of the Soviet Union that raised environmental concern and interest in soil quality (Davis and Miller, 2000). Historical forces that have accelerated soil degradation fostered the emergence of soil quality as a distinct scientific object by the end of the twentieth century to pursue the simultaneous goals of increased or sustained production and the reduction or alleviation of environmental harm (Tilman et al., 2002; Schjønning et al., 2004). This objective arose in part to address perceived shortcomings of soil-fertility-based management that was focused too narrowly on soil chemistry and production (Letey et al., 2003). Soil quality and its attendant stewardship goals were quickly deemed to be essential for food security and successful ecological intensification needed to sustain or increase yield (Cassman, 1999). It was also referenced in association with other efforts to use soil properties to set standards for environmental cleanup (Siegrist, 1990), steward North American forests (Paige-Dumroese et al., 2000) and produce frameworks

Published by Burleigh Dodds Science Publishing Limited, 2022.

for 'Soil Care' (Stengel, 2000; Dumanski and Pieri, 2000). In such framings, soil quality can be seen as an addition to the well-established concept of soil capability classification. Soil quality adds consideration of land condition through the use of dynamic soil properties to try to optimize land use based on science (Steinhardt, 1995). Efforts have sought to use indicators associated with capability and condition at national and even global scales for decades (e.g. Gerasimov, 1983; Cox, 1995; Schipper and Sparling, 2000; Krüger et al., 2018).

The number of English language articles using soil quality and soil health in their titles has continued to increase at the pace of 100- and 500-fold during the last 20 years (Fig. 2). The number of peer-reviewed articles using 'soil' in their titles only doubled in the 90s and tripped in the 2000s in comparison with 1981–1990, suggesting publication access cannot explain the rapid increase in the use of these terms. A similar search for 'soil security' recovered only one paper published before 1950 and 30 between 2011–2020 using this term in the title. Increased international concern for soil quality and degradation resulted from increased awareness of the finiteness of arable land, projections for population growth and economic development and the recognition of the serious and immediate need to mitigate climate change (Gomiero, 2016; Karlen and Rice, 2015; Amelung et al., 2020). Interest in common global standards for indicators that began decades ago (Hortensius and Welling, 1996) has grown in kind (FAO, 2015).

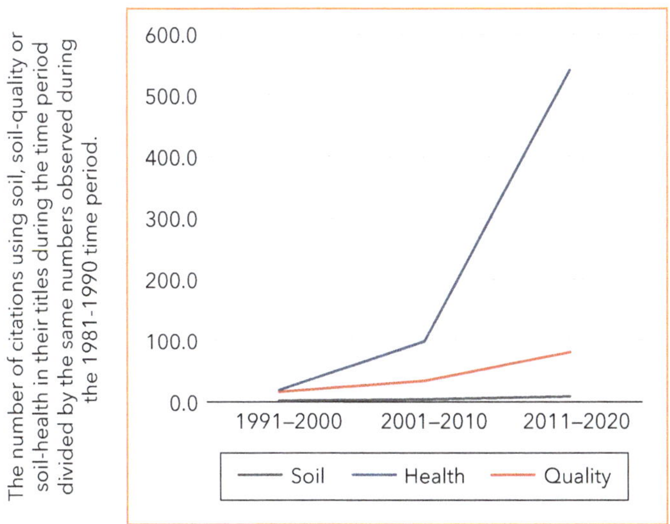

Figure 2 The increase in the numbers of manuscripts recovered using soil, soil-health or soil-quality in their title using Web of Science compared to the 1981–1990 period.

2 Evolving methodology

2.1 Multi-step process

Research efforts continue to address all aspects of indicator development including improvement of methods. Thiele-Bruhn et al. (2020) recently called for standardization of methods suitable for ecosystem service assessment endorsed by the FAO and UN, that consider soil microbial functions, including nutrient cycling and greenhouse gas emission, pest control and plant growth promotion, carbon cycling and sequestration, as well as soil structure development and filter functions. The criteria they outlined for indicator selection continued to include practicality, relevance and sensitivity judged subjectively by experts using methods that are generally consistent with the procedures and conceptual framing proposed by Andrews et al. (2002, 2004). Those procedures include indicator development, selection, interpretation and format for use-dependent applications with indicator scoring being related to performance or attributed based on inherent potential.

2.2 Indicators and key concepts

Reviewing a variety of approaches to develop and deploy indicators for agriculture Van der Werf and Petit (2002) separated soil properties or processes into 'means' (aka practice-based) indicators and 'effects'-based indicators associated with soil function. While some indicators, like water holding capacity or soil organic carbon, can be directly equated with services like water supply or carbon sequestration, the functions of many indicators are only broadly or indirectly associated with services. Soil organic matter and related qualities (e.g. C/N in Southeast Asian forest (Yamakura and Sahunalu, 1990), humus energy content (Kozin, 1990) and soil-dwelling arthropods (Reddy, 1986)) have long been recognized as useful integrative proxies that can aggregate information about multiple soil characteristics into one variable associated with shifts in carbon and nutrient cycles (Bailey et al., 2018). Emphasis on tight nutrient cycling is central to the soil health paradigm as it distinguishes soil health from fertility and associates these qualities with natural, undisturbed systems and biotic integrity (Wander, 2009). Accordingly, many biochemical assays, including extracellular enzymes and biotic indicators, are pursued as integrative assays for nutrient cycling, decomposition, pest suppression and plant resistance to stress (Burns et al., 2013; McDaniels et al., 2014; Ouyang et al., 2018; Larkin, 2015; Neher and Barbercheck, 2019; Tahat et al., 2020). These and other works continue to evaluate responsiveness to management and ways to establish functional value using indices that consider diversity and nutrient richness (Singh et al., 2012; Griffiths et al., 2018; Ney et al., 2019) as well as resistance and resilience (De Vries et al., 2012). Despite their proven value biota other than bacteria remain underused in soil health applications (Geisen et al., 2017). Much of the emphasis remains on the development of high-throughput microbial assays and 'OMICS' with work remaining in the comparative mode to understand the assay by drawing samples from locations under contrasting management or land use. Besides developing targeted approaches for the isolation of microorganisms from the soil, which allows a classical taxonomic assignment of genotypic and phenotypic traits, novel approaches integrating metagenomic datasets with other types of data such as metabolomics and abiotic factors will add insight into the workings of the microbiome (Feng et al., 2016). Unfortunately, many studies continue to rely on extensional definitions asserting relationships between diversity, richness and evenness or community composition resulting from management with nutrient cycling, plant growth promotion (Hartmann et al., 2015; De Curato et al., 2020). Ideally one could audit the community status of the system in relation to restoration targets and the effectiveness of management interventions and, shifts in the community or indicator status would be related to performance or enhancement of the rate of recovery of degraded systems (Harris, 2009). While this concept and aspiration are widely embraced, few works have meaningfully related microbiome attributes to system-level phenomena and relationships between omics, including sequence-relative abundance and function, unfortunately, remain obscure (Bailey et al., 2018). Efforts continue to strive to overcome taxonomic barriers and categorize soil microorganisms based on their ecological strategies associated with functional attributes but contend with the fact that despite the ever-growing sequence databases, marker gene, genomic and metagenomic analyses, most metagenomic reads cannot be assigned to a function and most soil microorganisms remain undescribed (Fierer, 2017).

Typically, the quantity, composition or process rate representing the 'effects' of management are commonly associated with soil services including nutrient cycling through correlation (Doran and Parkin, 1994; Andrews et al., 2002). For example, a recently developed multi-enzyme assay (Acosta-Martinez et al., 2019) was validated by demonstrating it appropriately ranked nutrient cycling status of soils under aggrading or degrading management. While meta-analytic work can confirm broad relationships between indicators, organic matter and nutrient abundance (Wang et al., 2018; Chen et al., 2018) much work needs to be done to associate indicators with services in a trustworthy, site-specific manner. Effects-type indicators, that relate indicators to desired outcomes, are attractive proxies because these can allow growers to select among practices to determine how to achieve goals based on their farming system, location and market contexts (Wander et al., 2002; Van der Werf and Petit, 2002). A drawback of effects-based indicators relative to the more general means- or practice-based measures is that they require more intensive data collection and so are relatively costly and difficult to validate. Effects-based indicators have long been prioritized for hands-on applications by farmers or technical advisors. Efforts to develop applied tests that could help farmers relate soil quality to productivity that began in the late 1970s (e.g. Geyer et al., 1980) and identify a useful subset of indicators continue to be a priority (Mei et al., 2019; Xia and Wander, 2021). Whether or not the information they provide is of enough value to farmers remains an open question. It may be that transaction costs could be lowered enough to help farmers use soil information in the way envisioned many decades ago (Zusevics, 1979) by using technologyies that reflect land capability (Quandt et al., 2020).

Schjønning et al. (2004) prioritized interpretation of indicator response to management and asserted that management-based, not indicator-based, thresholds must be established to embed ethics into decision making to successfully implement the soil health concept. These types of means, or practice-based indicators, are well suited for policy-focused applications, conservation ranking and product valorization when adequately tied to services. Efforts attempting to relate indicator response to management use a wide variety of statistical techniques (Bunneman et al., 2018). Responsiveness to management is most often studied by comparing plots in replicated trials or multiple sites, including farm fields, or areas under different use or management using 'space for time substitution' that assumes qualities were initially similar. Samples taken from the same location separated in time are often assumed to be the best way to separate the influences of management and soil change or development. Unfortunately, this kind of comparison is much less common and can be problematic if care is not taken to prevent false equivalencies caused by slight differences in sampling, handling, processing or analytical methods. Meta-analytic studies comparing indicator status under contrasting management are commonly used to rank practice efficacy (Ugarte et al., 2018) and identify indicators that can discriminate among land-use histories (Lori et al., 2017). Unfortunately, few studies contextualize indicators or collect co-variates need to appropriately reward or target management through any mechanism by allowing users to separate management effects from uncontrollable or unchangeable system factors (Halberg et al., 2005). Extrapolation of services associated with management practices can be risky and when overextended can undermine the veracity of stewardship claims (Chabala et al., 2020).

2.3 Site specificity and resistance and resilience

Soil quality and health assessment struggle with the need to contextualize information and establish viable reference states. The environment or habitat provided by soils' physical and chemical structure is assumed to enable or constrain biological response to perturbation and thus, collectively determine soil stability, where resistance and resilience are governed by soil physicochemical structure acting through its effect on microbial community composition and physiology (Doran and Zeiss, 2000; Griffiths and Philippot, 2013). This conceptualization includes resistance and resilience as key criteria for soil quality and soil health assessments (Doran and Zeiss, 2000; Andrews and Carrol, 2001; Seybold et al., 1999). Related assessments commonly presume an economy of nature associated with tight nutrient cycling resulting from diversity and stability. This nature-based conceptualization treats non-managed or natural conditions as the ideal that serves as a baseline for comparison. Soils' intrinsic potential has often been assessed using data taken from soils maintained under natural or undisturbed land cover, typically grasslands or, by comparing the status of soils managed using best or better practices with controls managed using standard or conventional practices. Stability framings such as this are often challenged due to their vagueness and because they invoke the 'organismic theory of ecology' which relates communities' structure and function to idealized states (e.g. Clements' idea equilibrium state) that are not representative of reality and so widely criticized if not fully rejected by the scientific community (Ehrenfeld, 1992; Rapport et al., 1998). Microbial biomass-size, diversity or community evenness are examples of biotic indicators routinely assumed to be positively related to soil function and health including soils' productivity function (Crowder et al., 2010; Wittebolle et al., 2009). Over-generalization of these kinds of interpretive assertions is a common problem as site, land-use history and scale are all needed to constrain the domain of validity of stability statements (Grimm and Wiessel, 1997). Despite these weaknesses and the fact that drivers of microbial community stability and ability to resist or recover from disturbance remain poorly understood, the association between resilience, microbial community composition and soil capacity to function persists (eg: Allison and Martiny, 2016; Shade et al., 2012).

2.4 Emerging concepts

Fortunately, we are gaining insights needed to manage ecological systems and their biotic components intentionally with more basic research. Functional genes have usefully revealed the influence of cover crop mixes and methods of termination on ammonia oxidizers and denitrifiers (Romdhane et al., 2019). We have modest confirmation of associations between cover crop biomass, taxonomic (species richness) or functional (legumes vs. non-legumes) diversity and cover crop mixtures (Finney et al., 2017). The growing interest in the use of functional taxa that endophytes, symbionts, pathogens, and plant-growth-promoting rhizobacteria provide useful information (Philippot et al., 2013) is paralleled by a rising interest in the larger community of soil microorganisms or soil microbiome (Chaparro et al., 2012). The plant microbiome has emerged as a fundamental trait that includes mutualism enabled through diverse biochemical mechanisms first revealed by studies of plant-growth-promoting and plant-health-promoting bacteria (Bulgarelli et al., 2013). Our ability to understand how plants recruit protective microorganisms and enhance microbial activity to suppress pathogens in the rhizosphere has begun to provide some mechanistic insights into theories proposed decades earlier (Berendsen et al., 2012; Philippot et al., 2013). This includes management of plant-growth-promoting rhizobacteria (PGPR) that colonize roots and mote phytohormones and other signals that enhance lateral root branching and development of root hairs (Vacheron et al., 2013). Communications between plants that modulate interactions favor association with beneficial microbes when grown under stressful conditions (Rosier et al., 2016). A rhizosphere-focused paradigm that has emerged posits that individual functions are coordinated by the community through biochemical networks connecting hosts and associated microbes where plants and animals are interdependent entities or 'holobionts' (Bordenstein et al., 2015). Research by Mendes et al. (2011) identified bacterial taxa and genes involved in the suppression of a fungal root pathogen by coupling PhyloChip-based metagenomics of the rhizosphere with culture-dependent functional analyses a decade ago suggests that indicators might reveal the status of the holobiont. While this kind of knowledge is useful to have, it is unlikely to lead to soil quality indicators as they are currently conceived because molecular communications fluctuate in space and time with plant growth stage, interactions between plants and other species, management techniques and edaphic factors (Chaparro et al., 2012). In many ways, the soil-testing approach used to assess and manage soils is incompatible with understanding or management of these fine-scape processes or associated (Vestergaard et al., 2017; Baveye et al., 2018). Appreciation of co-evolution of crops and their holobionts (Peiffer et al., 2013) has made crop breeding and selection more promising targets for information use. Intentional efforts to determine how to select microbiome–host interactions employ basic principles of quantitative genetics and community ecology. Mueller and Sachs (2015) outline core concepts and suggest how to manage soils to facilitate coadaptation between plants and microbes to enhance useful productivity along with resistance and resilience of the desired community.

3 Soil health assessment

3.1 Scoring

After efforts have identified responsive indicators and selected them for use, the indicators are scored or evaluated in relation to one or more functions or in relation to an acceptable reference state. Individual variables are commonly transformed to simultaneously rate and convert values into unitless scores that can then be combined into an integrated health index (Doran and Parkin, 1994; Liu et al., 2018). For example, Mukherjee and Lal (2014) followed standard steps when they developed a scoring function for soil productivity using the weighted additive method (Karlen and Stott, 1994; Amacher et al., 2007; Fernandes et al., 2011) that considered root development, water storage and nutrient supply in relation to crop productivity by deriving indicator scores using principal component analysis. The maximum yield observed was used as the reference state for functional scoring and this single dimension score does not distinguish quality from productivity and does very little to address health. Whether biotic indicators are judged through correlation in relation to the soil's intrinsic potential (optimum or maximum state based on inherent potential) or in relation to an established reference state, one must identify some kind of benchmark and interpret indicator status using objective criteria that are holistic. Thiele-Bruhn et al. (2020) suggest explicit consideration of indicator constancy and dynamic equilibrium and use of endpoints that represent a potential function of soil microorganisms (implying a reference) rather than the use of actual activity levels could improve indexing. Regionalized references tailored to cropping systems can provide reference states. For example, Krüger et al. (2018) used regional references to assess spatial and seasonal variability of biological indicators (soil respiration potential, microbial biomass carbon, microbial C/N ratio, net nitrogen mineralization, metabolic potential of soil bacteria, earthworm abundance, microbial quotient and metabolic quotient) in grassland and cropland soils in Belgium. That and other works suggest population-based approach may be superior to the use of natural or idealized reference states. Ideally, reference ranges for indicators can be established within cropland or other land-use types that consider not only spatial variability but also time and of course, depth of sampling.

3.2 Combining

Once vetted, individual indicators are often combined into indexes after transformation using a variety of statistical techniques. For example, Ritz et al. (2009) used a semi-objective logical sieve to reduce 183 possible candidate indicators to 17 genotypic-, phenotypic- and functional- variables for use in national-scale soil monitoring with scoring by experts and stakeholders performed using scientific and technical criteria. Repeated iterations allowed indicator scores and weightings to account for end-user requirements and expert opinion. Scoring methods can recognize the need for site-specific scoring that considers both inherent and dynamic properties. For example,

Vogel et al. (2018, 2019) propose a systemic modeling framework to couple reductionist yet observable indicators for soil functions with detailed process understanding. They propose that functional attributes can be explained by a network of interacting processes derived from scientific evidence, with the non-linear character of interactions leading to stability and resilience of soil function. Integration steps have begun to attempt to quantify interdependencies and tradeoffs between functions. For example, Vrebos et al. (2020) used DayCent to estimate model indicators to build crop-specific Bayesian belief networks (BBNs) to estimate productivity, climate protections, water protections and nutrient use efficiency and tradeoffs in function under current European Union conditions. Only relationships found to be plausible were retained after 'validation' by expert opinion. Algorithms used by BBNs can help capture the inductive inferences of stakeholders (Bonawitz et al., 2011) and provide a computational framework needed to compare the expected consequences of different types of management actions (Dorazio and Johnson, 2003). While, In theory, BBNs, which are a graphical network of nodes linked by probabilities, can help users make more informed and disclosed decisions about resource management (McCann et al., 2006), caution must be applied when they are based on limited or synthetic data that is also not proven (Marcot et al., 2006). Some fear the subjective aspect of BBNs can potentially obfuscate and embed bias in a way that is difficult to remove (Schweizer, 2019); however, this is one of the few ways to enable valid scientific quantification of perceptions of soil functions /services (including uncertainties) (Baveye et al., 2016). Baveye et al. (2016) recommend the use of BBNs with deliberative decision-making methods and note that the literacy of participants is critical to effectively and ethically rank services. Correlation and co-occurrence analyses and multi-criteria decision support systems are examples of useful techniques (Zwetsloot et al., 2020). Before decisions can be made, estimates of services and preferences should be provided (Coker et al., 2019) to parties engaged in ranking and weighting activities but unfortunately, location and system-specific information about services supplied by practices or policies are commonly inadequate.

4 Conclusion

Even though soil quality and health continue to receive academic critiques of their utility and scientific validity, the public and private sectors have rallied around the soil health concept as a way to help address food security, environmental degradation and climate change. Whether or not indicators of quality or soil health assessments prove to be useful or, instead consume resources better spent on other approaches may depend upon whether proponents can better clarify their objectives to apply indicators with functional relevance for specific biophysical and civic contexts. In order for effects-based indicators to find commercial success and be useful for valorization of individual fields or farms, extensional definitions that assert nebulous notions of diversity and stability must be replaced with appropriate references states and well-articulated relationships between indicator status and nutrient use efficiency, plant health and pest suppression. These understandings must be combined with up-to-date agronomic information and integrated nutrient management strategies that improve not only soil quality (Wu and Ma, 2015) but also productivity and other services. Improving data access through digital platforms and new technologies may be able to lower both the transaction and monetary costs associated with historical land capability classification to make information useful for site-specific applications. Calls for standardized methods and national or even global commitment to this approach to support ecosystem service markets first need to overcome our limited or incomplete understanding of inherent sensitivity, responsiveness to management and functional relevance of many indicators (Römbke et al., 2018). Very few efforts proceed to establish benchmarks that provide contextual information, delimited by space and land use in sufficient detail to reflect the inherent potential of a site in its optimum status and allow comparison with the present condition. Calls for policies that promote soil stewardship through practices that build soil organic matter and presumably other soil health attributes (e.g. Gomiero, 2016; Amelung et al., 2020) need to more than just acknowledge the need for contextualization. Efforts to implement policy or set targets often fall back on expert opinions and assertions about the benefits of different farming practices despite full knowledge of site-based variability. Analytical approaches like BBNs and deliberative frameworks might include procedures for a clear record of the decision-making process, flexibility in prioritization of function, and the ability to accommodate the inclusion of new or additional methods or indicators into the framework (Ritz et al., 2009; Baveye et al., 2016).

5 Where to look for further information

- Nature Reviews: Plant microbiome. https://www.nature.com/collections/jcbagaigaa.
- Standards by ISO 190: Soil Quality: https://www.iso.org/committee/54328/x/catalogue.
- Natural Resources Conservation Service Soil Health: https://www.nrcs.usda.gov/wps/portal/nrcs/main/soils/health/.

6 References

Acosta-Martinez, V., Pérez-Guzmána, L. and Johnson, J. M. F. 2019. Simultaneous determination of β-glucosidase, β-glucosaminidase, acid phosphomonoesterase, and arylsulfatase activities in a soil sample for a biogeochemical cycling index. *Applied Soil Ecology* 142:72–80.

Allison, S. D. and Martiny, J. B. H. 2016. Resistance, resilience, and redundancy in microbial communities. *PNAS* 105(Suppl. 1):11512–11519. DOI:10.1073/pnas.0801925105.

Amacher, M. C., O'Neill, K. P. and Perry, C. H. 2007. Soil vital signs: a new soil quality index (SQI) for assessing forest soil health. *Research Papers*. DOI:10.2737/RMRS-RP-65.

Amelung, W., Bossio, D., de Vries, W. , Kögel-Knabner, I., Lehmann, J., Amundson, R., Bol, R., Collins, C., Lal, R., Leifeld, J., Minasny, B., Pan, G., Paustian, K., Rumpel, C., Sanderman, J., van Groenigen, W. J., Mooney, S., van Wesemael, B., Wander, M. and Chabbi, A. 2020. Towards a global-scale soil climate mitigation strategy. *Nature Communications* 11:5427. DOI:10.1038/s41467-020-18887-7.

Andrews, S. S. and Carroll, C. R. 2001. Designing a soil quality assessment tool for sustainable agroecosystem management. *Ecological Applications* 11(6):1573–1585.

Andrews, S. S., Karlen, D. L. and Mitchell, J. P. 2002. A comparison of soil quality indexing methods for vegetable production systems in Northern California. *Agriculture, Ecosystems and Environment* 90(1):25–45.

Andrews, S. S., Flora, C. B., Mitchell, J. P. and Karlen, D. L. 2003. Growers' perceptions and acceptance of soil quality indices. *Geoderma* 114(3–4):187–213. DOI:10.1016/S0016-7061(03)00041-7.

Andrews, S. S., Karlen, D. L. and Cambardella, C. A. 2004. The soil management assessment framework: a quantitative soil quality evaluation method. *Soil Science Society of America Journal* 68(6):1945–1962.

Bailey, V. L., Bond-Lamberty, B., DeAngelis, K., Grandy, A. S., Hawkes, C. V., Heckman, K., Lajtha, K., Phillips, R. P., Sulman, B. N., Todd-Brown, K. E. O. and Wallenstein, M. D. 2018. Soil carbon cycling proxies: understanding their critical role in predicting climate change feedbacks. *Global Change Biology* 24(3):895–905. DOI:10.1111/gcb.13926.

Baveye, P. C. 2020. Bypass and hyperbole in soil research: a personal view on plausible causes and possible remedies. *European Journal of Soil Science* 72(1):21–28. DOI:10.1111/ejss.12940.

Baveye, P. C., Baveye, J. and Gowdy, J. 2016. Soil "ecosystem" services and natural capital: critical appraisal of research on uncertain ground. *Frontiers in Environmental Science* 4:49. DOI:10.3389/fenvs.2016.00041.

Baveye, P. C., Otten, W., Kravchenko, A., Balseiro-Romero, M., Beckers, É, Chalhoub, M., Darnault, C., Eickhorst, T., Garnier, P., Hapca, S., Kiranyaz, S., Monga, O., Mueller, C. W., Nunan, N., Pot, V., Schlüter, S., Schmidt, H. and Vogel, H. J. 2018. Emergent properties of microbial activity in heterogeneous soil microenvironments: different research approaches are slowly converging, yet major challenges remain. *Frontiers in Microbiology* 9:1929. DOI:10.3389/fmicb.2018.01929.

Berendsen, R. L., Pieterse, C. M. J. and Bakker, P. A. H. M. 2012. The rhizosphere microbiome and plant health. *Trends in Plant Science* 17(8):478–486. DOI:10.1016/j.tplants.2012.04.001.

Bonawitz, E., Denison, S., Chen, A., Gopnik, A. and Griffiths, T. 2011. A simple sequential algorithm for approximating bayesian inference. Proceedings of the Annual Meeting of the Cognitive Science Society. Available at: https://escholarship.org/uc/item/40t6n43w (vol. 33).

Bordenstein, S. R., Theis, K. R., Theis, S. R. and Waldor, K. R. 2015. Host biology in light of the microbiome: ten principles of holobionts and hologenomes. *PLoS Biology* 13:e1002226.

Bouma, J. and McBratney, A. 2013. Framing soils as an actor when dealing with wicked environmental problems. *Geoderma* 200-201:130–139.

Breeuwsma, A., Wösten, J. H. M., Vleeshouwer, J. J., van Slobbe, A. M. and Bouma, J. 1986. Derivation of land qualities to assess environmental problems from soil surveys. *Soil Science Society of America Journal* 50(1):186–190.

Brevick, E. C., Steffan, J. J., Rodrigo-Comino, J., Neubert, D., Burgess, L. C. and Cerdà, A. 2019. Connecting the public with soil to improve human health. *European Journal of Soil Science* 70(4):898-910. DOI:10.1111/ejss.12764.

Bulgarelli, D., Schlaeppi, K., Spaepen, S., Ver Loren van Themaat, E. and Schulze-Lefert, P. 2013. Structure and functions of the bacterial microbiota of plants. *Annual Review of Plant Biology* 64:807–838.

Bunemann, E. K., Bongiorno, G., Bai, Z. G., Creamer, R. E., De Deyn, G., de Goede, R., Fleskens, L., Geissen, V., Kuyper, T. W., Mäder, P., Pulleman, M., Sukkel, W., van Groenigen, J. W. and Brussaard, L. 2018. Soil quality-a critical review. *Soil Biology and Biochemistry* 120:105–125. DOI:10.1016/j.soilbio.2018.01.030.

Burns, R. G., DeForest, J. L., Marxsen, J., Sinsabaugh, R. L., Stromberger, M. E., Wallenstein, M. D., Weintraub, M. N. and Zoppini, A. 2013. Soil enzymes in a changing environment: current knowledge and future directions. *Soil Biology and Biochemistry* 58:216–234. DOI:10.1016/j.soilbio.2012.11.009.

Cassman, K. G. 1999. Ecological intensification of cereal production systems: yield potential, soil quality, and precision agriculture. *Proceedings of the National Academy of Sciences of the United States of America* 96(11):5952-5959. DOI:10.1073/pnas.96.11.5952.

Chabala, L. M., Chimungu, J. G., Lark, R. M., Mtambanengwe, F., Nalivata, P. C., Phiri, E. and Sakala, G. M. 2020. Eliciting experts' tacit models for the interpretation of soil information, an example from the evaluation of potential benefits from conservation agriculture. *Geoderma* 376:114545.

Chaparro, J. M., Sheflin, A. M., Manter, D. K. and Vivanco, J. M. 2012. Manipulating the soil microbiome to increase soil health and plant fertility. *Biology and Fertility of Soils* 48(5):489–499.

Chen, H., Li, D., Zhao, J., Zhang, W., Xiao, K. and Wang, K. 2018. Nitrogen addition aggravates microbial carbon limitation: evidence from ecoenzymatic stoichiometry. *Geoderma* 329:61-64. DOI:10.1016/j.geoderma.2018.05.019.

Coker, G., Richard, M. G., Bayne, K. M., Smaill, S. K., Garrett, L. S., Matson, A. L., Wakelin, S. A. and Martínez-Paz, S. 2019. Stakeholder valuation of soil ecosystem services from New Zealand's planted forests. *PLoS ONE* 14(8):e0221291.

Cox, C. 1995. Soil quality–goals for national policy. *Journal of Soil and Water Conservation* 50(3):223–223.

Crowder, D. W., Northfield, T. D., Strand, M. R. and Snyder, W. E. 2010. Organic agriculture promotes evenness and natural pest control. *Nature* 466:109–112.

Dale, T. and Carter, V. G. 1955. *Topsoil and Civilization*. Norman, OK: University of Oklahoma Press.

Davis, C. B. and Miller, F. P. 2000. Soil, sustainability, and security; the importance of ecosystem integrity. In: Wilson, M. J. and Maliszewska-Kordybach, B. (Eds) *Soil Quality, Sustainable Agriculture and Environmental Security in Central and Eastern Europe Springer-Science*. DOI:10.1007/978-94-011-4181-9.

Debailleul, G. 1990. Evolution of U.S. soil conservation policies since World War II: lessons for Canada. *Canadian Journal of Agricultural Economics/Revue Canadienne d'Agroeconomie* 38(4):757–769.

De Curato, U. 2020. Soil microbiota manipulation and its role in suppressing soil-borne plant pathogens in organic farming systems under the light of microbiome-assisted strategies. *Chemical and Biological Technologies in Agriculture* 7(1):17. DOI:10.1186/s40538-020-00183-7.

De Vries, F. T., Liiri, M. E., Bjørnlund, L., Bowker, M. A., Christensen, S., Setälä, H. M. and Bardgett, R. D. 2012. Land use alters the resistance and resilience of soil food webs to drought. *Nature Climate Change* 2(4):276–280.

Doran, J. W. and Parkin, T. B. 1994. *Defining Soil Quality for a Sustainable Environment*. SSSA Special Publication. Madison, WI: Madison Book Company.

Doran, J. W. and Zeiss, M. R. 2000. Soil health and sustainability: managing the biotic component of soil quality. *Applied Soil Ecology* 15(1):3–11.

Dorazio, R. M. and Johnson, F. A. 2003. Bayesian inference and decision theory - A framework for decision making in natural resource management. *Ecological Applications* 13(2):556–563.

Dumanski, J. and Pieri, C. 2000. Land quality indicators: research plan. *Agriculture, Ecosystems and Environment* 81(2):93–102.

Dumroese, P., Jurgensen, M., Elliot, W., Rice, T., Nesser, J., Collins, T. and Meurisse, R. 2000. Soil quality standards and guidelines for forest sustainability in northwestern North America. *Forest Ecology and Management* 138(1):445–462.

Ehrenfeld, D. 1992. Ecosystem health and ecological theories. In: Costanza, R., Norton, B. and Haskell, B. (Eds) *Ecosystem Health–New Goals for Environmental Management*. Washington, DC: Island Press. pp. 135–143.

Feng, Q., Liu, Z., Zhong, S., Li, R., Xia, H., Jie, Z.,Wen, B., Chen, X., Yan, W., Fan, Y., Guo, Z., Meng, N., Chen, J., Yu, X., Zhang, Z., Kristiansen, K., Wang, J., Xu, X., He, K. and Li, G. 2016. Integrated metabolomics and metagenomics analysis of plasma and urine identified microbial metabolites associated with coronary heart disease. *Scientific Reports*. DOI:10.1038/srep22525.

Fernandes, J. C., Gamero, C. A., Rodrigues, J. G. L. and Mirás-Avalos, J. M. 2011. Determination of the quality index of a Paleudult under sunflower culture and different management systems. *Soil and Tillage Research* 112(2):167–174.

Fierer, N. 2017. Embracing the unknown: disentangling the complexities of the soil microbiome. *Nature Reviews. Microbiology* 15(10):579–590.

Finney, D. M., Buyer, J. S. and Kaye, J. P. 2017. Living cover crops have immediate impacts on soil microbial community structure and function. *Journal of Soil and Water Conservation* 72(4):361–373; DOI:10.2489/jswc.72.4.361.

Food and Agricultural Organization 2015. *Status of the World's Soil Resources: Main Report*. Rome Italy. ISBN 978-92-5-109004-6. p. 650.

Forbes, A. C. 1928. Quality of soil in relation to food and timber supply. *Nature* 122(3063):54–55.

Foucault, M. 1970. The order of discourse. 1981. In: McLeod, I. (Trans.) Young, R. (Ed.) *Untying the Text: A Post-Structuralist Reader*. London: Routledge. pp. 48–78.

Geisen, S., Mitchell, E. A. D., Wilkinson, D. M., Adl, S., Bonkowski, M., Brown, M. W., Fiore-Donno, A. M., Heger, T. J., Jassey, V. E. J., Krashevska, V., Lahr, D. J. G., Marcisz, K., Mulot, M., Payne, R., Singer, D., Anderson, O. R., Charman, D. J., Ekelund, F., Griffiths, B. S., Rønn, R., Smirnov, A., Bass, D., Belbahri, L., Berney, C., Blandenier, Q., Chatzinotas, A., Clarholm, M., Dunthorn, M., Feest, A., Fernández, L. D., Foissner, W., Fournier, B., Gentekaki, E., Hájek, M., Helder, J., Jousset, A., Koller, R., Kumar, S., La Terza, A., Lamentowicz, M., Mazei, Y., Santos, S. S., Seppey, C. V. W., Spiegel,

F. W., Walochnik, J., Winding, A. and Lara, E. 2017. Soil protistology rebooted: 30 fundamental questions to start with. *Soil Biology and Biochemistry* 111:94–103.

Gerasimov, I. P. 1983. Land resources of the world; their use and reserves. Definition of land quality and agricultural potential by soil survey maps. *Geoforum* 14(4):427–439.

Geyer, W. A., Marquard, R. D. and Barber, J. F. 1980. Black walnut site quality in relation to soil and topographic characteristics in northeastern Kansas. *Journal of Soil and Water Conservation* 35(3):135–137.

Giampietro, M. and, Funtowicz, S. O. 2020. From elite folk science to the policy legend of the circular economy. *Environmental Science and Policy* 109:64–72.

Gomiero, T. 2016. Soil degradation, land scarcity and food security: reviewing a complex challenge. *Sustainability* 8(3):1–1.

Griffiths, B. S. and Philippot, L. 2013. Insights into the resistance and resilience of the soil microbial community. *FEMS Microbiology Reviews* 37(2):112–129. DOI:10.1111/j.1574-6976.2012.00343.x.

Griffiths, B. S., Faber, J. and Bloem, J. 2018. Applying soil health indicators to encourage sustainable soil use: the transition from scientific study to practical application. *Sustainability* 10(9):3021.

Grimm, V. and Wissel, C. 1997. Babel, or the ecological stability discussions: an inventory and analysis of terminology and a guide for avoiding confusion. *Oecologia* 109(3):323–334. DOI:10.1007/s004420050090.

Halberg, N., Verschuur, G. and Goodlass, G. 2005. Farm level environmental indicators: are they useful? An overview of green accounting systems for European farms. *Agriculture, Ecosystems and Environment* 105(1–2):195–212.

Harris, J. 2009. Soil microbial communities and restoration ecology: facilitators or followers? *Science* 325(5940):573–574.

Hartmann, M., Frey, B., Mayer, J., Mäder, P., Widmer, F. and Maeder, P. 2015. Distinct soil microbial diversity under long-term organic and conventional farming. *The ISME Journal* 9(5):1177–1194.

Hillel, D. 1991. *Out of the Earth: Civilization and the Life of the Soil*. Berkeley, CA : University of California Press.

Hoon, R. C. and Dhawan, C. L. 1950. Natural flora as an index of soil quality, II. *Indian Journal of Agricultural Sciences* 20:185–195.

Hortensius, D. and Welling, R. 1996. International standardization of soil quality measurements. *Communications in Soil Science and Plant Analysis* 27(3–4):387–402.

Karlen, D. L. and Rice, C. W. 2015. Soil degradation: will humankind ever learn? *Sustainability* 7(9):12490–12501.

Karlen, D. L. and Stott, D. E. 1994. A framework for evaluating physical and chemical indicators of soil quality. In: Doran, J. W., Coleman, D. C., Bezdicek, D. F. and Stewart, B. A. (Eds) *Defining Soil Quality for a Sustainable Environment*. Madison, WI: Soil Science Society of America. pp. 53–72.

Karlen, D. L., Mausbach, M. J., Doran, J. W., Cline, R. G., Harris, R. F. and Schuman, G. E. 1997. Soil quality: a concept, definition, and framework for evaluation. *SSSAJ* 61:4. DOI:10.10.2136/sssaj1997.03615995006100010001x.

Kozin, V. K. 1990. Energy supply of humus as a criterion for soil quality evaluation. *Soviet Soil Science* 22(6):122–125.

Krüger, I., Chartin, C., van Wesemael, B. and Carnol, M. 2018. Defining a reference system for biological indicators of agricultural soil quality in Wallonia, Belgium. *Ecological Indicators* 95:568–578.

Landa, G. and Angel, J. 2014. Notes on 'the order of discourse'. Available at: https://ssrn.com/abstract=2439772 or DOI:10.2139/ssrn.2439772.

Larkin, R. P. 2015. Soil health paradigms and implications for disease management. *Annual Review of Phytopathology* 53:199–221. DOI:10.1146/annurev-phyto-080614-120357.

Lekkerkerk, L. J. A., Vanveen, J. A. and Vandegeijn, S. C. 1990. Influence of climatic change on soil quality: consequences of increased atmospheric CO2 concentration on carbon input and turnover in agroecosystems. *Greenhouse Effect and Primary Productivity in European Agro-Ecosystems. Wageningen. Pudoc*:46–47.

Letey, J., Sojka, R. E., Upchurch, D. R., Cassel, D. K., Olson, K., Payne, B., Petrie, S., Price, G., Reginato, R. J., Scott, H. D., Smethurst, P. and Triplett, G. 2003. Deficiencies in the soil quality concept and its application. *Journal of Soil and Water Conservation* 58:180–187.

Liu, J., Wu, L., Chen, D., Yu, Z. and Wei, C. 2018. Development of a soil quality index for Camellia oleifera forestland yield under three different parent materials in Southern China. *Soil and Tillage Research* 176:45–50.

Liu, Y., Wu, K. and Zhao, R. 2020. Bibliometric analysis of research on soil health from 1999 to 2018. *Journal of Soils and Sediments* 20(3):1513–1525.

Lori, M., Symnaczik, S., MaÈder, P., De Deyn, G. and Gattinger, A. 2017. Organic farming enhances soil microbial abundance and activity a meta-analysis and meta-regression. *PLoS ONE* 12(7):e0180442. DOI:10.1371/journal.pone.0180442.

Marcot, B. G., Steventon, J. D., Sutherland, G. D. and McCann, R. K. 2006. Guidelines for developing and updating Bayesian belief networks applied to ecological modeling and conservation. *Canadian Journal of Forest Research* 36(12):3063-3074.

McCann, R. K., Marcot, B. G. and Ellis, R. 2006. Bayesian belief networks: applications in ecology and natural resource. *Canadian Journal of Forest Research* 36: 3053–3062. DOI: 10.1 139R06-238.

McDaniel, M. D., Grandy, A. S., Tiemann, L. K. and Weintraub, M. N. 2014. Crop rotation complexity regulates the decomposition of high and low quality residues. *Soil Biology and Biochemistry* 78:243–254.

Mei, N., Yang, B., Tian, P., Jiang, Y., Sui, P., Sun, D., Zhang, Z. and Qi, H. 2019. Using a modified soil quality index to evaluate densely tilled soils with different yields in Northeast China. *Environmental Science and Pollution Research International* 26(14):13867–13877.

Mendes, R., Kruijt, M., de Bruijn, I., Dekkers, E., van der Voort, M., Schneider, J. H., Piceno, Y. M., DeSantis, T. Z., Andersen, G. L., Bakker, P. A. and Raaijmakers, J. M. 2011. Deciphering the rhizosphere microbiome for disease-suppressive bacteria. *Science* 332(6033):1097–1100.

Mueller, U. G. and Sachs, J. L. 2015. Engineering microbiomes to improve plant and animal health trends in microbiology 23:10. DOI:10.1016/j.tim.

Mukherjee, A. and Lal, R. 2014. Comparison of soil quality index using three methods. *PLoS ONE* 9(8):e105981. DOI:10.1371/journal.pone.0105981.

Neher, D. A. and Barbercheck, M. E. 2019. Soil microarthropods and soil health: intersection of decomposition and pest suppression in agroecosystems. *Insects* 10(12):414. DOI:10.3390/insects10120414.

Ney, L., Franklin, D., Mahmud, K., Cabrera, M., Hancock, D., Habteselassie, M., Newcomer, Q., Dahal, S. and Subedi, A. 2019. Sensitivity of nematode community analysis to agricultural management practices and inoculation with local effective microorganisms in the Southeastern United States. *Soil Systems* 3(2):41.

Ouyang, Y., Reeve, J. R. and Norton, J. M. 2018. Soil enzyme activities and abundance of microbial functional genes involved in nitrogen transformations in an organic farming system. *Biology and Fertility of Soils* 54(4):437-450. DOI:10.1007/s00374-018-1272-y.

Patzel, N., Sticher, H. and Karlen, D. L. 2000. Soil fertility–phenomenon and concept. *Journal of Plant Nutrition and Soil Science* 163(2):129-142.

Peiffer, J. A., Spor, A., Koren, O., Jin, Z., Tringe, S. G., Dangl, J. L., Buckler, E. S. and Ley, R. E. 2013. Diversity and heritability of the maize rhizosphere microbiome under field conditions. *Proceedings of the National Academy of Sciences of the United States of America* 110(16):6548-6553.

Peters, B. C. 1973. Changing ideas about the use of vegetation as an indicator of soil quality: example of New York and Michigan. *Journal of Geography* 72(2):18-28. DOI:10.1080/00221347308981021.

Philippot, L., Raaijmakers, J. M., Lemanceau, P. and van der Putten, W. H. 2013. Going back to the roots: the microbial ecology of the rhizosphere. *Nature Reviews Microbiology* 11:789-799.

Quandt, A., Herrick, J., Peacock, G., Salley, S., Buni, A., Mkalawa, C. C. and Neff, J. 2020. A standardized land capability classification system for land evaluation using mobile phone technology. *Journal of Soil and Water Conservation* 75(5):579-589:DOI:10.2489/jswc.2020.00023.

Rapport, D. J., Costanza, R. and McMichael, A. J. 1998. Assessing ecosystem health. *Tree* 13:397-401.

Reddy, M. V. 1986. Soil-inhabiting arthropods as indicators of environmental quality. *Acta Biologica Hungarica* 37(1):79-84.

Ritz, K., Black, H. I. J., Campbell, C. D., Harris, J. A. and Wood, C. 2009. Selecting biological indicators for monitoring soils: a framework for balancing scientific and technical opinion to assist policy development. *Ecological Indicators* 9(6):1212-1221.

Römbke, J., Bernard, J. and Martin-Laurent, F. 2018. Standard methods for the assessment of structural and functional diversity of soil organisms: a review. *Integrated Environmental Assessment and Management* 14(4):463-479.

Romdhane, S., Spor, A., Busset, H., Falchetto, L., Martin, J., Bizouard, F., Bru, D., Breuil, M. C., Philippot, L. and Cordeau, S. 2019. Cover crop management practices rather than composition of cover crop mixtures affect bacterial communities in no-till agroecosystems. *Frontiers in Microbiology* 10:1618-1618.

Rosier, A., Bishnoi, U., Lakshmanan, V., Sherier, D. J. and Bais, H. P. 2016. A perspective on inter-kingdom signaling in plant-beneficial microbe interactions. *Plant Molecular Biology* 90(6):537-548. DOI:10.1007/s11103-016-0433-3.

Schipper, L. A. and Sparling, G. P. 2000. Performance of soil condition indicators across taxonomic groups and land uses. *Soil Science Society of America Journal* 64(1):300-311.

Schipper, L. A. and Sparling, G. P. 2002. Soil quality at a national scale in New Zealand. *Journal of Environmental Quality* 31(6):1848-1857.

Schjønning, P., Elmholt, S. and Christensen, B. T. 2004. Soil quality management–concepts and terms. Chapter: 1 In: Schjønning, P., Elmholt, S. and Christensen, B. T. (Eds) *Managing Soil Quality–Challenges in Modern Agriculture*. CABI Publishing. DOI:10.1079/9780851996714.0001.

Schultz, A. 1925. Studies on the vegetation in western Nyland (South Finland) and its relationship with the quality of the soil. *Petermanns Mitteilungen* 43:71.9-10:223-223.

Schweizer, M. 2019. De-biasing role induced bias using Bayesian networks. *Law, Probability and Risk* 18:255–273. DOI:10.1093/lpr/mgz015.

Seybold, C. A., Herrick, J. E. and Brejda, J. J. 1999. Soil resilience: a fundamental component of soil quality. *Soil Science* 164(4):224–234.

Shade, A., Peter, H., Allison, S. D., Baho, D. L., Berga, M., Burgmann, H., Huber, D. H., Langenheder, S., Lennon, J. T., Martiny, J. B. H., Matulich, K. L., Schmidt, T. M. and Handelsman, J. 2012. Fundamentals of microbial community resistance and resilience. *Frontiers in Microbiology* 3:417. DOI:10.3389/fmicb.2012.00417.

Siegrist, R. L. 1990. Development and implementation of soil quality and cleanup criteria for contaminated sites. *Contaminated Soil* 90(1 and 2): 149–156.

Singh, A. N., Fu, S., Chonghui, S., Wang, S., Yan, S., Singh, A. N., Li, Y., Cui, Y. and Hu, L. 2012. A soil fauna index for assessing soil quality. *Soil Biology and Biochemistry* 47:158. DOI:10. 016/j.soilbio.2011.11.014.

Steinhardt, G. C. 1995. Soil quality: a new idea that includes and old one. *Journal of Soil and Water Conservation* 50(3):222.

Stengel, P. 2000. Relations plante/sol/climat Connaître le sol : nouveaux enjeux pour la gestion des territoires ruraux. *Oléagineux, Corps Gras, Lipides* 7(6):485–489.

Tahat, M. M, Alananbeh, K. M., Othman, Y. A. and Leskovar, D. I. 2020. Soil health and sustainable agriculture. *Sustainability* 12(12):4859.

Thiele-Bruhn, S., Schloter, M., Wilke, B., Beaudette, L. A., Martin Laurent, F., Cheviron, N., Mougin, C. and Römbke, J. 2020. Identification of new microbial functional standards for soil quality assessment. *SOIL* 6(1):17–34.

Tilman, D., Cassman, K. G., Matson, P. A., Naylor, R. and Polasky, S. 2002. Agricultural sustainability and intensive production practices. *Nature* 418(6898):671-677.

Ugarte, C. M., Kwon, H.-Y. and Wander, M. M. 2018. Conservation management and ecosystem services in Midwest agricultural systems. *Journal of Soil and Water Conservation* 73(4):422–433. DOI:10.2489/jswc.73.4.422.

Urzelai, A., Vega, M. and Angulo, E. 2000. Deriving ecological risk-based soil quality values in the Basque Country. *Science of the Total Environment* 247(2–3):279–284.

Vacheron, J., Desbrosses, G., Bouffaud, M. L., Touraine, B., Moënne-Loccoz, Y., Muller, D., Legendre, L., Wisniewski-Dyé, F. and Prigent-Combaret, C. 2013. Plant growth-promoting rhizobacteria and root system functioning. *Frontiers in Plant Science* 4:356–356.

Van der Werf, H. M. G. and Petit, J. 2002. Evaluation of the environmental impact of agriculture at the farm level: a comparison and analysis of 12 indicator-based methods. *Agriculture, Ecosystems and Environment* 93(1–3):131–145.

Vestergaard, G., Schulz, S., Schöler, and Schloter, M. 2017. Making big data smart–how to use metagenomics to understand soil quality. *Biology and Fertility of Soils* 53:479–484. DOI:10.1007/s00374-017-1191-3.

Vogel, H., Bartke, S., Daedlow, K., Helming, K., Kögel Knabner, I., Lang, B., Rabot, E., Russell, D., Stößel, B., Weller, U., Wiesmeier, M. and Wollschläger, U. 2018. A systemic approach for modeling soil functions. *SOIL* 4(1):83-92.

Vogel, H., Eberhardt, E., Franko, U., Lang, B., Ließ, M., Weller, U., Wiesmeier, M., Wollschläger, U. and Liess, M. 2019. Quantitative evaluation of soil functions: potential and state. *Frontiers in Environmental Science* 22. DOI:10.3389/fenvs.2019.00164.

Vrebos, D., Jones, A., Lugato, E., O'Sullivan, L., Schulte, R., Staes, J. and Meire, P. 2020. Spatial evaluation and trade-off analysis of soil functions through Bayesian networks. *European Journal of Soil Science* 72(4):1575–1589. DOI:10.1111/ejss.13039.

Wander, M. M. 2009. Agroecosystem integrity and the internal cycling of nutrients. In: Bohlen, P. and House, G. (Eds) *Sustainable Agroecosystem Management*. Boca Raton, Fl: Taylor and Francis. pp. 137-166.

Wander, M. M., Walter, G. L., Nissen, T. M., Bollero, G. A., Andrews, S. S. and Cavanaugh-Grant, D. A. 2002. Soil quality. *Agronomy Journal* 94(1):23-32.

Wander, M. M., Cihacek, L. J., Coyne, M., Drijber, R. A., Grossman, J. M., Gutknecht, J. L. M., Horwath, W. R., Jagadamma, S., Olk, D. C., Ruark, M., Snapp, S. S., Tiemann, L. K., Weil, R. and Turco, R. F. 2019. Developments in agricultural soil quality and health: reflections by the research committee on soil organic matter management. *Frontiers in Environmental Science* 7. DOI:10.3389/fenvs.2019.00109.

Wang, C., Liu, D. and Baia, E. 2018. Decreasing soil microbial diversity is associated with decreasing microbial biomass under nitrogen addition. *Soil Biology and Biochemistry* 120:126-133. DOI:10.1016/j.soilbio.2018.02.003.

Wilson, M. J. 2000. Soil quality, sustainable development and environmental security. Introduction to general concepts. In: *Soil Quality, Sustainable Agriculture and Environmental Security in Central and Eastern Europe*. Dordrecht. B. Maliszewska-Kordybach: Springer. pp 3-16.

Wittebolle, L., Marzorati, M., Clement, L., Balloi, A., Daffonchio, D., Heylen, K., De Vos, P., Verstraete, W. and Boon, N. 2009. Initial community evenness favors functionality under selective stress. *Nature* 458(7238):623-626.

Wolff, W. 1939. The science of soil basis of German soil quality evaluation. *Naturwissenschaften* 27(20-21):374-376.

Worster, D. 2004. *Dust Bowl. The Southern Plains in the 1930s*. New York, NY: Oxford University Press:.

Wu, W. and Ma, B. 2015. Integrated nutrient management (INM) for sustaining crop productivity and reducing environmental impact: a review. *Science of the Total Environment* 512-513:415-427.

Xia, Y. and Wander, M. 2021. Responses of β-glucosidase, permanganate oxidizable carbon, and fluorescein diacetate hydrolysis to conservation practices. *Soil Science Society of America Journal*, 1-14. DOI:10.1002/saj2.20261.

Yamakura, T. and Sahunalu, P. 1990. Soil carbon nitrogen ratio as a site quality index for some South-East Asian forests. *Journal of Tropical Ecology* 6(3):371-377.

Yassoglou, N. J. and Kollias, V. J. 1989. Computer-assisted soil mapping for the evaluation of soil-erosion risk and land quality in Greece. In: *Application of Computerized EC Soil Map and Climate Data*. Luxembourg: Commission European Communities. pp. 237-246.

Zimmerman, C. C. 1939. Division of Esthonian soil by quality. *Rural Sociology* 4(4):490-491.

Zusevics, A. J. 1979. Field determination of soil quality based on morphological properties. *Agrochimica* 23(3-4):235-244.

Zwetsloot, M. J., Leeuwen, J., Hemerik, L., Martens, H., Simó Josa, I., Broek, M., Debeljak, M., Rutgers, M., Sandén, T., Wall, D. P., Jones, A. and Creamer, R. E. 2020. Soil multifunctionality: synergies and trade-offs across European climatic zones and land uses. *European Journal of Soil Science* 72(4):1640-1654. DOI:10.1111/ejss.13051.

Chapter 3

Understanding biological processes in soil

Regina O'Kelley and David D. Myrold, Oregon State University, USA

1 Introduction

Soil health is the manifestation of the interactions among a soil's physical, chemical, and biological properties. Because health is associated with living things, we need to view soil as a living ecosystem if we are to consider the health of a soil and design practices that enhance soil health.

The soil ecosystem consists of the myriad of organisms that live within the soil matrix. The activities of these organisms are central to determining the health of a soil. These activities can be grouped into those associated with: (1) gathering resources (e.g. organic matter decomposition, N mineralization), (2) altering the soil habitat (e.g. aggregation, C sequestration), and (3) interactions with other organisms (e.g. symbioses with plants, suppression of pathogens). Of course, some activities span these groupings: Nematodes feeding on bacteria are an interaction that results in the production of CO_2 and NH_4+. In this chapter, we will focus on the role of the microorganisms – collectively, the microbiome – in resource gathering because this has been how biology has been represented in measurements of soil health. We will first review the history of the biological indicators of soil health, concentrating on the literature prior to 2000, when the term 'soil health' emerged (Lehmann et al., 2020). Second, we will describe current efforts to refine those historic indicators and evaluate other microbial properties that may reflect the health status of a soil. Finally, we will consider the potential of more recently

http://dx.doi.org/10.19103/AS.2021.0094.02

developed molecular tools to improve our description of the microbial contribution to soil health.

2 Pre-2000

There has always been an appreciation of the importance of soil organisms in the turnover of organic matter and provisioning of nutrients for plant growth. For example, returning plant residues and the use of legumes in crop rotations are centuries-old management practices used to maintain the fertility. Although these practices continued, the advent of the widespread use of chemical fertilizers after the Second World War shifted the emphasis of soil fertility from biology to chemistry as a way of managing plant nutrients (Fig. 1). This shift led to the development of a suite of extraction methods (Bray and Kurtz, 1945; Olsen et al., 1954; Schollenberger and Simon, 1945) and the calibration of their use to determine the fertilizer requirements of various crops. Measuring the capacity of a soil to provide plant available N was more problematic (see (Keeney, 1983)), primarily because it is mediated by biological activity, which is subject to soil properties and environmental conditions.

By the end of the 1980s, there was the desire to expand beyond soil as a medium for plant growth and include additional ecosystem services, such as environmental quality and plant and animal health. This led to the idea of soil quality and the adoption of a range of soil quality indicators (Doran et al., 1994). A defined set of indicators was never established; however, they were

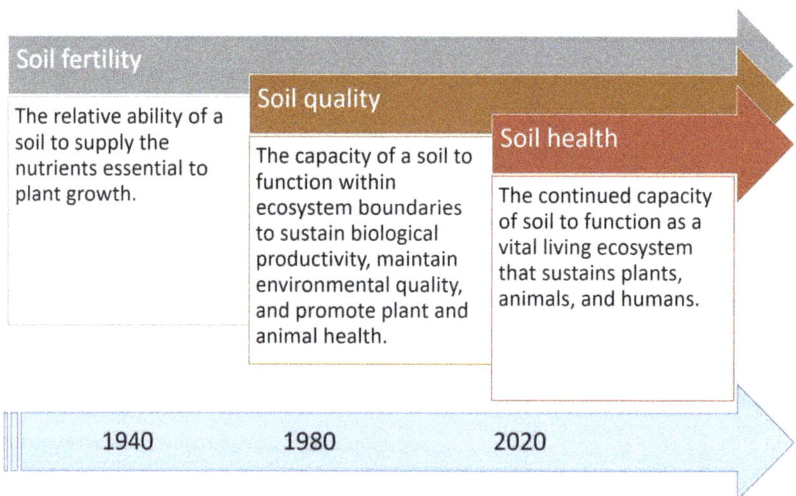

Figure 1 Timeline of development of the soil health concept and associated definitions from its origins in soil fertility to the current appreciation of soil as a living ecosystem (NRCS-USDA, n.d.; Soil Science Society of America, 2008).

dominated by soil physical and, especially, soil chemical properties (Doran and Jones, 1997; Lehmann et al., 2020). Apart from microbial biomass C – measured by the fumigation-extraction method – the other widely used biological indicators of soil quality – respiration and N mineralization assays – are reflective of microbial activity and, to some extent, nutrient availability.

During the soil quality period, we were limited in our ability to study and quantify microbial community structure and diversity. The measurement of lipids extracted from soil (PLFA–phospholipid fatty acids) was being developed (Federle, 1986). The PLFA method can differentiate between the microbial communities of soils under different management regimes (e.g. (Bossio et al., 1998)) but it provides only coarse taxonomic categorization and lacks the resolution necessary to quantify microbial diversity at the genus or species level. These shortcomings were overcome only after the advent of high-throughput sequencing a decade later (e.g. (Roesch et al., 2007)).

3 2000-Present day

Although the term 'soil health' was used earlier (Warkentin, 1995), it took another decade for its widespread adoption and efforts to refine and develop indicators and metrics of soil health (Fig. 1). Current indicators aim to characterize biologically relevant soil properties and the microbiome itself. Indicators can be categorized as reflecting either: (1) the resources available to microbes, (2) the functional capabilities of soil microbes, or (3) the activity resulting from interacting effects of resources, microbial function potential, and environmental constraints (Fig. 2). Environmental factors (i.e. temperature, pH) can largely be controlled in laboratory analyses, so indicators generally fall on a spectrum from those that are resource driven to those that reflect the functional potential of the microbiome (Fig. 3). Indicators closer to the middle, like carbon mineralization, are better tools in terms of predicting activity in the field because they show how multiple factors will interact with one another. In this section, we will discuss the properties that currently available indicators reflect, as well as practical considerations that inform their utility as soil health indicators.

3.1 Reframing carbon sequestration for ecosystem savvy management

A major goal of soil conservation has long been to restore soil organic carbon (C), originally for the benefit of fertility and erosion reduction. Fresh interest in soil C sequestration as a climate solution began to gain momentum around the same time as soil health research. However, this siloed thinking created a

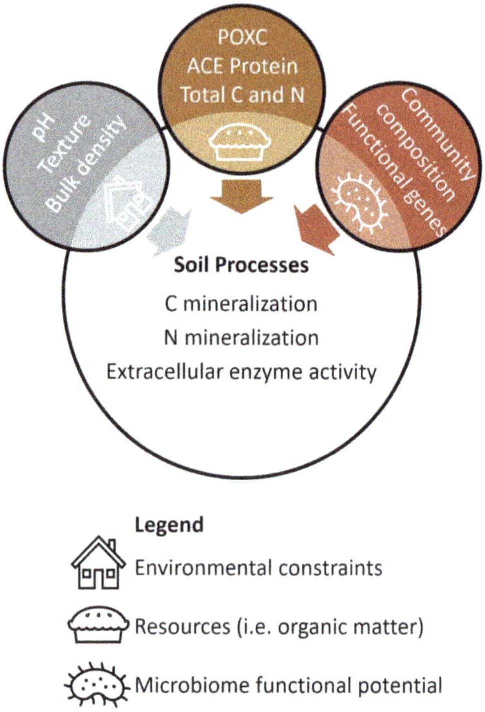

Figure 2 Conceptual diagram showing that the microbiome functional potential, soil resources, and environmental constraints contribute to the rates of measurable soil processes. Soil health indicators that are reflective of these factors or processes are listed. Abbreviations: Carbon (C), nitrogen (N) permanganate oxidizable carbon (POXC) (Weil et al., 2003), autoclaved citrate extractable (ACE) protein index (Moebius-Clune et al., 2016).

dilemma. More than a decade ago, Janzen (2006) put it succinctly: 'The soil carbon dilemma: Shall we hoard it or use it?' The central challenge arises from the conflicting goals of increasing soil C stocks and maintaining the ecosystem functionality that arises from its decomposition. Organic C in the soil is useful when it functions as a substrate for microbes that drive nutrient cycles. Seeking to increase the stock of organic C for the sole purpose of sequestration from the atmosphere would be shortsighted with respect to other ecosystem properties. Importantly, Janzen does not argue that soil C recovery is unwise, only that we should seek to go about it in a way that recognizes soil C as a 'flow' rather than a static pool, and manage decomposition to offer the greatest benefit to crops.

Soil health indicators used in recent years to evaluate organic matter decomposition reflect increasing recognition of the multifaceted functions of the soil ecosystem. Concern with total microbial biomass and organic C has somewhat yielded attention on the activities of microbes and what types of

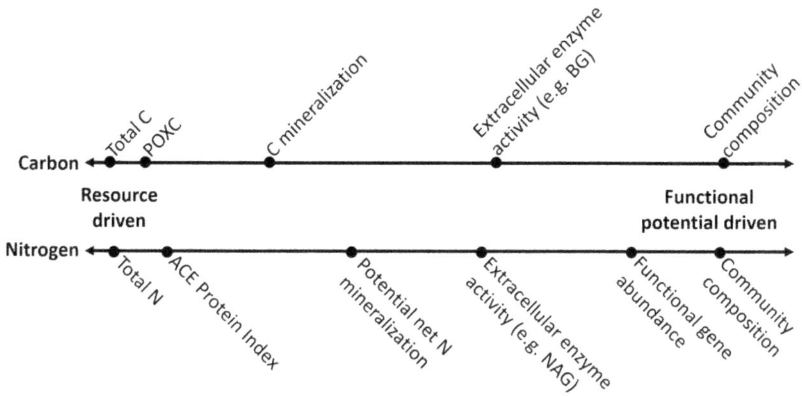

Figure 3 Conceptual diagram showing biological soil health indicators related to carbon and nitrogen acquisition along a spectrum of resource-driven to functional potential-driven properties. Abbreviations: Carbon (C), nitrogen (N) permanganate oxidizable carbon (POXC) (Weil et al., 2003), autoclaved citrate extractable (ACE) protein index (Moebius-Clune et al., 2016), β-glucosidase (BG), N-acetylglucosaminidase (NAG).

organic matter they can most easily access. In other words, the framework around C cycling has shifted to investigating C resources, microbial functional potential, and how these interact to influence C mineralization and storage (Fig. 3). This shift has contributed to refining soil health indicators to include biologically relevant C fractions, extracellular enzyme activities, and community composition data.

Efforts to identify indicators that responded more quickly to management changes in the early 2000s led to great interest in measuring the microbially active fraction of C. One candidate measure of active C, permanganate oxidizable C (POXC) (Weil et al., 2003), gained popularity as a practical, sensitive indicator. Although the nickname 'active C' implies a fraction with a fast turnover time, POXC has been reported to reflect relatively stable C and organic matter accumulation processes (Culman et al., 2012; Hurisso et al., 2016). Methodology for mineralizable C has also been refined for greater reliability in soil health frameworks. The flush of CO_2 from rewetted soil during a 1–3 day incubation is sensitive to management and closely related to other microbial variables (Franzluebbers et al., 2000; Hurisso et al., 2016; Ladoni et al., 2015). Soil health frameworks incorporating mineralizable C (Moebius-Clune et al., 2016) have largely moved to a 4-day incubation, which increases the precision of the measurement. Hurisso et al. (2016) postulated that C mineralization and POXC may have complementary roles in soil health, with POXC representing C accumulation processes and mineralization representing short-term nutrient availability.

Potential extracellular enzyme activity assays add detail to studies of C decomposition because they allow us to compare decomposition activity at a substrate level, rather than a more general index of activity such as C mineralization. Cellulolytic extracellular enzymes are an important class of enzymes in soil health research because of their role in decomposing plant litter. Endoglucanases and cellobiohydrolases (CBH) catalyze the initial steps, with β-glucosidase (BG) cleaving the resulting short glucose chains into monomers. Potential activity of BG is the most measured because it catalyzes the final, glucose releasing step. However, both BG and CBH have shown sensitivity to management and environmental changes (Fan et al., 2012; Hammerl et al., 2019; Luo et al., 2017). Oxidative enzymes (such as phenol oxidase) tend to be given greater attention in forest soil research due to their role in the decomposition of woody material. However, evidence for their apparent interaction with hydrolytic enzymes in agricultural and natural ecosystems (Chen et al., 2018; Freeman et al., 2001) suggests they may be worth further consideration in soil health research.

A significant challenge throughout soil microbial ecology is describing the role of biodiversity. It is often assumed that greater taxonomic diversity confers greater functional diversity of the microbiome, but properties of the soil ecosystem call this assumption into question. Due to the high degree of diversity present in the soil, it is theoretically possible to have sufficient redundancy to buffer functionality against changes in taxonomic diversity. On the other hand, the high degree of heterogeneity in the soil could create a high degree of niche specialization that limits the existence of "true" functional redundancy (Nunan et al., 2020).

Researchers have studied the effect of experimentally manipulating biodiversity on the total soil ecosystem functional capacity (de Graaff et al., 2015). To investigate this link in C cycle metabolism paths, de Graaff et al. (2015) conducted a meta-analysis of studies manipulating biodiversity or tracking it in situ. They found that decreasing biodiversity of soil biota negatively impacted C mineralization rates and plant tissue decomposition, but did not find evidence that it directly controls soil C pools. Microbial biodiversity decreases were more strongly associated with lowered C mineralization rates, while faunal biodiversity decreases were the driving factor behind reduced plant tissue decomposition. Counter to the conventional model of high functional redundancy, loss of diversity within body size classes of fauna (i.e. meso and macrofauna) affected decomposition rates as much as loss of diversity across multiple body size classes. In most experiments, species richness was orders of magnitude lower than in natural systems, obscuring conclusions of the meta-analysis.

Another aspect of the C cycle concerns possible climate and environmental perturbations to decomposition processes. Lehmann et al. (2020) identified sustainable planetary health as a soil health goal that distinguishes it from soil quality, despite the lack of soil health indicators explicitly addressing greenhouse gas emissions. The soil is the largest pool of organic C in the terrestrial environment (Friedlingstein et al., 2020). Characterizing C storage and fluxes in the soil is of paramount importance not only to quantifying resources in the soil environment, but also to research in the global C cycle and climate change processes. Global change, which is more broadly defined than climate change, is expected to impact the soil ecosystem through shifts in weather patterns, land use, and nutrient cycles. This is a broad topic that cannot be comprehensively addressed in this chapter. Instead, we will discuss how the soil C cycle may respond to two well-studied facets of global change: soil warming and long-term N addition.

Soil warming was initially thought to result in greater mineralization of soil organic C (SOC) by the microbial community (Jenkinson et al., 1991), and some short-term studies seemed to support this hypothesis. Despite efforts to demonstrate a mechanistic explanation for the phenomenon – typically reductions in C use efficiency (CUE) or increased cell turnover – results of long-term warming studies do not consistently support the predicted continuous loss of soil C under warming (Bradford et al., 2016; Lu et al., 2013; Melillo et al., 2017). Results are inconsistent with respect to the magnitude and direction (let alone mechanisms) of the warming effect on CUE (Frey et al., 2013; Li et al., 2019). Nonetheless, climate models frequently continue to assume a reduction in CUE due to soil warming that contributes to positive feedbacks. Hence, the uncertainty in the SOC response to warming contributes to substantial uncertainty in predicting climate feedback effects (Bradford et al., 2016). Despite decades of research in this area, open questions remain.

The results of long-term N addition experiments suggest that fertilization affects C cycling dynamics via shifts in the microbial community. By alleviating nutrient limitation in the soil for soil biota, N addition can alter C allocation patterns. These can lead to shifts in the microbial community composition, and the activity of N fertilization also lowers soil pH, which is another component of changes to the microbial community. Surprisingly, long-term fertilization can have a greater impact on soil microbial communities than plant species (Semenov et al., 2020; Soman et al., 2017), although this is not a universal observation.

Greater C storage after N fertilization appears to be mediated by extracellular enzyme activities. Potential enzyme activity response to N fertilization has been highly variable across the many studies reporting links between the two. A meta-analysis of 65 N fertilization experiments found that this treatment on the whole increased hydrolase activities and suppressed oxidase activities (Jian et al. 2016). This is consistent with other meta-analyses showing that suppressed oxidase activity mediated the accumulation of C under N addition (Chen et al., 2018; Liu et al., 2016). Chen et al. (2018) reported that the oxidase activity effect greatly outweighed the effect of hydrolase (β-glucosidase (BG), cellobiohydrolase (CBH), and N-acetylglucosaminidase

(NAG))activities, and other environmental variables. Another meta-analysis of N-addition experiments found that N stimulated accumulation of the chemically recalcitrant substances with which the oxidative enzymes are associated (Liu et al., 2016). Together, these meta-analyses provide strong evidence that oxidative enzymes will interact with N fertilization to influence decomposition rates.

3.2 Linking the microbial community to mineralizable macronutrients

Nitrogen (N) is a critical component of proteins, so it is an important nutrient for plants and other soil biota. The vast majority of soil N is in an organic form that must be decomposed before being taken into the cell (Bremner, 1965), so estimating the size and availability of this pool is vital to understanding ecosystem functionality. There are a number of soil health indicators available that aim to estimate the size of the organic N pool and how microbes are using it (Fig. 3).

Autoclaved citrate extractable (ACE) protein index is based on a method that was initially developed to quantify glomalin content in the soil (Wright and Upadhyaya, 1996), but has subsequently been shown to detect a wide variety of protein sources in the soil (Hurisso et al., 2018; Rosier et al., 2006). Soil protein (still referred to as glomalin in many studies) tends to decrease with increased tillage, and increase with organic amendment inputs (Moebius-Clune et al., 2008). Soil protein is also positively related to aggregate stability, although the causality of this relationship is likely complex. In a study investigating protein resources on a range of tillage intensities, soil protein accumulated in macroaggregates in the no-till system only (Wright et al., 2007). Wright et al. proposed that the extracted protein likely functioned as a glue that stabilized the aggregates. It is plausible that aggregates in turn physically occlude soil protein from microbial access, slowing decomposition.

The ACE protein index is primarily an indicator of the organic N available, while potential net N mineralization reflects both accessible N and the capacity of microbes to mineralize it (Fig. 3). Organic N fractions, such as ACE protein, have been suggested as a possible predictor for N mineralization rates, but the differing properties influencing these indicators suggests that they are not interchangeable. In soil samples from 57 field sites across the northern half of California, ACE protein was positively related to N mineralization over a 28-day aerobic incubation (Geisseler et al., 2019). However, the authors point out that this relationship was relatively weak compared to that of N mineralization and total N.

As outlined previously, many soil health indicators focus on measuring activities or changing nutrient pools. It has long been appreciated that microbes are responsible for driving many of the reactions, yet direct measurement of the microbial community is not common in soil health indicator suites. As sequencing technology becomes faster and more affordable, researchers are increasingly able to collect datasets identifying microorganisms. Identified microorganisms may be linked with their putative functions using the numerous bioinformatics pipelines and tools such as PICRUSt (Phylogenetic Investigation of Communities by Reconstruction of Unobserved States) (Langille et al., 2013). Yet, it is unclear how often metabolic capacity can be linked to activity in the field, and whether this provides significant insight over traditional activity measurements.

Loss of N to denitrification and leaching processes is a top concern in agroecosystems. Nitrogen loss from the system is not only costly but also can lead to polluting reactive N losses. Managing soil to minimize this loss is critical and requires a nuanced understanding of the biological processes driving mineralization, assimilation, and loss. The response of these processes to management is variable across studies and processes. Ouyang et al. (2018b) studied enzyme activity and microbial functional genes under compost and manure in organically farmed crops to understand the potential for N transformation (Ouyang et al., 2018b). Amendments, especially manure, increased the potential for processes associated with N loss (nitrification and denitrification), as assessed by both functional gene abundance and potential enzyme activities.

The rate and net products of N transformation processes depend not only on the available N, but also on the functional potential and nutrient needs of the microbial community. Increasingly, research suggests that these factors synergistically, not singularly, influence process rates. Microbial community structure can inform the functional potential and resource demand because different taxa will have different life history strategies. A meta-analysis of studies comparing community structure and activity rates found that changes in structure were not as strongly linked with nitrification and denitrification as when specifically linked genes were tested (Bier et al., 2015).

The connection between gene abundance and activity varies among processes. Rocca et al. (2015) reviewed the correlation between gene abundance and corresponding process rates in a meta-analysis and found variable support for this connection. Correlations were significant for ammonia mono-oxygenases and nitroreductases, but not for other genes related to N transformation. It is probable that consumption of the products and alternative sources for the product blurs this link, especially for processes with a broad phylogenetic distribution (Rocca et al., 2015). Both Bier et al. (2015) and Rocca et al. (2015) observed that gene abundance for narrowly distributed processes, such as nitrification, had a stronger relationship with the corresponding process rate. This observation highlights the need to differentiate the specific purposes of analysis for targeted work. Measuring gene abundance in the absence of measuring activity may not be appropriate if the goal is to predict net availability of a product when accounting for all processes. Alternatively, gene abundance may be of greater interest if the research questions are related to the microbial potential for a particular process.

In soil health research, studying how quickly a potential indicator changes is a priority. One potential benefit of measuring gene abundance is that it is plausible that the functional potential of the microbiome could shift before the process rate itself. In a meta-analysis of N-fertilization experiments, Ouyang et al. (2018a) reported that gene abundance of nirS, nosZ, and AOB amoA responded positively to fertilization within 5 years, with the effect typically increasing over time. The fertilization treatment needed to be greater than 5 years to detect an increase in nirK and AOA amoA abundance (Ouyang et al., 2018a). The slower response of some genes to treatment suggests some limitation in this strategy for soil health research.

The activity of phosphorus (P) mineralization enzymes, acid, and alkaline phosphatases are commonly measured as soil health indicators (Acosta-Martínez and Tabatabai, 2011). They estimate how fast P could potentially be liberated from organic P sources in the soil, but do not address mobilization from mineral or other occluded sources. Assessing phosphatase gene abundance may influence the P cycle, and more research is needed to understand how gene regulation is to management. Perturbations to the N supply may interact with the P cycle and vice versa to affect phosphatase gene abundance. For example, in fields with maize growing for nearly 30 years, alkaline phosphatase activity and phoD gene abundance decreased after N amendment (Chen et al., 2019). Microbial transformations of the N and P cycle are also linked to the nutrient demands and environmental limitations of the microbiome (Margalef et al., 2017). Identifying the dominant P solubilizing and mineralizing taxa under different environmental conditions is another important area of research.

Most of the sulfur in the soil is in organic forms that must be mineralized to sulfate, the most important plant available form. This mineralization is accomplished via arylsulfatases (ARS), whose activity is often incorporated into soil health research. Emissions regulations have substantially reduced atmospheric deposition of sulfur (gaseous forms are a common constituent of air pollution). Additionally, triple superphosphate, which contains only small amounts of sulfur, is now used in place of sulfur-containing single superphosphate fertilizer. The frequency of sulfur deficiency problems in agriculture has increased in recent decades as a result (Dick et al., 2008; Scherer, 2009). This has increased interest in monitoring and predicting sulfate release from organic matter in the soil. Authors have argued for the need for improved spatial-temporal modelling of variability in sulfatase activity to better constrain rates at the field and annual scales. Further, investigation into ARS at multiple spatial scales may resolve conflicting reported relationships between ARS activity rates and other soil properties (Baldrian, 2014; Chen et al., 2019).

3.3 Integrating extracellular enzyme results

Although there is great complexity in the genetic diversity of enzymes and the microbes that make them (Caldwell, 2005), researchers would benefit from a parsimonious way to study them. Soil health indicators must be suitable for high throughput studies, so a minimum dataset is clearly desirable. Extracellular enzyme activities are considered soil health indicators because they have been shown to have good sensitivity to management. However, fertilization, tillage, and cover crop treatments often only affect a few of a suite of enzymes measured, and sometimes effects are only observed when multiple management variables are altered. For example, Singh and Kumar (2021) found stronger interactive effects of tillage by cover type on urease activity than that of either alone. Effects of management on enzyme activities also varied seasonally and by substrate, which underscores the need for targeted sampling and a suite of enzymes rather than single representatives (Singh and Kumar, 2021).

Various soil health assessment systems include suites of representative enzyme assays, but they are not in complete agreement. The process of developing soil health indicator methodology will be discussed later in more detail. However, possible paths forward for simplifying enzyme assays also function as a framework for understanding biological processes. Determining which enzymes are best to measure in a soil health study begs the essential questions: Which extracellular enzymes are most important in organic matter decomposition and nutrient availability, and which are most responsive to management changes?

Broadly, this problem can be addressed with two approaches; the most common is to calculate an index from multiple enzyme activity rates. Reducing the number of activity rates measured in the first place is an alternative approach. The latter has the obvious advantage of saving time and money that could be reallocated to other analyses. The trade-off with this approach is the possibility of losing important nuance by excluding individual activity rates from the analysis.

Several indices are available to analyze extracellular enzymes. The geometric means of several enzyme activities are commonly calculated as an integrative index of soil quality because it is sensitive to management differences and simple to interpret (García-Ruiz et al., 2008; Raiesi and Kabiri, 2016; Raiesi and Salek-Gilani, 2018). However, nutrient-specific resource demand must be assessed separately.

Multivariate analyses such as principal coordinate analysis may also be used to relate enzyme activities to environmental variables (Pandey et al., 2014) and to the microbial community (Caldwell, 2005). In a long-term fertilization experiment, redundancy analysis found that extracellular enzyme activity rates were better explained by environmental variables than by differences in bacterial community composition (Ling et al., 2014). Another study using path analysis and ordination to follow long-term fertilizer inputs found that changes to the bacterial (but not fungal) community mediated increases in hydrolytic enzyme activities (Ai et al., 2018). This study also found taxon-specific changes. For example, bacteria from the Cytophagaceae had greater relative abundance in fertilized plots. It is not clear, however, what the contribution of individual enriched taxa are to changes in extracellular enzyme activities.

Ecoenzymatic stoichiometry is a way to synthesize potential activities by calculating ratios among various enzymes. Ecological stoichiometry is considered a 'unifying theory of ecology', (Sterner and Elser, 2002) that describes ecosystem resources explicitly in terms of the balance of chemical elements and associated biological demand. Ecoenzymatic stoichiometry was proposed as an application of ecological stoichiometry, where the ratio of enzyme activities indicates if C, N, or P is most in-demand (Sinsabaugh et al., 2009). For example, the ratio of β-1,4-glucosidase (BG) to N-acetylglucosaminidase (NAG) + leucine aminopeptidase (LAP) was proposed to describe the C:N demand, where greater NAG and LAP indicates N limitation. Sinsabaugh et al. (2009) proposed measuring the activity of BG, NAG, LAP, and acid phosphatase (AP) for this work, while other groups have recommended augmenting the measurements with other enzymes. For instance, one study argued that urease should be included because it catalyzes the hydrolysis of urea, a common agricultural fertilizer (Lin et al., 2017). Moorhead et al. (2013) proposed calculating enzyme activity vectors from the ratios of BG/AP and BG/NAG to assess relative nutrient limitation from a single plot. Plotting the BG/AP ratio on the horizontal axis and BG/NAG on the vertical axis creates an enzyme activity space, where a vector can be plotted from the plot origin to a point with coordinates (BG/AP, BG/NAG). The length of the vector is related to C limitation, and the vector angle increases with greater phosphorus limitation to nitrogen limitation. Enzyme activity vectors also appear to change with resource demand independent of individual enzyme activities and total enzyme activity (Moorhead et al., 2013).

Acosta-Martínez proposed an alternative assay simultaneously measuring enzyme activity rates by adding two to four substrates in the same assay. Measured rates were highly correlated to the sum of individual enzyme activities and showed some sensitivity to management across several soils (Acosta-Martínez et al., 2018, 2019). This approach is much faster and more cost-effective than individual assays, but more work is needed to assess the limitations of this technique. Although it may be highly beneficial to commercial laboratories, the combined assay would not be appropriate for research questions concerning specific substrates because the contribution of individual enzymes is unknown.

4 Future trends in research

There are a number of more recently developed tools that may prove valuable to soil health research in the future but are not currently included in soil health frameworks. These have the potential to: (1) characterize biologically relevant fractions of organic matter, (2) provide protein-level microbial activity data, and (3) clarify which populations are responsible for which activities. Here we will discuss a few of these techniques and their possible roles in soil health.

4.1 Beyond the grave: Appreciating the role of soil necromass

One facet of C persistence is understanding what types of organic inputs are likely to remain in the soil. The oft-quoted adage, 'The microbial biomass is the eye of the needle through which all organic matter must pass' (Jenkinson, 1977), quickly illustrates the magnitude, but not the complexity, of the problem. Soil organic matter must be decomposed by soil organisms before its nutrients are remobilized for other uses, yet its path is not so simple as passing through a needle. A modern viewpoint highlights the myriad paths a given atom of C may take on its way 'through' the microbial biomass (Waring et al., 2020).

While it was once thought that old organic matter primarily arose by resisting decomposition through chemical recalcitrance, modern frameworks instead emphasize that high-quality organic matter is more likely to become microbial biomass, which is more likely to gain mineral protection (Cotrufo et al., 2013). Similarly, the soil microbial C pump concept illustrates the process of incorporating SOC into reduced forms in the microbial biomass, which are more likely to become mineral associated and protected from further decomposition (Zhu et al., 2020). Soil microbial biomass is a small percentage of the soil organic C pool, but its derivatives (residues and necromass) have an outsized role in the generation of slow cycling soil organic matter (Liang et al., 2019). Characterizing the role of necromass in slow cycling organic matter is important to soil health because we are interested in what gives rise to soil organic matter that is likely to persist longer in the soil.

Distinguishing SOC of microbial origin is challenging. Amino sugar biomarkers are a proposed proxy for measuring the soil necromass because they are not found in plants (Liang et al., 2012; Zhu et al., 2020). Variability in conversion factors from amino sugars to microbial C across microorganisms and soils adds uncertainty to estimates of necromass in the soil (Liang et al., 2019). Analogous uncertainties plague quantification of microbial biomass, which remains routine in soil biology research (Geyer et al., 2020). Necromass may offer useful insights in soil health research because it is well correlated to SOC yet responds independently to management. In an investigation of SOC accumulation 6 years after planting bioenergy crops, Zhu et al. (2020) estimated necromass comprised up to 93% of added SOC. However, necromass does not universally increase with SOC. Microbial necromass was unaffected by 7 years of cover cropping corn silage with winter rye, despite increases in SOC and extracellular enzyme activity rates. These amino sugar biomarkers are not currently used in soil health research routinely, but they may prove useful in situations where the goal is to accumulate SOC.

4.2 Novel tools for exploring the microbiome

Harnessing 'omics' data may offer insight into competing processes in nutrient cycles, with N as an example. Although microbial community composition measurements are a popular proxy for estimating N cycling capacity, they often do not discriminate between active and dormant microbes, the latter of which can form a significant proportion of microbiome richness (Blagodatskaya and Kuzyakov, 2013). Furthermore, identifying the presence of a microbe believed to have the genetic capacity for a functional trait does not guarantee that it is expressing the relevant genes (Prosser, 2015). Focusing on the abundance of transcribed functional genes associated with the N cycle could be a way forward because it circumvents both the challenges of competing processes and inactive microbes. However, the correlation strength between gene abundance and activity (reviewed by Bier et al. (2015)) is variable, although the connection is difficult to evaluate given the aforementioned competing processes. Numerous approaches to the problem – stable isotope probing (SIP), metagenomic, transcriptomic, and proteomic approaches are all strategies that can be applied to N cycle questions - but no consensus on which are most appropriate for routine soil health monitoring.

4.3 Who's who in extracellular enzymes

Management decisions and environmental changes can alter extracellular enzyme activity, but it is not often clear whether potential activity is lower because of shifts in the microbial community or because of altered activity from the same cells as before. More information about 'who is doing what' could contribute to a mechanistic understanding of how edaphic properties affect extracellular enzyme activity and improve predictions of the consequences of disturbances and management choices (Schneider et al., 2012).

Metagenomics can capture the potential for microbes to produce proteins, but not their expression. The nutrient status of the soil largely drives biomass-specific production rates of extracellular enzymes, so the connectivity between phylogeny and function is variable. Deeper insights into the metabolic profile of the microbiome can be gained from metaproteomics - sequencing proteins in the soils and annotating them with functional and taxonomic characteristics. Metaproteomics is a more direct measurement because it measures the proteins that are present in the soil. This technique has been used to investigate the major groups responsible for litter decomposition (Keiblinger et al., 2016; Schneider et al., 2012) and identify plant-microbe interactions that influence crop yields and contribute to disease suppression (Starke et al., 2019). Although proteins may be linked to associated taxa upon identification, as Starke et al. (2019) point out, SIP is the most direct technique to relate functional activity to specific populations in the soil. However, given the cost of SIP, it is difficult to imagine this methodology as a practical soil health indicator.

In metaproteomics, proteins are extracted and isolated from the soil, sequenced by mass spectrometry, and then annotated using a reference database. Numerous tools and public databases are available for analysis of extracted proteins, but bias toward cultivatable organisms and extraction inefficiencies severely limit our annotation capacity. Extracellular enzymes are particularly difficult to extract efficiently because they tend to be stabilized by organic matter compounds and clay particles. As a consequence, few of these sequences are available in databases, relative to the more easily detected housekeeping proteins (Starke et al., 2019). Although the insights gained from metaproteomics may eventually improve interpretation of extracellular enzyme activities, metaproteomics is not considered an appropriate candidate for routine soil health monitoring due to the high cost and serious challenges with isolating and identifying proteins from soil (Abiraami et al., 2020). Advancements in technology and novel techniques (reviewed by Abiraami et al., 2020) could potentially improve extraction efficiencies and quantification in the future, although significant challenges remain for application to soil health.

5 Conclusion

Refining and developing new analysis techniques for soil health research has greatly improved insights into the biology of the soil ecosystem. Yet, just as the functional potential of the microbiome does not guarantee activity, new capacity to gather data does not automatically translate to practical guidance or theory. Many biological indicators still suffer from interpretation difficulty and methodological challenges (Fierer et al., 2021) that hinder the development of unified conceptual models. Prosser (2020) called for microbial ecology to return to practicing basic science, developing, and then rigorously testing hypotheses to arrive at theories. This argument was aimed at the field of microbial ecology, but they are relevant to soil health research because processes in the microbiome are fundamental to our understanding of soil ecosystem functionality. The integrative nature of soil health research is a challenge; however, with honest evaluations of what individual indicators can and cannot measure, this challenge can be transformed into our greatest strength.

6 Where to look for further information

For extended information about soil biological principles underlying the biological dimensions of soil health, the reader is referred to a standard text on soil microbiology (Paul, 2015). Full information on the text is listed in the References section.

For more detailed information on methodological considerations and interpretation of biological soil health indicators, the reader is directed to reviews by Fierer et al. (2021), Rocca et al. (2015), and Abiraami et al. (2020). Extracellular enzyme assay methodology and interpretation is a layered topic all its own, due to the variety of analysis options available. Methodological considerations for soil enzyme assays were previously detailed by Deng et al. (2017). Additionally, a useful discussion of the role of extracellular enzymes in soil biodiversity may be found in the review by Caldwell (2005).

6.1 Selected organizations

The following organizations are working to integrate up-to-date soil biology research into soil health management recommendations and dissemination of this research to the general public. All of the organizations below offer excellent resources for further information on the biological dimension of soil health.

- Natural Resource Conservation Service (NRCS) with the United States Department of Agriculture (USDA) (https://www.nrcs.usda.gov/wps/portal/nrcs/main/or/soils/health/).
- Soil and Water Conservation Society (https://www.swcs.org/).
- Soil Health Institute (https://soilhealthinstitute.org/).

7 References

Abiraami, T. V., Singh, S. and Nain, L. (2020). Soil metaproteomics as a tool for monitoring functional microbial communities: Promises and challenges. *Reviews in Environmental Science and Bio/Technology* 19(1): 73–102. DOI: 10.1007/s11157-019-09519-8.

Acosta-Martínez, V., Cano, A. and Johnson, J. (2018). Simultaneous determination of multiple soil enzyme activities for soil health-biogeochemical indices. *Applied Soil Ecology* 126: 121–128. DOI: 10.1016/j.apsoil.2017.11.024.

Acosta-Martínez, V., Pérez-Guzmán, L. and Johnson, J. M. F. (2019). Simultaneous determination of β glucosidase, β-glucosaminidase, acid phosphomonoesterase, and arylsulfatase activities in a soil sample for a biogeochemical cycling index. *Applied Soil Ecology* 142: 72–80. DOI: 10.1016/j.apsoil.2019.05.001.

Acosta-Martínez, V. and Tabatabai, M. A. (2011). Phosphorus cycle enzymes. In: *Methods of Soil Enzymology*. New York: John Wiley & Sons, Ltd: 161–183. DOI: 10.2136/sssabookser9.c8.

Ai, C., Zhang, S., Zhang, X., Guo, D., Zhou, W. and Huang, S. (2018). Distinct responses of soil bacterial and fungal communities to changes in fertilization regime and crop rotation. *Geoderma* 319: 156–166. DOI: 10.1016/j.geoderma.2018.01.010.

Baldrian, P. (2014). Distribution of extracellular enzymes in soils: Spatial heterogeneity and determining factors at various scales. *Soil Science Society of America Journal* 78(1): 11–18. DOI: 10.2136/sssaj2013.04.0155dgs.

Bier, R. L., Bernhardt, E. S., Boot, C. M., Graham, E. B., Hall, E. K., Lennon, J. T., Nemergut, D. R., Osborne, B. B., Ruiz-González, C., Schimel, J. P., Waldrop, M. P. and Wallenstein, M. D. (2015). Linking microbial community structure and microbial processes: An empirical and conceptual overview. *FEMS Microbiology Ecology* 91(10): fiv113. DOI: 10.1093/femsec/fiv113.

Blagodatskaya, E. and Kuzyakov, Y. (2013). Active microorganisms in soil: Critical review of estimation criteria and approaches. *Soil Biology and Biochemistry* 67: 192–211. DOI: 10.1016/j.soilbio.2013.08.024.

Bossio, D. A., Scow, K. M., Gunapala, N. and Graham, K. J. (1998). Determinants of soil microbial communities: Effects of agricultural management, season, and soil type on phospholipid fatty acid profiles. *Microbial Ecology* 36(1): 1–12. DOI: 10.1007/s002489900087.

Bradford, M. A., Wieder, W. R., Bonan, G. B., Fierer, N., Raymond, P. A. and Crowther, T. W. (2016). Managing uncertainty in soil carbon feedbacks to climate change. *Nature Climate Change* 6(8): 751–758. DOI: 10.1038/nclimate3071.

Bray, R. H. and Kurtz, L. T. (1945). Determination of total, organic, and available forms of phosphorus in soils. *Soil Science* 59(1): 39–46. DOI: 10.1097/00010694-194501000-00006.

Bremner, J. M. (1965). Total nitrogen. In: *Agronomy Monographs*. New York: John Wiley & Sons, Ltd: 1149–1178. DOI: 10.2134/agronmonogr9.2.c32.

Caldwell, B. A. (2005). Enzyme activities as a component of soil biodiversity: A review. *Pedobiologia* 49(6): 637–644.

Chen, X., Jiang, N., Condron, L. M., Dunfield, K. E., Chen, Z., Wang, J. and Chen, L. (2019). Soil alkaline phosphatase activity and bacterial phoD gene abundance and diversity under long-term nitrogen and manure inputs. *Geoderma* 349: 36–44. DOI: 10.1016/j.geoderma.2019.04.039.

Chen, J., Luo, Y., Groenigen, K. J. van, Hungate, B. A., Cao, J., Zhou, X. and Wang, R. W. (2018). A keystone microbial enzyme for nitrogen control of soil carbon storage. *Science Advances* 4(8): eaaq1689.

Cotrufo, M. F., Wallenstein, M. D., Boot, C. M., Denef, K. and Paul, E. (2013). The Microbial Efficiency-Matrix Stabilization (MEMS) framework integrates plant litter decomposition with soil organic matter stabilization: Do labile plant inputs form stable soil organic matter? *Global Change Biology* 19(4): 988–995. DOI: 10.1111/gcb.12113.

Culman, S. W., Snapp, S. S., Freeman, M. A., Schipanski, M. E., Beniston, J., Lal, R., Drinkwater, L. E., Franzluebbers, A. J., Glover, J. D., Grandy, A. S., Lee, J., Six, J., Maul, J. E., Mirksy, S. B., Spargo, J. T. and Wander, M. M. (2012). Permanganate oxidizable carbon reflects a processed soil fraction that is sensitive to management. *Soil Science Society of America Journal* 76(2): 494–504. DOI: 10.2136/sssaj2011.0286.

de Graaff, M.-A., Adkins, J., Kardol, P. and Throop, H. L. (2015). A meta-analysis of soil biodiversity impacts on the carbon cycle. *Soil* 1(1): 257–271. DOI: 10.5194/soil-1-257-2015.

Deng, S., Dick, R., Freeman, C., Kandeler, E. and Weintraub, M. N. (2017). Comparison and standardization of soil enzyme assay for meaningful data interpretation. *Journal of Microbiological Methods* 133: 32–34. DOI: 10.1016/j.mimet.2016.12.013.

Dick, W. A., Kost, D. and Chen, L. (2008). Availability of sulfur to crops from soil and other sources. In: *Agronomy Monographs*. Sulfur: A Missing Link between Soils. John Wiley & Sons, Ltd: 59–82. DOI: 10.2134/agronmonogr50.c5.

Doran, J. W., Coleman, D. C., Bezdicek, D. F. and Stewart, B. A. (Eds.). (1994). *Defining Soil Quality for a Sustainable Environment. SSSA Special Publications*. Madison, WI: Soil Science Society of America and American Society of Agronomy. DOI: 10.2136/sssaspecpub35.

Doran, J. W. and Jones, A. J. (Eds.). (1997). *Methods for Assessing Soil Quality. SSSA Special Publications*. Madison, WI: Soil Science Society of America. DOI: 10.2136/sssaspecpub49.

Fan, F., Li, Z., Wakelin, S. A., Yu, W. and Liang, Y. (2012). Mineral fertilizer alters cellulolytic community structure and suppresses soil cellobiohydrolase activity in a long-term fertilization experiment. *Soil Biology and Biochemistry* 55: 70–77. DOI: 10.1016/j.soilbio.2012.06.008.

Federle, T. (1986). Microbial distribution in soil–New techniques. In: Megusar, F. and Gantar, M. (Eds.), *Perspectives in Microbial Ecology*. Ljubljana: Slovene Society for Microbiology: 493–498.

Fierer, N., Wood, S. A. and Bueno de Mesquita, C. P. (2021). How microbes can, and cannot, be used to assess soil health. *Soil Biology and Biochemistry* 153: 108111. DOI: 10.1016/j.soilbio.2020.108111.

Franzluebbers, A. J., Haney, R. L., Honeycutt, C. W., Schomberg, H. H. and Hons, F. M. (2000). Flush of carbon dioxide following rewetting of dried soil relates to active organic pools. *Soil Science Society of America Journal* 64(2): 613–623. DOI: 10.2136/sssaj2000.642613x.

Freeman, C., Ostle, N. and Kang, H. (2001). An enzymic 'latch' on a global carbon store. *Nature* 409(6817): 149–149. DOI: 10.1038/35051650.

Frey, S. D., Lee, J., Melillo, J. M. and Six, J. (2013). The temperature response of soil microbial efficiency and its feedback to climate. *Nature Climate Change* 3(4): 395–398. DOI: 10.1038/nclimate1796.

Friedlingstein, P., O'Sullivan, M., Jones, M. W., Andrew, R. M., Hauck, J., Olsen, A., Peters, G. P., Peters, W., Pongratz, J., Sitch, S., Le Quéré, C., Canadell, J. G., Ciais, P., Jackson, R. B., Alin, S., Aragão, L. E. O. C., Arneth, A., Arora, V., Bates, N. R., Becker, M., Benoit-Cattin, A., Bittig, H. C., Bopp, L., Bultan, S., Chandra, N., Chevallier, F., Chini, L. P., Evans, W., Florentie, L., Forster, P. M., Gasser, T., Gehlen, M., Gilfillan, D., Gkritzalis, T., Gregor, L., Gruber, N., Harris, I., Hartung, K., Haverd, V., Houghton, R. A., Ilyina, T., Jain, A. K., Joetzjer, E., Kadono, K., Kato, E., Kitidis, V., Korsbakken, J. I., Landschützer, P., Lefèvre, N., Lenton, A., Lienert, S., Liu, Z., Lombardozzi, D., Marland,

G., Metzl, N., Munro, D. R., Nabel, J. E. M. S., Nakaoka, S., Niwa, Y., O'Brien, K., Ono, T., Palmer, P. I., Pierrot, D., Poulter, B., Resplandy, L., Robertson, E., Rödenbeck, C., Schwinger, J., Séférian, R., Skjelvan, I., Smith, A. J. P., Sutton, A. J., Tanhua, T., Tans, P. P., Tian, H., Tilbrook, B., van der Werf, G., Vuichard, N., Walker, A. P., Wanninkhof, R., Watson, A. J., Willis, D., Wiltshire, A. J., Yuan, W., Yue, X. and Zaehle, S. (2020). Global carbon budget 2020. *Earth System Science Data* 12(4): 3269–3340. DOI: 10.5194/essd-12-3269-2020.

García-Ruiz, R., Ochoa, V., Hinojosa, M. B. and Carreira, J. A. (2008). Suitability of enzyme activities for the monitoring of soil quality improvement in organic agricultural systems. *Soil Biology and Biochemistry* 40(9): 2137–2145.

Geisseler, D., Miller, K., Leinfelder-Miles, M. and Wilson, R. (2019). Use of soil protein pools as indicators of soil nitrogen mineralization potential. *Soil Science Society of America Journal* 83(4): 1236–1243. DOI: 10.2136/sssaj2019.01.0012.

Geyer, K., Schnecker, J., Grandy, A. S., Richter, A. and Frey, S. (2020). Assessing microbial residues in soil as a potential carbon sink and moderator of carbon use efficiency. *Biogeochemistry* 151(2–3): 237–249. DOI: 10.1007/s10533-020-00720-4.

Hammerl, V. B., Grant, K., Pritsch, K., Jentsch, A., Schloter, M., Beierkuhnlein, C. and Gschwendtner, S. (2019). Seasonal effects of extreme weather events on potential extracellular enzyme activities in a temperate grassland soil. *Frontiers in Environmental Science* 6: 157. DOI: 10.3389/fenvs.2018.00157.

Hurisso, T. T., Culman, S. W., Horwath, W. R., Wade, J., Cass, D., Beniston, J. W., Bowles, T. M., Grandy, A. S., Franzluebbers, A. J., Schipanski, M. E., Lucas, S. T. and Ugarte, C. M. (2016). Comparison of permanganate-oxidizable carbon and mineralizable carbon for assessment of organic matter stabilization and mineralization. *Soil Science Society of America Journal* 80(5): 1352–1364. DOI: 10.2136/sssaj2016.04.0106.

Hurisso, T. T., Moebius-Clune, D. J., Culman, S. W., Moebius-Clune, B. N., Thies, J. E. and Es, H. M. (2018). Soil protein as a rapid soil health indicator of potentially available organic nitrogen. *Agricultural and Environmental Letters* 3(1): 180006. DOI: 10.2134/ael2018.02.0006.

Janzen, H. H. (2006). The soil carbon dilemma: Shall we hoard it or use it? *Soil Biology and Biochemistry* 38(3): 419–424. DOI: 10.1016/j.soilbio.2005.10.008.

Jenkinson, D. S. (1977). The soil microbial biomass. *New Zealand Soil News* 25: 213–218.

Jenkinson, D. S., Adams, D. E. and Wild, A. (1991). Model estimates of CO_2 emissions from soil in response to global warming. *Nature* 351(6324): 304–306. DOI: 10.1038/351304a0.

Jian, S., Li, J., Chen, J., Wang, G., Mayes, M. A., Dzantor, K. E., Hui, D. and Luo, Y. (2016). Soil extracellular enzyme activities, soil carbon and nitrogen storage under nitrogen fertilization: A meta-analysis. *Soil Biology and Biochemistry* 101: 32–43. DOI: 10.1016/j.soilbio.2016.07.003.

Keeney, D. R. (1983). Nitrogen–Availability indices. In: *Agronomy Monographs*. New York: John Wiley & Sons, Ltd: 711–733. DOI: 10.2134/agronmonogr9.2.2ed.c35.

Keiblinger, K. M., Fuchs, S., Zechmeister-Boltenstern, S. and Riedel, K. (2016). Soil and leaf litter metaproteomics–A brief guideline from sampling to understanding. *FEMS Microbiology Ecology* 92(11): fiw180. DOI: 10.1093/femsec/fiw180.

Ladoni, M., Basir, A. and Kravchenko, A. (2015). Which soil carbon fraction is the best for assessing management differences? A statistical power perspective. *Soil Science Society of America Journal* 79(3): 848–857. DOI: 10.2136/sssaj2014.10.0426.

Langille, M. G., Zaneveld, J., Caporaso, J. G., McDonald, D., Knights, D., Reyes, J. A., Clemente, J. C., Burkepile, D. E., Vega Thurber, R. L., Knight, R., Beiko, R. G. and

Huttenhower, C. (2013). Predictive functional profiling of microbial communities using 16S rRNA marker gene sequences. *Nature Biotechnology* 31(9): 814–821. DOI: 10.1038/nbt.2676.

Lehmann, J., Bossio, D. A., Kögel-Knabner, I. and Rillig, M. C. (2020). The concept and future prospects of soil health. *Nature Reviews Earth and Environment* 1(10): 544–553. DOI: 10.1038/s43017-020-0080-8.

Li, J., Wang, G., Mayes, M. A., Allison, S. D., Frey, S. D., Shi, Z., Hu, X. M., Luo, Y. and Melillo, J. M. (2019). Reduced carbon use efficiency and increased microbial turnover with soil warming. *Global Change Biology* 25(3): 900–910. DOI: 10.1111/gcb.14517.

Liang, C., Amelung, W., Lehmann, J. and Kästner, M. (2019). Quantitative assessment of microbial necromass contribution to soil organic matter. *Global Change Biology* 25(11): 3578–3590. DOI: 10.1111/gcb.14781.

Liang, C., Read, H. W. and Balser, T. C. (2012) GC-based detection of aldononitrile acetate derivatized glucosamine and muramic acid for microbial residue determination in soil. *Journal of Visualized Experiments : JoVE* 63(63): e3767. DOI: 10.3791/3767.

Lin, S., Wang, S., Si, Y., Yang, W., Zhu, S. and Ni, W. (2017). Variations in eco-enzymatic stoichiometric and microbial characteristics in paddy soil as affected by long-term integrated organic-inorganic fertilization. *PLoS ONE* 12(12): e0189908. DOI: 10.1371/journal.pone.0189908.

Ling, N., Sun, Y., Ma, J., Guo, J., Zhu, P., Peng, C., Yu, G., Ran, W., Guo, S. and Shen, Q. (2014). Response of the bacterial diversity and soil enzyme activity in particle-size fractions of Mollisol after different fertilization in a long-term experiment. *Biology and Fertility of Soils* 50(6): 901–911. DOI: 10.1007/s00374-014-0911-1.

Liu, J., Wu, N., Wang, H., Sun, J., Peng, B., Jiang, P. and Bai, E. (2016). Nitrogen addition affects chemical compositions of plant tissues, litter and soil organic matter. *Ecology* 97(7): 1796–1806. DOI: 10.1890/15-1683.1.

Lu, M., Zhou, X., Yang, Q., Li, H., Luo, Y., Fang, C., Chen, J., Yang, X. and Li, B. (2013). Responses of ecosystem carbon cycle to experimental warming: A meta-analysis. *Ecology* 94(3): 726–738. DOI: 10.1890/12-0279.1.

Luo, L., Meng, H. and Gu, J. D. (2017). Microbial extracellular enzymes in biogeochemical cycling of ecosystems. *Journal of Environmental Management* 197: 539-549. DOI: 10.1016/j.jenvman.2017.04.023.

Margalef, O., Sardans, J., Fernández-Martínez, M., Molowny-Horas, R., Janssens, I. A., Ciais, P., Goll, D., Richter, A., Obersteiner, M., Asensio, D. and Peñuelas, J. (2017). Global patterns of phosphatase activity in natural soils. *Scientific Reports* 7(1): 1337. DOI: 10.1038/s41598-017-01418-8.

Melillo, J. M., Frey, S. D., DeAngelis, K. M., Werner, W. J., Bernard, M. J., Bowles, F. P., Pold, G., Knorr, M. A. and Grandy, A. S. (2017). Long-term pattern and magnitude of soil carbon feedback to the climate system in a warming world. *Science* 358(6359): 101–105. DOI: 10.1126/science.aan2874.

Moebius-Clune, B. N., Moebius-Clune, D. J., Gugino, B. K., Idowu, O. J., Schindelbeck, R. R., van Es, H. M., Thies, J. E., Shayler, H. A., McBride, M. B., Wolfe, D. W. and Abawi, G. S. (2016). *Comprehensive Assessment of Soil Health: The Cornell Framework Manual*. Geneva, NY: Cornell University: 3.2.

Moebius-Clune, B. N., van Es, H. M., Idowu, O. J., Schindelbeck, R. R., Moebius-Clune, D. J., Wolfe, D. W., Abawi, G. S., Thies, J. E., Gugino, B. K. and Lucey, R. (2008). Long-term effects of harvesting maize Stover and tillage on soil quality. *Soil Science Society of America Journal* 72(4): 960–969. DOI: 10.2136/sssaj2007.0248.

Moorhead, D. L., Rinkes, Z. L., Sinsabaugh, R. L. and Weintraub, M. N. (2013). Dynamic relationships between microbial biomass, respiration, inorganic nutrients and enzyme activities: Informing enzyme-based decomposition models. *Frontiers in Microbiology* 4: 223. DOI: 10.3389/fmicb.2013.00223.

NRCS-USDA. (n.d.). Soil. Available at: https://www.nrcs.usda.gov/wps/portal/nrcs/main/soils/health/ (accessed 5 April 2021). Health Communications.

Nunan, N., Schmidt, H. and Raynaud, X. (2020). The ecology of heterogeneity: Soil bacterial communities and C dynamics. *Philosophical Transactions of the Royal Society of London. Series B, Biological Sciences* 375(1798): 20190249. DOI: 10.1098/rstb.2019.0249.

Olsen, S. R., Cole, C. V. and Watanabe, F. S. (1954). *Estimation of Available Phosphorus in Soils by Extraction with Sodium Bicarbonate*. Washington, DC: United States Department of Agriculture.

Ouyang, Y., Evans, S. E., Friesen, M. L. and Tiemann, L. K. (2018a). Effect of nitrogen fertilization on the abundance of nitrogen cycling genes in agricultural soils: A meta-analysis of field studies. *Soil Biology and Biochemistry* 127: 71–78. DOI: 10.1016/j.soilbio.2018.08.024.

Ouyang, Y., Reeve, J. R. and Norton, J. M. (2018b). Soil enzyme activities and abundance of microbial functional genes involved in nitrogen transformations in an organic farming system. *Biology and Fertility of Soils* 54(4): 437–450. DOI: 10.1007/s00374-018-1272-y.

Pandey, D., Agrawal, M. and Bohra, J. S. (2014). Effects of conventional tillage and no tillage permutations on extracellular soil enzyme activities and microbial biomass under rice cultivation. *Soil and Tillage Research* 136: 51-60. DOI: 10.1016/j.still.2013.09.013.

Paul, E. A. (Ed.). (2015). *Soil Microbiology, Ecology, and Biochemistry (5th ed)*. Academic Press.

Prosser, J. I. (2020). Putting science back into microbial ecology: A question of approach. *Philosophical Transactions of the Royal Society B: Biological Sciences* 375(1798): 20190240. DOI: 10.1098/rstb.2019.0240.

Prosser, J. I. (2015). Dispersing misconceptions and identifying opportunities for the use of 'omics' in soil microbial ecology. *Nature Reviews. Microbiology* 13(7): 439–446. DOI: 10.1038/nrmicro3468.

Raiesi, F. and Kabiri, V. (2016). Identification of soil quality indicators for assessing the effect of different tillage practices through a soil quality index in a semi-arid environment. *Ecological Indicators* 71: 198–207. DOI: 10.1016/j.ecolind.2016.06.061.

Raiesi, F. and Salek-Gilani, S. (2018). The potential activity of soil extracellular enzymes as an indicator for ecological restoration of rangeland soils after agricultural abandonment. *Applied Soil Ecology* 126: 140–147. DOI: 10.1016/j.apsoil.2018.02.022.

Rocca, J. D., Hall, E. K., Lennon, J. T., Evans, S. E., Waldrop, M. P., Cotner, J. B., Nemergut, D. R., Graham, E. B. and Wallenstein, M. D. (2015). Relationships between protein-encoding gene abundance and corresponding process are commonly assumed yet rarely observed. *ISME Journal* 9(8): 1693–1699. DOI: 10.1038/ismej.2014.252.

Roesch, L. F. W., Fulthorpe, R. R., Riva, A., Casella, G., Hadwin, A. K., Kent, A. D., Daroub, S. H., Camargo, F. A., Farmerie, W. G. and Triplett, E. W. (2007). Pyrosequencing enumerates and contrasts soil microbial diversity. *The ISME Journal* 1(4): 283–290. DOI: 10.1038/ismej.2007.53.

Rosier, C. L., Hoye, A. T. and Rillig, M. C. (2006). Glomalin-related soil protein: Assessment of current detection and quantification tools. *Soil Biology and Biochemistry* 38(8): 2205–2211. DOI: 10.1016/j.soilbio.2006.01.021.

Scherer, H. W. (2009). Sulfur in soils. *Journal of Plant Nutrition and Soil Science* 172(3): 326–335. DOI: 10.1002/jpln.200900037.

Schneider, T., Keiblinger, K. M., Schmid, E., Sterflinger-Gleixner, K., Ellersdorfer, G., Roschitzki, B., Richter, A., Eberl, L., Zechmeister-Boltenstern, S. and Riedel, K. (2012). Who is who in litter decomposition? Metaproteomics reveals major microbial players and their biogeochemical functions. *The ISME Journal* 6(9): 1749–1762. DOI: 10.1038/ismej.2012.11.

Schollenberger, C. J. and Simon, R. H. (1945). Determination of exchange capacity and exchangeable bases in soil–Ammonium acetate method. *Soil Science* 59(1): 13–24. DOI: 10.1097/00010694-194501000-00004.

Semenov, M. V., Krasnov, G. S., Semenov, V. M. and van Bruggen, A. H. C. (2020). Long-term fertilization rather than plant species shapes rhizosphere and bulk soil prokaryotic communities in agroecosystems. *Applied Soil Ecology* 154: 103641. DOI: 10.1016/j.apsoil.2020.103641.

Singh, J. and Kumar, S. (2021). Seasonal changes of soil carbon fractions and enzyme activities in response to winter cover crops under long-term rotation and tillage systems. *European Journal of Soil Science* 72(2): 886–899. DOI: 10.1111/ejss.13028.

Sinsabaugh, R. L., Hill, B. H. and Follstad Shah, J. J. (2009). Ecoenzymatic stoichiometry of microbial organic nutrient acquisition in soil and sediment. *Nature* 462(7274): 795–798. DOI: 10.1038/nature08632.

Soil Science Society of America. (2008). *Glossary of Soil Science Terms: ASA, CSSA, and SSSA Books*. Madison, WI: American Society of Agronomy and Soil Science Society of America. DOI: 10.2136/2008.glossarysoilscienceterms.

Soman, C., Li, D., Wander, M. M. and Kent, A. D. (2017). Long-term fertilizer and crop-rotation treatments differentially affect soil bacterial community structure. *Plant and Soil* 413(1–2): 145–159. DOI: 10.1007/s11104-016-3083-y.

Starke, R., Jehmlich, N. and Bastida, F. (2019). Using proteins to study how microbes contribute to soil ecosystem services: The current state and future perspectives of soil metaproteomics. *Journal of Proteomics* 198: 50–58. DOI: 10.1016/j.jprot.2018.11.011.

Sterner, R. W. and Elser, J. J. (2002). Ecological stoichiometry: The biology of elements from molecules to the biosphere. Available at: http://www.jstor.org/stable/j.ctt1jktrp3 (accessed 7 December 2020).

Waring, B. G., Sulman, B. N., Reed, S., Smith, A. P., Averill, C., Creamer, C. A., Cusack, D. F., Hall, S. J., Jastrow, J. D., Jilling, A., Kemner, K. M., Kleber, M., Liu, X. A., Pett-Ridge, J. and Schulz, M. (2020). From pools to flow: The PROMISE framework for new insights on soil carbon cycling in a changing world. *Global Change Biology* 26(12): 6631–6643. DOI: 10.1111/gcb.15365.

Warkentin, B. P. (1995). The changing concept of soil quality. *Journal of Soil and Water Conservation* 50(3): 226–228.

Weil, R. R., Islam, K. R., Stine, M. A., Islam, K. R., Stine, M. A., Gruver, J. B., Samson-Liebig, S. E. and Weil, R. R. (2003). Estimating active carbon for soil quality assessment: A simplified method for laboratory and field use. *American Journal of Alternative Agriculture* 18(1). Cambridge University Press 18(1): 3–17.

Wright, S., Green, V. and Cavigelli, M. (2007). Glomalin in aggregate size classes from three different farming systems. *Soil and Tillage Research* 94(2): 546–549. DOI: 10.1016/j.still.2006.08.003.

Wright, S. F. and Upadhyaya, A. (1996). Extraction of an abundant and unusual protein from soil and comparison with hyphal protein of arbuscular mycorrhizal fungi. *Soil Science* 161(9): 575–586.

Zhu, X., Jackson, R. D., DeLucia, E. H., Tiedje, J. M. and Liang, C. (2020). The soil microbial carbon pump: From conceptual insights to empirical assessments. *Global Change Biology* 26(11): 6032–6039. DOI: 10.1111/gcb.15319.

Chapter 4

Mycorrhizae and soil health

Muhammad Adnan and Fazli Wahid, The University of Swabi, Pakistan; Shah Fahad, The University of Haripur, Pakistan; Muhammad Arif, The University of Agriculture Peshawar, Pakistan; Songmei Shi, Yunnan Agricultural University, China; and Xinhua He, Southwest University, China, University of California-Davis, USA and University of Western Australia, Australia

1 Introduction

Soil is a complex environment that supports the largest and most diverse microbial community on the planet. It has been suggested that bacterial DNA comprises 97%, archaeal DNA 1%, fungal DNA 0.3% and other eukaryotic DNA 1.2% within the overall soil microbial community (the remainder is assigned to viruses or unknown groups) (Hirsch, 2018). However, soil fungi have been recognized as playing a key role in areas such as nutrient cycling in soil. Through enzyme release, fungi are very efficient organic matter decomposers, thus playing a key role in soil carbon (C) and cycling of nutrients such as nitrogen (N) and phosphorus (P) (Wahid et al., 2020; Žifčáková et al., 2016). They improve plant growth by increasing the uptake of mineral nutrients (Hoeksema et al., 2010; Bagyaraj and Ashwin, 2017). As a result, they have been described as 'ecosystem engineers' in soil (Swift, 2005; Gardi and Jeffery, 2009; Sun et al., 2010; Bücking and Kafle, 2015). Fungi also contribute

http://dx.doi.org/10.19103/AS.2021.0094.05

to plant nitrogen (N) fixation (Jayne and Quigley, 2014; Muhammad et al., 2019) and are actively involved in soil carbon sequestration as well (Parniske, 2008).

This chapter explores the particular role of mycorrhizae in improving soil health via mechanisms such as aggregate formation and stabilization, remediation of soil heavy metals and reclamation of degraded soils, soil organic matter (SOM) formation, nutrient cycling and carbon sequestration. The chapter also looks at the ways mycorrhizae can alleviate both abiotic stresses such as nutrient deficiency and biotic stresses such as disease.

2 Mycorrhizae and their role in soil health

A key group of fungal symbioses is the mycorrhizae. The term 'mycorrhizae' comes from the Greek words for 'fungus' and 'root'. The term refers to the mutually beneficial relationship of host plants with soil fungi (Smith and Read, 2008). Mycorrhizae have the ability to form a symbiotic relationship with the roots of ~80% of higher plants (Wang and Qiu, 2006; Smith and Read, 2008; Brundrett, 2009; Brundrett and Tedersoo, 2018). In this relationship, plant roots provide essential nutrients to the fungi (carbohydrates) whilst the large mass of fungal hyphae (long filamentous structures) acts as an extended root system for the plant, increasing its ability to take up water as well as macro- and micronutrients, particularly immobile mineral nutrients such as NH_4^+, P, Cu and Zn (Friese and Allen, 1991; Marschner and Dell, 1994; Marschner and Rengel, 2012; Meena et al., 2016; Dotaniya et al., 2016; Adnan et al., 2020; Huq et al., 2020).

Mycorrhizae differ in their ability to form extraradical hyphae (Jilling et al., 2018). In some conditions, mycorrhizae can account for up to half of standing microbial biomass (Bååth et al., 2004; Högberg and Högberg, 2002). The extraradical hyphal biomass in Sweden forests e.g. ranges from 4.8 g m^{-2} to 5.8 g m^{-2} over the upper 70 cm soil profile in pure spruce and mixed spruce-oak stands (Wallander et al., 2004) and 5.0 g m^{-2} to 14.0 g m^{-2} in the organic horizon of a pine forest (Hagenbo et al., 2017). The hyphae of endomycorrhizae grow within plant cells and out into the soil whilst the less common ectomycorrhizae form a sheath around plant roots called a Hartig net from which hyphae develop (Miller and Jastrow, 2000; Smith and Read, 2008; Schwendemann et al., 2011).

The association of arbuscular mycorrhizal fungi (AMF) with plant roots has been recognized as one of the oldest and most widespread types of symbiotic relationships between plants and fungi (Paszkowski, 2006; Martin et al., 2007). AMF are a type of endomycorrhizae (from the phylum Glomeromycota) in which the fungus penetrates root cells and forms intercellular structures called arbuscules (Douds and Millner, 1999; Schüßler et al., 2001; Wang and Qiu, 2006; Smith and Read, 2008; Mian et al., 2019). Arbuscles provide the sites for nutrient exchange between fungus and plant (Ezawa et al., 2002; Smith and Read, 2008). Beyond plant roots, AMF form elaborate hyphal networks that can extend 5–15 cm beyond the root zone and exploit nutrients in bulk soil (i.e. soil outside the rhizosphere) for the benefit of both fungus and plant (Jakobsen et al., 1992; Khan et al., 2019; Brady and Weil, 2002; Rouphael et al., 2015). This relationship supports plant growth in conditions of limited nutrient supply (Borowicz, 2001; Smith and Read, 2008).

Mycorrhizae not only increase plant growth directly by enhancing uptake of nutrients but indirectly through their non-nutritional roles by promoting plant tolerance to biotic and abiotic stresses as well as by improving soil structure and water retention (Smith et al., 2009). Like other fungi, they also are important in areas such as soil C sequestration and remediation of soil pollutants (Hodge et al., 2001; Hodge and Fitter, 2010; Banegas et al., 2022; Singh et al., 2022).

As noted earlier, as well as promoting plant health, mycorrhizae play an important, broader role in soil health (Harris, 2009; Kumar et al., 2015). Soil health has been defined as the continued capacity of soil to function as a living ecosystem able to sustain plants, animals and humans, primarily by delivering ecosystem services in such areas as: providing physical stability and support (e.g. for plant roots), filtering and buffering pollutants, nutrient cycling, regulating air quality, water regulation and as a source of nutrients for plants and animals (Doran and Zeiss, 2000; Bardgett, 2005; Adnan et al., 2019).

Mycorrhizae play a key role in enhancing soil ecosystem services. Given their role in nutrient cycling and organic matter decomposition and accumulation in soil, they affect C and N cycling on a global scale (Smith and Read, 2008). They are also involved e.g. in the breakdown of pollutants (Torstensson et al., 1998). There is thus a strong association between soil biota, plant and soil health (Altieri and Nicholls, 2003; Giller et al., 2005). Healthy populations of soil biota like mycorrhizae are linked to crop yield and quality, nutrient cycling, overall plant and soil health (Al-Karaki et al., 2007; Wahid et al., 2019). The following sections discuss the different roles of mycorrhizae in promoting soil health in more detail.

3 Mycorrhizae and soil composition

AMF create a branching network reaching up to 6.5–19.4 m of hyphae per g^{-1} soil (Cavagnaro, 2008; Shen et al., 2016). This mycelial network (a rootlike structure produced by fungi) is a substantial portion of the total soil microbial biomass and can contribute up to 50% of fungal mycelia in soil (Rillig and Mummey, 2006; Leake et al., 2004). This complex network improves soil structure due to its binding action of organic matter that assists in forming and stabilizing aggregates (Miller and Jastrow, 1992a; Bedini et al., 2007).

AMF deposit slow cycling organic products such as chitin and glomalin into soil (Wright and Upadhyaya, 1996; Rillig et al., 2005) which supports soil aggregation (Bedini et al., 2007; Singh et al., 2022). Glomalin, also referred to as glomalin-related soil protein (GRSP), is a glycoprotein produced by AMF and plays an important role in soil structure and C sequestration (Wright and Upadhyaya, 1996; Wright et al., 2007; Zhang et al., 2017; Banegas et al., 2022; Holátko et al., 2021). The formation of soil aggregates also promotes aeration which enhances nutrient availability (Veresoglou et al., 2012). Soil aggregates also provide good habitats for other microbes in soil (Zamin et al., 2019). By improving water retention, they help control soil moisture-sensitive processes like nitrification (Avrahami and Bohannan, 2007), denitrification (Saleem et al., 2020) and leaching (Tran et al., 2021).

As has been noted, mycorrhizae such as AMF can also play a role in managing polluted soils. AMF are able to sequester and extract potentially toxic heavy metals from contaminated soils (Hutchinson et al., 2003). Whilst AMF promote plant micronutrient uptake, they have been found to have a buffer function in protecting plants from excessive and toxic uptake of heavy metals such as Al, Cu, Zn, Fe and Mn as well as chemical pollutants (Cumming and Weinstein, 1990; Vogel-Mikuš et al., 2005). This has given them a role in reclaiming soils polluted from industrial activities such as mining (Wang et al., 2017; Malcová et al., 2003; Festin et al., 2019; Singh et al., 2022).

4 Mycorrhizae and soil organic matter formation

The main route through which C enters soil is probably due to the release of low molecular weight substances by mycorrhizal roots, including amino acids, enzymes, fatty acids, glucose, organic acids, phenolics, vitamins and sterol sugars (Dakora and Phillips, 2002; Kuzyakov, 2002; Pausch and Kuzyakov, 2018). Root exudates are adsorbed on mineral surfaces, allowing the formation and stabilization of SOM which sequesters C (Keiluweit et al., 2015; Kleber et al., 2015; Zhang et al., 2019). Mycorrhizae can also release these exudates directly (Toljander et al., 2007; Kaiser et al., 2015).

Compared to the pool of total SOM, the pool of living biomass represented by soil microorganisms is small at about 1–3% of the total C within the soil. However, given rapid turnover, microbial residues (known as necromass) can play a major role in contributing to total SOM in comparison to the pool of standing microbial biomass (Grandy and Neff, 2008; Cotrufo et al., 2013; Schmidt et al., 2011). Mycorrhizae produce SOM which, when stabilized, becomes a resource for heterotrophic microorganisms to further contribute to total SOM (Clemmensen et al., 2013; Schweigert et al., 2015). It has also been hypothesized that competition of mycorrhiza with free-living saprotrophs slows organic matter decomposition and hence helps its accumulation. Research thus suggests that mycorrhizal necromass plays a key role in overall SOM formation and stabilization.

Although mycorrhiza play a key role in the formation, stabilization and C storage, their activities can also cause loss and destabilization of SOM. Like saprotrophic fungi, AMF metabolize SOM for C or energy acquisition (Zak et al., 2019; Lindahl and Tunlid, 2015). The destabilization and loss of mycorrhizal SOM result in an enzymatic release of C and nutrients, especially N and P, which are then available for utilization by plants and mycorrhizal fungi (Read, 1991; Hodge et al., 2001; Hodge and Fitter, 2010).

5 Mycorrhizae, nutrient cycling and carbon sequestration

As has been noted, mycorrhizae have been found to play a key role in promoting both nutrient availability and nutrient cycling in soil (Rillig, 2004; Smith and Read, 2008; Fellbaum et al., 2012). Arbuscular mycorrhizal fungi can mineralize soil organic matter (Cheng et al., 2012; Hodge and Fitter, 2010) and organic compounds (Whiteside et al., 2012) which reduce N leaching (Bender and van der Heijden, 2015). This process of mineralization results in a large, stabilized pool of organic N on the surface of soil mineral surfaces (Jilling et al., 2018). AMF also have the ability to reduce P leaching (Jansa et al., 2013; Bender et al., 2015). By promoting N cycling, AMF have been found to play a role in reducing N_2O emissions (Cavagnaro et al., 2012; Bender et al., 2014; Lazcano et al., 2014). Similarly, by improving soil water-holding capacity, AMF can indirectly contribute to limiting greenhouse gases emissions (Augé, 2004; Cavagnaro et al., 2006).

Ectomycorrhizae play a significant role in the regulation of the C cycle through mineralization of soil organic matter and C stabilization (Fernandez and Kennedy, 2015; Lindahl and Tunlid, 2015). They can increase soil carbon sequestration by slowing down decomposition of soil organic matter (SOM) via a process known as the Gadgil effect (Gadgil and Gadgil, 1971). Whilst the Gadgil effect has not been observed with AMF, they may still play an important role in C sequestration e.g. in stabilizing SOM or indirectly by promoting formation of soil aggregates with greater C storage potential (Hodge et al., 2001; Koide and Wu, 2003; Hodge and Fitter, 2010; Orwin et al., 2011; Cheng et al., 2012; Lindahl and Tunlid, 2015; Carteron et al., 2021). Research has also shown the ways mycorrhizae either facilitate C uptake by plants or use C themselves in locking up C within plants or soil (Barrett et al., 2010; Courty et al., 2011; Eriksson and Kainulainen, 2011; Gartner et al., 2012; Stöckel et al., 2014; van der Heijden et al., 2015).

AMF extraradical hyphae act as an important component in the formation and stabilization of SOM content. Extracellular compounds are produced and secreted by hyphae that help in aggregating clay particles that leads to the stabilization of C and mineral nutrients on particle surfaces and within soil particles (Jilling et al., 2018). The interaction between mineral nutrient formation and organic matter has been suggested as a key factor in storing soil C due to a reduced turnover of SOM (Kleber et al., 2015). The enmeshing of soil particles is caused by the adherent properties of fungal hyphae that can generate stable aggregates and thus slow the decomposition of SOM (Six et al., 1998). Extraradical hyphae also increase soil aggregation and the creation of water-stable aggregates (Singh et al., 2022; Lehmann et al., 2017).

6 Mycorrhizae, abiotic and biotic stress

Abiotic stresses such as drought, heat, salinity, nutrient deficiencies or heavy metal toxicity cause significant losses in agricultural productivity (Evelin et al., 2009). AMF can play a significant beneficial role in stressed environments (Lal, 2009). Mycorrhizal plants are generally more resistant to environmental stresses than non-mycorrhizal plants. AMF e.g. have been shown to be resistant to drought and to promote plant tolerance to drought stress, although the precise mechanisms are still not clear (Sylvia et al., 1993; Adnan et al., 2017).

The hyphal network of AMF significantly increases the capacity of plants to access and absorb available water (Safir and Nelson, 1985). By increasing the capacity to access water and nutrients (particularly nutrients like P that need water to diffuse effectively), mycorrhizae have the ability to improve plant tolerance to heat, drought, unfavorable pH, salinity, metal toxicity and low soil fertility (Trappe, 1962, 1977; Adnan et al., 2017; Tarraf et al., 2017). For example, salinity stress in olive plants in Spain has been alleviated by promoting AMF in soils (Porras-Soriano et al., 2009). Marulanda et al. (2009) e.g. observed a 42% reduction in water requirement in cultivation of *Retama sphaerocarpa* by inoculation with a native drought-tolerant strain of the AMF *Rhizophagus irregularis*. Indigenous AMF isolated from dry soils in Mediterranean ecosystems have been observed to improve soil quality and re-vegetation of degraded soils (Marulanda-Aguirre et al., 2008). AMF have also been proposed as part of the solution to reverse desertification which affects around 20% of terrestrial ecosystems (Zhan et al., 2019; Jambhulkar and Kumar, 2019; Stefanowicz et al., 2020).

Mycorrhizae can also play a role in competing with and suppressing pathogens in the rhizosphere that can cause disease in plants (Hakeem et al., 2016). Research suggests that AMF directly and/or indirectly suppress plant pathogens (Akhtar and Siddiqui, 2008). AMF both stimulate plant defenses and compete with pathogens for resources (Secilia and Bagyaraj, 1987). After inoculation with AMF, plants release plant defense-related compounds such as chitinases, peroxidases, β-1,3-glucanases, phenolics, callose, proteins and phytoalexins (Dehne and Schonbeck, 1979; Gianinazzi and Gianinazzi-Pearson,

1992; Gianinazzi-Pearson et al., 1994). Examples of pathogen reduction using AMF include reducing *Phytophthora cinnamon* in tomato plants (Meyer and Linderman, 1986), preventing *Fusarium solani* infection in beans (Al-Askar and Rashad, 2010) and *Ralstonia solanacerum*-induced wilt disease in tomato (Vailleau et al., 2007). *Glomus etinicatunium* and *G. margarita* inoculation has been used to fight Verticillium disease (Matsubara et al., 1995).

7 Mycorrhizal interaction with other soil microorganisms

Beneficial microorganisms in soil include mycorrhizal fungi, N_2-fixing rhizobia (NFB) and plant growth-promoting rhizobacteria (PGPR). PGPR promote nutrient cycling and can enhance plant resistance to disease (Probanza et al., 2002; Dobbelaere et al., 2001), NFB contribute to increasing soil N (Postgate, 1998). Other bacteria promote the availability of chemicals such as P and Mg (Toro et al., 1997; Nogueira et al., 2007). Through changes in the rhizosphere, improving nutrient cycling and soil structure, mycorrhizae can create a favorable environment for these other microorganisms (Edwards et al., 1998; Rambelli, 1973; Linderman, 1994).

It has been found that different bacteria can work synergistically in promoting the growth and health of plants such as poplars (Quoreshi and Khasa, 2008). These synergies have been found e.g. for AMF and PGPR (Garbaye, 1994; Jäderlund et al., 2008; Vestergård et al., 2008). Other bacteria in the rhizosphere have e.g. been found to enhance root colonization by AMF and their ability to process Fe, Mn and other nutrients (Kothari et al., 1990; Berthelin et al., 1991; Mozafar et al., 1992; Bending et al., 2006). In turn, AMF help to activate Mn-oxidizing bacteria in reducing Mn toxicity to plants in acidic soils (Arines et al., 1992). AMF also stimulate activity in phosphate-solubilizing bacteria (Zhang et al., 2016; Wahid et al., 2016, 2019, 2020).

Research has also investigated interactions between mycorrhizal fungi and other fungal taxa including Colletotrichum, Mucoromycotina and Sebacina (Ogura-Tsujita et al., 2012; Hiruma et al., 2016; Kong et al., 2015). The complex interaction of mycorrhizae with other soil microbes has been linked to C cycling and sequestration processes (Gryndler et al., 2014; Bukovská et al., 2016; Wang et al., 2022).

8 Conclusion and future trends

Arbuscular mycorrhizal fungi (AMF) are an active component of soil and play a critical role in maintaining soil health and plant productivity. The symbiotic interactions of mycorrhiza with other microbes enhance nutrient availability and uptake, organic matter decomposition, soil aggregation, C cycling and water holding capacity. Mycorrhizae also protect plants from diseases and abiotic stresses such as drought and heat stress.

AMF and other mycorrhizae are vulnerable to poor soil conditions with impacts on soil health, plant growth and the wider functioning of terrestrial ecosystems (Uddin et al., 2016; Lehmann et al., 2017; Shafi et al., 2020). This is particularly the case in the upper layer of soil (0-30cm) which, whilst it represents <0.5% of the total soil volume, contains the highest concentration of total soil organic matter, available nutrients and biological activity. The amount and diversity of AMF can be adversely affected by intensive agronomic practices acting on this layer of soil in particular such as tillage or chemical fertilizer application (Njeru et al., 2015; Lal, 2015). In contrast, plant-mycorrhizal symbiotic relationships can be improved by agronomic practices like zero-tillage, continuous cropping systems, cover crops and reduction of chemical fertilizer inputs (Rillig et al., 2005; Basir et al., 2016; Singh et al., 2022).

Future research should focus on mycorrhizal interactions with other soil microorganisms at different soil depths or layers. More research is also needed on the mechanisms underlying the effects of mycorrhizae in such areas as aggregate stability, nutrient cycling, plant root function and plant growth, including the role of specific genes and proteins in mycorrhizal activity. More work needs to be done to develop mass production and testing of stable mycorrhizal inoculums as biofertilizers for use on a variety of crops in different agro-ecological environments in promoting plant and soil health.

9 Where to look for further information

For both research progresses and practical applications regarding mycorrhizae and soil health, please refer to the following two websites of https://mycorrhizas.org/ and https://mycorrhizae.com/, as well as corresponding mycological or microbial societies in your home country.

10 Acknowledgements

Research is supported by a Free Exploration Program at Southwest University (SWU116002), National Natural Science Foundation of China - Youth Program (4111800096), Key Laboratory of Eco-environments in Three Gorges Reservoir Region of Ministry of Education, Chongqing Key Laboratory of Plant Resource Conservation and Germplasm Innovation, and National Base of International S&T Collaboration on Water Environmental Monitoring and Simulation in the Three Gorges Reservoir Region, China.

11 References

Adnan, M., Fahad, S., Khan, I. A., Saeed, M., Ihsan, M. Z., Saud, S., Riaz, M., Wang, D. and Wu, C. (2019). Integration of poultry manure and phosphate solubilizing bacteria improved availability of Ca bound P in calcareous soils, *Biotech* 9(10), 368.

Adnan, M., Fahad, S., Zamin, M., Shah, S., Mian, I. A., Danish, S., Zafar-Ul-Hye, M., Battaglia, M. L., Naz, R. M. M., Saeed, B., Saud, S., Ahmad, I., Yue, Z., Brtnicky, M., Holatko, J. and Datta, R. (2020). Coupling phosphate-solubilizing bacteria with phosphorus supplements improve maize phosphorus acquisition and growth under lime induced salinity stress, *Plants (Basel)* 9(7), 900.

Adnan, M., Shah, Z., Fahad, S., Arif, M., Alam, M., Khan, I. A., Mian, I. A., Basir, A., Ullah, H., Arshad, M. and Rahman, I. U. (2017). Phosphate-solubilizing bacteria nullify the antagonistic effect of soil calcification on bioavailability of phosphorus in alkaline soils, *Sci. Rep.* 7(1), 16131.

Akhtar, M. S. and Siddiqui, Z. A. (2008). Arbuscular mycorrhizal fungi as potential bioprotectants against plant pathogens. In: Siddiqui, Z., Akhtar, M. and Futai, K. (Eds.), *Mycorrhizae: Sustainable Agriculture and Forestry*. Dordrecht: Springer, pp. 61-97.

Al-Askar, A. A. and Rashad, Y. M. (2010). Arbuscular mycorrhizal fungi: A biocontrol agent against common bean fusarium root rot disease, *Plant Pathol. J.* 9(1), 31-38.

Al-Karaki, G. N. O., Othman, Y. and Al-Ajmi, A. (2007). Effects of mycorrhizal fungi inoculation on landscape turf establishment under Arabian Gulf region conditions, *Arabgulf J. Sci. Res.* 25, 147-152.

Altieri, M. A. and Nicholls, C. I. (2003). Soil fertility management and insect pests: Harmonizing soil and plant health in agroecosystems, *Soil Till. Res.* 72(2), 203-211.

Arines, J., Porto, M. E. and Vilarino, A. (1992). Effect of manganese on vesicular-arbuscular mycorrhizal development in red clover plants and on soil Mn-oxidizing bacteria, *Mycorrhiza* 1(3), 127-131.

Augé, R. M. (2004). Arbuscular mycorrhizae and soil/plant water relations, *Can. J. Soil Sci.* 84(4), 373-381.

Avrahami, S. and Bohannan, B. J. (2007). Response of *Nitrosospira* sp. strain AF-like ammonia oxidizers to changes in temperature, soil moisture content, and fertilizer concentration, *Appl. Environ. Microbiol.* 73(4), 1166-1173.

Bååth, E., Nilsson, L. O., Göransson, H. and Wallander, H. (2004). Can the extent of degradation of soil fungal mycelium during soil incubation be used to estimate ectomycorrhizal biomass in soil?, *Soil Biol. Biochem.* 36(12), 2105-2109.

Bagyaraj, D. J. and Ashwin, R. (2017). Soil biodiversity: Role in sustainable horticulture, *Biodivers. Hortic. Crops* 5, 1-18.

Banegas, N., Dos Santos, D. A., Molina, F. G., Albanesi, A. and Pedraza, R. (2022). Glomalin contribution to soil organic carbon under different pasture managements in a saline soil environment, *Arch. Agron. Soil Sci.* 68(3), 340-354.

Bardgett, R. (2005). *The Biology of Soil: A Community and Ecosystem Approach*. Oxford: Oxford University Press.

Barrett, C. F., Freudenstein, J. V., Lee Taylor, D. L. and Kõljalg, U. (2010). Range wide analysis of fungal associations in the fully mycoheterotrophic *Corallorhiza striata* complex (Orchidaceae) reveals extreme specificity on ectomycorrhizal Tomentella (Thelephoraceae) across North America, *Am. J. Bot.* 97(4), 628-643.

Basir, A., Jan, M. T., Alam, M., Shah, A. S., Afridi, K., Adnan, M., Ali, K. and Mian, I. A. (2016). Impacts of tillage, stubble management, and nitrogen on wheat production and soil properties, *Can. J. Soil Sci.* 97, 133–140.

Bedini, S., Avio, L., Argese, E. and Giovannetti, M. (2007). Effects of long-term land use onarbuscular mycorrhizal fungi and glomalin-related soil protein, *Agr. Ecosyst. Environ.* 120(2–4), 463–466.

Bender, S. F., Conen, F. and van der Heijden, M. G. A. (2015). Mycorrhizal effects on nutrient cycling, nutrient leaching and N_2O production in experimental grassland, *Soil Biol. Biochem.* 80, 283–292.

Bender, S. F., Plantenga, F., Neftel, A., Jocher, M., Oberholzer, H. R., Köhl, L., Giles, M., Daniell, T. J. and van der Heijden, M. G. A. (2014). Symbiotic relationships between soil fungi and plants reduce N_2O emissions from soil, *ISME J.* 8, 1336–1345.

Bender, S. F. and van der Heijden, M. G. A. (2015). Soil biota enhance agricultural sustainability by improving crop yield, nutrient uptake and reducing nitrogen leaching losses, *J. Appl. Ecol.* 52(1), 228–239.

Bending, G. D., Aspray, T. J. and Whipps, J. M. (2006). Significance of microbial interactions in the mycorrhizosphere, *Adv. Appl. Microbiol.* 60, 97–132.

Berthelin, J., Leyval, C., Laheurte, F. and Giudici, P. D. (1991). Involvement of roots and rhizosphere microflora in the chemical weathering of soil minerals, *Plant Root Growth Regul.*, 187–210.

Borowicz, V. A. (2001). Do arbuscular mycorrhizal fungi alter plant-pathogen relations?, *Ecology* 82(11), 3057–3068.

Brady, N. C. and Weil, R. R. (2002). *The Nature and Properties of Soils*. Upper Saddle River, NJ: Prentice-Hall Inc.

Brundrett, M. C. (2009). Mycorrhizal associations and other means of nutrition of vascular plants: Understanding the global diversity of host plants by resolving conflicting information and developing reliable means of diagnosis, *Plant Soil* 320(1–2), 37–77.

Brundrett, M. C. and Tedersoo, L. (2018). Evolutionary history of mycorrhizal symbioses and global host plant diversity, *New Phytol.* 220(4), 1108–1115.

Bücking, H. and Kafle, A. (2015). Role of arbuscular mycorrhizal fungi in the nitrogen uptake of plants: Current knowledge and research gaps, *Agronomy* 5(4), 587–612.

Bukovská, P., Gryndler, M., Gryndlerová, H., Püschel, D. and Jansa, J. (2016). Organic nitrogen-driven stimulation of arbuscular mycorrhizal fungal hyphae correlates with abundance of ammonia oxidizers, *Front. Microbiol.* 7, 711.

Carteron, A., Cichonski, F. and Laliberte, E. (2021). Ectomycorrhizal stands accelerate decomposition to a greater extent than arbuscular mycorrhizal stands in a northern deciduous forest, *Ecosystems*. DOI: 10.1007/s10021-021-00712-x.

Cavagnaro, T. R. (2008). The role of arbuscular mycorrhizas in improving plant zinc nutrition under low soil zinc concentrations: A review, *Plant Soil* 304(1–2), 315–325.

Cavagnaro, T. R., Barrios-Masias, F. H. and Jackson, L. E. (2012). Arbuscular mycorrhizas and their role in plant growth, nitrogen interception and soil gas efflux in an organic production system, *Plant Soil* 353(1–2), 181–194.

Cavagnaro, T. R., Jackson, L. E., Six, J., Ferris, H., Goyal, S., Asami, D. and Scow, K. M. (2006). Arbuscular mycorrhizas, microbial communities, nutrient availability, and soil aggregates in organic tomato production, *Plant Soil* 282(1–2), 209–225.

Cheng, L., Booker, F. L., Tu, C., Burkey, K. O., Zhou, L., Shew, H. D., Rufty, T. W. and Hu, S. (2012). Arbuscular mycorrhizal fungi increase organic carbon decomposition under elevated CO_2, *Science* 337(6098), 1084–1087.

Clemmensen, K. E., Bahr, A., Ovaskainen, O., Dahlberg, A., Ekblad, A., Wallander, H., Stenlid, J., Finlay, R. D., Wardle, D. A. and Lindahl, B. D. (2013). Roots and associated fungi drive long-term carbon sequestration in boreal forest, *Science* 339(6127), 1615–1618.

Cotrufo, M. F., Wallenstein, M. D., Boot, C. M., Denef, K. and Paul, E. (2013). The microbial efficiency-matrix stabilization (MEMS) framework integrates plant litter decomposition with soil organic matter stabilization: Do labile plant inputs form stable soil organic matter?, *Glob. Change Biol.* 19(4), 988–995.

Courty, P. E., Walder, F., Boller, T., Ineichen, K., Wiemken, A., Rousteau, A. and Selosse, M. A. (2011). Carbon and nitrogen metabolism in mycorrhizal networks and mycoheterotrophic plants of tropical forests: A stable isotope analysis, *Plant Physiol.* 156(2), 952–961.

Cumming, J. R. and Weinstein, L. H. (1990). Aluminum-mycorrhizal interactions in the physiology of pitch pine-seedlings, Plant Soil 125(1), 7–18.

Dakora, F. D. and Phillips, D. A. (2002). Root exudates as mediators of mineral acquisition in low-nutrient environments. In: Adu-Gyamfi, J. J. (Ed.), *Food Security in Nutrient-Stressed Environments: Exploiting Plants' Genetic Capabilities*. Dordrecht: Springer, pp. 201–213.

Dehne, H. W. and Schonbeck, F. (1979). Influence of endotrophic mycorrhiza on plant diseases. II. Phenol metabolism and lignification, *J. Phytopathol.* 95(3), 210–216.

Dobbelaere, S., Croonenborghs, A., Thys, A., Ptacek, D., Vanderleyden, J., Dutto, P., Labandera-Gonzalez, C., Caballero-Mellado, J., Aguirre, J. F., Kapulnik, Y., Brener, S., Burdman, S., Kadouri, D., Sarig, S. and Okon, Y. (2001). Responses of agronomically important crops to inoculation with *Azospirillum*, *Funct. Plant Biol.* 28(9), 871–879.

Doran, J. W. and Zeiss, M. R. (2000). Soil health and sustainability: Managing the biotic component of soil quality, *Appl. Soil Ecol.* 15(1), 3–11.

Dotaniya, M. L., Meena, V. D., Basak, B. B. and Meena, R. S. (2016). Potassium uptake by crops: As well as microorganisms. In: Meena, V., Maurya, B., Verma, J. and Meena, R. (Eds.), Potassium Solubilizing Microorganisms for Sustainable Agriculture. New Delhi, India: Springer, pp. 67–280.

Douds Jr, D. D. and Millner, P. D. (1999). Biodiversity of arbuscular mycorrhizal fungi in agroecosystems, *Agr. Ecosyst. Environ.* 74(1–3), 77–93.

Edwards, D., Kerp, H. and Hass, H. (1998). Stomata in early land plants: An anatomical and ecophysiological approach, *J. Exp. Bot.* 49(Special), 255–278.

Eriksson, O. and Kainulainen, K. (2011). The evolutionary ecology of dust seeds, *Plant Ecol.* 13(2), 73–87.

Evelin, H., Kapoor, R. and Giri, B. (2009). Arbuscular mycorrhizal fungi in alleviation of salt stress: A review, *Ann. Bot.* 104(7), 1263–1280.

Ezawa, T., Smith, S. E. and Smith, F. A. (2002). P metabolism and transport in AM fungi, *Plant Soil* 244(1–2), 221–230.

Fellbaum, C. R., Mensah, J. A., Pfeffer, P. E., Kiers, E. T. and Bücking, H. (2012). The role of carbon in fungal nutrient uptake and transport: Implications for resource exchange in the arbuscular mycorrhizal symbiosis, *Plant Signal. Behav.* 7(11), 1509–1512.

Fernandez, C. W. and Kennedy, P. G. (2015). Moving beyond the black-box: Fungal traits, community structure, and carbon sequestration in forest soils, *New Phytol.* 205(4), 1378–1380.

Festin, E. S., Tigabu, M., Chileshe, M. N., Syampungani, S. and Odén, P. C. (2019). Progresses in restoration of post-mining landscape in Africa, *J. For. Res.* 30(2), 381–396.

Friese, C. F. and Allen, M. F. (1991). The spread of VA mycorrhizal fungal hyphae in the soil: Inoculum types and external hyphal architecture, *Mycologia* 83(4), 409–418.

Gadgil, R. L. and Gadgil, P. D. (1971). Mycorrhiza and litter decomposition, *Nature* 233(5315), 133–133.

Garbaye, J. (1994). Tansley review no. 76 helper bacteria: A new dimension to the mycorrhizal symbiosis, *New Phytol.* 128(2), 197–210.

Gardi, C. and Jeffery, S. (2009). *Soil Biodiversity*. Brussels: Office for Official Publications of the European Commission, EUR23759EN, pp. 1–29.

Gartner, T. B., Treseder, K. K., Malcolm, G. M. and Sinsabaugh, R. L. (2012). Extracellular enzyme activity in the mycorrhizospheres of a boreal fire chronosequence, *Pedobiologia* 55(2), 121–127.

Gianinazzi, S. and Gianinazzi-Pearson, V. (1992). Cytology, histochemistry and immunochemistry as tools for studying structure and function in endomycorrhiza. *Methods Microbiol.* 24, 109–139.

Gianinazzi-Pearson, V., Lemoine, M. C., Arnould, C., Gollotte, A. and Morton, J. B. (1994). Localization of β-glucans in spore and hyphal walls of fungi in the Glomales, *Mycologia* 86(4), 478–485.

Giller, K. E., Bignell, D. E. and Lavelle, P. A. T. (2005). Soil biodiversity in rapidly changing tropical landscapes: Scaling down and scaling up. In: De Ruiter, P. C., Neutel, A. M. and Moore, J. (Eds.), *Biological Diversity and Function in Soils*. Cambridge: Cambridge University Press, pp. 295–318.

Grandy, A. S. and Neff, J. C. (2008). Molecular C dynamics downstream: The biochemical decomposition sequence and its impact on soil organic matter structure and function, *Sci. Total Environ.* 404(2–3), 297–307.

Gryndler, M., Černá, L., Bukovská, P., Hršelová, H. and Jansa, J. (2014). *Tuber aestivum* association with non-host roots, *Mycorrhiza* 24(8), 603–610.

Hagenbo, A., Clemmensen, K. E., Finlay, R. D., Kyaschenko, J., Lindahl, B. D., Fransson, P. and Ekblad, A. (2017). Changes in turnover rather than production regulate biomass of ectomycorrhizal fungal mycelium across a *Pinus sylvestris* chronosequence, *New Phytol.* 214(1), 424–431.

Hakeem, K. R., Akhtar, M. S. and Abdullah, S. N. A. (Eds.). (2016). *Plant, Soil and Microbes*. Switzerland: Springer.

Harris, J. (2009). Soil microbial communities and restoration ecology: Facilitators or followers?, *Science* 325(5940), 573-574.

Hirsch, P. (2018). Soil microorganisms: Role in soil health. In: Reicosky, D. (Ed.), *Managing Soil Health for Sustainable Agriculture Volume 1: Fundamentals*. Cambridge: Burleigh Dodds Science Publishing.

Hiruma, K., Gerlach, N., Sacristán, S., Nakano, R. T., Hacquard, S., Kracher, B., Neumann, U., Ramírez, D., Bucher, M., O'Connell, R. J. and Schulze-Lefert, P. (2016). Root endophyte *Colletotrichum tofieldiae* confers plant fitness benefits that are phosphate status dependent, *Cell* 165(2), 464–474.

Hodge, A., Campbell, C. D. and Fitter, A. H. (2001) An arbuscular mycorrhizal fungus accelerates decomposition and acquires nitrogen directly from organic material, *Nature* 413(6853), 297–299.

Hodge, A. and Fitter, A. H. (2010). Substantial nitrogen acquisition by arbuscular mycorrhizal fungi from organic material has implications for N cycling, *Proc. Natl Acad. Sci. U. S. A.* 107(31), 13754–13759.

Hoeksema, J. D., Chaudhary, V. B., Gehring, C. A., Johnson, N. C., Karst, J., Koide, R. T., Pringle, A., Zabinski, C., Bever, J. D., Moore, J. C., Wilson, G. W., Klironomos, J. N. and Umbanhowar, J. (2010). A meta-analysis of context-dependency in plant response to inoculation with mycorrhizal fungi, *Ecol. Lett.* 13(3), 394-407.

Högberg, M. N. and Högberg, P. (2002). Extramatrical ectomycorrhizal mycelium contributes one-third of microbial biomass and produces, together with associated roots, half the dissolved organic carbon in a forest soil, *New Phytol.* 154(3), 791–795.

Holátko, J., Brtnický, M., Kùcerík, J., Kotianová, M., Elble, J., Antonín Kintl, A., Kynický, J., Benada, O., Rahul Datta, R. and Jansa, J. (2021). Glomalin - Truths, myths, and the future of this elusive soil glycoprotein, Soil Biol. Biochem. 153(2), 108116.

Huq, M. E., Fahad, S., Shao, Z., Sarven, M. S., Khan, I. A., Alam, M., Saeed, M., Ullah, H., Adnan, M., Saud, S., Cheng, Q., Ali, S., Wahid, F., Zamin, M., Raza, M. A., Saeed, B., Riaz, M. and Khan, W. U. (2020). Arsenic in a roundwater environment in Bangladesh: Occurrence and mobilization, *J. Environ. Manage.* 262, 110318.

Hutchinson, S. L., Schwab, A. P. and Banks, M. K. (2003). Biodegradation of petroleum hydrocarbons in the rhizosphere. In: McCutcheon, S. C. and Schnoor, J. L. (Eds.), *Phytoremediation: Transformation and Control of Contaminants.* Hoboken, NJ: Wiley, pp. 355–386.

Jäderlund, L., Arthurson, V., Granhall, U. and Jansson, J. K. (2008). Specific interactions between arbuscular mycorrhizal fungi and plant growth-promoting bacteria: As revealed by different combinations, *FEMS Microbiol. Lett.* 287(2), 174–180.

Jakobsen, I., Abbott, L. K. and Robson, A. D. (1992). External hyphae of vesicular arbuscular mycorrhizal fungi associated with *Trifolium subterraneum* L. 1. Spread of hyphae and phosphorus inflow into roots, *New Phytol.* 120(3), 371–380.

Jambhulkar, H. P. and Kumar, M. S. (2019). Eco-restoration approach for mine spoil overburden dump through biotechnological route, *Environ. Monit. Assess.* 191(12), 772.

Jansa, J., Bukovská, P. and Gryndler, M. (2013). Mycorrhizal hyphae as ecological niche for highly specialized hypersymbionts – Or just soil free-riders?, *Front. Plant Sci.* 4, 134.

Jayne, B. and Quigley, M. (2014). Influence of arbuscular mycorrhiza on growth and reproductive response of plants under water deficit: A meta-analysis, *Mycorrhiza* 24(2), 109–119.

Jilling, A., Keiluweit, M., Contosta, A. R., Frey, S., Schimel, J., Schnecker, J., Smith, R. G., Tiemann, L. and Grandy, A. S. (2018). Minerals in the rhizosphere: Overlooked mediators of soil nitrogen availability to plants and microbes, *Biogeochemistry* 139(2), 103–122.

Kaiser, C., Kilburn, M. R., Clode, P. L., Fuchslueger, L., Koranda, M., Cliff, J. B., Solaiman, Z. M. and Murphy, D. V. (2015). Exploring the transfer of recent plant photosynthates to soil microbes: Mycorrhizal pathway vs direct root exudation, *New Phytol.* 205(4), 1537–1551.

Keiluweit, M., Bougoure, J. J., Nico, P. S., Pett-Ridge, J., Weber, P. K. and Kleber, M. (2015). Mineral protection of soil carbon counteracted by root exudates, *Nat. Clim. Change* 5(6), 588–595.

Khan, A., Fahad, S., Khan, A., Saud, S., Adnan, M., Wahid, F., Noor, M., Nasim, W., Hammad, H. M., Bakhat, H. F., Ahmad, S., Habib ur Rehman, M., Wang, D. and Sönmez, O. (2019). Managing tillage operation and manure to restore soil carbon stocks in wheat-maize cropping system, *Agron. J.* 111(5), 2600-2609.

Kleber, M., Eusterhues, K., Keiluweit, M., Mikutta, C., Mikutta, R. and Nico, P. S. (2015). Mineral-organic associations: Formation, properties, and relevance in soil environments, *Adv. Agron.* 130, 1-140.

Koide, R. T. and Wu, T. (2003). Ectomycorrhizas and retarded decomposition in a *Pinus resinosa* plantation, *New Phytol.* 158(2), 401-407.

Kong, A., Cifuentes, J., Estrada-Torres, A., Guzmán-Dávalos, L., Garibay-Orijel, R. and Buyck, B. (2015). Russulaceae associated with mycoheterotroph *Monotropa uniflora* (Ericaceae) in Tlaxcala, Mexico: A phylogenetic approach, *Cryptogam. Mycol.* 36(4), 479-512.

Kothari, S. K., Marschner, H. and Römheld, V. (1990). Direct and indirect effects of VA mycorrhizal fungi and rhizosphere microorganisms on acquisition of mineral nutrients by maize (*Zea mays* L.) in a calcareous soil, *New Phytol.* 116(4), 637-645.

Kumar, V., Teotia, P., Bisht, S. and Sharma, S. (2015). Plant-microbe interactions for land and sustainable agriculture development. In: Arora, N. K. (Ed.), *Plant Microbes Symbiosis: Applied Facets*. New Delhi: Springer, pp. 77-94.

Kuzyakov, Y. (2002). Review: Factors affecting rhizosphere priming effects, *J. Plant Nutr. Soil Sci.* 165(4), 382-396.

Lal, R. (2009). Soil degradation as a reason for inadequate human nutrition, *Food Secur.* 1(1), 45-57.

Lal, R. (2015). Restoring soil quality to mitigate soil degradation, *Sustainability* 7(5), 5875-5895.

Lazcano, C., Barrios-Masias, F. H. and Jackson, L. E. (2014). Arbuscular mycorrhizal effects on plant water relations and soil greenhouse gas emissions under changing moisture regimes, *Soil Biol. Biochem.* 74, 184-192.

Leake, J., Johnson, D., Donnelly, D., Muckle, G., Boddy, L. and Read, D. (2004). Networks of power and influence: The role of mycorrhizal mycelium in controlling plant communities and agroecosystem functioning, *Can. J. Bot.* 82(8), 1016-1045.

Lehmann, A., Zheng, W. S. and Rillig, M. C. (2017). Soil biota contributions to soil aggregation, Nature Ecol. Evol. 1(12), 1828-1835.

Lindahl, B. D. and Tunlid, A. (2015). Ectomycorrhizal fungi-potential organic matter decomposers, yet not saprotrophs, *New Phytol.* 205(4), 1443-1447.

Linderman, R. G. (1994). Role of VAM fungi in biocontrol. In: Pfleger, F. L. and Linderman, R. G. (Eds.), *Mycorrhizae and Plant Health*. St Paul, MN: APS Publishing, pp. 1-26.

Malcová, R., Vosátka, M. and Gryndler, M. (2003). Effects of inoculation with *Glomus intraradices* on lead uptake by *Zea mays* L. and *Agrostisca pillaris* L., *Appl. Soil Ecol.* 23(1), 55-67.

Marschner, H. and Dell, B. (1994). Nutrient uptake in mycorrhizal symbiosis, *Plant Soil* 159(1), 89-102.

Marschner, P. and Rengel, Z. (2012). Nutrient availability in soils. In: Marschner, P. (Ed.), *Marschner's Mineral Nutrition of Higher Plants*. London: Academic Press, pp. 315-330.

Martin, F., Perotto, S. and Bonfante, P. (2007). Mycorrhizal fungi: A fungal community at the interface between soil and roots. In: Pinton, R., Varanini, Z. and Nannipieri, P. (Eds.),

The Rhizosphere: Biochemistry and Organic Substances at the Soil-Plant Interface (2nd edn.). Boca Raton, FL: CRC Press, pp. 201-236.

Marulanda, A., Barea, J. M. and Azcón, R. (2009). Stimulation of plant growth and drought tolerance by native microorganisms (AM fungi and bacteria) from dry environments: Mechanisms related to bacterial effectiveness, *J. Plant Growth Regul.* 28(2), 115-124.

Marulanda-Aguirre, A., Azcón, R., Ruiz-Lozano, J. M. and Aroca, R. (2008). Differential effects of a *Bacillus megaterium* strain on *Lactuca sativa* plant growth depending on the origin of the arbuscular mycorrhizal fungus coinoculated: Physiologic and biochemical traits, *J. Plant Growth Regul.* 27(1), 10-18.

Matsubara, Y. I., Harada, T. and Yakuwa, T. (1995). Effect of inoculation density of vesicular-arbuscular mycorrhizal fungal spores and addition of carbonized material to bed soil on growth of Welsh onion seedlings, *J. Jap. Soc. Hort. Sci.* 64(3), 549-554.

Meena, R. S., Bohra, J. S., Singh, S. P., Meena, V. S., Verma, J. P., Verma, S. K. and Sihag, S. K. (2016). Towards the prime response of manure to enhance nutrient use efficiency and soil sustainability a current need: A book review, *J. Cleaner Prod.* 112, 1258-1260.

Meyer, J. R. and Linderman, R. G. (1986). Selective influence on populations of rhizosphere or rhizoplane bacteria and actinomycetes by mycorrhizas formed by *Glomus fasciculatum*, *Soil Biol. Biochem.* 18(2), 191-196.

Mian, I. A., Muhammad, M. W., Fahad, S. and Adnan, M. (2019). Seed priming improves yield and nitrogen uptake in wheat under varying nitrogen and moisture regimes, *Pure Appl. Biol.* 8(3), 1925-1934.

Miller, R. M. and Jastrow, J. D. (1992). The application of VA mycorrhizae to ecosystem restoration and reclamation. In: Allen, M. F. (Ed.), *Mycorrhizal Functioning: An Integrative Plant-Fungal Process*. New York: Chapman and Hall, pp. 438-467.

Miller, R. M. and Jastrow, J. D. (2000). Mycorrhizal fungi influence soil structure. In: Kapulnik, Y. and Dounds, D. D. (Eds.), *Arbuscular Mycorrhizas: Physiology and Function*. Dordrecht: Springer, pp. 3-18.

Mozafar, A., Duss, F. and Oertli, J. J. (1992). Effect of *Pseudomonas fluorescens* on the root exudates of two tomato mutants differently sensitive to Fe chlorosis, *Plant Soil* 144(2), 167-176.

Muhammad, B., Adnan, M., Munsif, F., Fahad, S., Saeed, M., Wahid, F., Arif, M., Amanullah, J., Wang, D., Saud, S., Noor, M., Zamin, M., Subhan, F., Saeed, B., Raza, M. A. and Mian, I. A. (2019). Substituting urea by organic wastes for improving maize yield in alkaline soil, *J. Plant Nutr.* 42(19), 2423-2434.

Njeru, E. M., Avio, L., Bocci, G., Sbrana, C., Turrini, A., Bàrberi, P., Giovannetti, M. and Oehl, F. (2015). Contrasting effects of cover crops on 'hot spot'arbuscular mycorrhizal fungal communities in organic tomato, *Biol. Fertil. Soils* 51(2), 151-166.

Nogueira, M. A., Nehls, U., Hampp, R., Poralla, K. and Cardoso, E. J. B. N. (2007). Mycorrhiza and soil bacteria influence extractable iron and manganese in soil and uptake by soybean, *Plant Soil* 298(1-2), 273-284.

Ogura-Tsujita, Y., Yokoyama, J., Miyoshi, K. and Yukawa, T. (2012). Shifts in mycorrhizal fungi during the evolution of autotrophy to mycoheterotrophy in Cymbidium (Orchidaceae), *Am. J. Bot.* 99(7), 1158-1176.

Orwin, K. H., Kirschbaum, M. U., St John, M. G. and Dickie, I. A. (2011). Organic nutrient uptake by mycorrhizal fungi enhances ecosystem carbon storage: A model-based assessment, *Ecol. Lett.* 14(5), 493-502.

Parniske, M. (2008). Arbuscular mycorrhiza: The mother of plant root endosymbioses, *Nat. Rev. Microbiol.* 6(10), 763-775.

Paszkowski, U. (2006). A journey through signaling in arbuscular mycorrhizal symbioses 2006, *New Phytol.* 172(1), 35–46.

Pausch, J. and Kuzyakov, Y. (2018). Carbon input by roots into the soil: Quantification of rhizodeposition from root to ecosystem scale, *Glob. Change Biol.* 24(1), 1–12.

Porras-Soriano, A., Soriano-Martín, M. L., Porras-Piedra, A. and Azcón, R. (2009). Arbuscular mycorrhizal fungi increased growth, nutrient uptake and tolerance to salinity in olive trees under nursery conditions, *J. Plant Physiol.* 166(13), 1350–1359.

Postgate, J. (1998). *Nitrogen Fixation*. Cambridge: Cambridge University Press.

Probanza, A., Lucas García, J. A., Ruiz Palomino, M., Ramos, B. and Gutiérrez Mañero, F. J. (2002). *Pinus pinea* L. seedling growth and bacterial rhizosphere structure after inoculation with PGPR *Bacillus* (*B. licheniformis* CECT 5106 and *B. pumilus* CECT 5105), *Appl. Soil Ecol.* 20(2), 75–84.

Quoreshi, A. M. and Khasa, D. P. (2008). Effectiveness of mycorrhizal inoculation in the nursery on root colonization, growth, and nutrient uptake of aspen and balsam poplar, *Biomass Bioenerg.* 32(5), 381–391.

Rambelli, A. (1973). The rhizosphere of mycorrhizae. In: Marks, G. L. and Koslowski, T. T. (Eds.), *Ectomycorrhizae*. New York: Academic Press, pp. 299–343.

Read, D. J. (1991). Mycorrhizas in ecosystems, *Experientia* 47(4), 376–391.

Rillig, M. C. (2004). Arbuscular mycorrhizae and terrestrial ecosystem processes, *Ecol. Lett.* 7(8), 740–754.

Rillig, M. C., Lutgen, E. R., Ramsey, P. W., Klironomos, J. N. and Gannon, J. E. (2005). Microbiota accompanying different arbuscular mycorrhizal fungal isolates influence soil aggregation, *Pedobiologia* 49(3), 251–259.

Rillig, M. C. and Mummey, D. L. (2006). Mycorrhizas and soil structure, *New Phytol.* 171(1), 41–53.

Rouphael, Y., Franken, P., Schneider, C., Schwarz, D., Giovannetti, M., Agnolucci, M., Pascale, S. D., Bonini, P. and Colla, G. (2015). Arbuscular mycorrhizal fungi act as biostimulants in horticultural crops, *Sci. Hortic.* 196, 91–108.

Safir, G. R. and Nelson, C. E. (1985). VA-mycorrhizas plant and fungal water relations. In: Molina, R. (Ed.), *Proceedings of the 6th North American Conference on Mycorrhiza*, Corvallis, p. 471.

Saleem, M. H., Fahad, S., Adnan, M., Ali, M., Rana, M. S., Kamran, M., Ali, Q., Hashem, I. A., Bhantana, P., Ali, M. and Hussain, R. M. (2020). Foliar application of gibberellic acid endorsed phytoextraction of copper and alleviates oxidative stress in jute (*Corchorus capsularis* L.) plant grown in highly copper-contaminated soil of China, *Environ. Sci. Pollut. Res.* 37, 121–133.

Schmidt, M. W., Torn, M. S., Abiven, S., Dittmar, T., Guggenberger, G., Janssens, I. A., Kleber, M., Kögel-Knabner, I., Lehmann, J., Manning, D. A., Nannipieri, P., Rasse, D. P., Weiner, S. and Trumbore, S. E. (2011). Persistence of soil organic matter as an ecosystem property, *Nature* 478(7367), 49–56.

Schüβler, A., Schwarzott, D. and Walker, C. (2001). A new fungal phylum, the Glomeromycota: Phylogeny and evolution, *Mycol. Res.* 105(12), 1413–1421.

Schweigert, M., Herrmann, S., Miltner, A., Fester, T. and Kästner, M. (2015). Fate of ectomycorrhizal fungal biomass in a soil bioreactor system and its contribution to soil organic matter formation, *Soil Biol. Biochem.* 88, 120–127.

Schwendemann, A. B., Decombeix, A. L., Taylor, T. N., Krings, E. L. and Krings, M. (2011). Morphological and functional stasis in mycorrhizal root nodules as exhibited by a Triassic conifer, *Proc. Natl Acad. Sci. U. S. A.* 108(33), 13630–13634.

Secilia, J. and Bagyaraj, D. J. (1987). Bacteria and actinomycetes associated with pot cultures of vesicular-arbuscular mycorrhizas, *Can. J. Microbiol.* 33(12), 1069-1073.

Shafi, M. I., Adnan, M., Fahad, S., Wahid, F., Khan, A., Yue, Z., Danish, S., Zafar-ul-Hye, M., Brtnicky, M. and Datta, R. (2020). Application of single superphosphate aestivum L. with humic acid improves the growth, yield and phosphorus uptake of wheat (Triticum in calcareous soil), *Agronomy* 10, 1224.

Shen, Q., Kirschbaum, M. U., Hedley, M. J. and Camps Arbestain, M. (2016). Testing an alternative method for estimating the length of fungal hyphae using photomicrography and image processing, *PLoS ONE* 11(6), e0157017.

Singh, A. K., Zhu, X. A., Chen, C. F., Wu, J. N., Yang, B., Zakari, S., Jiang, X. J., Singh, N. and Liu, W. (2022). The role of glomalin in mitigation of multiple soil degradation problems, *Crit. Rev. Env. Sci. Technol.* 52(9), 1604-1638.

Six, J., Elliott, E. T., Paustian, K. and Doran, J. W. (1998). Aggregation and soil organic matter accumulation in cultivated and native grassland soils, *Soil Sci. Soc. Am. J.* 62(5), 1367-1377.

Smith, F. A., Grace, E. J. and Smith, S. E. (2009). More than a carbon economy: Nutrient trade and ecological sustainability in facultative arbuscular mycorrhizal symbioses, *New Phytol.* 182(2), 347-358.

Smith, S. E. and Read, D. J. (2008). *Mycorrhizal Symbiosis* (3rd edn.). San Diego: Academic Press.

Stefanowicz, A. M., Kapusta, P., Zubek, S., Stanek, M. and Woch, M. W. (2020). Soil organic matter prevails over heavy metal pollution and vegetation as a factor shaping soil microbial communities at historical Zn-Pb mining sites, *Chemosphere* 240, 124922.

Stöckel, M., Těšitelová, T., Jersáková, J., Bidartondo, M. I. and Gebauer, G. (2014). Carbon and nitrogen gain during the growth of orchid seedlings in nature, *New Phytol.* 202(2), 606-615.

Sun, C., Johnson, J. M., Cai, D., Sherameti, I., Oelmüller, R. and Lou, B. (2010). *Piriformospora indica* confers drought tolerance in Chinese cabbage leaves by stimulating antioxidant enzymes, the expression of drought-related genes and the plastid-localized CAS protein, *J. Plant Physiol.* 167(12), 1009-1017.

Swift, M. J. (2005). Human impacts on biodiversity and ecosystem services: An overview. In: Dighton, J., White, J. F. and Oudemans, P. (Eds.), *The Fungal Community Its Organization and Role in Ecosystems* (3rd edn). Boca Raton, FL: CRC Press, pp. 627-641.

Sylvia, D. M., Hammond, L. C., Bennett, J. M., Haas, J. H. and Linda, S. B. (1993). Field response of maize to a VAM fungus and water management, *Agron. J.* 85(2), 193-198.

Tarraf, W., Ruta, C., Tagarelli, A., De Cillis, F. and De Mastro, G. (2017). Influence of arbuscular mycorrhizae on plant growth, essential oil production and phosphorus uptake of *Salvia officinalis* L., *Ind. Crops Prod.* 102, 144-153.

Toljander, J. F., Lindahl, B. D., Paul, L. R., Elfstrand, M. and Finlay, R. D. (2007). Influence of arbuscular mycorrhizal mycelial exudates on soil bacterial growth and community structure, *FEMS Microbiol. Ecol.* 61(2), 295-304.

Toro, M., Azcon, R. and Barea, J. (1997). Improvement of arbuscular mycorrhiza development by inoculation of soil with phosphate-solubilizing rhizobacteria to improve rock phosphate bioavailability (Sup32) and nutrient cycling, *Appl. Environ. Microbiol.* 63(11), 4408-4412.

Torstensson, L., Pell, M. and Stenberg, B. (1998). Need of a strategy for evaluation of arable soil quality, *AMBIO* 27, 4–8.

Tran, C. T. K., Watts-Williams, S. J., Smernik, R. J. and Cavagnaro, T. R. (2021). Root and arbuscular mycorrhizal effects on soil nutrient loss are modulated by soil texture, Appl. Soil Ecol. 167(11), 104097.

Trappe, J. M. (1962). Fungus associates of ectotrophic mycorrhizae, *Bot. Rev.* 28(4), 538–606.

Trappe, J. M. (1977). Selection of fungi for ectomycorrhizal inoculation in nurseries, *Annu. Rev. Phytopathol.* 15(1), 203–222.

Uddin, M. N., Ali, M., Muhammad, M. F., Farooq, M., Ahmad, N., Jamil, J., Kalsoom, K., Adnan, M., Shah, N. and Khan, A. (2016). Characterizing microbial populations in petroleum-contaminated soils of Swat district, Pakistan, *Pol. J. Environ. Stud.* 25(4), 1721–1727.

Vailleau, F., Sartorel, E., Jardinaud, M. F., Chardon, F., Genin, S., Huguet, T., Gentzbittel, L. and Petitprez, M. (2007). Characterization of the interaction between the bacterial wilt pathogen *Ralstonia solanacearum* and the model legume plant *Medicago truncatula*, *Mol. Plant Microbe Interact.* 20(2), 159–167.

van der Heijden, M. G. A., Martin, F. M., Selosse, M. A. and Sanders, I. R. (2015). Mycorrhizal ecology and evolution: The past, the present, and the future, *New Phytol.* 205(4), 1406–1423.

Veresoglou, S. D., Menexes, G. and Rillig, M. C. (2012). Do arbuscular mycorrhizal fungi affect the allometric partition of host plant biomass to shoots and roots? A meta-analysis of studies from 1990 to 2010, *Mycorrhiza* 22(3), 227–235.

Vestergård, M., Henry, F., Rangel-Castro, J. I., Michelsen, A., Prosser, J. I. and Christensen, S. (2008). Rhizosphere bacterial community composition responds to arbuscular mycorrhiza, but not to reductions in microbial activity induced by foliar cutting, *FEMS Microbiol. Ecol.* 64(1), 78–89.

Vogel-Mikuš, K., Drobne, D. and Regvar, M. (2005). Zn, Cd and Pb accumulation and arbuscular mycorrhizal colonisation of pennycress *Thlaspi praecox* Wulf. (Brassicaceae) from the vicinity of a lead mine and smelter in Slovenia, *Environ. Pollut.* 133(2), 233–242.

Wahid, F., Fahad, S., Danish, S., Adnan, M., Yue, Z., Saud, S., Siddiqui, M. H., Brtnicky, M., Hammerschmiedt, T. and Datta, R. (2020). Sustainable management with mycorrhizae and phosphate solubilizing bacteria for enhanced phosphorus uptake in calcareous soils, *Agriculture* 10(8), 334.

Wahid, F., Sharif, M., Fahad, S., Adnan, M., Khan, I. A., Aksoy, E., Ali, A., Sultan, T., Alam, M., Saeed, M., Ullah, H., Basir, A., Noor, M. and Khan, N. A. (2019). Arbuscular mycorrhiza fungi improve the growth and phosphorus uptake of bean plants fertilized with compost rock phosphate fed dung, *J. Plant Nutr.* 42(15), 1760–1769.

Wahid, F., Sharif, M., Khan, M. A. and Khan, M. J. (2016). Status and phosphorus solubilization potential of bacteria and arbuscular mycorrhizal fungi isolated from various locations of Khyber Pakhtunkhwa Province, *Pak. J. Bot.* 48, 2121–2130.

Wallander, H., Göransson, H. and Rosengren, U. (2004). Production, standing biomass and natural abundance of ^{15}N and ^{13}C in ectomycorrhizal mycelia collected at different soil depths in two forest types, *Oecologia* 139(1), 89–97.

Wang, B. and Qiu, Y. L. (2006). Phylogenetic distribution and evolution of mycorrhizas in land plants, *Mycorrhiza* 16(5), 299–363.

Wang, W., Shi, J., Xie, Q., Jiang, Y., Yu, N. and Wang, E. (2017). Nutrient exchange and regulation in arbuscular mycorrhizal symbiosis, *Mol. Plant* 10(9), 1147–1158.

Wang, Y., He, X. and Yu, F. (2022). Non-host plants: Are they mycorrhizal networks players?, *Plant Divers.* 44(2), 127-134. DOI: 10.1016/j.pld.2021.06.005.

Whiteside, M. D., Garcia, M. O. and Treseder, K. K. (2012). Amino acid uptake in arbuscular mycorrhizal plants, *PLoS ONE* 7(10), e47643.

Wright, S. F., Green, V. S. and Cavigelli, M. A. (2007). Glomalin in aggregate size classes from three different farming systems, Soil Till. Res. 94, 546-549.

Wright, S. F. and Upadhyaya, A. (1996). Extraction of an abundant and unusual protein from soil and comparison with hyphal protein of arbuscular mycorrhizal fungi, Soil Sci. 161, 575-586.

Zak, D. R., Pellitier, P. T., Argiroff, W., Castillo, B., James, T. Y., Nave, L. E., Averill, C., Beidler, K. V., Bhatnagar, J., Blesh, J., Classen, A. T., Craig, M., Fernandez, C. W., Gundersen, P., Johansen, R., Koide, R. T., Lilleskov, E. A., Lindahl, B. D., Nadelhoffer, K. J., Phillips, R. P. and Tunlid, A. (2019). Exploring the role of ectomycorrhizal fungi in soil carbon dynamics, *New Phytol.* 223(1), 33-39.

Zamin, M., Fahad, S., Khattak, A., Adnan, M., Wahid, F., Raza, A., Wang, D., Saud, S., Noor, M., Bakhat, H. F. and Mubeen, M. (2019). Developing the first halophytic turf grasses for the urban landscape from native Arabian Desert grass, *Environ. Sci. Pollut. Res.* 27, 39702-39716.

Zhan, F., Li, B., Jiang, M., Li, T., He, Y., Li, Y. and Wang, Y. (2019). Effects of arbuscular mycorrhizal fungi on the growth and heavy metal accumulation of Bermudagrass [*Cynodondactylon* (L.) Pers.] grown in a lead-zinc mine wasteland, *Int. J. Phytoremediation* 21(9), 849-856.

Zhang, J., Tang, X. L., Zhong, S. Y., Yin, G. C., Gao, Y. F. and He, X. (2017). Recalcitrant carbon components in glomalin-related soil protein facilitate soil organic carbon preservation in tropical forests, *Sci. Rep.* 7(1), 2391.

Zhang, L., Xu, M., Liu, Y., Zhang, F., Hodge, A. and Feng, G. (2016). Carbon and phosphorus exchange may enable cooperation between an arbuscular mycorrhizal fungus and a phosphate-solubilizing bacterium, *New Phytol.* 210(3), 1022-1032.

Zhang, Z., Phillips, R. P., Zhao, W., Yuan, Y., Liu, Q. and Yin, H. (2019). Mycelia-derived C contributes more to nitrogen cycling than root-derived C in ectomycorrhizal alpine forests, *Funct. Ecol.* 33(2), 346-359.

Žifčáková, L., Větrovský, T., Howe, A. and Baldrian, P. (2016). Microbial activity in forest soil reflects the changes in ecosystem properties between summer and winter, *Environ. Microbiol.* 18(1), 288-301.

Part 2

Cultivation practices and soil health

Chapter 5

Agricultural traffic management systems and soil health

Paula A. Misiewicz, Magdalena Kaczorowska-Dolowy, David R. White, Edward Dickin and Richard J. Godwin, Harper Adams University, UK

1 Introduction

Since 1950, the annual increase in the global human population has varied from a peak of 1.98% in 1960 to a current rate of 1.20%. The global population is currently around 7.8 billion and is projected to reach 9.9 billion by 2050, an increase of more than 27% (UN population data, 2021: www.worldometer.info). In addition, dietary requirements are also rapidly changing particularly in developing nations (Ray et al., 2013; Godfray and Garnett, 2014). This dramatic, long-term increase in population and demand for food has prompted an equally dramatic increase in food production, made possible by developments such as advances in high-yielding varieties, the development and use of a new range of fertilisers and pesticides, and major advances in technology to allow crop planting, management, and harvesting to occur on a much larger scale. From 1960 to 2014, e.g. UK wheat production increased 240%, barley 196%, and oats 226% (Ritchie and Roser, 2019) (Fig. 1).

http://dx.doi.org/10.19103/AS.2021.0094.20

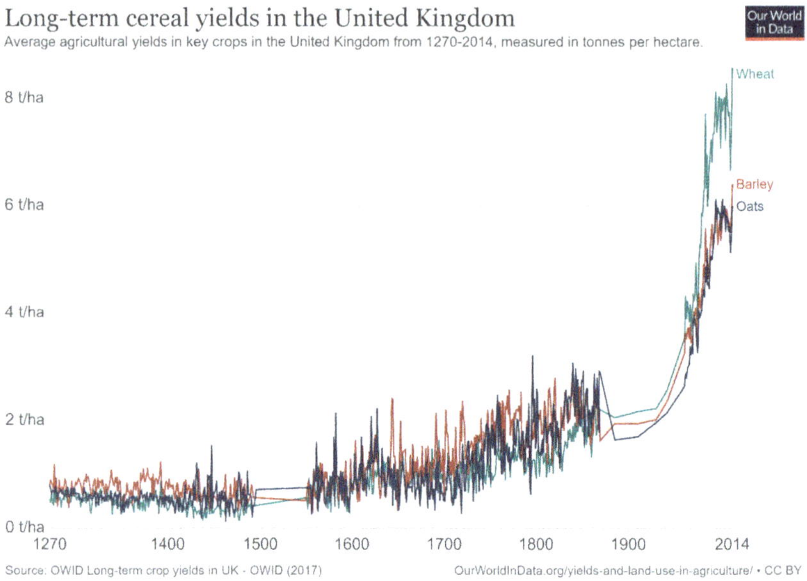

Figure 1 Long-term cereal yields in the United Kingdom after OWID, (2017).

The continued improvement in the sophistication and size of farm machinery to support this growth in agricultural output has resulted in larger, more efficient farms with less labour but much greater productivity (Cavallo et al., 2014). However, increases in agricultural machinery power and capability have also been accompanied by steadily increasing machinery weight. The size and weight of farm machinery has increased significantly in the last 50 years. Chamen et al. (2006) reported that, since 1966, the average weight of farm vehicles has approximately tripled while the average wheel load has increased six times. Schjønning et al. (2015) reported that the average mass of fully loaded agricultural equipment increased by a factor of 6, from 4.3 Mg in 1958 to an average of about 25 Mg in 2009. It has been estimated that average wheel loads of tractors increased from about 1.5 Mg in 1960 to 4 Mg in 2000; wheel loads of combine harvesters increased from 1.5 Mg in 1958 to almost 9 Mg by 2009 (Keller et al., 2019).

One consequence of increasing machinery weight is its link to the problem of soil compaction. Soil compaction has been identified as a factor in stagnating yields more recently observed in wheat and other crops (Knight et al., 2012). Soil compaction has been defined as 'the process by which the soil grains are rearranged to decrease void space and bring them into closer contact with one another, thereby increasing bulk density' (SSSA, 2008). Soil compaction occurs when the application of a load or stress provided by a vehicle on the soil surface at a given depth exceeds the soil strength (Sohne, 1958). The process of densification results in a wide range of changes to a range of soil characteristics such as porosity, permeability, and strength (Soane and van Ouwerkerk, 1994). Compaction disrupts soil structure and in turn soil biology. Morales et al. (2015) suggested that changes in soil moisture and nutrient availability resulting from soil compaction can influence microbial diversity and consequently the ecosystem services delivered by soil, including carbon (C) sequestration. Kaczorowska-Dolowy (2022) also reported a significant negative effect of soil compaction on Collembola population and soil fauna feeding activity. Soil compaction can also accelerate wind and water soil erosion (Houšková and Montanarella, 2008) and has been linked to problems such as surface run-off and resulting pollution of surface water (Soane and van Ouwerkerk, 1994; Bagarello et al., 2004; Kroulik et al., 2009).

Soil compaction has also been linked to lower crop yields by a number of researchers (Raghavan et al., 1979; Horn et al., 2003; Hula et al., 2009; Chamen 2011), which in turn has been identified as a factor in stagnating yields more recently observed in wheat and other crops (Knight et al., 2012). Yield reduction on trafficked soils has been related to restricted root growth and poorer access to water and nutrients as a result of increased bulk density (BD) and soil strength, and reduced pore size (Kaspar et al., 2001; McHugh et al., 2009; Nawaz et al., 2013; Ngo-Cong et al., 2021). As an example, soil compaction has been associated with changes in root structure, leading to shorter, thicker roots with, potentially, less overall surface area for water and nutrient uptake (Hettiaratchi, 1990; Materechera et al., 1991; Głąb, 2008; Chen et al., 2014; Kaczorowska-Dolowy et al., 2019).

Soil compaction is also seen as a factor in overall soil degradation which is estimated to affect over 30% of arable land in Europe (Oldeman, 1998; Keller et al., 2019). This leads to a conclusion that soil compaction has a detrimental effect on 'soil health', a term that integrates soil physical, chemical, and biological properties. Soil health refers to the general quality of the soil resource and embraces both the provision for agricultural crop production and the provision of other ecosystem services (Kibblewhite et al., 2008). Doran (2002) defined soil health as 'the capacity of a living soil to function, within natural or managed ecosystem boundaries, to sustain plant and animal productivity, maintain or enhance water and air quality, and promote plant and animal health'. Improved soil health is vital for resilience and adaptability which in turn is essential for future production, particularly in the face of climate change (Andriuzzi, 2015; Congreves et al., 2015).

This chapter examines the following aspects of agricultural traffic and soil compaction. It begins by reviewing research on how agricultural traffic affects soil compaction as well as ways of measuring soil compaction and its effects. It then discusses a range of potential techniques to avoid/reduce soil compaction.

These include controlled traffic farming (CTF), low ground pressure tyre systems as well as tracks and gantry systems. The chapter also discusses the relationship between different tillage practices and soil compaction. It includes a case study based on research conducted by the authors.

2 How agricultural traffic affects soil compaction

It is important to distinguish between compaction of natural origin (e.g. hardsetting and glacial overburden) from compaction induced by agricultural traffic and trampling by grazing animals (Chartres et al., 1990; Raper and Kirby, 2006; Botta et al., 2020). Soil compaction caused by farm traffic has been related to three main factors (Raper et al., 1995; Soane and van Ouwerkerk, 1994):

1 wheel load;
2 tyre inflation pressure;
3 contact area.

Other relevant factors include soil consistency (mainly influenced by soil water content, which influences soil strength and bearing capacity) and the characteristics of the traffic itself (e.g. number of passes, speed) (Soane et al., 1980). This section examines each of these three main factors.

Sohne (1958) reported that higher wheel loads cause stresses reaching progressively greater depths below the soil surface as loads increase (Fig. 2). At the same time, the pressure reaches greater depths with increasing soil moisture content. Soil compaction due to wheel traffic has been found to deteriorate soil structure both in topsoil and deeper in the soil profile (Domżał et al., 1991). Increased loads have thus been shown to increase compaction both in the surface layer and in the subsoil (Kroulik et al., 2009). Obour et al.

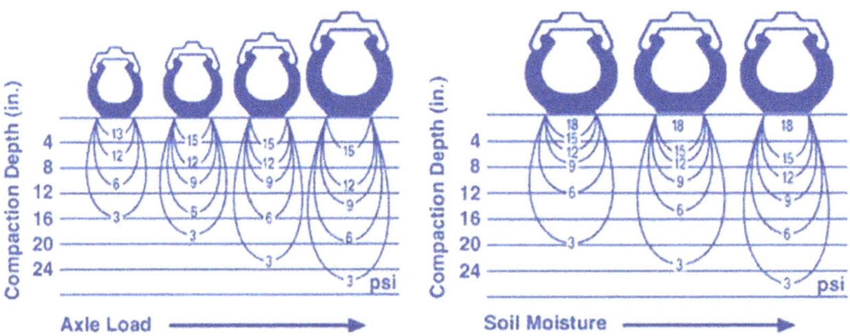

Figure 2 Curves of pressure under a range of axle load (on the left) and increasing moisture condition (on the right). Source: After Sohne (1958).

(2017) found that four-wheel passes by a tractor-trailer combination with a wheel load of 8 Mg on a sandy loam soil had a significant effect on soil structural properties down to 0.65 m. However, he found that 3 Mg wheel load and five-wheel passes did not affect soil structure, highlighting the importance of load as a factor in soil compaction.

Increased tyre inflation pressure increases contact pressure as a result of reduced contact area defined as the area in contact between the tyre and the soil. Raper et al. (1995) reported that rut depth increased with an increasing tyre inflation pressure, confirming the relationship between inflation pressure and vertical impact on the soil profile. Antille et al. (2013), based on the work of Ansorge and Godwin (2007, 2008), investigated three different sizes of combine harvester tyres at three inflation pressures; they reported higher soil BD under higher inflation pressure tyres with a smaller contact area. Spoor et al. (2003) related contact pressure P_c with load and tyre contact area:

$$P_c = \frac{W}{A}$$

where W is the wheel load and A is contact area.

This equation does not, however, account for tyre carcass stiffness or tyre inflation pressure, thus underestimating actual pressure. Misiewicz et al. (2016) accounted for these factors as follows:

$$P_c = P_i + P_{cs}$$

where P_c is calculated contact pressure, P_i is tyre inflation pressure, and P_{cs} is tyre carcass stiffness.

A third variable affecting compaction is contact area. As presented in the equation by Spoor et al. (2003) the greater the contact area, the lower the contact pressure for a given load. The increase of the contact area can be obtained by mounting additional tyres (Alakukku et al., 2003), reducing tyre inflation pressure (Antille et al., 2013), the use of larger tyres (increased tyre width and/or diameter) (Koolen and Kuipers, 2012), or use of tracks (Ansorge and Godwin, 2007).

The total area of a field over which the compaction is applied is determined by the intensity of farming traffic (Kroulik et al., 2009), which, in turn, depends on crop and agronomic practices (Botta et al., 2022). The trafficked area of a field, that is the area covered by wheels each time a crop is produced, can be as high as 90% (Soane et al., 1980) in conventional (grain) farming systems and up to 120–150% in horticulture and root crop systems, but it can be significantly reduced (e.g. to less than 20%) with CTF (Antille et al., 2016a, 2019a,b; Tullberg, 2010).

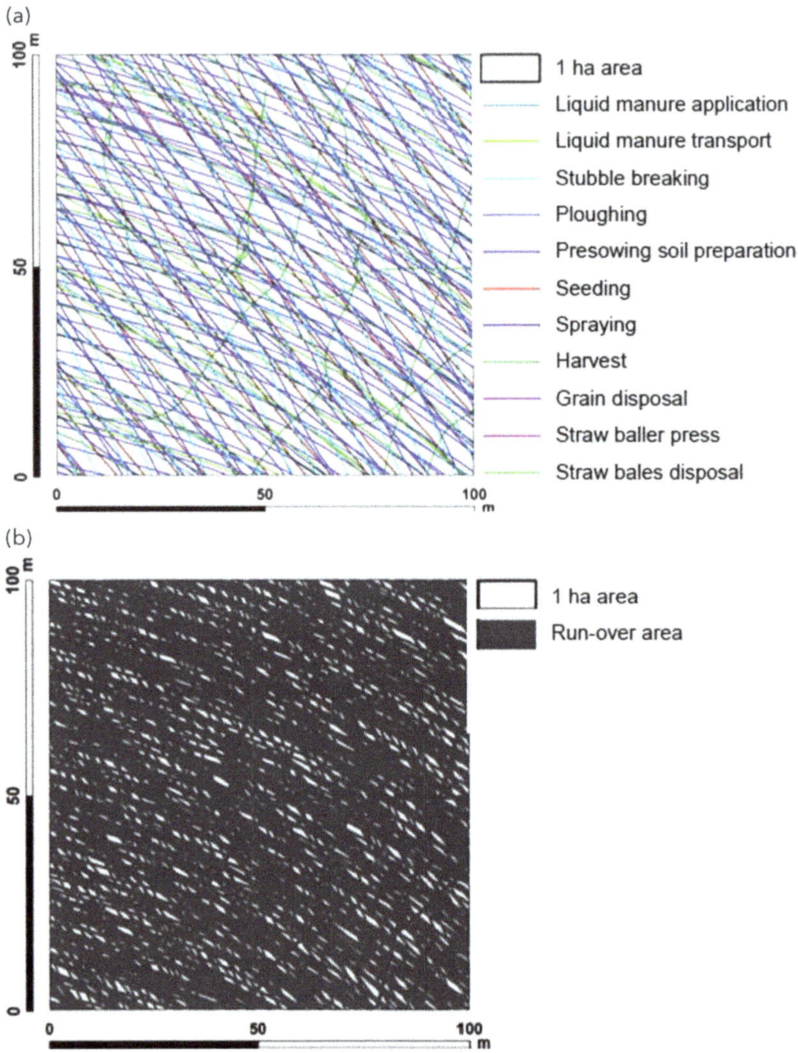

Figure 3 Graphic representation of (a) machinery trajectories, and (b) total trafficked area for RTF with conventional mouldboard ploughing. Source: Reproduced from Kroulik et al. (2009).

Figure 3a illustrates the intensity of traffic on a 1-hectare field resulting from conventional (random) traffic (RTF) with a mouldboard ploughing tillage system during one season of cereal production (Kroulik et al., 2009). This leads to 86% of the field being subjected to at least one wheel pass every growing year (Fig. 3b). Traffic intensity is in part determined by different cultivation techniques and increases with more intensive tillage practices. Reducing tillage intensity leads to a decrease in the trafficked area as a consequence of the lower number of field operations required to prepare the seed bed. The use of shallow tillage and zero tillage reduces the total wheeled area to around 65% and 43%, respectively, as shown in Figs. 4 and 5 (Kroulik et al., 2009).

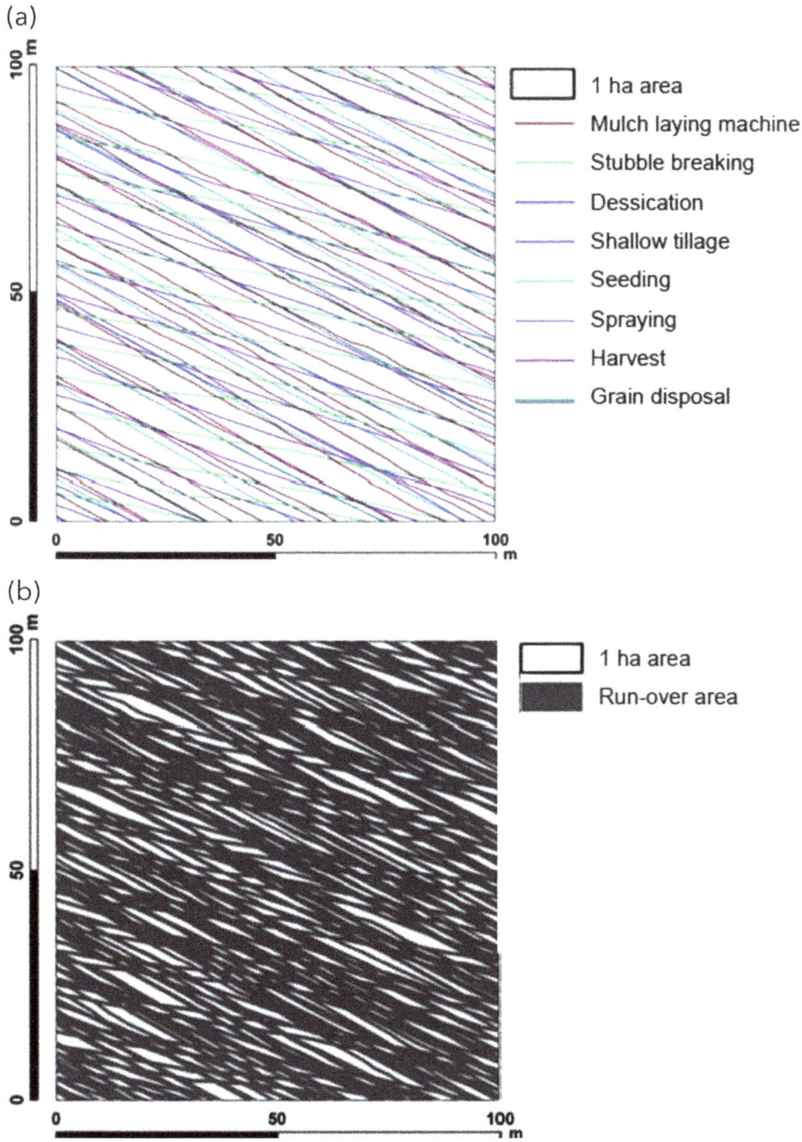

Figure 4 Graphic representation of (a) machinery trajectories, and (b) total trafficked area for RTF with shallow tillage. Source: Reproduced from Kroulik et al. (2009).

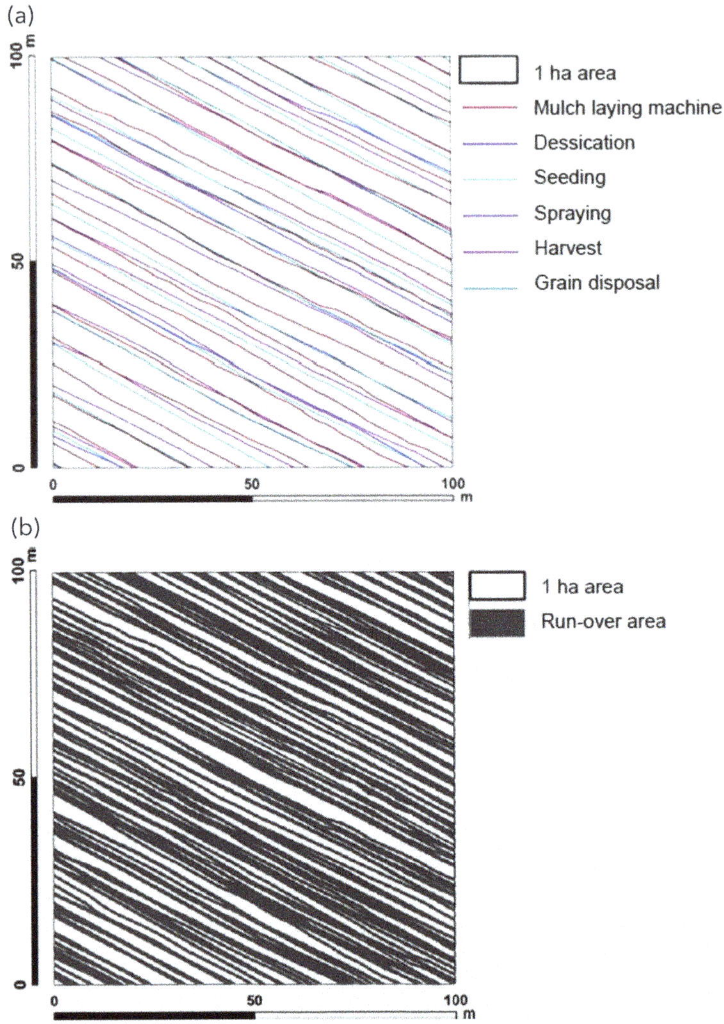

Figure 5 Graphic representation of (a) machinery trajectories, and (b) total trafficked area for RTF with zero tillage. Source: Reproduced from Kroulik et al. (2009).

3 Measuring soil compaction and its effects

Soil compaction occurs when soil particles are brought together, thus reducing the soil pore space and pore size distribution (Berissso et al., 2012). Heavily compacted soils contain few large pores, less total pore volume and, consequently, have greater density. One way, therefore, to measure compaction is by measuring soil BD. Given higher density and disrupted pore structure, water flows and transmission within the soil is retarded or impeded (Ngo-Cong et al., 2021). A compacted soil thus has a reduced rate of both water infiltration and internal drainage. For similar reasons, gas exchange between the soil and atmosphere is also reduced, potentially reducing soil aeration.

Soil compaction has been linked to increased soil BD (Millington, 2019; Shaheb, 2019; Smith, 2017; Lipiec et al., 2003; Evans, 1996), though increases in soil BD are not linear with degree of compaction. As a result, the first pass of a vehicle has the greatest effect on soil compaction (Inns and Kilgour, 1978; Antille et al., 2013; Stranks, 2006). Ahmadi and Ghaur (2015) reported that an increase in the number of passes (from 0 to 4) increased soil BD by 13% and had detrimental effects on root growth. In silt and silt loam soils, Logsdon and Karlen (2004) found that the threshold of soil BD at which roots experience restrictions to growth is at 1.55 Mg m^3 whereas, for sandy and sandy loam soils, this critical threshold is increased to 1.6 Mg m^3 (Huber et al., 2008).

As well as increased soil density, compaction also increases soil strength which can be measured by penetration resistance (PR). Domżał et al. (1991) observed soil compaction as characterised by increased PR, shear resistance, increased cohesion, and crushing strength of soil aggregates. In their research on cotton, Taylor and Gardner (1963) concluded that the most critical factor for root penetration in the sandy soils of the Southern Great Plains of the United States was soil strength rather than soil BD. They concluded that PR greater than 2.96 MPa is a limiting value for root penetration, regardless of whether soil strength is caused by a decrease in soil moisture or by an increase in BD (though other research has suggested different values, depending on factors such as plant species (Bengough, 2006) and soil texture). Arvidsson (2001) reported a significant increase of PR after 2–4 years of traffic in comparison to undisturbed soil. Increased strength as measured by PR means plant roots must exert greater force to penetrate the compacted layer.

Apart from soil BD and PR, another characteristic of the physical condition of soil is porosity which is defined as the fraction of the total soil volume that is taken up by the pore space, i.e. a measurement of the amount of space available to water or other fluids (Nimmo, 2013). Changes in pore size distribution resulting from compaction can be inferred from water retention curves, obtained when saturated soil samples are drained (Dexter and Bird, 2001). However, this method does not identify the complex changes to soil pore structure resulting from different stresses due to compaction (Peth et al., 2010). Another method to analyse pore structure is 2-D image analysis of thin slices of undisturbed soil samples as used by Pagliai et al. (2003). Dried samples are impregnated by polyester resin, sliced into thin sections and then analysed with software such as Image Pro-plus. However, this method of porosity analysis is time-consuming, costly, and requires specialist training (Lipiec and Hatano, 2003). To overcome these limitations, a number of scientists have investigated X-ray computer tomography (CT) to quantify pore size and its distribution in soil (Udawatta and Anderson, 2008; Rab et al., 2014; Beckers et al., 2014; Millington et al., 2017; Shaheb, 2019; Kaczorowska-Dolowy, 2022).

Kim et al. (2010) used X-ray CT to investigate the effect of compaction on a silty loam soil in Missouri (USA) and found that porosity in compacted soil was reduced by 64% in comparison to uncompacted soil, while the number of pores decreased by 71%. The same authors also reported an 8% increase in BD. Kim et al. (2010), Czyz (2004), and Berisso et al. (2012) found that soil compaction reduces the number of larger pores and thus affects total porosity. Shaheb (2019) reported almost two-fold higher macro porosity in untrafficked soil in comparison to trafficked soil, which agrees with observations made by Antille et al. (2016b) and Ebrahimi et al. (2022) based on model-derived data. Millington et al. (2016) found that reduced pore size from soil compaction results in anaerobic conditions, reducing plant establishment and root dry mass of winter barley in compacted areas. Millington et al. (2017) also reported that soil compaction resulting from farm traffic results in reduced porosity, confirming the findings of Soane and van Ouwerkerk (1994) who suggested that farm traffic results in homogenisation of pore systems. The size and distribution of pores within soil is essential for the transport of air, water, and nutrients necessary for the growing plant (Eden et al., 2011). Reduced porosity increases waterlogging and anaerobic conditions, leading to denitrification and a resulting reduction in root growth (Rab et al., 2014), thus reducing crop yield (Czyz, 2004).

The rate at which water enters the soil from the surface to the profile is known as infiltration (Parr and Betrand, 1960). The higher the infiltration rate, the less susceptible the soil is to surface runoff and erosion, particularly when soil is saturated (Barthes and Roose, 2002). Bagarello et al. (2004) argue that 'the hydraulic conductivity (K_{fs}) of saturated soil is one of the most important soil properties controlling water infiltration and surface runoff'. The value of K_{fs} is governed by cracks, root holes as well as by aggregate stability (Kirkham, 2014). It is also dependent on soil structure and texture (Bagarello et al., 2004), as well as on organic matter content and populations of earthworms which build the vertical holes facilitating water movement down the soil profile (Unger, 1996).

Field traffic has been suggested as having a detrimental effect on soil saturated hydraulic conductivity. Ankeny et al. (1995) reported that wheel traffic reduced ponded infiltration rates regardless of tillage system (chisel plough or no-till) by 33–64%, depending on location. Chyba (2012) concluded that the first pass of traffic is responsible for the greatest decrease in surface water infiltration rate (by approximately 82%). Silburn and Glanville (2002) found that rates of water infiltration were 29% higher on untrafficked soils when compared to trafficked soil. Chamen (2011) also reported a 400% difference in infiltration rates on untrafficked compared to trafficked soil. Arvidsson (2001) found that heavy traffic affected saturated hydraulic conductivity at 0.3 and 0.5 m depth.

In addition to soil physical properties, some soil biology attributes have been investigated to assess the effects of farming practices on soil health. Smith (2016) investigated the effects of traffic and tillage systems on earthworm density and found that zero tillage was associated with significantly increased earthworm numbers than deep tillage (250 mm) and shallow tillage (100 mm) measured over the winter period. No significant effects were found for alternative traffic systems. Morales et al. (2015) suggested that changes in soil moisture and nutrient availability resulting from soil compaction can influence microbial diversity and in turn ecosystem services delivered by soil. Rodgers et al. (2018) reported that field traffic significantly decreased Collembola abundance compared to untrafficked soils with highly significant differences in the spring. Thompson et al. (2020) in a study of the effects of wheeled traffic on soil microbial communities in grassland found that direct wheeled traffic had no significant effect on the abundance, diversity, or community structure of either the bacterial or archaeal communities (primitive, single-celled procaryote organisms). In contrast, traffic wheeling increased fungal communities in the loamy soil in comparison to unwheeled soil.

Various measures of soil properties have been used to assess the effects of farm traffic and resulting soil compaction and its consequences on crop growth and yield. Most research supports the view that heavy levels of traffic have a negative effect on crop growth and yields. Chamen (2011) reported a 16% decrease in winter wheat yield under heavily trafficked soil in comparison to untrafficked soil. Dickson and Ritchie (1996) reported that winter wheat, winter barley, and oilseed rape yields increased by 19% in uncompacted soil when compared to soil subject to compaction from farm traffic. In Australia, Li et al. (2007) found that, without traffic, winter wheat yields increased by 9% in comparison to soil exposed to one pass of a tractor. Demmel et al. (2015) also observed higher wheat and rye yields from untrafficked compared to trafficked zones in a study in Germany.

However, Raghavan and McKyes (1978) found that moderate levels of compaction had less effects on yields. Arvidsson and Håkansson (2014) investigated the effect of traffic-induced compaction on crop yields in an experiment across 13 sites in Sweden. They concluded that, with moderate compaction (one pass with low tyre inflation pressures), wheat showed relative yield increases of up to 12% compared to untrafficked and previously loosened soil. However, a further increase in traffic intensity (to three passes with high tyre inflation pressures) led to bulk densities of 1.40–1.45 Mg m^3 which resulted in crop yield decreases of 1–21.3% for winter wheat (*Triticum aestivum*) and horse bean (*Vicia faba*), respectively. Seehusen et al. (2014) studied the effects of load and wheel pass intensities of two different farming vehicle combinations at a total load of 16 Mg and 36 Mg, respectively. They concluded that a single pass at 16 Mg and 36 Mg resulted in a 23% and 28% yield reduction, respectively.

The same authors also observed that 10 passes at 36 Mg resulted in total crop loss. Research by Ball and Ritchie (1999) has highlighted that wet conditions play an important role in the effects of soil compaction, decreasing yields by 24% under trafficked compared to untrafficked soil. The impact on yield is greater in drier years, because of the effect of compaction on plant available water: in wet years, the crop is not constrained by water and thus yield is not affected to the same extent than in drier years (Hussein et al., 2021a,b).

4 Techniques to avoid soil compaction: controlled traffic farming

One technique to reduce soil compaction from farm traffic to a very small area of the field is CTF where field operations follow pre-determined traffic lanes and equipment widths and wheel track spacing are matched as far as possible to mimimise trafficked area (Raper, 2005; Tullberg et al., 2007). Confining farming traffic to permanent, pre-selected traffic lanes in a field requires the use of technology such as navigation and auto-steering systems. Use of RTK-GPS (Real Time Kinematic Global Position System) technology provides errors in positional accuracy to less than 20 mm, making it possible to drive farm vehicles on the same permanent traffic lanes every year, leaving the crop zones in between free of traffic and the risk of compaction (Gasso et al., 2013; Bochtis et al., 2010; Raper, 2005). Compacted permanent traffic lanes also improve energy use efficiency (by reducing wheel slip and rolling resistance) and have even made it possible to start work in fields after heavy rainfall at an earlier stage than would be possible if the compacted lanes were not in place (Taylor, 1994; McPhee et al., 2013; Dickson and Ritchie, 1996; Tullberg, 2000; Luhaib et al., 2017).

Bell et al. (2003) have suggested that the greatest benefits of CTF systems are achieved when the width of all track gauges match, that is, when the distance from wheel centre to wheel centre across all equipment is the same. Figure 6 illustrates one preferred ratio for equipment width, called a '3:1 ratio' layout (Australian terminology) or 'ComTrac' system (European terminology). This layout is suitable for equipment less than 12 m wide using a single wheel track (Isbister et al., 2013; Antille et al., 2019b). There is no universal CTF track layout in Europe, given the diverse range of agricultural machinery and road traffic restrictions in different countries. However, as the combine harvester is usually the largest and most expensive piece of equipment, the most popular solution is an 'OutTrac' system where all other machines are adapted or replaced to run on the same track gauge while the combine harvester runs on its own track gauge. Other potential systems are TwinTrac, where a tractor with a narrower gauge straddles adjacent passes of vehicles with a wider gauge (as shown in Fig. 7), and AdTrac, where one track of the wider gauge, i.e. a combine

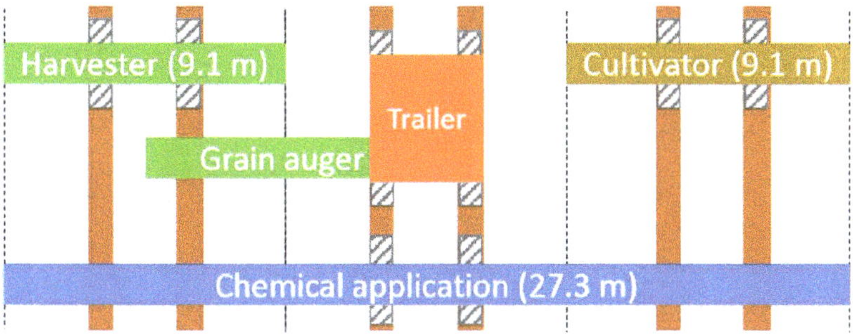

Figure 6 Layout of a '3:1 ratio' or 'Com Trac' CTF system. Source: Adapted from Chamen (2011).

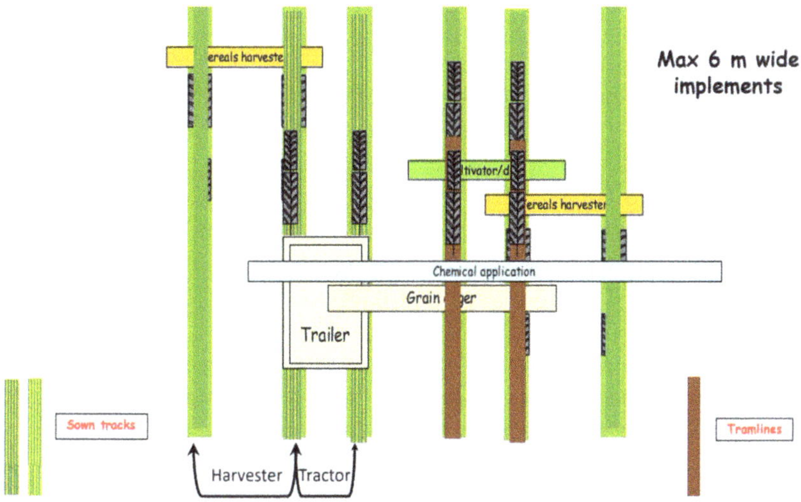

Figure 7 Twin Track CTF system where tractor straddle harvester passes. Source: Adapted from Chamen (2011).

harvester, coincides with the narrower gauge of a tractor (as shown in Fig. 8) (Hargreaves et al., 2016). The uptake of CTF systems has been limited by factors such as technology costs and compatibility of equipment widths (Tullberg, 2010; Chamen, 2015). The economics of CTF have been investigated by a number of authors, including Tullberg et al. (2007) and Godwin et al. (2017, 2022).

Compared with conventional (random) traffic farming (RTF), CTF can reduce the trafficked area from up to 85%, in the case of random traffic mouldboard ploughing operations (Kroulik, 2012), to only 10–20% of the total field area using CTF (Antille et al., 2015, 2016a,b; Gasso et al., 2013; Soane and van Ouwerkerk, 1994; Tullberg, 2010;). By reducing the area exposed to

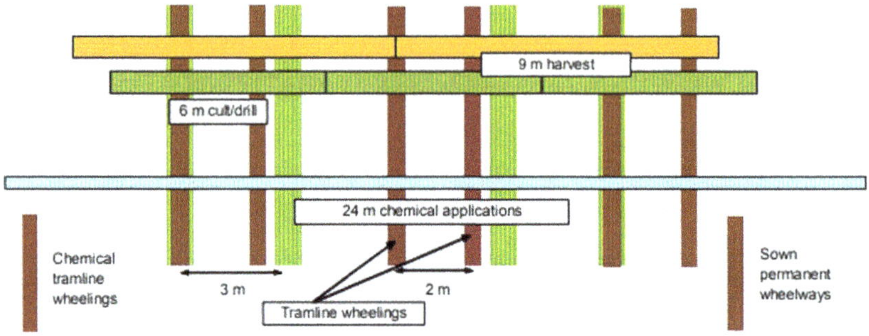

Figure 8 Ad2Trac CTF system with two track and two implement widths. One wheel of the narrower track (e.g. tractor) coincides with the harvester track. Source: Adapted from Chamen (2011).

farm traffic and thus limiting soil compaction, CTF systems have been seen to improve overall soil health and crop yields (Godwin *at al.*, 2015, 2017). By implementing a CTF system using existing commercially available, unmodified equipment, Galambošová et al. (2017) showed that the field trafficked area could be reduced from 64% to 45% using a 6-m wide CTF system, resulting in 0.5 Mg ha^1 increase in crop yields, which they attributed to the overall improvement in the soil physical and mechanical conditions after field traffic was reorganised. Chamen et al. (1992b) reported that CTF increased yields of potatoes, sugar beet, onions, and ryegrass by between 4% and 14%, with improvements in wheat and barley yields varying between −9% to +21% depending on factors such as soil type and weather conditions. Chamen (2011) reported yield improvements between 7% and 35% using CTF, while Godwin et al. (2015) reported yield increases of 7–10%. Hussein et al. (2021a,b) further showed that cereal yields could be increased by 9–30% in crops grown in CTF systems, compared to non-CTF systems. A detailed long-term study by Harper Adams University in the UK also found CTF delivered higher crop yields than random traffic using standard tyre pressures (STP) (Kaczorowska-Dolowy et al., 2020; Godwin et al., 2022).

Since compaction affects soil aeration, CTF has been reported to have additional environmental benefits in addition to beneficial effects on crop yields. These include reduction of both powerful greenhouse gases: nitrous oxide fluxes (21–45%) and methane (372–2100%), as well as water runoff (27–42%) from CTF in comparison to random traffic operations (Gasso et al., 2013). In a 3-year Australian study of CTF, Tullberg et al. (2018) found even greater reduction of nitrous oxide (N_2O) emissions from CTF in comparison to those from trafficked soils which were lower by a factor of 2.2. They also reported a decrease of methane emissions from untrafficked soils and concluded that the adoption of CTF might reduce overall soil emissions by 30–50%.

5 Techniques to avoid soil compaction: low ground pressure tyre systems

As noted earlier, increased tyre inflation pressure increases contact pressure. A number of studies have therefore investigated the use of low ground pressure tyres as a way to avoid soil compaction (Tijink et al., 1995; Alakukku et al., 2003; Antille et al., 2013; Chamen, 2015). A range of improved flexion tyres and very high flexion tyres has been developed for many agricultural machines. These tyres feature uniform load distribution thanks to a wider footprint of the wheel which reduces compaction and, at the same time, improves fuel efficiency and tyre life (Michelin, 2014).

The effects of tyre inflation pressure (in the range of 150–300 kPa) on PR were investigated by Schjønning et al. (2016) who found that a lower tyre pressure (LTP) produced lower stresses in the upper soil profile, resulting in lower PR though, for deeper layers, PR was more correlated with vehicle load. Antille et al. (2013) investigated the effects of a single pass of three sizes of combine harvester tyres at a fixed vertical load of 10.5 Mg; they reported that the least change in soil BD and vertical soil displacement was found when the combine was fitted with a larger tyre size with the lowest safe tyre inflation pressure.

On the other hand, when Millington (2019) and Smith (2017) investigated the effects of tyre pressures in the range of 70–100 kPa on the front axle and 80–90 kPa on the rear axle for LTP and STP, respectively on PR, they found no significant differences between those two tyre pressures systems.

Shaheb (2019) found that LTP resulted in a 4.6% increase in corn (*Zea mays*) yield (2-year average: 2017 and 2018) in a silty clay loam soil; in the first year of his experiment, there was no difference in yields; however, in the second year, the low inflation pressure tyres brought a noticeable increase in yields compared to tyres with standard inflation pressures. Chamen et al. (1990) investigated the effects of different tyre pressures on winter wheat yields and found no significant effects of low pressures. Kaczorowska-Dolowy et al. (2020) found that the effects of reduced tyre inflation pressures depend on the depth of tillage: deep tilled sandy loam soil (250 mm) benefits most from low inflation pressure tyres, resulting in an average 4% crop yield increase. For zero and shallow tillage, there was no significant effect using different tyre pressures.

6 Techniques to avoid soil compaction: tracks and gantry systems

Another approach to minimise contact pressure is to equip the vehicle with tracks (Bashford and Kocher, 1999). The development of rubber tracked

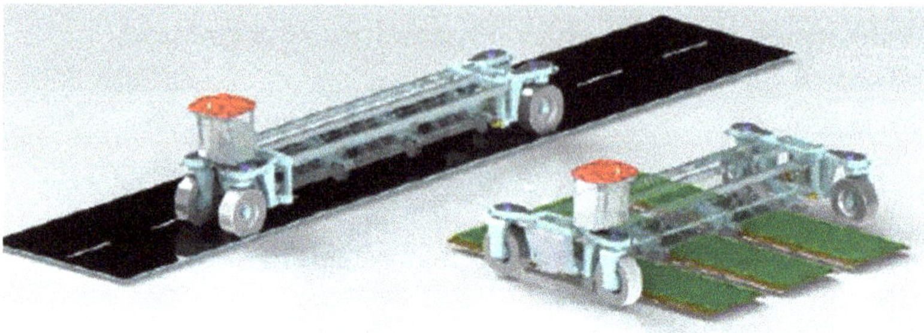

Figure 9 Wide-span gantry system transport on the highway and working in the field. Source: Adapted from CTF Europe (2020).

vehicles dates back to 1987 (Cousins et al., 2016). Blunden et al. (1994) studied the effects of a rubber tracked Cat Challenger on earthy sand and found no significant difference in the maximum stress from the rubber tracked vehicle and dual tyres at the depth of 0.3 m. However, deeper in the soil profile (i.e. at 0.4 and 0.5 m), the tyres resulted in significantly higher maximum stress, suggesting that stresses under the tracked vehicle do not extend as deep in the soil profile. Alakukku et al. (2003) also found that the uneven load distribution from tracked vehicles only affected soil surface layers. On the other hand, Bashford et al. (1988) found no significant differences in the effects of tyres and tracks on soil BD in the topsoil and subsoil.

Ansorge and Godwin (2007, 2008) found that wheels caused greater soil deformation than tracks. They also suggested that cone penetrometer resistance is minimal using tracks, though pressure distribution using tracks is not constant. Arvidsson (2014) also found out that stresses were more variable along the length of the track in comparison to wheels, though the highest soil stress was caused by single wheel tyre vehicles.

Another potential solution to the problem of soil compaction has been identified as the development of wide-span gantry systems. These systems rely on a frame mounted on a wide track gauge with equipment attached onto sections that are able to move independently of each other, as shown in Fig. 9. Chamen et al. (1992a) investigated the effects of a partial 12-m-wide gantry system and found that it reduced the trafficked area by 50%, reduced fuel consumption by up to 44% and increased yield by 19% in comparison to conventional systems. Pedersen et al. (2013) found that a 9.6 m prototype gantry system reduced the trafficked area from 21% to just 6%.

7 Tillage practices and soil compaction

Tillage is defined as the mechanical manipulation of soil for crop production. The role of tillage is to provide favourable conditions for good crop establishment, including providing a suitable seedbed, incorporating the residues of previous crops back into the soil and suppressing weeds (Godwin, 2014). Alskaf et al. (2020) reports that conventional tillage (i.e. mouldboard plough) is still practised on 45% of arable land in England.

Primary tillage includes the use of different types of plough (e.g. mouldboard or disk ploughs) to turn over the soil after a crop is harvested in order to prepare a seedbed for the next crop. Subsequent tillage operations to prepare the seedbed are known as secondary tillage. They include the use of tandem or off-set disks, field cultivators, and harrows to break up larger aggregates left from mouldboard ploughing (Morris et al., 2010; Hallett and Bengough, 2013). The purpose of these operations includes improving soil aeration, water infiltration, and facilitating root development (Sommer and Zach, 1992; Hallett and Bengough, 2013).

Deep tillage (operating at depths of 200+ mm) has been related to heavy traffic and consequently increased soil compaction (Kroulik et al., 2009; Foley et al., 2011). As a result of running two tractor wheels in the open furrow during ploughing, a compacted stratum is created in the soil at about 200–350 mm from the surface; subsoiling can alleviate this but again poses a risk of re-compaction from subsequent traffic (Spoor and Voorhees, 1986; Morris et al., 2010). Alakukku (1996) found that the effects of soil compaction below 0.1 m on clay soil can be long-lasting, despite annual ploughing to 0.2 m. Soane et al. (1986) also found significant re-compaction as a result of trafficking of soil that had been deep loosened and then ploughed. Millington et al. (2017) found porosity decreased with depth in deep tilled soils across the analysed soil horizon down to 250 mm. Dal Ferro et al. (2014) also found that tillage had significant effects on macroporosity (enhancing the pore class in the range 54–250 μm) but did not have a significant effect on microporosity in the analysed soil horizon down to 300 mm.

The effects of conventional tillage on soil BD are variable. Alvarez and Steinbach (2009) in their review of 35 papers, reported higher BD from reduced or zero tillage in comparison to conventional mouldboard ploughing in the upper 0–200 mm. However, the levels of BD and cone PR in the 0–20 cm layer were often below the thresholds which inhibit root development. Franzluebbers et al. (1995) investigated BD down to 200 mm and reported decreased BD shortly after tillage; however, over the growing season, the differences between tilled and no-tilled soil decreased, regardless of the crop (wheat, sorgum, and soya bean). Other studies have found that tillage systems did not significantly affect BD, including Taboada et al. (1998), who did not observe significant differences down to 400 mm of a silty clay soil profile before harvest. Logsdon and Cambardella (2000), who investigated fine loamy soils down to 300 mm in Iowa (USA), reported that none of the observed bulk densities were significantly different between the zero tillage and cultivated fields. Jabro et al. (2016) reported that soil BD of sandy loam in North Dakota (USA) was not significantly influenced by tillage (deep, shallow, and zero) in the 0–40 cm layer in 3 out of 4 years. Similarly, Millington (2019) did not find significant differences in BD measured down to 250 mm soil profile several months after tillage on sandy loam in the West Midlands (UK).

The effects of tillage on macroporosity have been investigated by many researchers but with often contradictory results. Capowiez et al. (2009) reported that mouldboard ploughing leads to a significant decrease in the total number and distribution of pores in comparison to reduced tillage. However, other research suggests higher hydraulic conductivity and macroporosity from conventionally ploughed soils compared to zero-tilled soils (Ferreras et al., 2000; Fabrizzi et al., 2005; Sasal et al., 2006; Villarreal et al., 2020).

There is similar inconsistency in studies on the effects of tillage on water infiltration rates and hydraulic conductivity. Nielsen et al. (2005) reported an increase of infiltration rates under zero-tillage, but Rasmussen (1999) found greater water infiltration in tilled soil. Esser (2017) also found higher infiltration and less runoff under conventional tillage as well as higher soil moisture in the top soil under conventionally tilled soil than under Conservation Agriculture. These conflicting results can be attributed in some cases to temporal variability in soil infiltration rate as suggested by Strudley et al. (2008) who found that infiltration was highest immediately after tillage but decreased with time, whilst water infiltration under zero-tillage can be higher than in tilled soil after the first wetting-drying cycle.

Wuest (2010) found that tillage to depths of 100 mm and 150 mm preserved up to 0.01 kg^{1} greater water content than zero-tillage or shallow tillage (to 50 mm). However, Oorts et al. (2007) found no significant differences in soil moisture between different tillage systems. Smith (2017) also did not find any significant differences in gravimetric soil moisture between contrasting tillage and traffic treatments. Other studies suggest that soil moisture is higher under zero tillage but with significant variability in results, depending on factors such as soil conditions, depth, crop, and season (Rasmussen, 1999; Fuentes et al., 2003; Alvarez and Steinbach, 2009; Gruber et al., 2011; Slawinski et al., 2012). Wuest (2010) have suggested it is often difficult to discern significant differences between tillage practices.

Because it leaves the soil bare until the next crop is planted, conventional tillage has been linked to greater risk of erosion by wind and water (Lal, 2007; Huggins and Reganold, 2008; Bogunovic et al., 2018). Since nutrients and pesticides from eroded soils may be attached to sediment, the transport of soil particles to surface waters possess a significant pollution risk (Rickson, 2014; Melland et al., 2017). For these and other reasons, reduced tillage (non-inversion tillage) and zero-tillage have been suggested as alternatives to conventional mouldboard ploughing (Gauer et al., 1982; Tullberg et al., 2007;

Verhulst et al., 2010; Derpsch et al., 2010, 2014; Warner et al., 2016). Zero-tillage is often combined with other techniques such as continuous soil cover and the use of rotations in what has become known as 'Conservation Agriculture', actively promoted by the FAO as promoting soil health and more sustainable agricultural practices (FAO, 2013). This makes it difficult to distinguish the benefits of zero-tillage on soil properties from other Conservation Agriculture practices which may have a more important impact on soil health or crop yields (Esser, 2017; Bogunovic et al., 2018).

Joergensen and Emmerling (2006) in their review of methods for evaluating human impact on soil microorganisms suggested that soil microorganisms are sensitive to farm management practices. Pelosi et al. (2014) in his study into the effects of three main cultivations types: ploughing (250–300 mm), reduced tillage (80 mm), and zero-tillage on earthworm populations concluded that the tillage intensity did not affect the number of observed species nor abundance though different cultivation techniques did affect some functional traits (e.g. body length, body mass/length ratio, cocoon diameter, vertical distribution). That experiment however contained varied crops across the experimental plots (i.e. sugar beet, wheat, and flax). Crittenden et al. (2015), in a 2-year experiment, found that non-inversion tillage significantly increased total earthworm density by 34% and total earthworm biomass by 15% compared with inversion tillage. The same authors also reported that anecic earthworms particularly suffered from deep tillage which resulted in very low numbers of these organisms. Springtails are also suggested to be sensitive to different soil use practices. The effects of two deep tillage systems on Collembola communities were investigated by Petersen (2002), who reported that in the uppermost 4 cm stratum conventional ploughing reduced their population more than the non-inverting deep tillage while, in the deepest stratum (28–32 cm), the immediate effect was opposite. Hu et al. (2014) reported that the soil microbial biomass is determined by soil organic carbon, which is influenced by plant inputs. However, there are no coherent conclusions on the effects of farming traffic or tillage on microbial populations. Some studies show that tillage negatively affects the size of microbial populations (Sun et al., 2016; Wright et al., 2008), others reported on little effect of tillage on microbial biomass (Calderón et al., 2000; Jackson et al., 2003). Kaiser et al. (2014) suggested that those inconsistencies might derive from different soil physical properties across different studies as well as from varying tillage intensities. Some studies have found changes between microbial taxa but with total biomass unchanged. Campbell et al. (1991), who investigated crop rotation and influence of fertilisers on soil microbial biomass, suggested that fungal communities can benefit from a decrease in bacterial communities and vice versa. In agreement, Sun et al. (2018) found that susceptibility to tillage intensity varied across different microbial communities and reduced tillage had a greater effect on fungal communities while bacterial communities were more affected by mouldboard ploughing.

Some researchers have reported an increase in crop yields under zero-tillage compared to mouldboard ploughing (e.g. Lal, 1997; Singh et al., 2016). However, other studies have reported no differences in crop yields between these two tillage systems (e.g. Shipitalo and Edwards, 1998; Logsdon and Karlen, 2004; Alvarez and Steinbach, 2009), while some have found reduced yields under zero-tillage compared to conventional tillage (Clutterbuck and Hodgson, 1984; Rusu, 2005; Rieger et al., 2008; Alvares and Steinbach, 2009; Smith, 2017). A possible reason for this apparent lack of consistency in reported crop yield data from zero-tillage studies is the fact that field traffic management practices are not always reported, which therefore makes interpretation of such data rather cumbersome.

Researchers have suggested zero-till may be more suited to semi-arid regions where it is more effective in water retention (Buschiazzo et al., 1998; De Vita et al., 2007) whereas, in wet conditions, deep tillage can deliver higher crop yields due to better infiltration (Alvarez and Steinbach, 2009). As an example, Godwin et al. (2015) concluded that chalk limestone soils and well-drained loamy soils might provide yields of both autumn and spring cereals under zero-till similar to those from conventionally cultivated crops. On calcareous clays and clayey or loamy soils over clay with improved drainage, only winter cereal crop yields are likely to be similar to those from conventionally cultivated soil, whilst spring crops yields are likely to be lower. The least suitable soils for zero-till (with a substantial risk of lower yields) are sandy soils with low organic matter content, silty soils, wet alluvial soils, and poorly drained clayey soils. Another complication is that the benefits to soil health of a transition from conventional cultivation to Conservation Agriculture may take a number of years to materialise with an initial dip, e.g. in yields followed by longer-term improvements in soil properties, ultimately leading to increased crop yields which equal or exceed yields under conventional tillage (Rhoton, 2000; Chamen, 2011; Jemai et al., 2013; Kaczorowska-Dolowy et al., 2020; Godwin et al., 2022).

8 Case study

The previous sections suggest the effects of traffic on soil compaction vary, though high traffic levels generally have a significant impact on compaction. Compaction leads to higher soil BD, soil strength, a decreased number of pores and lower pore size, as well as poorer infiltration. Research suggests that CTF can play a significant role in reducing compaction and its effects, whilst the benefits of other measures such as LTP systems are less clear. The research on differing types of tillage shows widely varying effects on soil characteristics such as BD, pore number and size, infiltration, and overall moisture levels.

Some of these issues have been the basis of a long-term experiment established at Harper Adams University in the UK (Smith, 2017; Millington, 2019; Kaczorowska-Dolowy, 2022). The experiment was started in 2011 by Smith (2017) on a uniform sandy loam field called Large Marsh within the Harper Adams University campus in the county of Shropshire (52°46'58.0"N 2°25'43.9"W). The aim of the experiment was to examine the effects of three traffic systems:

1 random traffic with standard tyre inflation pressure (STP);
2 random traffic with low tyre inflation pressure (LTP);
3 CTF.

These were subject to three tillage depths:

1 deep tillage: 250 mm;
2 shallow tillage: 100 mm;
3 zero tillage.

These different systems were applied across a typical crop rotation for the UK with cereals as main crop, including wheat, barley, and oats with field beans as a break crop. A starting point was the research conducted by Chamen (2011) who suggested that the effects of reduced trafficking on soil properties, yields, and profitability requires systematic investigation. The study examined effects of the different systems on soil properties, crop growth, yield, and farming economics.

The first 5 years of the experiment by Smith (2017) and Millington (2019) focused on the effects of the traffic and tillage systems on soil physical properties related to compaction (BD, PR, moisture, hydraulic conductivity, total porosity, and pore distribution) and crop yields. Initial published results from the study found that yields for all crops across the 5-year rotation were significantly lower using zero tillage than those for deep and shallow tillage, with an average reduction of 1.0 t ha^{1} (Godwin et al., 2017). Nevertheless, as from the sixth year, no yield penalties have been observed under zero tillage. Moreover, the yield under zero tillage exceeded the yield from deep tillage in year 7 and 8 whereas, in comparison to shallow tillage, it was significantly greater in year 7 (Godwin et al., 2022).

Compared to random traffic using standard tyre inflation pressure, CTF with trafficked areas of 30% and 15% produced 4% and 7% greater yields worth £39 ha^{-1} and £74 ha^{-1}, respectively. The study also investigated the use of LTP of 0.7 bar for both the front and rear types of the tractor, compared to conventional pressures of 1.2 bar in front tyres and 1.5 bar in rear tyres. The beneficial effect of low inflation pressure tyres on crop yields was found for the deep tillage treatment where the crop yields were 3.9% greater than those of the STP systems giving a financial gain of £39 ha^{-1} (Godwin et al., 2022).

Millington (2019) continued the study by focusing on analysing the long-term effects of traffic and tillage on soil physical properties using X-ray tomography. This has confirmed that vehicular traffic decreases soil macroporosity down to 0–250 mm of the soil profile. Moreover, this effect is enhanced under deep tillage, resulting in a reduction in the ability of soil to support vehicular traffic which leads to significant soil recompaction. Research by Kaczorowska-Dolowy (2022) was focused on the effects of traffic and tillage systems on soil biological properties, including soil fauna feeding activity, soil organic matter, soil microbial carbon, earthworm, and Collembola abundance. She reported that vehicular traffic decreased invertebrate feeding activity (in the 0–80 mm soil horizon), whereas the effect of tillage depended on the year and under zero tillage, it was significantly lower than under both tilled systems in 2019, whereas in 2020 under shallow and zero tillage, the soil fauna feeding activity was significantly greater than under deep tillage. Similarly, the abundance of Collembola in 2019 decreased under compacted soil in comparison to unwheeled area, at the same time Collembola abundance was significantly greater under shallow tillage than zero tillage. Reduced tillage (shallow and zero) increased soil organic matter over deep tillage in the 0–100 mm soil horizon. Earthworms' abundance increased with decreased tillage intensity and so under deep tillage, it was significantly lower than under zero and shallow tillage whereas under zero tillage, it was significantly greater than under shallow tillage (Kaczorowska-Dolowy, 2022).

9 Conclusion

Agricultural intensification has been accompanied by an increase in machinery size and weight, resulting in soil compaction and leading to soil degradation. Compacted soil inhibits root development, nutrient uptake, water movement, and availability leading to yield losses. Remedial actions such as subsoiling are expensive, time-consuming, not always effective at greater depths, and require burning additional fossil fuels. Maintaining the soil in a good (uncompacted) condition is key to improving aspects of soil health such as water infiltration, crop growth, and health as well as ecosystem services such as carbon sequestration. The effects of alternative tillage practices on soil properties and crop growth have been well reported. The effects of alternative traffic management systems generally indicate their benefits for soil health though more research is needed, especially effects on soil biology.

10 Where to look for further information

Further information can be obtained from the following sources:

- Australian Controlled Traffic Farming Association Inc.: http://www.actfa.net/.
- Controlled Traffic Farming Europe Ltd.: http://controlledtrafficfarming.com/Home/Default.aspx.
- Godwin R. J., 2014. Potential of "No-till" Systems for Arable Farming. The Worshipful Company of Farmers. London, UK. http://cdn.harper-adams.ac.uk/document/project/Potential-of-No-till-Systems-for-Arable-Farming-Report.pdf.
- McPhee, J. E., Antille, D. L., Tullberg, J. N., Doyle, R. B. and Boersma, M. (2020). Managing soil compaction - a choice of low-mass autonomous vehicles or controlled traffic? Biosystems Engineering 195: 227-241. https://doi.org/10.1016/j.biosystemseng.2020.05.006.
- The National Centre for Precision Farming, Harper Adams University: http://www.harper-adams.ac.uk/research/ncpf/.
- The Soil and Water Management Centre: http://www.soilandwater.org.uk/.
- Traffic and Tillage Research Project, Harper Adams University: http://www.harper.ac.uk/tillage.

11 References

Ahmadi, I. and Ghaur, H. 2015. Effects of soil moisture content and tractor wheeling intensity on traffic-induced soil compaction. *Journal of Central European Agriculture* 16(4), 489-502.

Alakukku, L. 1996. Persistence of soil compaction due to high axle load traffic. I. Short-term effects on the properties of clay and organic soils. *Soil and Tillage Research* 37(4), 211-222.

Alakukku, L., Weisskopf, P., Chamen, W. C. T., Tijink, F. G. J., Van der Linden, J. P., Pires, S., Sommer, C. and Spoor, G. 2003. Prevention strategies for field traffic-induced subsoil compaction: a review: part 1. *Soil and Tillage Research* 73(1-2), 145-160.

Alskaf, K., Sparkes, D. L., Mooney, S. J., Sjogersten, S. and Wilson, P. 2020. The uptake of different tillage practices in England. *Soil Use and Management* 36(1), 27-44. https://doi.org/10.1111/sum.12542.

Alvarez, R. and Steinbach, H. S. 2009. A review of the effects of tillage systems on some soil physical properties, water content, nitrate availability and crops yield in the argentine pampas. *Soil and Tillage Research* 104(1), 1-15.

Andriuzzi, W. S. 2015. Ecological functions of earthworms in soil. Doctoral dissertation, Wageningen University and Research.

Ankeny, M. D., Kaspar, T. C. and Prieksat, M. A. 1995. Traffic effects on water infiltration in chisel-plow and no-till systems. *Soil Science Society of America Journal* 59(1), 200-204.

Ansorge, D. and Godwin, R. J. 2007. The effect of tyres and a rubber track at high axle loads on soil compaction, part 1: single axle-studies. *Biosystems Engineering* 98(1), 115–126.

Ansorge, D. and Godwin, R. J. 2008. The effect of tyres and a rubber track at high axle loads on soil compaction, part 2: multi-axle studies. *Biosystems Engineering* 99(3), 338–347.

Antille, D. L., Ansorge, D., Dresser, M. L. and Godwin, R. J. 2013. Soil displacement and soil bulk density changes as affected by tire size. *Transactions of the ASABE* 56(5), 1683–1693.

Antille, D. L., Chamen, W. C. T., Tullberg, J. N. and Lal, R. 2015. The potential of controlled traffic farming to mitigate greenhouse gas emissions and enhance carbon sequestration in arable land: a critical review. *Transactions of the ASABE* 58(3), 707–731.

Antille, D. L., Bennett, J. M. and Jensen, T. A. 2016a. Soil compaction and controlled traffic considerations in Australian cotton-farming systems. *Crop and Pasture Science* 67(1), 1–28.

Antille, D. L., Huth, N. I., Eberhard, J., Marinoni, O., Cocks, B., Poulton, P. L., Macdonald, B. C. T. and Schmidt, E. J. 2016b. The effects of coal seam gas infrastructure development on arable land in southern Queensland, Australia: field investigations and modeling. *Transactions of the ASABE* 59(4), 879-901. https://doi.org/10.13031/trans.59.11547.

Antille, D. L., Peets, S., Galambošová, J., Botta, G. F., Rataj, V., Macak, M., Tullberg, J. N., Chamen, W. C. T., White, D. R., Misiewicz, P. A., Hargreaves, P. R., Bienvenido, J. F. and Godwin, R. J. 2019a. Review: soil compaction and controlled traffic farming in arable and grass cropping systems. *Agronomy Research* 17(3), 653–682.

Antille, D. L., Chamen, T., Tullberg, J. N., Isbister, B., Jensen, T. A., Chen, G., Baillie, C. P. and Schueller, J. K. 2019b. Chapter 10: controlled traffic farming in precision agriculture. In: Stafford, J. V. (Ed.), *Precision Agriculture for Sustainability, Part 2: Delivery Systems. Burleigh Dodds Series in Agricultural Science No.: 52*. Cambridge: Burleigh Dodds Science Publishing Limited, pp. 239–270.

Arvidsson, J. 2001. Subsoil compaction caused by heavy sugarbeet harvesters in southern Sweden. I. soil physical properties and crop yield in six field experiments. *Soil and Tillage Research* 60(1), 67–78.

Arvidsson, J. 2014. Soil stresses under tracks and tyres – measurements and model development. *Zurich*. [Online]. Presented at the International Conference of Agricultural Engineering, 6–10 July 2014. AgEng Available at: http://www.geyseco.es/geystiona/adjs/comunicaciones/304/C06210001.pdf.

Arvidsson, J. and Håkansson, I. 2014. Response of different crops to soil compaction–short-term effects in Swedish field experiments. *Soil and Tillage Research* 138, 56–63.

Bagarello, V., Iovino, M. and Elrick, D. 2004. A simplified falling-head technique for rapid determination of field-saturated hydraulic conductivity. *Soil Science Society of America Journal* 68(1), 66–73.

Barthes, B. and Roose, E. 2002. Aggregate stability as an indicator of soil susceptibility to runoff and erosion; validation at several levels. *CATENA* 47(2), 133–149.

Bashford, L. L., Jones, A. J. and Mielke, L. N. 1988. Comparison of bulk density beneath a belt track and tire. *Applied Engineering in Agriculture* 4(2), 122–125.

Bashford, L. L. and Kocher , M. F. 1999. Belts vs tires, belts vs belts, tires vs tires. *Applied Engineering in Agriculture* 15(3), 175–181.

Beckers, E., Plougonven, E., Roisin, C., Hapca, S., Léonard, A. and Degré, A. 2014. X-ray microtomography: a porosity-based thresholding method to improve soil pore network characterization? *Geoderma* 219, 145-154.

Ball, B. C. and Ritchie, R. M. 1999. Soil and residue management effects on arable cropping conditions and nitrous oxide fluxes under controlled traffic in Scotland: 1. soil and crop responses. *Soil and Tillage Research* 52(3-4), 177-189.

Bell, M. J., Halpin, N. V., Garside, A. L., Moody, P. W., Stirling, G. R. and Robotham, B. J. 2003. PK editorial services. *Proceedings of the-Australian Society of Sugar Cane Technologists* (vol. 1999).

Bengough, A. G. 2006. Mechanical impedance to root growth: a review of experimental techniques and root growth responeses. *European Journal of Soil Science* 41(3), 341-358.

Berisso, F. E., Schjønning, P., Keller, T., Lamandé, M., Etana, A., de Jonge, L. W., Iversen, B. V., Arvidsson, J. and Forkman, J. 2012. Persistent effects of subsoil compaction on pore size distribution and gas transport in a loamy soil. *Soil and Tillage Research* 122, 42-51.

Blunden, B. G., McBride, R. A., Daniel, H. and Blackwell, P. S. 1994. Compaction of an earthy sand by rubber tracked and tired vehicles. *Soil Research* 32(5), 1095-1108.

Bochtis, D. D., Sørensen, C. G., Green, O., Moshou, D. and Olesen, J. 2010. Effect of controlled traffic on field efficiency. *Biosystems Engineering* 106(1), 14-25.

Bogunovic, I., Pereira, P., Kisic, I., Sajko, K. and Sraka, M. 2018. Tillage management impacts on soil compaction, erosion and crop yield in stagnosols (Croatia). *CATENA* 160, 376-384.

Botta, G. F., Antille, D. L., Bienvenido, F., Rivero, D., Avila-Pedraza, E. A., Contessotto, E. E., Ghelfi, D. G., Nistal, A. I., Pelizzari, F. M., Rocha-Meneses, L. and Ezquerra Canalejo, A. 2020. Effect of cattle trampling and farm machinery traffic on soil compaction of an Entic Haplustoll in a semiarid region of Argentina. *Agronomy Research* 18(2), 1163-1176. https://doi.org/10.15159/AR.20.063.

Botta, G. F., Antille, D. L., Nardón, G. F., Rivero, D., Bienvenido, F., Contessotto, E. E., Ezquerra-Canalejo, A. and Ressia, J. M. 2022. Zero and controlled traffic improved soil physical conditions and soybean yield under no-tillage. *Soil and Tillage Research* 215, Article number: 105235. https://doi.org/10.1016/j.still.2021.105235.

Buschiazzo, D. E., Panigatti, J. L. and Unger, P. W. 1998. Tillage effects on soil properties and crop production in the subhumid and semiarid Argentinean pampas. *Soil and Tillage Research* 49(1-2), 105-116.

Calderón, F. J., Jackson, L. E., Scow, K. M. and Rolston, D. E. 2000. Microbial responses to simulated tillage in cultivated and uncultivated soils. *Soil Biology and Biochemistry* 32(11-12), 1547-1559.

Campbell, C. A., Biederbeck, V. O., Zentner, R. P. and Lafond, G. P. 1991. Effect of crop rotations and cultural practices on soil organic matter, microbial biomass and respiration in a thin *Black Chernozem. Canadian Journal of Soil Science* 71(3), 363-376.

Capowiez, Y., Cadoux, S., Bouchant, P., Ruy, S., Roger-Estrade, J., Richard, G. and Boizard, H. 2009. The effect of tillage type and cropping system on earthworm communities, macroporosity and water infiltration. *Soil and Tillage Research* 105(2), 209-216.

Chamen, T. 2015. Controlled traffic farming - from worldwide research to adoption in Europe and its future prospects. *Acta Technologica Agriculturae* 18(3), 64-73.

Chamen, W. C. T. 2011. The effects of low and controlled traffic systems on soil physical properties, yields and the profitability of cereal crops on a range of soil types. PhD thesis. Cranfield University.

Chamen, W. C. T., Watts, C. W., Leeds, P. R. and Longstaff, D. J. 1992a. Assessment of a wide span vehicle (gantry), and soil and cereal crop responses to its use in a zero-traffic regime. *Soil and Tillage Research* 24(4), 359–380.

Chamen, W. C. T., Vermeulen, G. D., Campbell, D. J. and Sommer, C. 1992b. Reduction of traffic-induced soil compaction: a synthesis. *Soil and Tillage Research* 24(4), 303–318.

Chamen, W. C. T., Horn, R., Fleige, H., Peth, S. and Peng, X. 2006. Controlled traffic farming on a field scale in the UK. *Advances in Geoecology* 39, 251–260.

Chamen, W. C. T., Chittey, E. T., Leede, P. R., Goss, M. J. and Howse, K. R. 1990. The effect of tyre/soil contact pressure and zero traffic on soil and crop responses when growing winter wheat. *Journal of Agricultural Engineering Research* 47(1), 1–21.

Cavallo, E., Ferrari, E., Bollani, L. and Coccia, M. 2014. Strategic management implications for the adoption of technological innovations in agricultural tractor: the role of scale factors and environmental attitude. *Technology Analysis and Strategic Management* 26(7), 765–779.

Chartres, C. J., Kirby, J. M. and Raupach, M. 1990. Poorly ordered silica and aluminosilicates as temporary cementing agents in hard-setting soils. *Soil Science Society of America Journal* 54(4), 1060–1067. https://doi.org/10.2136/sssaj1990.03615995005400040022x.

Chen, Y. L., Palta, J., Clements, J., Buirchell, B., Siddique, K. H. M. and Rengel, Z. 2014. Root architecture alteration of narrow-leafed lupin and wheat in response to soil compaction. *Field Crops Research* 165, 61–70.

Chyba, J. 2012. *The Influence of Traffic Intensity and Soil Texture on Soil Water Infiltration Rate*. Newport: Harper Adams University College.

Clutterbuck, B. J. and Hodgson, D. R. 1984. Direct drilling and shallow cultivation compared with ploughing for spring barley on a clay loam in northern England. The Journal of Agricultural Science, 102 (1), pp. 127–134.

Congreves, K. A., Hayes, A., Verhallen, E. A. and Van Eerd, L. L. 2015. Long-term impact of tillage and crop rotation on soil health at four temperate agroecosystems. *Soil and Tillage Research* 152, 17–28.

Cousins, D., Andrews, J. and Mark, O. 2016. Lamma. 2016: 7 big kit highlights from the ag machinery show. *Farmers Weekly*.

Crittenden, S. J., Huerta, E., De Goede, R. G. M. and Pulleman, M. M. 2015. Earthworm assemblages as affected by field margin strips and tillage intensity: an on-farm approach. *European Journal of Soil Biology* 66, 49–56.

CTF Europe [online], available from: http://ctfeurope.dk/2020/wide-span-gantry-tractor-english/ [accessed 10.11.2020].

Czyz, E. A. 2004. Effects of traffic on soil aeration, bulk density and growth of spring barley. *Soil and Tillage Research* 79(2), 153–166.

Dal Ferro, N., Sartori, L., Simonetti, G., Berti, A. and Morari, F. 2014. Soil macro-and microstructure as affected by different tillage systems and their effects on maize root growth. *Soil and Tillage Research* 140, 55–65.

De Vita, P., Di Paolo, E., Fecondo, G., Di Fonzo, N. and Pisante, M. 2007. No-tillage and conventional tillage effects on durum wheat yield, grain quality and soil moisture content in southern Italy. *Soil and Tillage Research* 92(1–2), 69–78.

Demmel, M. R., Kirchmeier, H., Brandhuber, R., Kupke, S. and Blumental, B. 2015. Controlled traffic farming in Germany-Technical and Organizational Realization and Results. In *2015 ASABE Annual International Meeting (p. 1)*. American Society of Agricultural and Biological Engineers.

Derpsch, R., Friedrich, T., Kassam, A. and Li, H. 2010. Current status of adoption of no-till farming in the world and some of its main benefits. *International Journal of Agricultural and Biological Engineering* 3(1), 1-25.

Derpsch, R., Franzluebbers, A. J., Duiker, S. W., Reicosky, D. C., Koeller, K., Friedrich, T., Sturny, W. G., Sá, J. and Weiss, K. 2014. Why do we need to standardize no-tillage research? Soil and Tillage Research, 137 pp. 16-22.

Dexter, A. R. and Bird, N. R. A. 2001. Methods for predicting the optimum and the range of soil water contents for tillage based on the water retention curve. *Soil and Tillage Research* 57(4), 203-212.

Dickson, J. W. and Ritchie, R. M. 1996. Zero and reduced ground pressure traffic systems in an arable rotation 2. Soil and crop responses. *Soil and Tillage Research* 38(1-2), pp.89-113.

Domżał, H., Gliński, J. and Lipiec, J. 1991. Soil compaction research in Poland. *Soil and Tillage Research* 19(2-3), 99-109.

Doran, J. W. 2002. Soil health and global sustainability: translating science into practice. *Agriculture, Ecosystems and Environment* 88(2), 119-127.

Ebrahimi, E., Tekeste, M. Z., Huth, N. I., Antille, D. L., Archontoulis, S. V. and Horton, R. 2022. Measured and modelled maize and soybean growth and water use on pipeline disturbed land. *Soil and Tillage Research* 220, Article number: 105340. https://doi.org/10.1016/j.still.2022.105340.

Eden, M., Schjønning, P., Moldrup, P. and De Jonge, L. W. 2011. Compaction and rotovation effects on soil pore characteristics of a loamy sand soil with contrasting organic matter content. *Soil Use and Management* 27(3), 340-349.

Esser, K. B. 2017. Water infiltration and moisture in soils under conservation and conventional agriculture in agro-ecological zone IIa, Zambia. *Agronomy* 7(2).

Evans, R. 1996. *Soil Erosion and Its Impacts in England in Wales*. Friends of the Earth.

Fabrizzi, K. P., Garcıa, F. O., Costa, J. L. and Picone, L. I. 2005. Soil water dynamics, physical properties and corn and wheat responses to minimum and no-tillage systems in the southern pampas of Argentina. *Soil and Tillage Research* 81(1), 57-69.

FAO, 2013. *Climate Smart Agriculture*: Sourcebook. 203-204, online source [https://www.fao.org/3/i3325e/i3325e.pdf] accessed on 20.12.2021.

Ferreras, L. A., Costa, J. L., Garcia, F. O. and Pecorari, C. 2000. Effect of no-tillage on some soil physical properties of a structural degraded petrocalcic paleudoll of the southern "Pampa" of Argentina. *Soil and Tillage Research* 54(1-2), 31-39.

Foley, J. A., Ramankutty, N., Brauman, K. A., Cassidy, E. S., Gerber, J. S., Johnston, M., Mueller, N. D., O'Connell, C., Ray, D. K., West, P. C., Balzer, C., Bennett, E. M., Carpenter, S. R., Hill, J., Monfreda, C., Polasky, S., Rockström, J., Sheehan, J., Siebert, S., Tilman, D. and Zaks, D. P. 2011. Solutions for a cultivated planet. *Nature* 478(7369), 337-342.

Franzluebbers, A. J., Hons, F. M. and Zuberer, D. A. 1995. Tillage and crop effects on seasonal dynamics of soil CO_2 evolution, water content, temperature, and bulk density. *Applied Soil Ecology* 2(2), 95-109.

Fuentes, J. P., Flury, M., Huggins, D. R. and Bezdicek, D. F. 2003. Soil water and nitrogen dynamics in dryland cropping systems of Washington State, USA. Soil and Tillage Research 71(1), 33-47.

Galambošová, J., Macák, M., Rataj, V., Antille, D. L., Godwin, R. J., Chamen, W. C. T., Žitnák, M., Vitázková, B., Dudák, J. and Chlpík, J. 2017. Field evaluation of controlled traffic farming in central Europe using commercially available machinery. *Transactions of the ASABE* 60(3), 657-669.

Gasso, V., Sørensen, C. A. G., Oudshoorn, F. W. and Green, O. 2013. Controlled traffic farming: a review of the environmental impacts. *European Journal of Agronomy* 48, 66-73.

Gauer, E., Shaykewich, C. F. and Stobbe, E. H. 1982. Soil temperature and soil water under zero tillage in manitoba. *Canadian Journal of Soil Science* 62(2), 311-325.

Głąb, T. 2008. Effects of tractor wheeling on root morphology and yield of lucerne (Medicago sativa L.). Grass and Forage Science, 63 (3), pp 398-406.

Godfray, H. C. J. and Garnett, T. 2014. Food security and sustainable intensification. *Philosophical Transactions of the Royal Society of London. Series B, Biological Sciences* 369(1639), 20120273.

Godwin, R. J. 2014. *Potential of "No-Till" Systems for Arable Farming. The Worshipful Company of Farmers*. London.

Godwin, R. J., Misiewicz, P. A., White, D., Smith, E., Chamen, T., Galambošová, J. and Stobart, R. 2015. Results from recent traffic systems research and the implications for future work. *Acta Technologica Agriculturae* 18(3), 57-63.

Godwin, R. J., Misiewicz, P. A., Smith, E. K., Millington, W. A., White, D. R., Dickin, E. T. and Chaney, K. 2017. *2017 ASABE Annual International Meeting*. American Society of Agricultural and Biological Engineers.

Godwin, R. J., White, D. R., Dickin, E. T., Kaczorowska-Dolowy, M., Millington, W. A. J., Pope, E. K. and Misiewicz, P. A. 2022. The effects of traffic management systems on the yield and economics of crops grown in deep, shallow and zero tilled sandy loam soil over eight years. *Soil and Tillage Research* 223, 105465.

Gruber, S., Möhring, J. and Claupein, W. 2011. On the way towards conservation tillage-soil moisture and mineral nitrogen in a long-term field experiment in Germany. *Soil and Tillage Research* 115-116, 80-87.

Hallett, P. D. and Bengough, A. G. 2013. Managing the soil physical environment for plants. *Soil Conditions and Plant Growth*. Oxford: Blackwell Publishing, pp. 238-268.

Hargreaves, P., Peets, S., Chamen, W., Misiewicz, P. A., White, D. R. and Godwin, R. J. 2016. Controlled traffic farming: methods applied to grassland silage management, AHDB dairy report. Stoneleigh Park, Kenilworth, Warwickshire, UK.

Hettiaratchi, D. 1990. Soil compaction and plant root growth. *Philosophical Transactions of the Royal Society B: Biological Sciences* 329(1255), 343-355.

Horn, R., Way, T. and Rostek, J. 2003. Effect of repeated tractor wheeling on stress/strain properties and consequences on physical properties in structured arable soils. *Soil and Tillage Research* 73(1-2), 101-106.

Houšková, B. and Montanarella, L. 2008. The natural susceptibility of European soils to compaction. *Threats to Soil Quality in Europe Eur.* 23438, 23-35.

Hu, Y., Xiang, D., Veresoglou, S. D., Chen, F., Chen, Y., Hao, Z., Zhang, X. and Chen, B. 2014. Soil organic carbon and soil structure are driving microbial abundance and community composition across the arid and semi-arid grasslands in northern China. *Soil Biology and Biochemistry* 77, 51-57.

Huber, S., Prokop, G., Arrouays, D., Banko, G., Bispo, A., Jones, R. J. A., Kibblewhite, M. G., Lexer, W., Moller, A., Rickson, R. J. and Shishkov, T. 2008. Environmental Assessment of Soil for Monitoring. Volume I: Indicators & Criteria (No. JRC 47184). European Communities.

Huggins, D. R. and Reganold, J. P. 2008. No-till: the quiet revolution. *Scientific American* 299(1), 70-77.

Hůla, J., Kroulík, M. and Kovaříček, P. 2009. Effect of repeated passes over the soil on the degree of compaction (Vliv opakovaných přejezdů po půdě nastupeň zhutnění půdy). GPS autopiloty v zemědělství, pp. 39-44.

Hussein, M. A., Antille, D. L., Kodur, S., Chen, G. and Tullberg, J. N. 2021a. Controlled traffic farming delivers improved agronomic performance of wheat as a result of enhanced rainfall and fertilizer nitrogen use efficiency. *Acta Agriculturae Scandinavica, Section B - Soil and Plant Science* 71(5), 377-398. https://doi.org/10.1080/09064710.2021.1903984.

Hussein, M. A., Antille, D. L., Kodur, S., Chen, G. and Tullberg, J. N. 2021b. Controlled traffic farming effects on productivity of grain sorghum, rainfall and fertiliser nitrogen use efficiency. *Journal of Agriculture and Food Research* 3, Article number: 100111. https://doi.org/10.1016/j.jafr.2021.100111.

Inns, F. M. and Kilgour, J. 1978. *Agricultural Tyres*. London, UK: Dunlop Ltd.

Isbister, B., Blackwell, P., Riethmuller, G., Davies, S., Whitlock, A. and Neale, T. 2013. Controlled Traffic Farming Technical Manual. Department of Agriculture and Food, Perth.

Jabro, J. D., Iversen, W. M., Stevens, W. B., Evans, R. G., Mikha, M. M. and Allen, B. L. 2016. Physical and hydraulic properties of a sandy loam soil under zero, shallow and deep tillage practices. *Soil and Tillage Research* 159, 67-72.

Jackson, L. E., Calderon, F. J., Steenwerth, K. L., Scow, K. M. and Rolston, D. E. 2003. Responses of soil microbial processes and community structure to tillage events and implications for soil quality. *Geoderma* 114(3-4), 305-317.

Jemai, I., Aissa, N. B., Guirat, S. B., Ben-Hammouda, M. and Gallali, T. 2013. Impact of three and seven years of no-tillage on the soil water storage, in the plant root zone, under a dry subhumid Tunisian climate. *Soil and Tillage Research*, 126, 26-33.

Joergensen, R. G. and Emmerling, C. 2006. Methods for evaluating human impact on soil microorganisms based on their activity, biomass, and diversity in agricultural soils. *Journal of Plant Nutrition and Soil Science* 169(3), 295-309.

Kaczorowska-Dolowy, M., Godwin, R. J., Dickin, E. T., White, D. R. and Misiewicz, P. A. 2019. Controlled traffic farming delivers better crop yield of winter bean as a result of improved root development. *Agronomy Research* 17(3), 725-740.

Kaczorowska-Dolowy, M. A., Misiewicz, P. A., Dickin, E., White, D. R., Millington, W. A. J., Smith, E. K. and Godwin, R. J. 2020. The effect of traffic and tillage on soil biology and crop yields in a long-term experiment on sandy loam. *Paper Delivered to the ASABE Annual International Meeting: 12-15th July, 2020*, Omaha - Nebraska, USA.

Kaczorowska-Dolowy, M. A. 2022. Traffic and tillage effects on soil health and crop growth. PhD Thesis. Newport: Harper Adams University.

Kaiser, M., Piegholdt, C., Andruschkewitsch, R., Linsler, D., Koch, H. J. and Ludwig, B. 2014. Impact of tillage intensity on carbon and nitrogen pools in surface and sub-surface soils of three long-term field experiments. *European journal of soil science*, 65(4), 499-509.

Kaspar, T. C., Radke, J. K. and Laflen, J. M. 2001. Small grain cover crops and wheel traffic effects on infiltration, runoff, and erosion. *Journal of Soil and Water Conservation* 56(2), 160-164.

Keller, T., Sandin, M., Colombi, T., Horn, R. and Or, D. 2019. Historical increase in agricultural machinery weights enhanced soil stress levels and adversely affected soil functioning. *Soil and Tillage Research* 194, 104293.

Kibblewhite, M. G., Ritz, K. and Swift, M. J. 2008. Soil health in agricultural systems. *Philosophical Transactions of the Royal Society of London. Series B, Biological Sciences* 363(1492), 685–701.

Kim, H., Anderson, S. H., Motavalli, P. P. and Gantzer, C. J. 2010. Compaction effects on soil macropore geometry and related parameters for an arable field. *Geoderma* 160(2), 244–251.

Kirkham, M. B. 2014. *Principles of Soil and Plant Water Relations*. Academic Press, Burlington, MA.

Knight, S., et al. 2012. Desk study to evaluate the contributory causes of the current 'yield plateau' in wheat and oilseed rape. Project Report No. 502. *AHDB-HGCA* Available at: https://ahdb.org.uk/desk-study-to-evaluate-contributory-causes-of-the-current-yield-plateau-in-wheat-and-oilseed-rape.

Koolen, A. J. and Kuipers, H. 2012. *Agricultural Soil Mechanics*. Springer Science & Business Media, Berlin.

Kroulik, M. 2012. Benefits of GPS agricultural guidance for sustainable agriculture. *Agrociencia Uruguay* 16(3), 107–116.

Kroulik, M., Kumhala, F., Hula, J. and Honzik, I. 2009. The evaluation of agricultural machines field trafficking intensity for different soil tillage technologies. *Soil and Tillage Research* 105(1), 171–175.

Lal, R. 1997. Degradation and resilience of soils. *Philosophical Transactions of the Royal Society of London. Series B: Biological Sciences* 352(1356), 997–1010.

Lal, R. 2007. Anthropogenic influences on world soils and implications to global food security. *Advances in Agronomy*, 93, 69–93.

Li, Y. X., Tullberg, J. N. and Freebairn, D. M. 2007. Wheel traffic and tillage effects on runoff and crop yield. *Soil and Tillage Research*, 97(2), 282–292.

Lipiec, J., Arvidsson, J. and Murer, E. 2003. Review of modelling crop growth, movement of water and chemicals in relation to topsoil and subsoil compaction. *Soil and Tillage Research* 73(1–2), 15–29.

Lipiec, J. and Hatano, R. 2003. Quantification of compaction effects on soil physical properties and crop growth. *Geoderma* 116(1–2), 107–136.

Logsdon, S. D. and Karlen, D. L. 2004. Bulk density as a soil quality indicator during conversion to no-tillage. *Soil and Tillage Research* 78(2), 143–149.

Logsdon, S. D. and Cambardella, C. A. 2000. Temporal changes in small depth-incremental soil bulk density. *Soil Science Society of America Journal* 64(2), 710–714.

Luhaib, A. A. A., Antille, D. L., Tullberg, J. N., Hussein, M. A. and Chen, G. 2017. Effect of controlled traffic farming on energy saving in Australian grain cropping systems. ASABE Paper No.: 1700583. St. Joseph, MI: American Society of Agricultural and Biological Engineers. https://doi.org/10.13031/aim.201700583.

Materechera, S. A., Dexter, A. R. and Alston, A. M. 1991. Penetration of very strong soils by seedling roots of different plant species. *Plant and Soil (Netherlands)*, 135(1), 31–41.

McHugh, A. D., Tullberg, J. N. and Freebairn, D. M. 2009. Controlled traffic farming restores soil structure. *Soil and Tillage Research* 104(1), 164–172.

McPhee, J. E., Neale, T. and Aird, P. L. 2013. Controlled traffic for vegetable production: Part 2. layout considerations in a complex topography. *Biosystems Engineering*, 116(2), 171–178.

Melland, A. R., Antille, D. L. and Dang, Y. P. 2017. Effects of strategic tillage on short-term erosion, nutrient loss in runoff and greenhouse gas emissions. *Soil Research* 55(3), 201–214. https://doi.org/10.1071/SR16136.

Michelin. 2014. 10 years of Michelin Ultraflex Technology. [Online] Available at: https://www.michelin.com/eng/content/download/14637/202334/version/2/file/2014_DP+10+ans_Ultraflex+Technologies+EN.pdf [accessed 30 October 2018].

Millington, W. A. J., Misiewicz, P. A., Dickin, E. T., White, D. R. and Godwin, R. J. 2016. An investigation into the effect of soil compaction and tillage on plant growth and yield of Winter Barley (Hordeum vulgare L.), ASABE Annual International Meeting. Orlando, FL, 17–20 July 2016, Paper Number: 162461725. Available at: https://doi.org/10.13031/aim.20162461725.

Millington, W. A. J., Misiewicz, P. A., White, D. R., Dickin, E. T., Mooney, S. J. and Godwin, R. J. 2017. An investigation into the effect of traffic and tillage on soil properties using X-ray computed tomography. ASABE Annual International Meeting. Spokane, WA, 16-19 July, 2017, Paper Number: 1700380.

Millington, W. A. J. 2019. The effect of low ground pressure and controlled traffic farming systems on soil properties and crop development for three tillage systems. PhD Thesis. Newport: Harper Adams University.

Misiewicz, P. A., Richards, T. E., Blackburn, K. and Godwin, R. J. 2016. Comparison of methods for estimating the carcas stiffness of agricultural tyres on hard surfaces. *Biosystems Engineering* 147, 183–192.

Morales, D., Vargas, M. M., Oliveira, M. P. D., Taffe, B. L., Comin, J., Soares, C. R. and Lovato, P. 2015. Response of soil microbiota to nine-year application of swine manure and urea. *Ciência Rural*, 46, 260–266.

Morris, N. L., Miller, P. C. H., Orson, J. H. and Froud-Williams, R. J. 2010. The adoption of non-inversion tillage systems in the United Kingdom and the agronomic impact on soil, crops and the environment–a review. *Soil and Tillage Research* 108(1-2), 1-15.

Nawaz, M. F., Bourrie, G. and Trolard, F. 2013. Soil compaction impact and modelling: a review. *Agronomy for Sustainable Development* 33(2), 291–309.

Ngo-Cong, D., Antille, D. L., van Genuchten, M. T., Nguyen, H. Q., Tekeste, M. Z., Baillie, C. P. and Godwin, R. J. 2021. A modeling framework to quantify the effects of compaction on soil water retention and infiltration. *Soil Science Society of America Journal* 85(6), 1931–1945. https://doi.org/10.1002/saj2.20328.

Nielsen, D. C., Unger, P. W. and Miller, P. R. 2005. Efficient water use in dryland cropping systems in the great plains. *Agronomy Journal* 97(2), 364–372.

Nimmo, J. R. 2013. *Porosity and Pore Size Distribution, Reference Module in Earth Systems and Environmental Sciences,* Elsevier.

Obour, P. B., Schjønning, P., Peng, Y. and Munkholm, L. J. 2017. Subsoil compaction assessed by visual evaluation and laboratory methods. *Soil and Tillage Research* 173, 4–14.

Oldeman, L. R. 1998. Soil degradation: a threat to food security. Paper prepared for International Conference on Time Ecology: "Time for Soil Culture- temporal perspectives on Sustainable Use of Soil" - April 6-9, 1997, Tutzing

Oorts, K., Laurent, F., Mary, B., Thiebeau, P., Labreuche, J. and Nicolardot, B. 2007. Experimental and simulated soil mineral N dynamics for long-term tillage systems in northern France. *Soil and Tillage Research*, 94(2), 441–456.

Pagliai, M., Marsili, A., Servadio, P., Vignozzi, N. and Pellegrini, S. 2003. Changes in some physical properties of a clay soil in central Italy following the passage of rubber tracked and wheeled tractors of medium power. *Soil and Tillage Research* 73(1-2), 119-129.

Parr, J. F. and Bertrand, A. R. 1960. Water infiltration into soils. *Advances in Agronomy* 12(C), 311-363. https://doi.org/10.1016/S0065-2113(08)60086-3.

Pedersen, H. H., Sorensen, C. G., Oudshoorn, F. W. and McPhee, J. E. 2013. International commission of agricultural and biological engineers. *CIOSTA XXXV Conference "From Effective to Intelligent Agriculture and Forestry", Billund, Denmark (section V).*

Pelosi, C., Pey, B., Hedde, M., Caro, G., Capowiez, Y., Guernion, M., Peigné, J., Piron, D., Bertrand, M. and Cluzeau, D. 2014. Reducing tillage in cultivated fields increases earthworm functional diversity. *Applied Soil Ecology* 83, 79-87.

Petersen, H. 2002. Effects of non-inverting deep tillage vs. conventional ploughing on Collembola populations in an organic wheat field. *European Journal of Soil Biology* 38(2), 177-180.

Peth, S., Nellesen, J., Fischer, G. and Horn, R. 2010. Non-invasive 3D analysis of local soil deformation under mechanical and hydraulic stresses by μCT and digital image correlation. *Soil and Tillage Research* 111(1), 3-18.

Rab, M. A., Haling, R. E., Aarons, S. R., Hannah, M., Young, I. M. and Gibson, D. 2014. Evaluation of X-ray computed tomography for quantifying macroporosity of loamy pasture soils. *Geoderma* 213, 460-470.

Raghavan, G. S. V. and McKyes, E. 1978. Effect of vehicular traffic on soil moisture content in corn (maize) plots. *Journal of Agricultural Engineering Research* 23(4), 429-439.

Raghavan, G. S. V., McKyes, E., Amir, I., Chasse, M. and Broughton, R. S. 1979. Prediction of soil compaction due to off-road vehicle traffic. *Transactions of the ASAE. American Society of Agricultural Engineers,* 19(4), pp.610-0613.

Raper, R. L. 2005. Agricultural traffic impacts on soil. *Journal of Terramechanics* 42(3-4), 259-280.

Raper, R. L., Bailey, A. C., Burt, E. C., Way, T. R. and Liberati, P. 1995. Inflation pressure and dynamic load effects on soil deformation and soil-tire interface stresses. *Transactions of the ASAE* 38(3), 685-689.

Raper, R. L. and Kirby, J. M. 2006. Soil compaction: how to do it, undo it, or avoid doing It. ASAE distinguished Lecture No: 30. Agricultural Equipment Technology Conference, 12-14 February 2006. KY: Louisville. ASABE Publication Number 913C0106, pp. 1-14.

Rasmussen, K. J. 1999. Impact of ploughless soil tillage on yield and soil quality: a Scandinavian review. *Soil and Tillage Research* 53(1), 3-14.

Ray, D. K., Mueller, N. D., West, P. C. and Foley, J. A. 2013. Yield trends are insufficient to double global crop production by 2050. *PLoS ONE* 8(6), e66428.

Rhoton, F. E. 2000. Influence of time on soil response to no-till practices. *Soil Science Society of America Journal* 64(2), 700-709.

Rickson, R. J. 2014. Can control of soil erosion mitigate water pollution by sediments? *Science of the Total Environment* 468-469, 1187-1197.

Rieger, S., Richner, W., Streit, B., Frossard, E. and Liedgens, M. 2008. Growth, yield, and yield components of winter wheat and the effects of tillage intensity, preceding crops, and N fertilisation. *European Journal of Agronomy*, 28(3), 405-411.

Ritchie, H. and Roser, M. 2019. *Crop Yields*, Published online at OurWorldInData.org. Retrieved from: 'https://ourworldindata.org/crop-yields' [Online Resource].

Rodgers, D., McPhee, J., Aird, P. and Corkrey, R. 2018. Soil arthropod responses to controlled traffic in vegetable production. *Soil and Tillage Research* 180, 154-163.

Rusu, T. 2005. The influence of minimum tillage systems upon the soil properties, yield and energy efficiency in some arable crops. *Journal of Central European Agriculture*, 6(3), 287–294.

Sasal, M. C., Andriulo, A. E. and Taboada, M. A. 2006. Soil porosity characteristics and water movement under zero tillage in silty soils in Argentinian pampas. *Soil and Tillage Research* 87(1), 9–18.

Schjønning, P., van den Akker, J. J. H., Keller, T., Greve, M. H., Lamandé, M., Simojoki, A., Stettler, M., Arvidsson, J. and Breuning-Madsen, H. 2015. Chapter five - driver-pressure-state-impact-response (DPSIR) analysis and risk assessment for soil Compaction–A European perspective. *Advances in Agronomy,* 133, 183–237.

Schjønning, P., Lamandé, M., Munkholm, L. J., Lyngvig, H. S. and Nielsen, J. A. 2016. Soil precompression stress, penetration resistance and crop yields in relation to differently-trafficked, temperate-region sandy loam soils. *Soil and Tillage Research* 163, 298–308.

Shaheb, R. 2019. A study on the effect of tyre inflation pressure on soil properties, growth and yield of maize and soybean in central Illinois. PhD Thesis.

Shipitalo, M. J. and Edwards, W. M. 1998. Runoff and erosion control with conservation tillage and reduced-input practices on cropped watersheds. *Soil and Tillage Research* 46(1–2), 1–12.

Seehusen, T., Riley, H., Riggert, R., Fleige, H., Børresen, T., Horn, R. and Zink, A. 2014. Traffic-induced soil compaction during manure spreading in spring in south-east Norway. *Acta Agriculturae Scandinavica, Section B-Soil and Plant Science*, 64(3), 220–234.

Silburn, D. M. and Glanville, S. F. 2002. Management practices for control of runoff losses from cotton furrows under storm rainfall. I. runoff and sediment on a black vertosol. *Soil Research* 40(1), 1–20.

Singh, V. K., Yadvinder-Singh, Dwivedi, B. S., Singh, S. K., Majumdar, K., Jat, M. L., Mishra, R. P. and Rani, M. 2016. Soil physical properties, yield trends and economics after five years of conservation agriculture-based rice-maize system in north-western India. *Soil and Tillage Research* 155, 133–148.

Slawinski, C., Cymerman, J., Witkowska-Walczak, B. and Lamorski, K. 2012. Impact of diverse tillage on soil moisture dynamics. *International Agrophysics* 26(3), 301-309.

Smith, E. K. 2017. The effect of agricultural traffic and tillage on soil physical properties and crop yields. PhD Thesis. Newport: Harper Adams University.

Smith, V. L. 2016. The effect of tillage depth and compaction level on earthworm density. BSc Dissertation. Newport: Harper Adams University.

Soane, B. D., Blackwell, P. S., Dickson, J. W. and Painter, D. J. 1980. Compaction by agricultural vehicles: a review II. compaction under tyres and other running gear. *Soil and Tillage Research* 1, 373–400.

Soane, G. C., Godwin, R. J. and Spoor, G. 1986. Influence of deep loosening techniques and subsequent wheel traffic on soil structure. *Soil and Tillage Research* 8, 231–237.

Soane, B. D. and van Ouwerkerk, C. 1994. *Soil Compaction in Crop Production*. Amsterdam: Elsevier.

Sohne, W. 1958. Fundamentals of pressure distribution and soil compaction under tractor tires. *Agricultural Engineering,* 39, 290.

Sommer, C. and Zach, M. 1992. Managing traffic-induced soil compaction by using conservation tillage. *Soil and Tillage Research*, 24(4), pp. 319–336.

Spoor, G. and Voorhees, W. B. 1986. Deep tillage. *Soil and Tillage Research* 8(C), 317–319. https://doi.org/10.1016/0167-1987(86)90345-4.

Spoor, G., Tijink, F. G. J. and Weisskopf, P. 2003. Subsoil compaction: risk, avoidance, identification and alleviation. *Soil and Tillage Research* 73(1-2), 175-182.

SSSA 2008. Glossary of soil science terms. Available at: https://www.soils.org/publications/soils-glossary/. Soil Science Society of America.

Stranks, S. N. 2006. *The Effects of Tyre Systems on the Depth and Severity of Compaction*. In: Soane, B. D. and van Ouwerkerk, C. (Eds), Soil Compaction in Crop Production. Elsevier, Amsterdam, pp. 521-537

Strudley, M. W., Green, T. R. and Ascough, J. 2008. Tillage effects on soil hydraulic properties in space and time: state of the science. *Soil and Tillage Research* 99(1), 4-48.

Sun, H., Koal, P., Liu, D., Gerl, G., Schroll, R., Gattinger, A., Joergensen, R. G. and Munch, J. C. 2016. Soil microbial community and microbial residues respond positively to minimum tillage under organic farming in Southern Germany. *Applied Soil Ecology* 108, 16-24.

Sun, R., Li, W., Dong, W., Tian, Y., Hu, C. and Liu, B. 2018. Tillage changes vertical distribution of soil bacterial and fungal communities. *Frontiers in Microbiology* 9, 699.

Taboada, M. A., Micucci, F. G., Cosentino, D. J. and Lavado, R. S. 1998. Comparison of compaction induced by conventional and zero tillage in two soils of the rolling pampa of Argentina. *Soil and Tillage Research* 49(1-2), 57-63.

Taylor, J. H. 1994. Chapter 22: development and benefits of vehicle gantries and controlled-traffic systems. In: Soane, B. D. and van Ouwerkerk, C. (Eds), *Soil Compaction in Crop Production*. Elsevier, pp. 521-537.

Taylor, H.M. and Gardner, H. R. 1963. Penetration of cotton seedling taproots as influenced by bulk density, moisture content, and strength of soil. *Soil Science*, 96(3), 153-156.

Thompson, K. A., Bent, E., James, K., Carlyle, C. N., Quideau, S. and Bork, E. W. 2020. Access mats partially mitigate direct traffic impacts on soil microbial communities in temperate grasslands. *Applied Soil Ecology* 145, 103353.

Tijink, F. G. J., Döll, H. and Vermeulen, G. D. 1995. Technical and economic feasibility of low ground pressure running gear. *Soil and Tillage Research* 35(1-2), 99-110.

Tullberg, J. N. 2010. Tillage, traffic and sustainability–a challenge for ISTRO. *Soil and Tillage Research* 111(1), 26-32.

Tullberg, J. N. 2000. Wheel traffic effects on tillage draught. *Journal of Agricultural Engineering Research* 75(4), 375-382. https://doi.org/10.1006/jaer.1999.0516.

Tullberg, J. N., Antille, D. L., Bluett, C., Eberhard, J. and Scheer, C. 2018. Controlled traffic farming effects on soil emissions of nitrous oxide and methane. *Soil and Tillage Research* 176, 18-25.

Tullberg, J. N., Yule, D. F. and McGarry, D. 2007. Controlled traffic farming–from research to adoption in Australia. *Soil and Tillage Research* 97(2), 272-281.

Udawatta, R. P. and Anderson, S. H. 2008. CT-measured pore characteristics of surface and subsurface soils influenced by agroforestry and grass buffers. *Geoderma* 145(3-4), 381-389.

Unger, P. W. 1996. Soil bulk density, penetration resistance, and hydraulic conductivity under controlled traffic conditions. *Soil and Tillage Research*, 37(1), 67-75.

Verhulst, N., Govaerts, B., Verachtert, E., Castellanos-Navarrete, A., Mezzalama, M., Wall, P., Deckers, J. and Sayre, K. D. 2010. Conservation agriculture, improving soil quality for sustainable production systems. *Advances in Soil Science*, Food security and soil quality. CRC Press, Boca Raton, FL, USA, 137-208.

Villarreal, R., Lozano, L. A., Salazar, M. P., Bellora, G. L., Melani, E. M., Polich, N. and Soracco, C. G. 2020. Pore system configuration and hydraulic properties. Temporal variation during the crop cycle in different soil types of Argentinean pampas region. *Soil and Tillage Research* 198, 104528.

Warner, D., Stobart, R., Morris, N., Tzilivakis, J., Green, A. and Lewis, K. 2016. Crop specific implications of yield and energy use efficiency in non-inversion tillage systems, Paper presented at Crop Protection in Northern Britain 2016, Dundee, United Kingdom, 23/02/16 - 24/02/16 pp. 17-22.

Wuest, S. B. 2010. Tillage depth and timing effects on soil water profiles in two semiarid soils. *Soil Science Society of America Journal* 74(5), 1701-1711. https://doi.org/10.2136/sssaj2010.0046.

Wright, A. L., Hons, F. M., Lemon, R. G., McFarland, M. L. and Nichols, R. L. 2008. Microbial activity and soil C sequestration for reduced and conventional tillage cotton. Applied Soil Ecology 38(2), 168-173.

Chapter 6

Assessing the effects of no-till cultivation practices on soil health

Alison Hamm and Daniel K. Manter, USDA-ARS, USA

1 Introduction

Tillage has been used for thousands of years to prepare the seed bed, mix nutrients, and control weeds in crops. However, tillage can reduce soil health by altering its physical, chemical, and biological properties. No-till (NT) is 'a conservation farming system in which seeds are placed into otherwise untilled soil by opening a narrow slot, trench, or hole of only sufficient width and depth to obtain proper seed placement and coverage', resulting in minimal soil disturbance (Derpsch et al., 2011). In the United States, NT adoption increased in reaction to the Dust Bowl of the 1930s and the desire to prevent wind (Fig. 1) and water soil erosion. Since then, NT has increased in popularity due to its potential to reduce farm inputs, especially labor and fuel, and improve soil health (Karlen et al., 1994, Lal et al., 2007, Zhang et al., 2007, Pittelkow et al., 2015). In 2017, 104 million acres, or 26% of cropland in the United States, was managed under NT with an additional 98 million acres in reduced tillage (USDA, 2017).

In addition to its success in controlling erosion, NT has clear benefits for conserving water (Fig. 2) and reducing energy costs (Grandy et al., 2006, Obour et al., 2017, Lewis et al., 2018). In comparison to conventional tillage (CT), NT can double water infiltration rate and increase soil water content during sowing and flowering periods (Alvarez and Steinbach, 2009). Absence of tilling practices also reduces fuel and labor costs (Soane et al., 2012). However, challenges

http://dx.doi.org/10.19103/AS.2021.0094.07
Published by Burleigh Dodds Science Publishing Limited, 2023.

Figure 1 Wind erosion in a conventional till farm. Photo source: https://www.nrcs.usda.gov/wps/portal/nrcs/detail/national/about/history/?cid=nrcs143_021392.

such as weed management (Dang et al., 2015), soil compaction (Moreno et al., 2010, Blanco-Canqui and Ruis, 2018), and other potentially unwanted changes in soil health (e.g. nutrient leaching and stratification) may arise due to NT. Stacking conservation practices such as crop rotations and cover cropping can mitigate such changes, and can help increase crop yields relative to CT (Unger and McCalla, 1980). These challenges are often seen in systems lacking in one or more of the three pillars of Conservation Agriculture: reduced disturbance, permanent soil cover, and diversified crop rotation. The combination of these practices can mitigate negative effects of NT and help increase crop yields. Further information on Conservation Agriculture can be found in a recent three-volume review 'Advances in Conservation Agriculture' (Kassam, 2020a, Kassam, 2020b, Kassam, 2022).

It is important to note that while we focus mainly on the role of NT in this chapter, tillage methods are varied and can result in a wide range of disturbance depending upon the type of operation, its frequency, and depth. The moldboard plow, also known as inversion tillage, was the standard method until the 1940s when its effect on soil health came into question (Faulkner, 1943). Its use is still prevalent today and is almost always the implement used in long-term NT versus CT studies. Alternative implements with less soil disturbance, including chisel and disc plows, have increased in popularity alongside other conservation agriculture practices.

Figure 2 Side-by-side comparison of water infiltration differences in no-till (left) and conventional tillage (right). Photo source: Courtesy of Neil Sass (Decorah, IA).

Tillage operations are defined as follows:

- No-till: From harvest to planting, soil is left undisturbed. Planting or drilling is done with disc openers, coulter(s), row cleaners, in-row chisels, or rototillers. Weed control includes crop protection products (herbicides), but minimal cultivation can be used for emergencies.
- Strip-till: Same as NT, except strips up to one-third of the row width are cultivated including residue and/or soil disturbance.
- Reduced-till: One or more tillage passes disturbing all of the soil surface, prior to and/or during planting. After planting, there is 15–30% of residue cover. Weed control includes crop protection products and/or soil cultivation.
- Conventional-till: Multiple tillage passes disturbing all of the soil surface, prior to and/or during planting. Less than 15% residue cover remains after planting. Weed control includes crop protection products and/or soil cultivation. Often uses a moldboard plow implement for inversion tillage that flips over the top 6–12″ of soil.

The goal of this chapter is to assess the effect of NT on soil health. However, it is important to note that the concept of soil health is relatively ambiguous and remains poorly defined. It may be best defined as 'the vitality of a soil in sustaining the socio-ecological functions of its enfolding lands' (Janzen et al.,

2021). Definitions of soil health are context dependent, differing according to the objectives and scale being considered (Bennett et al., 2010, Bouma and McBratney, 2013, Norris et al., 2020), and may differ across locations (Ng and Zhang, 2019). However, most definitions define soil health in terms of desirable physical properties, chemical composition, and type/degree of biological activity. As a result, we focus our discussion on the effects of NT on soil health by examining its impact on the physical, chemical, and biological properties generally accepted as promoting the function and vitality of soil, and which are often tightly interconnected. Table 1 summarizes some of the known changes in key soil health indicators in response to NT in the top 0–15 cm of soil.

Table 1 Common soil health indicators and their trends with NT versus CT

	Indicator	Effect under NT	References
Physical	Aggregate stability	↑	Andruschkewitsch et al. (2014); Aziz et al. (2013)
	Bulk density	↓↑ –	Blanco-Canqui et al. (2017); Dam et al. (2005); Fiorini et al. (2018); Sekaran et al. (2021)
	Compaction	↓↑	Blanco-Canqui and Ruis (2018); Moreno et al. (2010)
Chemical	Soil organic carbon	↑	Aziz et al. (2013); Chellappa et al. (2021); Sekaran et al. (2021)
	Total nitrogen	↑	Aziz et al. (2013); Gál et al. (2007)
	Nitrate leaching and runoff	↑	Blanco-Canqui and Ruis (2018); Daryanto et al. (2017)
	Nutrient stratification	↑	Angers et al. (1997); Franzluebbers and Hons (1996); Garcia et al. (2007)
Biological	Glomalin-related soil protein	↑	González-Chávez et al. (2010); Sekaran et al. (2021)
	Phospholipid fatty acids (PLFA)	↑	Feng et al. (2003); Helgason et al. (2010)
	Fatty acid methyl ester (FAME)	↑	Drijber et al. (2000); González-Chávez et al. (2010)
	Soil enzymes: β-glucosidase (C cycling); β-glucosaminidase, urease, arylamidase (N cycling); phosphodiesterase, alkaline phosphatase (P cycling); arylsulphatase (S cycling)	↑	Bergstrom et al. (1998); Deng and Tabatabai (1997); Mbuthia et al. (2015); Sekaran et al. (2021)

2 Soil physical properties

The physical structure of soil is a key component of soil health, influencing both chemical and biological conditions. Changes in soil physical properties are highly dependent on time and management. While it is generally agreed upon that NT improves soil physical health by increasing aggregate stability and porosity, there are mixed results concerning compaction and bulk density (BD), especially during the first few years after adopting NT. Building soil physical structure takes time, with the first 3–6 years being a transitional period rather than true NT (Grigar et al., 2018). Decaying roots and earthworms create vertical biopores after several years of adoption of NT, and aggregate stabilization can take decades. NT also results in lower soil temperatures which can have positive and negative effects on plant growth.

One frequent concern regarding NT adoption is soil compaction. Without mechanically breaking up the topsoil, compaction will occur through natural forces (rain, gravity) and mechanical force from farm equipment (Moreno et al., 2010). Soil compaction results in poor seed emergence, decreased root growth, and overall decreased yields (Unger and Kaspar, 1994). Tilling will temporarily alleviate compaction and mix in fertilizers for optimum plant germination. However, tilling breaks down aggregates into finer soil particles more prone to compaction, necessitating the need for more tilling which can then lead to further compaction (Hamza and Anderson, 2005, Batey, 2009).

Soil compaction is often defined as an increase in BD and decrease in porosity and penetration resistance (PR). The effects of NT on these indicators are mixed in the literature, and often confounded by other variables such as climate, time under NT, machinery traffic, etc. (Dam et al., 2005, Moreno et al., 2010, Blanco-Canqui et al., 2017, Sekaran et al., 2021). In 62 studies reviewed by Blanco-Canqui and Ruis (2018), NT either increased, decreased, or did not change BD and PR in the topsoil. Such varied results are attributed to study differences, especially length in time under NT, since BD can increase during early transition years (Blanco-Canqui and Ruis, 2018). In a review of NT agriculture in Europe, the authors conclude NT generally increases BD in the short term; however, its effects are diminished by increased vertical porosity and earthworm abundance (Soane et al., 2012). Soils under long-term NT can have twice as many pores with higher microporosity and pore connectivity (Galdos et al., 2019), supporting higher root density despite a higher measured BD (Fiorini et al., 2018).

It is widely agreed that NT increases aggregate size and stability which, in turn, helps lower BD and compaction (Sekaran et al., 2021). When soil is left undisturbed, macroaggregates and microaggregates are under a dynamic equilibrium of balanced mineralization and formation, enhancing soil physical structure (Andruschkewitsch et al., 2014). The distribution of aggregates, particularly macroaggregates, creates macropores essential for root growth, water and nutrient transmission and storage, and gas exchange (Lipiec and Stępniewski, 1995). Microaggregates within larger macroaggregates house mesopores and micropores important for water, soil organic carbon (SOC), and nutrient storage. Soils left undisturbed under NT retain aggregate and pore structure providing growing roots with space, water, and nutrients to grow (Voorhees and Lindstrom, 1984).

Aggregates in undisturbed topsoil under NT have a higher SOC content, important for maintaining aggregate stability and the soil physical structure (Chellappa et al., 2021). In a long-term (>19 years) moldboard plow CT system compared to NT in three sites in the Central Great Plains loam/silt loam, 0–5 cm depth, soils showed that long-term NT had significantly lower BD, higher critical water content (CWC), and more SOC at all sites. Accumulation of SOC in stable aggregates helps decrease BD and increase CWC enabling soils to resist compaction. The authors conclude 'NT soils can be trafficked at relatively higher water contents and have less susceptibility to compaction than tilled soils at the same water content'. In fact, plowed soils at water contents below the CWC were more susceptible to compaction due to lack of stable aggregates and the macropores and micropores they form (Blanco-Canqui et al., 2009). These findings are supported by Sekaran et al. (2021), who measured increased SOC, reduced BD, higher glomalin-related soil protein (GRSP), higher soil water retention, increased proportion of fine and coarse mesopores, and higher proportion of large >2 mm aggregates. These differences were more pronounced in silty clay loam versus silt loam, and reached statistical significance in soils under NT for 25 years, but not always in soils under NT for only 7 years, demonstrating that improvements in soil health may take a decade or more (Sekaran et al., 2021).

In general, the effect of NT on moderating the soil temperature is beneficial to soil health. NT is associated with the retention of crop residues that have a high solar reflectivity and thermal conductivity compared to soil (Van Wijk et al., 1959). NT crop residues thus act as an insulating layer which moderates fluctuations in temperature, influencing the soil energy balance and reducing average temperature compared to CT (Shen et al., 2018). The lack of an insulating layer from crop residues under CT also exposes soils to freeze-thaw and wet-dry cycles that can further result in the loss of macroaggregates and increase in nutrient mineralization (Six et al., 2004). However, concern is often raised regarding its potential impact on delaying seed germination and reducing crop yields. Several studies suggest that cooler soil temperatures at planting in NT systems in the arid west of the United States can delay early crop development and final yield (Swan et al., 1987, Sims et al., 1998, Halvorson et al., 2006). However, research from more humid areas suggests that any potential delays in crop development did not have a detrimental effect on crop yield (Mehdi et al., 1999, Wolfe and Eckert, 1999). A global meta-analysis of 610 studies (5463 paired yield observations) showed NT management from diverse cropping systems had equivalent yields for oilseeds, cotton, and legume crops but reduced yields in cereals, horticultural, and root crops. Except for maize, no negative effect on crop yield was observed for NT systems that were 10+ years old (Pittelkow et al., 2015).

Soil physical processes are slow, but careful management strategies can help mitigate short-term compaction while enhancing aggregate formation and stability. Stacking conservation practices such as crop rotations and cover crops help improve the soil physical structure through root penetration and creation of biopores. Intermittent tillage is also effective at mixing the soil and lowering BD, with limited detriment to aggregate stability. In addition, crop residues help buffer soil temperatures from extremes that can influence the physical, chemical, and biological health of soils.

3 Soil chemical properties

NT increases nutrient stocks, particularly SOC and total nitrogen (TN), partly through being stabilized by the physical and biological properties of soil associated with NT compared to CT. Aggregates store nutrients as organic matter, organo-mineral complexes, and microbial biomass (Rice and Smith, 1984, Lynch and Bragg, 1985, Karami et al., 2012). Tilling results in aggregate destabilization and mineralization, resulting in long-term depletion of nutrient stocks, particularly SOC, as well as microbial enzymes involved in nutrient cycling which promote nutrient uptake by plants (Piazza et al., 2020).

SOC is a key component of physical, chemical, and biological soil functions and a common soil health indicator measured by farmers and researchers alike. SOC improves soil health by increasing aggregate turnover, water and nutrient retention, and microbial growth (Islam and Weil, 2000, Six et al., 2000, Bronick and Lal, 2005). SOC is protected in aggregates as microbial biomass and microaggregates as organic matter (Jastrow et al., 1996, Six et al., 1998). In turn, SOC helps stabilize aggregates, reducing their mineralization and nutrient loss (Gál et al., 2007).

Protection of aggregate stability under NT systems is thought to contribute to C sequestration, supported by consistent findings of increased SOC in the topsoil of NT versus CT in long-term research plots (Aziz et al., 2013, Chellappa et al., 2021, Sekaran et al., 2021). However, most studies only sampled topsoil and subsoil (<30 cm) whilst the CT system used for comparison was the inversion (moldboard) plow (Gál et al., 2007, Blanco-Canqui et al., 2011, Blanco-Canqui et al., 2021). Depending on plow type and depth, tillage may support deeper root growth and therefore carbon deposition at deeper depths for long-term storage (Baker et al., 2007). More recently, in a long-term >34-year study, Blanco-Canqui et al. (2021) found that, with deep sampling to 100 cm, NT increased SOC in the whole soil profile compared to use of a moldboard plow, but no difference was seen with a disc or chisel plow (Blanco-Canqui et al., 2021).

However, without the mixing of soil associated with CT, nutrient stratification can occur in NT and limit plant growth (Angers et al., 1997, Garcia et al., 2007). Tillage can increase nutrient availability in the short term by mixing soil and speeding up the decomposition of crop residues and releasing stored C, N, macronutrients, and micronutrients from aggregates (Blanco-Canqui and Wortmann, 2020). Depending on the depth type of plow used, tilling also mixes plant residues and nutrients deeper into the soil profile. Under NT, nutrients, including N, P, K, Zn, Fe Mn, and Cu, accumulate at the surface along with lower pH (Franzluebbers and Hons, 1996, Holanda et al., 1998, Chellappa et al., 2021). Comparing tillage system effects on C and N stocks in the 0–60 cm profile, Angers et al. (1997) found that NT and use of a disc plow had higher C and N in the top 0–10 cm of soil, while the reverse was true in moldboard plow. However, there was no difference in the cumulative C and N along the whole 0–60 cm profile (Angers et al., 1997). However, accumulation of nutrients in the topsoil under NT may lessen fertilizer requirements. In a 48-year NT versus moldboard plow study on well-drained soils, NT improved N stocks, mineralization, and efficient fertilizer use: NT yielded 12% higher than CT despite equal N inputs (Feng et al., 2003).

The improved physical characteristics of soil under NT can also reduce the loss of applied fertilizer through runoff and leaching (Zhang et al., 2007, Daryanto et al., 2017). Up to 70% of nitrogen fertilizers can be lost due to volatilization, denitrification, runoff, and leaching (Malhi et al., 2001). Other fertilizers such as phosphorus are prone to loss through runoff and leaching. Nutrient leaching through the subsurface can occur under NT due to its increased SOC and aggregate stability, especially due to undisturbed macropores (Daryanto et al., 2017).

Cover crops, intermittent tillage, and modified fertilizer application rates and timing can help improve nutritional stocks of soils while limiting nutrient losses to the environment (Soane et al., 2012, Frasier et al., 2017). Injection of nitrate fertilizer below the surface can reduce runoff, increasing plant N and yields (Mengel et al., 1982). Similar results are seen with P where injected P had 33% less dissolved reactive P lost to runoff compared to P broadcast on the soil surface (Carver et al., 2022). Reduced rates and split applications can also mitigate fertilizer runoff and leaching, as can use of cover crops (Blanco-Canqui and Ruis, 2018). Cover crops improve soil structure, fill vertical macropores, and decrease sediment loss, reducing mean annual dissolved nutrient losses up to 77% (Zhu et al., 1989). Intermittent tilling can also mitigate nutrient losses by disrupting vertical macropores (Daryanto et al., 2017).

In summary, NT increases SOC and TN which, in turn, support the physical and biological health of soils. However, nutrient stratification can occur, potentially reducing availability of nutrients to plant roots. Applied fertilizers are subject to losses from leaching and runoff where soils have high SOC, aggregate stability, and high porosity. Cover crops and intermittent tillage can help reduce nutrient stratification, disrupt vertical biopores and lower BD to reduce losses and improve soil chemical health.

4 Soil biological properties

Changes to the physical and chemical environment of soils alter the biology of soils which then, in turn, influence soil processes (Schimel, 1995). Soil organisms from bacteria to earthworms serve important soil functions such as nutrient storage and cycling, physical aggregation, and plant stress tolerance. Earthworm, bacterial, and fungal biomass is particularly quick to change in response to environmental changes and are often used as soil health indicators (Deng and Tabatabai, 1997, Aziz et al., 2013, Melman et al., 2019). The microbial biomass itself is also an important source and sink of nutrients and provides nutrients at least one order of magnitude faster than the remaining organic matter (Dalal, 1998). This section of the chapter looks at the effects of CT on macro- and microorganisms such as bacteria and fungi, comparing these with the effects of NT on soil biological activity.

Tilling physically disrupts the soil structure, incorporates plant matter, and introduces oxygen. These physical and chemical habitat disruptions result in a decrease in overall macro- and microorganism abundance and biodiversity. It has been shown that tillage lowers the abundance of soil macroorganisms: invertebrates such as earthworms, termites, and ants (Karlen et al., 1994). Macroorganisms are often referred to as 'soil engineers', influencing physical, chemical, and biological properties of soils through burrowing and translocation of surface organic matter (Bottinelli et al., 2015). Invertebrate burrows (biopores) increase gas exchange and oxygen availability at lower depths (Uteau et al., 2013), increase water in the subsurface for future crops (Gaiser et al., 2012), increase nutrient availability directly through invertebrate excretions and translocation of organic matter from the surface to deeper depths, and provide niche habitats for beneficial bacteria and fungi (Lavelle, 1988, Tiunov and Scheu, 1999). In an inoculation study in barley, earthworms resulted in more root biomass and higher crop yields, attributed to the ability of roots to grow in biopore spaces lined with a higher concentration of nutrients than the surrounding soil (Edwards and Lofty, 1980). Left undisturbed, biopores are re-used by macroorganisms (and roots), further increasing their abundance over time (Pagenkemper et al., 2015). Comparing NT versus CT with or without residue retention, earthworm abundance was five times greater in NT with residue, and correlated with increase in water infiltration rate and aggregate stability (Melman et al., 2019).

The loss of pore structures and aggregates associated with tillage also disrupts microhabitats for fungi and bacteria, resulting in lower microbial diversity (Lupwayi et al., 1998, Gupta and Germida, 2015) which has been suggested to be associated with soil resilience and plant resistance to stress (Van Der Heijden et al., 1998, Kennedy, 1999). Microbial diversity is important for community structure and function and their roles in maintaining the chemical and physical properties of soil. Increased microbial biomass and diversity broadens the scope of functions provided to the soil, such as physical scaffolding and enzyme production for nutrient cycling (Wagg et al., 2014, Reicosky, 2018). Bacteria also secrete exopolysaccharides that act as glues, holding together soil particles forming microaggregates (Sandhya and Ali, 2015). After tillage, the bacterial community shifts towards fast-growing 'competitors', different from slow growing 'stress tolerators' (Schmidt et al., 2018). Increased abundance of aerobic bacteria with high metabolic rates (Kim et al., 2020) break down roots organic matter and fungal hyphae (Pankhurst et al., 2002). AMF fungi are root symbionts with many beneficial functions, particularly pathogen resistance (Wehner et al., 2010) and nutrient acquisition (Veresoglou et al., 2012). AMF also helps structure the soil by hyphal growth and production of the glycoprotein glomalin which contributes to the formation of aggregates (Wright and Upadhyaya, 1996, Piotrowski et al., 2004). During tilling, fungal hyphae are mechanically broken, disrupting growth and disentangling the hyphal network which holds soil aggregates together, breaking them down further (Beare et al., 1997) AMF biomass, as measured by glomalin-related soil protein (GRSP), is consistently shown to be lower under CT versus NT (González-Chávez et al., 2010, Singh et al., 2013), with a decreased diversity of species (Brito et al., 2012). Even a single tilling event significantly changes the fungal community structure. Two years after a single tillage in a NT system, Wortmann et al. (2008) measured 22% less AMF and 6% more non-AMF fungi (Wortmann et al., 2008), not only lowering overall biomass but community structure as well.

In contrast to the negative effects of CT, research suggests that NT preserves and promotes soil biological activity. Abundance and diversity of macro- and microfauna are consistently lower in CT compared to NT, which has implications for the chemical and physical health of soil. Other conservation practices such as cover crops (Korucu et al., 2018) and crop diversification (Rodríguez et al., 2020) help increase microorganism biomass and diversity under NT systems.

5 Case studies

5.1 Case study 1 (Aziz et al., 2013)

Soil scientists use physical, chemical, and biological indicators to quantify soil health in response to different types of farm managements. Aziz et al. (2013)

used such indicators to measure the effects of 5 years of NT or CT used in continuous corn cultivation in silt loam soil. At the start and end of 5 years, soil biological (SBQ), chemical (SCQ), and physical measurements (SPQ) were calculated into a single soil quality index (SQI). After 5 years of NT, topsoil showed statistically significant higher SBQ, SCQ, and SPQ scores and, therefore, an improved overall SQI score. After 5 years of CT, scores neither improved nor declined. Similar results were found by Karlen et al. (1994) in 12-year NT versus CT plots on silt loam soils with NT having a higher SQI score than plow or chisel tillage (Karlen et al., 1994).

Soil biological, chemical, and physical indicators were calculated for quality scores, as summarized in Table 2. The SCQ was 10% higher under NT than CT with total carbon (TC), active carbon (AC), and total nitrogen (TN) increasing under NT but not under CT. The rate of change in SCQ under NT significantly increased 2.1% per year, whilst it decreased 0.4% per year under CT. SPQ was significantly higher (5%) in NT versus CT. Aggregate stability (AS) and particulate organic matter (POM) were significantly higher under NT. This is likely due to lack of physical disturbance to the soil, but more likely

Table 2 Significant changes in biological, chemical, and physical soil quality indicators in soil under NT or CT for 5 years

	Indicator	CT over 5 years	NT over 5 years	NT versus CT
Biological	Microbial biomass (C_{mic})	–	↑	↑
	Basal respiration (BR)	↑	↑	–
	Microbial metabolic quotient (qCO_2)	↑	↓	↓
	Soil Biological Quality Index (SBQ)	–	↑	↑
Chemical	Total carbon (TC)	–	↑	↑
	Active carbon (AC)	–	↑	↑
	Total nitrogen (TN)	–	↑	–
	Soil Chemical Quality Index (SCQ)	–	↑	↑
Physical	Total porosity (f_t)	–	–	–
	Aggregate stability (AS)	–	↑	↑
	Particulate organic matter (POM)	–	↑	↑
	Soil Physical Quality Index (SPQ)	–	↑	↑
Total	Soil Quality Index (SQI)	–	↑	↑

Source: Adapted from Aziz et al. (2013).

due to the increase in available labile C for microbial growth and production of organic binding agents, as well as slow microbial decomposition of protected POM. SBQ was 13% higher for NT versus CT. NT had a lower microbial metabolic quotient (qCO_2), indicating less carbon was used for respiration and decomposition and more carbon used for microbial growth and nutrient cycling. The authors state that 'an accumulation of labile C through enhanced microbial assimilation may be one of the most important and early mechanisms by which NT improved soil quality functions' and that biology is a 'sensitive and consistent indicator of soil quality in response to management practices'.

Significant changes in soil indicators were observed in just 5 years, a relatively short period to study effects of farm management on soil health. Comparing long term (>25 years) and short term (<7 years) tillage on three sites, Sekaran et al. (2021) found that topsoil aggregation, SOC, GRSP, soil water retention, and porosity improved under NT over 4–6 years but not significantly until after 7 years. Improvements were more pronounced in silty clay loam versus silt loam, highlighting the influence of soil type in response to management (Sekaran et al., 2021).

5.2 Case study 2 (Nunes et al., 2018)

Soil type is an important and often overlooked factor in tillage research (Derpsch et al., 2014). In a long-term study comparing three soil types, Nunes et al. (2018) examined the impact of long-term (20+ years) continuous NT in combination with cover crops on soil properties and corn yields on three different soils. A comprehensive group of soil health indicators was measured including:

- four biological indicators (soil organic matter, active carbon, respiration, and protein);
- four physiological indicators (available water capacity, water aggregation, penetration resistance, and water infiltration rate);
- three chemical indicators (plant available nutrients, pH, and TN).

Overall, NT performed best across these chemical, biological, and physical soil quality indicators. However, of the 16 measured soil properties, those influenced by tillage varied in degree depending on soil type, with a lesser effect in clay loam soil (Table 3). Corn yields in clay loam remained unchanged but were significantly higher in the loamy sand and silt loam soils. Soil biological indicators and wet aggregate stability positively correlated with

Table 3 Significant percent increases in soil health indicators vary by soil type

	Muskellunge clay loam	Honeoye-Lima silt loam	Stafford loamy fine sand
Organic matter	17%	15%	67%
Active carbon	-	29%	51%
Respiration	17%	33%	82%
Protein	65%	26%	49%
pH	-	-	-
Total nitrogen	-	17%	69%
P	95%	112%	-
K	-	12%	45%
Zn	66%	-	92%
Mg	-	-	-
Fe	-	-	-
Mn	-	-	-
Available water capacity	-	-	-
Water stable aggregates	76%	155%	92%
Penetration resistance (0-15 cm)	-	-	-
Water infiltration rate	-	67%	-

Source: Adapted from Nunes et al. (2018).

yield, especially for the silt loam site under CT and cover cropping. Chemical indicators that correlated with yield included TN, K, and P. In a similar 7-year study comparing five soil types, Fiorini et al. (2020) concluded improvements in soil quality indices attributed to aggregate stability was less pronounced where the SOC, clay, and silt fractions were high and crop residues low (Fiorini et al., 2020).

Nunes et al. also found that cover cropping improved soil health indicators in both CT and NT. Over the 4 years of the study, no cover crop (NC) or a grass-legume cover crop (CC) was inter-seeded with corn on the silt loam site. The soil health indicators (organic matter, microbial respiration, available water capacity, and water infiltration rate increased with the treatments ranking: CT-NC < CT-CC < NT-NC < NT-CC. Although yields declined the first year, presumably due to mechanical damage during inter-seeding, over the next 3 years cover cropping increased yields in both CT and NT to the same degree. Comparing plots in NT since 1962, yields are increased under long-term NT with cover crops compared to CT (Islam and Reeder, 2014).

6 Future trends and conclusion

Recent review articles have been quick to point out studies comparing CT and NT can miss important variables, making comparison difficult. A particular issue is that sampling depth varies across studies, making it difficult to draw conclusions; shallow sampling depths can miss important trends, such as changes in SOC stocks and nutrient stratification (Karlen et al., 1994, Gál et al., 2007, Aziz et al., 2013, Blanco-Canqui and Ruis, 2018, Blanco-Canqui et al., 2021). Shallow sampling means that conclusions drawn about carbon sequestration under NT as a method to mitigate climate change are controversial (Six et al., 2000, Lal, 2004, Gál et al., 2007, Luo et al., 2010, Ogle et al., 2012). To fully capture changes in soil health properties, sampling depths to at least 100 cm are recommended (Blanco-Canqui et al., 2021).

Over time, a single tillage operation (also known as strategic tillage or occasional tillage) can help alleviate compaction, incorporate crop residues, and control herbicide-resistant weeds without significantly impacting soil health or crop yields (Blanco-Canqui and Wortmann, 2020). Nunes et al. (2015) found that chiseling in NT reduced aggregate size in the 7–17 cm depth the first 6 months, but returned to pre-tillage levels after 12 months (Nunes et al., 2015). Incorporation of plant material in a single-tillage event can redistribute nutrients along the soil profile. Five years after a sub-till at 10 cm in a wheat-fallow system, Kettler et al. (2000) found reduction of SOC and TN at a depth of 0–7.5 cm, but an increase at 7.5–15 cm and, overall, no change in whole 0–30 cm profile. Weeds were also decreased 97% in the first crop, and 41% in the third, resulting in a 30% increase in wheat yields in the first crop, and 9% in the third crop (Kettler et al., 2000). In a similar cropping system of wheat-sorghum-fallow, after 6 years, a single 10-cm tillage did not affect crop yield, biomass, and available soil water, as compared to continuous NT (Schlegel et al., 2020).

Long-term NT research has provided the scientific community with a wealth of information, but methods are often outdated. For example, moldboard inversion plowing has largely been replaced by less invasive disc or chisel plow techniques. New and ongoing studies need to standardize methodologies including management practices and sampling depth, while including pertinent information such as soil type and climate data.

NT is a particularly effective method for controlling erosion and has been increasingly recognized as improving soil physical, chemical, and biological health without compromising yield. However, many challenges remain, including weed control, compaction, runoff, and leaching. Stacking conservation practices such as cover crops, crop rotations, and reduced and precise fertilizer applications can help mitigate challenges posed by NT as it develops further.

7 Where to look for further information

- USDA Census of Agriculture:
  - https://www.nass.usda.gov/Publications/AgCensus/2017.
  - Summary of agricultural activity for the United States and for each state, including number of farms by type and size, inventory and values for crops and livestock, producer characteristics, etc.
- Carbon Reduction Potential Evaluation Tool (CaRPE):
  - https://farmland.org/project/the-carpe-tool.
  - Mapping tool of the United States including current acres under conservation management with cover crops and tillage, as well as greenhouse gas potential, to the county level.
- USDA-NRCS soil health information homepage:
  - https://www.nrcs.usda.gov/wps/portal/nrcs/main/national/soils/health.
  - Educational materials on soil health, as well as soil health assessments and management guides.
- Bronick, C. J. and Lal, R. (2005). 'Soil structure and management: a review.' *Geoderma* 124(1-2): 3-22.
  - Review article overviewing the physical, chemical, and biological components of soil, including environmental impacts and management for enhanced soil structure.
- Janzen, H. H., Janzen, D. W. and Gregorich, E. G. (2021). 'The 'soil health' metaphor: Illuminating or illusory?' *Soil Biology and Biochemistry* 159: 108167.
  - Opinion paper challenging the definition of soil health as a metaphor for sustaining socioecological functions. The history, current trends, and future research pertaining to soil health is discussed.
- Blanco-Canqui, H. and Ruis, S. J. (2018). 'No-tillage and soil physical environment.' *Geoderma* 326: 164–200.
  - Review article on tillage and its effect on soil health indicators. Benefits of NT increase with time, and intermittent or single tillage is discussed.
- Hamza, M. and Anderson, W. (2005). 'Soil compaction in cropping systems: a review of the nature, causes and possible solutions.' *Soil and Tillage Research* 82(2): 121–145.
  - Review article on the nature of soil compaction and its detriment to crop productivity. Soil compaction is discussed to the molecular level in relation to soil health indicators. Ways to alleviate soil compaction are discussed.
- Pittelkow, C. M., Linquist, B. A., Lundy, M. E., Liang, X., Van Groenigen, K. J., Lee, J., Van Gestel, N., Six, J., Venterea, R. T. and Van Kessel, C. (2015).

'When does no-till yield more? A global meta-analysis.' *Field Crops Research* 183: 156–168.
  - Meta-analysis of 678 studies including 50 crops and 5005 paired observations. Yield losses or improvements under NT depend on edaphic and anthropologic factors.
- Reicosky, D. (ed). (2018). *Managing Soil Health for Sustainable Agriculture Volume 1: Fundamentals*, Burleigh Dodds Science Publishing, Cambridge, UK.
- Kassam, A. (ed). (2020). *Advances in Conservation Agriculture Volume 1: Systems and Science*, Burleigh Dodds Science Publishing, Cambridge, UK.
- Kassam, A. (ed). (2020). *Advances in Conservation Agriculture Volume 2: Practice and Benefits*, Burleigh Dodds Science Publishing, Cambridge, UK.
- Kassam, A. (ed). (2022). *Advances in Conservation Agriculture Volume 3: Adoption and Spread*, Burleigh Dodds Science Publishing, Cambridge, UK.

8 References

Alvarez, R. and Steinbach, H. S. (2009). A review of the effects of tillage systems on some soil physical properties, water content, nitrate availability and crops yield in the Argentine Pampas *Soil and Tillage Research* 104(1): 1–15.

Andruschkewitsch, R., Geisseler, D., Dultz, S., Joergensen, R. G. and Ludwig, B. (2014). Rate of soil-aggregate formation under different organic matter amendments–a short-term incubation experiment *Journal of Plant Nutrition and Soil Science* 177(2): 297–306.

Angers, D. A., Bolinder, M. A., Carter, M. R., Gregorich, E. G., Drury, C. F., Liang, B. C., Voroney, R. P., Simard, R. R., Donald, R. G., Beyaert, R. P. and Martel, J. (1997). Impact of tillage practices on organic carbon and nitrogen storage in cool, humid soils of eastern Canada *Soil and Tillage Research* 41(3–4): 191–201.

Aziz, I., Mahmood, T. and Islam, K. R. (2013). Effect of long term no-till and conventional tillage practices on soil quality *Soil and Tillage Research* 131: 28–35.

Baker, J. M., Ochsner, T. E., Venterea, R. T. and Griffis, T. J. (2007). Tillage and soil carbon sequestration–what do we really know? *Agriculture, Ecosystems and Environment* 118(1–4): 1–5.

Batey, T. (2009). Soil compaction and soil management–a review *Soil Use and Management* 25(4): 335–345.

Beare, M. H., Hu, S., Coleman, D. C. and Hendrix, P. F. (1997). Influences of mycelial fungi on soil aggregation and organic matter storage in conventional and no-tillage soils *Applied Soil Ecology* 5(3): 211–219.

Bennett, L. T., Mele, P. M., Annett, S. and Kasel, S. (2010). Examining links between soil management, soil health, and public benefits in agricultural landscapes: an Australian perspective *Agriculture, Ecosystems and Environment* 139(1–2): 1–12.

Bergstrom, D. W., Monreal, C. M. and King, D. J. (1998). Sensitivity of soil enzyme activities to conservation practices *Soil Science Society of America Journal* 62(5): 1286–1295.

Blanco-Canqui, H. and Ruis, S. J. (2018). No-tillage and soil physical environment *Geoderma* 326: 164–200.

Blanco-Canqui, H., Schlegel, A. J. and Heer, W. F. (2011). Soil-profile distribution of carbon and associated properties in no-till along a precipitation gradient in the central Great Plains *Agriculture, Ecosystems and Environment* 144(1): 107–116.

Blanco-Canqui, H., Shapiro, C., Jasa, P. and Iqbal, J. (2021). No-till and carbon stocks: is deep soil sampling necessary? Insights from long-term experiments *Soil and Tillage Research* 206: 104840.

Blanco-Canqui, H., Stone, L. R., Schlegel, A. J., Lyon, D. J., Vigil, M. F., Mikha, M. M., Stahlman, P. W. and Rice, C. W. (2009). No-till induced increase in organic carbon reduces maximum bulk density of soils *Soil Science Society of America Journal* 73(6): 1871–1879.

Blanco-Canqui, H., Wienhold, B. J., Jin, V. L., Schmer, M. R. and Kibet, L. C. (2017). Long-term tillage impact on soil hydraulic properties *Soil and Tillage Research* 170: 38–42.

Blanco-Canqui, H. and Wortmann, C. S. (2020). Does occasional tillage undo the ecosystem services gained with no-till? a review *Soil and Tillage Research* 198: 104534.

Bottinelli, N., Jouquet, P., Capowiez, Y., Podwojewski, P., Grimaldi, M. and Peng, X. (2015). Why is the influence of soil macrofauna on soil structure only considered by soil ecologists? *Soil and Tillage Research* 146: 118–124.

Bouma, J. and McBratney, A. (2013). Framing soils as an actor when dealing with wicked environmental problems *Geoderma* 200–201: 130–139.

Brito, I., Goss, M. J., de Carvalho, M., Chatagnier, O. and van Tuinen, D. (2012). Impact of tillage system on arbuscular mycorrhiza fungal communities in the soil under Mediterranean conditions *Soil and Tillage Research* 121: 63–67.

Bronick, C. J. and Lal, R. (2005). Soil structure and management: a review *Geoderma* 124(1–2): 3–22.

Carver, R. E., Nelson, N. O., Roozeboom, K. L., Kluitenberg, G. J., Tomlinson, P. J., Kang, Q. and Abel, D. S. (2022). Cover crop and phosphorus fertilizer management impacts on surface water quality from a no-till corn-soybean rotation *Journal of Environmental Management* 301: 113818.

Chellappa, J., Sagar, K. L., Sekaran, U., Kumar, S. and Sharma, P. (2021). Soil organic carbon, aggregate stability and biochemical activity under tilled and no-tilled agroecosystems *Journal of Agriculture and Food Research* 4: 100139.

CTIC (2002). Tillage type definitions Available at: www.ctic.org/resource_display/?id=322&title=Tillage+Type+Definitions.

Dalal, R. C. (1998). Soil microbial biomass–what do the numbers really mean? *Australian Journal of Experimental Agriculture* 38(7): 649–665.

Dam, R. F., Mehdi, B. B., Burgess, M. S. E., Madramootoo, C. A., Mehuys, G. R. and Callum, I. R. (2005). Soil bulk density and crop yield under eleven consecutive years of corn with different tillage and residue practices in a sandy loam soil in central Canada *Soil and Tillage Research* 84(1): 41–53.

Dang, Y. P., Moody, P. W., Bell, M. J., Seymour, N. P., Dalal, R. C., Freebairn, D. M. and Walker, S. R. (2015). Strategic tillage in no-till farming systems in Australia's northern grains-growing regions: II. Implications for agronomy, soil and environment *Soil and Tillage Research* 152: 115–123.

Daryanto, S., Wang, L. and Jacinthe, P. A. (2017). Impacts of no-tillage management on nitrate loss from corn, soybean and wheat cultivation: a meta-analysis *Scientific Reports* 7(1): 12117.

Deng, S. P. and Tabatabai, M. A. (1997). Effect of tillage and residue management on enzyme activities in soils: III. Phosphatases and arylsulfatase *Biology and Fertility of Soils* 24(2): 141–146.

Derpsch, R., Franzluebbers, A. J., Duiker, S. W., Reicosky, D. C., Koeller, K., Friedrich, T., Sturny, W. G., Sá, J. C. M. and Weiss, K. (2014). Why do we need to standardize no-tillage research? *Soil and Tillage Research* 137: 16–22.

Derpsch, R., Friedrich, T., Landers, J. N., Rainbow, R., Reicosky, D. C., Sá, J., Sturny, W., Wall, P., Ward, R. and Weiss, K. (2011). About the necessity of adequately defining no-tillage-a discussion paper. *Proceedings of the 5th World Congress of Conservation Agriculture.*

Díaz-Zorita, M. n., Duarte, G. A. and Grove, J. H. (2002). A review of no-till systems and soil management for sustainable crop production in the subhumid and semiarid Pampas of Argentina *Soil and Tillage Research* 65(1): 1–18.

Doran, J. W. (1980). Soil microbial and biochemical changes associated with reduced tillage *Soil Science Society of America Journal* 44(4): 765–771.

Drijber, R. A., Doran, J. W., Parkhurst, A. M. and Lyon, D. J. (2000). Changes in soil microbial community structure with tillage under long-term wheat-fallow management *Soil Biology and Biochemistry* 32(10): 1419–1430.

Edwards, C. A. and Lofty, J. R. (1980). Effects of earthworm inoculation upon the root growth of direct drilled cereals *Journal of Applied Ecology* 17(3): 533–543.

Faulkner, E. H. (1943). *Plowman's Folly*, University of Oklahoma Press.

Feng, Y., Motta, A. C., Reeves, D. W., Burmester, C. H., Van Santen, E. and Osborne, J. A. (2003). Soil microbial communities under conventional-till and no-till continuous cotton systems *Soil Biology and Biochemistry* 35(12): 1693–1703.

Fiorini, A., Boselli, R., Amaducci, S. and Tabaglio, V. (2018). Effects of no-till on root architecture and root-soil interactions in a three-year crop rotation *European Journal of Agronomy* 99: 156–166.

Fiorini, A., Boselli, R., Maris, S. C., Santelli, S., Perego, A., Acutis, M., Brenna, S. and Tabaglio, V. (2020). Soil type and cropping system as drivers of soil quality indicators response to no-till: a 7-year field study *Applied Soil Ecology* 155: 103646.

Franzluebbers, A. and Hons, F. (1996). Soil-profile distribution of primary and secondary plant-available nutrients under conventional and no tillage *Soil and Tillage Research* 39(3–4): 229–239.

Frasier, I., Noellemeyer, E., Amiotti, N. and Quiroga, A. (2017). Vetch-rye biculture is a sustainable alternative for enhanced nitrogen availability and low leaching losses in a no-till cover crop system *Field Crops Research* 214: 104–112.

Frey, S. D., Elliott, E. T. and Paustian, K. (1999). Bacterial and fungal abundance and biomass in conventional and no-tillage agroecosystems along two climatic gradients *Soil Biology and Biochemistry* 31(4): 573–585.

Gaiser, T., Perkons, U., Küpper, P. M., Puschmann, D. U., Peth, S., Kautz, T., Pfeifer, J., Ewert, F., Horn, R. and Köpke, U. (2012). Evidence of improved water uptake from subsoil by spring wheat following Lucerne in a temperate humid climate *Field Crops Research* 126: 56–62.

Gál, A., Vyn, T. J., Michéli, E., Kladivko, E. J. and McFee, W. W. (2007). Soil carbon and nitrogen accumulation with long-term no-till versus moldboard plowing overestimated with tilled-zone sampling depths *Soil and Tillage Research* 96(1-2): 42-51.

Galdos, M. V., Pires, L. F., Cooper, H. V., Calonego, J. C., Rosolem, C. A. and Mooney, S. J. (2019). Assessing the long-term effects of zero-tillage on the macroporosity of Brazilian soils using X-ray computed tomography *Geoderma* 337: 1126-1135.

Garcia, J. P., Wortmann, C. S., Mamo, M., Drijber, R. A. and Tarkalson, D. D. (2007). One-time tillage of no-till: effects on nutrients, mycorrhizae, and phosphorus uptake *Agronomy Journal* 99(4): 1093-1103.

González-Chávez, M. d. C. A., Aitkenhead-Peterson, J. A., Gentry, T. J., Zuberer, D., Hons, F. and Loeppert, R. (2010). Soil microbial community, C, N, and P responses to long-term tillage and crop rotation *Soil and Tillage Research* 106(2): 285-293.

Grandy, A. S., Robertson, G. P. and Thelen, K. D. (2006). Do productivity and environmental trade-offs justify periodically cultivating no-till cropping systems? *Agronomy Journal* 98(6): 1377-1383.

Grigar, J., Hatfield, J. L. and Reeder, R. (2018). Transitional no-till: what is it and how does it differ from 'true' no-till? *Crops & Soils* 51(6): 28-36.

Gupta, V. V. S. R. and Germida, J. J. (2015). Soil aggregation: influence on microbial biomass and implications for biological processes *Soil Biology and Biochemistry* 80: A3-A9.

Halvorson, A. D., Mosier, A. R., Reule, C. A. and Bausch, W. C. (2006). Nitrogen and tillage effects on irrigated continuous corn yields *Agronomy Journal* 98(1): 63-71.

Hamza, M. A. and Anderson, W. K. (2005). Soil compaction in cropping systems *Soil and Tillage Research* 82(2): 121-145.

Helgason, B. L., Walley, F. L. and Germida, J. J. (2010). No-till soil management increases microbial biomass and alters community profiles in soil aggregates *Applied Soil Ecology* 46(3): 390-397.

Holanda, F. S. R., Mengel, D. B., Paula, M. B., Carvaho, J. G. and Bertoni, J. C. (1998). Influence of crop rotations and tillage systems on phosphorus and potassium stratification and root distribution in the soil profile *Communications in Soil Science and Plant Analysis* 29(15-16): 2383-2394.

Islam, K. R. and Weil, R. (2000). Soil quality indicator properties in mid-Atlantic soils as influenced by conservation management *Journal of Soil and Water Conservation* 55(1): 69-78.

Islam, R. and Reeder, R. (2014). No-till and conservation agriculture in the United States: an example from the David Brandt farm, Carroll, Ohio *International Soil and Water Conservation Research* 2(1): 97-107.

Janzen, H. H., Janzen, D. W. and Gregorich, E. G. (2021). The 'soil health' metaphor: illuminating or illusory? *Soil Biology and Biochemistry* 159: 108167.

Jastrow, J. D., Miller, R. M. and Boutton, T. W. (1996). Carbon dynamics of aggregate-associated organic matter estimated by carbon-13 natural abundance *Soil Science Society of America Journal* 60(3): 801-807.

Karami, A., Homaee, M., Afzalinia, S., Ruhipour, H. and Basirat, S. (2012). Organic resource management: impacts on soil aggregate stability and other soil physico-chemical properties *Agriculture, Ecosystems and Environment* 148: 22-28.

Karlen, D. L., Wollenhaupt, N. C., Erbach, D. C., Berry, E. C., Swan, J. B., Eash, N. S. and Jordahl, J. L. (1994). Long-term tillage effects on soil quality *Soil and Tillage Research* 32(4): 313-327.

Kassam, A. (2020a). *Advances in Conservation Agriculture Volume 1: Systems and Science*, Burleigh Dodds Science Publishing, Cambridge, UK.

Kassam, A. (2020b). *Advances in Conservation Agriculture Volume 2: Practice and Benefits*, Burleigh Dodds Science Publishing, Cambridge, UK.

Kassam, A. (2022). *Adnances in Conservation Agriculture Volume 3: Adoption and Spread*, Burleigh Dodds Science Publishing, Cambridge, UK.

Kennedy, A. C. (1999). Bacterial diversity in agroecosystems *Agriculture, Ecosystems and Environment* 74(1-3): 65-76.

Kettler, T. A., Lyon, D. J., Doran, J. W., Powers, W. L. and Stroup, W. W. (2000). Soil quality assessment after weed-control tillage in a no-till wheat-fallow cropping system *Soil Science Society of America Journal* 64(1): 339-346.

Kim, N., Zabaloy, M. C., Riggins, C. W., Rodríguez-Zas, S. and Villamil, M. B. (2020). Microbial shifts following five years of cover cropping and tillage practices in fertile agroecosystems *Microorganisms* 8(11): 1773.

Kladivko, E. J. (2001). Tillage systems and soil ecology *Soil and Tillage Research* 61(1-2): 61-76.

Korucu, T., Shipitalo, M. J. and Kaspar, T. C. (2018). Rye cover crop increases earthworm populations and reduces losses of broadcast, fall-applied, fertilizers in surface runoff *Soil and Tillage Research* 180: 99-106.

Lal, R. (2004). Soil carbon sequestration impacts on global climate change and food security *Science* 304(5677): 1623-1627.

Lal, R., Reicosky, D. C. and Hanson, J. D. (2007). Evolution of the plow over 10,000 years and the rationale for no-till farming *Soil and Tillage Research*, Elsevier 93(1): 1-12.

Lavelle, P. (1988). Earthworm activities and the soil system *Biology and Fertility of Soils* 6(3): 237-251.

Lewis, K. L., Burke, J. A., Keeling, W. S., McCallister, D. M., DeLaune, P. B. and Keeling, J. W. (2018). Soil benefits and yield limitations of cover crop use in Texas High Plains cotton *Agronomy Journal* 110(4): 1616-1623.

Lipiec, J. and Stępniewski, W. (1995). Effects of soil compaction and tillage systems on uptake and losses of nutrients *Soil and Tillage Research* 35(1-2): 37-52.

Luo, Z., Wang, E. and Sun, O. J. (2010). Can no-tillage stimulate carbon sequestration in agricultural soils? A meta-analysis of paired experiments *Agriculture, Ecosystems and Environment* 139(1-2): 224-231.

Lupwayi, N. Z., Rice, W. A. and Clayton, G. W. (1998). Soil microbial diversity and community structure under wheat as influenced by tillage and crop rotation *Soil Biology and Biochemistry* 30(13): 1733-1741.

Lynch, J. M. and Bragg, E. (1985). Microorganisms and soil aggregate stability *Advances in Soil Science*, Springer:2: 133-171.

Malhi, S. S., Grant, C. A., Johnston, A. M. and Gill, K. S. (2001). Nitrogen fertilization management for no-till cereal production in the Canadian Great Plains: a review *Soil and Tillage Research* 60(3-4): 101-122.

Mathew, R. P., Feng, Y., Githinji, L., Ankumah, R. and Balkcom, K. S. (2012). Impact of no-tillage and conventional tillage systems on soil microbial communities *Applied and Environmental Soil Science* 2012: 1-10.

Mbuthia, L. W., Acosta-Martínez, V., DeBruyn, J., Schaeffer, S., Tyler, D., Odoi, E., Mpheshea, M., Walker, F. and Eash, N. (2015). Long term tillage, cover crop, and fertilization effects on microbial community structure, activity: implications for soil quality *Soil Biology and Biochemistry* 89: 24-34.

Mehdi, B. B., Madramootoo, C. A. and Mehuys, G. R. (1999). Yield and nitrogen content of corn under different tillage practices *Agronomy Journal* 91(4): 631-636.

Melman, D. A., Kelly, C., Schneekloth, J., Calderón, F. and Fonte, S. J. (2019). Tillage and residue management drive rapid changes in soil macrofauna communities and soil properties in a semiarid cropping system of Eastern Colorado *Applied Soil Ecology* 143: 98-106.

Mengel, D. B., Nelson, D. W. and Huber, D. M. (1982). Placement of nitrogen fertilizers for no-till and conventional till corn 1 *Agronomy Journal* 74(3): 515-518.

Moreno, F., Arrúe, J., Cantero-Martínez, C., López, M., Murillo, J., Sombrero, A., López-Garrido, R., Madejón, E., Moret, D. and Álvaro-Fuentes, J. (2010). *Conservation Agriculture under Mediterranean Conditions in Spain. Biodiversity, Biofuels, Agroforestry and Conservation Agriculture*, Springer: 175-193.

Naudin, K., Gozé, E., Balarabe, O., Giller, K. E. and Scopel, E. (2010). Impact of no tillage and mulching practices on cotton production in North Cameroon: a multi-locational on-farm assessment *Soil and Tillage Research* 108(1-2): 68-76.

Ng, E. L. and Zhang, J. (2019). The search for the meaning of soil health: lessons from human health and ecosystem health *Sustainability* 11(13): 3697.

Norris, C. E., Bean, G. M., Cappellazzi, S. B., Cope, M., Greub, K. L., Liptzin, D., Rieke, E. L., Tracy, P. W., Morgan, C. L. and Honeycutt, C. W. (2020). Introducing the North American project to evaluate soil health measurements *Agronomy Journal* 112(4): 3195-3215.

Nunes, M. R., Denardin, J. E., Pauletto, E. A., Faganello, A. and Pinto, L. F. S. (2015). Effect of soil chiseling on soil structure and root growth for a clayey soil under no-tillage *Geoderma* 259-260: 149-155.

Nunes, M. R., van Es, H. M., Schindelbeck, R., Ristow, A. J. and Ryan, M. (2018). No-till and cropping system diversification improve soil health and crop yield *Geoderma* 328: 30-43.

Obour, A. K., Mikha, M. M., Holman, J. D. and Stahlman, P. W. (2017). Changes in soil surface chemistry after fifty years of tillage and nitrogen fertilization *Geoderma* 308: 46-53.

Ogle, S. M., Swan, A. and Paustian, K. (2012). No-till management impacts on crop productivity, carbon input and soil carbon sequestration *Agriculture, Ecosystems and Environment* 149: 37-49.

Pagenkemper, S. K., Athmann, M., Uteau, D., Kautz, T., Peth, S. and Horn, R. (2015). The effect of earthworm activity on soil bioporosity-investigated with X-ray computed tomography and endoscopy *Soil and Tillage Research* 146: 79-88.

Pankhurst, C., Kirkby, B. H. and Harch, B. (2002). Impact of a change in tillage and crop residue management practice on soil chemical and microbiological properties in a cereal-producing red duplex soil in NSW, Australia *Biology and Fertility of Soils* 35(3): 189-196.

Piazza, G., Pellegrino, E., Moscatelli, M. C. and Ercoli, L. (2020). Long-term conservation tillage and nitrogen fertilization effects on soil aggregate distribution, nutrient stocks and enzymatic activities in bulk soil and occluded microaggregates *Soil and Tillage Research* 196: 104482.

Piotrowski, J. S., Denich, T., Klironomos, J. N., Graham, J. M. and Rillig, M. C. (2004). The effects of arbuscular mycorrhizas on soil aggregation depend on the interaction between plant and fungal species *New Phytologist* 164(2): 365–373.

Pittelkow, C. M., Linquist, B. A., Lundy, M. E., Liang, X., Van Groenigen, K. J., Lee, J., Van Gestel, N., Six, J., Venterea, R. T. and Van Kessel, C. (2015). When does no-till yield more? A global meta-analysis. *Field Crops Research* 183: 156–168.

Reicosky, D. (2018). *Managing Soil Health for Sustainable Agriculture Volume 1: Fundamentals*, Burleigh Dodds Science Publishing, Cambridge, UK.

Rice, C. W. and Smith, M. S. (1984). Short-term immobilization of fertilizer nitrogen at the surface of no-till and plowed soils *Soil Science Society of America Journal* 48(2): 295–297.

Rodríguez, M. P., Domínguez, A., Moreira Ferroni, M., Wall, L. G. and Bedano, J. C. (2020). The diversification and intensification of crop rotations under no-till promote earthworm abundance and biomass *Agronomy* 10(7): 919.

Sandhya, V. and Ali, S. Z. (2015). The production of exopolysaccharide by pseudomonas putida GAP-P45 under various abiotic stress conditions and its role in soil aggregation *Microbiology* 84(4): 512–519.

Schimel, J. (1995). *Ecosystem Consequences of Microbial Diversity and Community Structure. Arctic and Alpine Biodiversity: Patterns, Causes and Ecosystem Consequences*, Springer: 239–254.

Schlegel, A., Holman, J. D. and Assefa, Y. (2020). A single tillage in a long-term no-till system on dryland crop performance *Agronomy Journal* 112(4): 3174–3187.

Schmidt, R., Gravuer, K., Bossange, A. V., Mitchell, J. and Scow, K. (2018). Long-term use of cover crops and no-till shift soil microbial community life strategies in agricultural soil *PLoS ONE* 13(2): e0192953.

Sekaran, U., Sagar, K. L. and Kumar, S. (2021). Soil aggregates, aggregate-associated carbon and nitrogen, and water retention as influenced by short and long-term no-till systems *Soil and Tillage Research* 208: 104885.

Shen, Y., McLaughlin, N., Zhang, X., Xu, M. and Liang, A. (2018). Effect of tillage and crop residue on soil temperature following planting for a black soil in northeast China *Scientific Reports* 8(1): 4500.

Sims, A. L., Schepers, J. S., Olson, R. A. and Power, J. F. (1998). Irrigated corn yield and nitrogen accumulation response in a comparison of no-till and conventional till: tillage and surface-residue variables *Agronomy Journal* 90(5): 630–637.

Singh, P. K., Singh, M. and Tripathi, B. N. (2013). Glomalin: an arbuscular mycorrhizal fungal soil protein *Protoplasma* 250(3): 663–669.

Six, J., Bossuyt, H., Degryze, S. and Denef, K. (2004). A history of research on the link between (micro) aggregates, soil biota, and soil organic matter dynamics *Soil and Tillage Research* 79(1): 7–31.

Six, J., Elliott, E. T. and Paustian, K. (2000). Soil macroaggregate turnover and microaggregate formation: a mechanism for C sequestration under no-tillage agriculture *Soil Biology and Biochemistry* 32(14): 2099–2103.

Six, J., Elliott, E. T., Paustian, K. and Doran, J. W. (1998). Aggregation and soil organic matter accumulation in cultivated and native grassland soils *Soil Science Society of America Journal* 62(5): 1367–1377.

Soane, B. D., Ball, B. C., Arvidsson, J., Basch, G., Moreno, F. and Roger-Estrade, J. (2012). No-till in northern, western and south-western Europe: a review of problems and opportunities for crop production and the environment *Soil and Tillage Research* 118: 66-87.

Stirzaker, R. J., Passioura, J. B. and Wilms, Y. (1996). Soil structure and plant growth: impact of bulk density and biopores *Plant and Soil* 185(1): 151-162.

Swan, J. B., Schneider, E. C., Moncrief, J. F., Paulson, W. H. and Peterson, A. E. (1987). Estimating corn growth, yield, and grain moisture from air growing degree days and residue Cover1 *Agronomy Journal* 79(1): 53-60.

Taskin, E., Boselli, R., Fiorini, A., Misci, C., Ardenti, F., Bandini, F., Guzzetti, L., Panzeri, D., Tommasi, N., Galimberti, A., Labra, M., Tabaglio, V. and Puglisi, E. (2021). Combined impact of no-till and cover crops with or without short-term water stress as revealed by physicochemical and microbiological indicators *Biology* 10(1): 23.

Tiunov, A. V. and Scheu, S. (1999). Microbial respiration, biomass, biovolume and nutrient status in burrow walls of Lumbricus terrestris L.(Lumbricidae) *Soil Biology and Biochemistry* 31(14): 2039-2048.

Unger, P. W. and Kaspar, T. C. (1994). Soil compaction and root growth: a review *Agronomy Journal* 86(5): 759-766.

Unger, P. W. and McCalla, T. M. (1980). Conservation tillage systems *Advances in Agronomy* 33: 1-58.

USDA (2017). Census of agriculture Available at: www.nass.usda.gov/Publications/AgCensus/2017.

Uteau, D., Pagenkemper, S. K., Peth, S. and Horn, R. (2013). Root and time dependent soil structure formation and its influence on gas transport in the subsoil *Soil and Tillage Research* 132: 69-76.

Van Der Heijden, M. G. A., Klironomos, J. N., Ursic, M., Moutoglis, P., Streitwolf-Engel, R., Boller, T., Wiemken, A. and Sanders, I. R. (1998). Mycorrhizal fungal diversity determines plant biodiversity, ecosystem variability and productivity *Nature* 396(6706): 69-72.

Van Wijk, W. R., Larson, W. E. and Burrows, W. C. (1959). Soil temperature and the early growth of corn from mulched and unmulched soil *Soil Science Society of America Journal* 23(6): 428-434.

Veresoglou, S. D., Chen, B. and Rillig, M. C. (2012). Arbuscular mycorrhiza and soil nitrogen cycling *Soil Biology and Biochemistry* 46: 53-62.

Voorhees, W. B. and Lindstrom, M. J. (1984). Long-term effects of tillage method on soil tilth independent of wheel traffic compaction *Soil Science Society of America Journal* 48(1): 152-156.

Wagg, C., Bender, S. F., Widmer, F. and Van Der Heijden, M. G. (2014). Soil biodiversity and soil community composition determine ecosystem multifunctionality *Proceedings of the National Academy of Sciences of the United States of America* 111(14): 5266-5270.

Wehner, J., Antunes, P. M., Powell, J. R., Mazukatow, J. and Rillig, M. C. (2010). Plant pathogen protection by arbuscular mycorrhizas: a role for fungal diversity? *Pedobiologia* 53(3): 197-201.

Wolfe, A. M. and Eckert, D. J. (1999). Crop sequence and surface residue effects on the performance of no-till corn grown on a poorly drained soil *Agronomy Journal* 91(3): 363-367.

Wortmann, C. S., Quincke, J. A., Drijber, R. A., Mamo, M. and Franti, T. (2008). Soil microbial community change and recovery after one-time tillage of continuous no-till *Agronomy Journal* 100(6): 1681-1686.

Wright, S. F. and Upadhyaya, A. (1996). Extraction of an abundant and unusual protein from soil and comparison with hyphal protein of arbuscular mycorrhizal fungi *Soil Science* 161(9): 575–586.

Wright, S. F. and Upadhyaya, A. (1998). A survey of soils for aggregate stability and glomalin, a glycoprotein produced by hyphae of arbuscular mycorrhizal fungi *Plant and Soil* 198(1): 97–107.

Young, I. M. and Ritz, K. (2000). Tillage, habitat space and function of soil microbes *Soil and Tillage Research* 53(3–4): 201–213.

Zhang, G., Chan, K., Oates, A., Heenan, D. and Huang, G. (2007). Relationship between soil structure and runoff/soil loss after 24 years of conservation tillage *Soil and Tillage Research* 92(1–2): 122–128.

Zhu, J. C., Gantzer, C. J., Anderson, S. H., Alberts, E. E. and Beuselinck, P. R. (1989). Runoff, soil, and dissolved nutrient losses from no-till soybean with winter cover crops *Soil Science Society of America Journal* 53(4): 1210–1214.

Chapter 7

Cover crops for soil health

Sieglinde S. Snapp, Michigan State University, USA; Carmen M. Ugarte, University of Illinois at Urbana-Champaign, USA; Dane W. Hunter, Southern Illinois University, USA; and Michelle M. Wander, University of Illinois at Urbana-Champaign, USA

1 Introduction

Cover crops are widely heralded as a key approach to the sustainable management of row crops (Schipanski et al., 2014). They are an important tactic for climate-smart, sustainable agriculture (Paustian et al., 2019) because they bundle a wide range of functions (Table 1) into one technology. Among the multiple benefits of cover crop use, the most widely reputed outcomes include enhanced diversity, living cover and support for ecological processes occurring within the soil (Clark, 2008; Frasier et al., 2016). Cover crops are accessory crops that are integrated with cash crops through intentional seeding into or between cash cropping phases by relay cropping, overseeding, inter-seeding and double-cropping for purposes other than harvest. This means that the costs associated with their use and management must be justified in terms of the services provided. Even though the case for cover crops has been made over several decades, with foundational work establishing their value (Hartwig and Ammon, 2002; Hartwig and Hoffman, 1975), we still lack a detailed understanding of where and how to support particular cover crop functions within our various geographies, farming systems, institutions and markets (Daryanto et al., 2018).

http://dx.doi.org/10.19103/AS.2021.0094.11

Table 1 Key functional roles of cover crops, mechanisms involved and benefits at a farm level and a society level

Functional role	Mechanism	Farm	Society
Water and soil regulation	Improved structure, reduced crusting, hold soil in place, promote aggregation, intercept rainfall, enhanced water holding, increased non-crop season water use	Enhance crop productivity, reduce need for irrigation, retain nutrients, reduce need for edge of field infrastructure	Reduce downstream/infrastructure costs, reduce need for water treatment, reduce flash flooding and improves and protects water quality
Biomass, soil building, C sequestration and nutrient availability	Increase biomass added and prevent loss of topsoil	Enhance soil quality and thus soil function	Mitigate climate change, resist climate stress
Yield and economics	Enhanced water and nutrient retention, nutrient availability and supply, disease, weed or pest suppression	Reduce input costs, increase economic returns	Price stability, reduce indirect costs through crop insurance, need to invest in costly infrastructure
Biodiversity	Add vegetative complexity, disease and pest regulation, support pollinator and belowground foodwebs	Reduce input costs, increase economic returns	Benefit from pollinators, wildlife

Because cover crops fulfill several functional roles, the terms used to refer to these accessory crops vary based on the primary purpose for their use. The erosion reduction, water and soil regulation functions (Table 1) associated with the 'cover crop' label emphasize the use of vegetative cover to avoid or reduce the period of time that soil is bare. Their use to protect soils from erosion has been promoted in the United States at the National scale for almost a century. Early work by Williams et al. (1981) and Langdale et al. (1991) summarized findings from erosion control studies dating from the 1930s to the 1940s. This research is the basis used to develop the Universal Soil Loss Equation (USLE) that has influenced erosion models worldwide. The original core dataset included 8000 runoff-plot years from 21 states and 6 sites. This established that cover crops such as vetch and ryegrass grown in sequence with row crops such as cotton or corn provided erosion control similar to that achieved through rotation with meadows (Wischmeier, 1960). The Revised USLE (RUSLE) further updated cover crop benefits for erosion control (Langdale et al., 1991; Sojka et al., 1984).

Foundational research such as that used to develop RUSLE established predictive relationships among cover biomass, residue and reductions in wind erosion (Skidmore and Siddoway, 1978; Woodruff and Siddoway, 1965). The clear ties established between adaptive management of cover crops and factors used by dominant erosion models explain why erosion reductions are widely accepted as a cover crop service, even though relatively few studies measure erosion reductions directly. There is wide agreement in the literature that continuous soil cover is one the most effective means to reduce soil loss on agricultural land (Duniway et al., 2019). The protective function of continuous living cover that prevents erosion directly through raindrop interception and indirectly by improving soil structure to maintain soils' capacity to function may need to be better reflected in soil health evaluations to avoid degradation of soil and water resources by erosion that is currently widespread (Turner et al., 2018).

The large number of concomitant services derived from cover crop biomass produced by cover crop growth that is the primary basis for improvement of soil health is most often associated with *catch crops* and *green manures*. Covers are also commonly referred to as scavenger crops when nutrient capture and retention are the goals. Reviewing potential for what they term catch crops and green manures, Thorup-Kristensen et al. (2003) established that either by catching available nitrogen (N) in the soil and thereby preventing N leaching losses or providing N directly through biological fixation, cover crops can improve the nutrition of the subsequent cash crops. Reviews by Meisinger et al. (1991) and Di and Cameron (2002) provide evidence that cover crops can reduce N leaching losses using capture and fixation mechanisms. Cover crops are also effective means to prevent loss of reactive N and phosphorus

(P) through water pathways, to preserve high-quality water (Aryal et al., 2018; Thapa et al., 2018). While the nutrient supply and water quality protections provided by 'nitrogen husbandry' objectives described by Thorup-Kristensen et al. (2003) are direct and not always explicitly tied to soil health. Cover crop growth increases the time period during which organic residues are added and can thereby increase the amount of material returned, explaining why green manures are closely tied to 'soil building' and efforts to 'feed the soil' that underpin soil health (Idowu et al., 2009).

Cover crops are often commonly considered *soil conditioners*, which is a term that was widely used to refer to inputs used to improve soil physical condition by building organic matter (Wanic et al., 2018; Messiga et al., 2016). Organic matter accrual derived from cover crop use that is directly tied to delivery of carbon (C) sequestration and enhanced fertility functions can be seen in Table 1. Early work on cover crop contributions to soil organic carbon (SOC) by Kuo et al. (1997) demonstrated that gains were associated with cover biomass and suggested that grasses might be more successful than legume or brassica covers. Several recent reviews confirm cover crops' general ability to increase SOC (Poeplau and Don, 2015; Ugarte et al., 2014) and improve soil physical condition (Haruna et al., 2020).

Cover crop benefits are sometimes associated with tradeoffs among services. For example, Abdalla et al. (2019) conducted a global meta-analysis of cover crops and found significant decreases in N leaching and increased SOC sequestration, but reduced grain yield unless cover crops included a mixture of legume and non-legume species. Noting general differences between legume and grass cover crops associated with the source of N and rates of residue decay resulting from differences in composition (often reflected by C:N ratios), and their ability to mobilize insoluble nutrients or scavenge N, Fageria et al. (2005) argued that growing grass/legume mixtures may be the best strategy in obtaining maximum benefits from cover crops. Soil building and conditioning are tied to economic benefits of cover crops derived through improvements in yield and yield stability associated with prevention of erosion, or gained through the other mechanisms listed in Table 1. One of the primary motivations for cover crop adoption is the perceived yield benefit in the subsequent cash crop (Bergtold et al., 2019). The same is true for societal benefits derived from enhanced water protection, flood prevention, food security and climate mitigation.

Additional functions of *green manures* or cover crops used as 'smother' or 'rest crops' are intended to suppress weed and pest pressure through direct competition or indirect means associated with benefits derived from crop rotation where crops are grown intentionally to restore exhausted soils and break disease cycles (Fageria et al., 2005). Most studies of pest suppression by cover crops focus on direct interactions between the cover crops and pests with little or no attention to indirect effects through soil health improvements. These examples also suggest how important it is to understand how both direct and indirect factors benefit soil health feedback to deliver soil services. For example, an influential review by Abawi and Widmer (2000) summarized key mechanisms associated with pest suppression and used case studies to demonstrate how specific cover crop species suppress plant-parasitic nematodes and other pests. In a recent review of cover crop services, Daryanto et al. (2018) failed to quantify sufficient evidence to suggest suppression of plant-parasitic nematodes resulting from the use of cover crops, but it did not distinguish among cover crop species rated as poor- or non-hosts to plant-parasitic nematodes which is their primary mode of action. Early work also established the potential for allelopathic and allelochemical interactions facilitated by cover crops as conferring weed and pest control (Singh et al., 2003).

There are direct effects, but many mechanisms by which cover crops influence pest dynamics and crop health are indirect and through complex interactions of cover crops and soil health. This includes feedback mechanisms through soil properties that influence pest dynamics. Work by Kadziene et al. (2020) provides a good example of indirect cover crop benefits through general suppression of soil-borne diseases and weeds through enhanced soil health built up under cover crop practices. We need to gain a greater understanding of how to manage soil health to promote desired interactions. The importance of soil health interactions was already noted in early work by Liebman and Davis (2000) that established how resource competition and phytotoxic effects of cover crops on weeds varied with weed seed size and competition for N, and recently by Couëdel et al. (2019), who reported that the influences of brassica cover crops on pathogens, beneficial organisms and weeds arise not only from allelochemicals but also from non-host and plant cover effects, organic matter additions and improved soil structure.

Reviews promoting cover crop use have largely reiterated claims for potential benefits and soil health services, with few specifics that address climatic regions and management systems (Blanco-Canqui et al., 2015). We see the need to move beyond general reviews of cover crops that often rely upon asserted benefits. The approach taken here was to review primary literature and consider specifically where and how cover crops are being used in North America and resulting soil health–associated services. The chapter further explores how to integrate cover crops within cropping systems. This includes attention to the niche for growing cover crops and how this has been influenced by improvements in cultivation and planting equipment, herbicides and germplasm. Until recently, cover crops' greatest potential for use has been assumed across North America to be largely restricted to cool-season crops grown in warm areas with abundant precipitation. Soil moisture limitations were thought to preclude integration of cover crops in dryland crops in semiarid areas, as were cooler soil temperatures in high-altitude areas, which could retard early growth (Dabney et al., 2001; Unger and Vigil, 1998). Finally, we consider research gaps and future directions for enhancing soil health through cover crops.

2 Where and why to grow cover crops

2.1 Cover crop niches

Understanding how cover crops influence soil health and how to optimize such benefits requires an understanding of particular niches and which cover crop species can be integrated within regional crop rotations (Smith et al., 2008; Tiemann et al., 2015). Cropping diversity often correlates strongly with increased soil health and biomass, which is highly associated with most provisioning ecosystem services (Jarecki et al., 2009; O'Dea et al., 2015). Increasing niche space utilization can result in tradeoffs but also contribute to the optimization of management intensities within cropping systems (Franco et al., 2015). Spring and fall typically have the highest precipitation and the use of cover crops during these periods has the potential for nutrient and soil retention, but waning Growing Degree Days can limit fall establishment following cash crop harvest (Rusch et al., 2020). Exploitation of niches with low opportunity for biomass accumulation through the use of tactics like inter-seeding, for example, can enable cover crops to maximize growth.

Growing work in arid and semiarid regimes suggests covers can be valuable where previously thought unsuitable for cover crop growth due to water stress. For example, a long-term study in Spain by Gabriel et al. (2019) showed that benefits are achieved where water use by covers is compensated for by cover crop impacts on soil water retention and reductions in drainage losses. At the same time, in dry areas, tradeoffs may occur between cover crop maturity and a negative water balance. Many environmental services, if not all, are associated with cover crop vegetation presence, where termination occurs before maturity. Thus, early termination may be the most viable management approach to facilitate the adoption of cover crops in arid and semiarid locations.

Successful integration of cover crops into agronomic systems requires additional consideration of the socio-economic niches that encourage or discourage their use. Factors including farmers' agronomic preferences, access to equipment, production goals and societal priorities, may include value conflicts at farm and society level (Snapp et al., 2005). To succeed, cover crops must satisfy the farm-scale economic niche shaped by farm gate prices for goods and services and by societal scale incentives provided by policies and subsidies, to enable cover crop adoption (Bergtold et al., 2019; Schipanski et al.,

2014). Labor availability, access to appropriate seeds and societal incentives such as subsidies and insurance requirements all add to the complexity of socio-economic opportunities and constraints. The productivity function is more likely to be rewarded at the farm scale than is erosion prevention, which is more difficult to quantify and relate it to an economic value. Growth in educational programs, incentives and cost share promoting the use of cover crops provides evidence that the socio-economic niche for cover crops is expanding at the societal scale. This needs to be matched by access to cover crop seeds that are adapted and perform well in a given environment; thus market value chains must be developed along with institutional and educational incentives.

2.2 Key cover crop types by function

Cover crop services described in Table 1 are related to plant species functional traits. These are traits associated with plant families, or with species, or, in some cases, traits are associated with sub-species or cultivars. Functional traits include growth traits and properties of association, e.g. relative growth rate, length of the growth cycle, symbiosis with N fixing bacteria, winter hardiness, rooting depth and angle, and relative water use, among others (Clark, 2008). Common cover crop functional groups include legumes, grasses and brassicas (Table 2). The legume plant family generally is associated with Rhizobia and biologically fixes N. Thus, legume tissue properties are generally characterized by a narrow C:N ratio and labile C, properties that promote rapid, biologically mediated mineralization. Species in the grass family, on the other hand, tend to have a fast growth rate and moderate N content in tissues, which is often associated with slow N mineralization and can temporarily immobilize N to reduce losses to the environment (McSwiney et al., 2010; Snapp and Surapur, 2018). At the same time, tissue N is markedly influenced by plant maturity and the growth environment. The C:N ratio in grass plant tissue widens substantially during the shift from vegetative growth to the reproductive stage, as N is reallocated to seed tissue, and tissue properties are often in complex carbohydrates and lignin compounds, recalcitrant to mineralization. Brassicas also grow rapidly and take up inorganic N from the soil; however, their tissue characteristics are intermediate between legumes and grasses. For example, radish grown as a cover crop can provide a comparable amount of soil inorganic N as a N-fixing cover crop of soybean, grown prior to a summer cash crop (Holmes et al., 2019).

Broad generalizations are commonly made about cover crop types and species, allowing functional grouping by the plant family (Table 2). For example, legume cover crops are often planted to precede N-demanding cash crops such as tomatoes or cereals (Moore et al., 2014). Grass cover crops may positively impact systems when planted to temporarily immobilize residual inorganic N from a leguminous or potato cash crop. However, grass cover crops

Table 2 Dominant cover crops used in four North American regions, classified by functional group, services and niche

Family/ Common name	Seasonality	Services/functional traits	Cover niche	Common region	Comments
Leguminosae					
Field pea	Cool season annuals	• Biologically fix N • Narrow C:N ratio tissues • High N mineralization • Water infiltration	Summer & winter	N. Plains, South	• Water efficient • Relatively fast growth • Winterkills −10°C
Lentil			Winter	N. Plains	• Water efficient • Relative fast growth • Winterkills −6.1°C
Black medic			Summer	N. Plains	• Water efficient • Relative fast growth • Winterkills −5°C
Alfalfa	Winter and year long. Long-season annuals, biannuals and perennials	• Pollination • Biologically fix N • Narrow C:N ratio tissues • High N mineralization • Improve infiltration • Build soil N pools & organic matter • Decrease bulk density	Summer & winter	Widespread	• Slow to establish
Hairy vetch			Winter	Widespread	• Can offer weed suppression
Lana vetch			Winter	West	• Can offer weed suppression
Clovers			Winter	Regional varieties	• Growth habits vary
Soybean	Warm season	• Biologically fix N • Narrow C:N ratio tissues • High N mineralization • Improve infiltration	Summer	Widespread	• Accessible
Sun hemp			Summer	South, Midwest	• Fast growing
Cowpea			Summer	South	• Water efficient

Gramineae					
Cereal rye	Cool season	• Scavenge N • Wide C:N ratio* • Slow mineralization* • Enhance SOM • Improve infiltration • Decrease bulk density • Increase soil aggregation *When grown to the reproductive phase; low N content and lignin increases after boot stage.	Winter	Widespread	• Cold hardy • Easy to kill
Winter wheat			Winter or spring	Widespread	• Accessible • Used as cover crop or cash crop
Barley			Winter	Widespread	• Usage varies regionally • Quick growth • Water efficient • Winterkills at −8°C
Annual ryegrass			Winter	Widespread	• Penetrates fragipan • Winterkills at −22°C
Oats			Spring or fall	Widespread	• Quick growth • Good nurse crop • Winterkills between −7°C and −17°C
Sudangrass	Warm season		Summer	Midwest, South	• High Biomass
Sorghum			Summer	Midwest, South	• Water efficient
Brassicaceae					
Pennycress	Cool season	• Scavenge N • Narrow C:N ratio • Fast mineralization • Enhance SOM • Improve infiltration • Decrease bulk density	Winter	Midwest	• Used as cover crop or cash crop
Mustards			Spring or summer	Northwest, Northeast	• Winterkills between −4°C and −12°C
Canola			Spring or winter	N Plains, Northwest, Northwest	• Can winterkill
Radish	Cool & warm season		Spring, summer or fall	Widespread	• Winterkills between −3.9°C and −6.7°C

(*Continued*)

Table 2 (*Continued*)

Family/ Common name	Seasonality	Services/functional traits	Cover niche	Common region	Comments
Other					
Buckwheat	Cool and warm season	• Pollination • Scavenge N	Summer	Midwest, Northeast	• Fast growing
Phacelia		• Narrow C:N ratio • Fast mineralization • Improve infiltration	Spring, Fall	Northwest, Northeast, West	• Winterkills at −8°C

Sources: Clark, A. (2008), *Managing Cover Crops Profitably*, 3rd ed., Diane Publishing. USDA, NRCS, 2020. The PLANTS Database <http://plants.usda.gov>, 29 October 2020. National Plant Data Team, Greensboro, NC 27401-4901 USA.

can also immobilize N to the detriment of the cash crop, but this interaction can be avoided through adequate management. Continued use of N fixing and scavenging cover crops can build SOC pools to a point of creating conditions of N sufficiency or excess for the cash crop (Allen et al., 2011; Zentner et al., 2004) and in response to this, the combination of species with synergistic functional traits, like mixes of legume and grass species, have been recommended. By using this strategy, multiple services like C and N retention can be achieved (Fageria et al., 2005).

Many functional traits are shared by a number of species, such as winter hardiness. Such traits are commonly associated with species that are considered broadly analogous, such as winter wheat and rye, sharing traits for germination in cold temperatures and tolerance to cold, dry conditions (Clark, 2008; Snapp et al., 2005). Another plant family with common traits is spring establishing cereals, for example, oat and spring wheat. While spring or winter cereals fill similar niche space, each species has its own particular set of traits as well. Rye is representative of winter grasses, in that it is able to grow under very cold conditions, rapidly accumulating biomass. At the same time, its tissue has unique biochemical properties with allelopathic functions (Reiss et al., 2018). This trait can suppress weed growth but requires careful management to prevent damage to crops planted subsequent to incorporation of a rye cover crop. Vegetable growers often grow winter wheat as a cover crop, as small-seeded crops such as carrot are highly sensitive to rye allelopathic properties, thus wheat as an analogous winter cereal to rye is substituted (Clark, 2008).

There are also examples of species-specific mechanisms and plant traits. These are often not well described, with many involving belowground traits and rhizosphere functions, which are challenging to quantify. For example, root system traits of annual ryegrass are associated with ameliorating a hardened soil layer (Karathanasis et al., 2014). Root exudates may enhance sparingly soluble P either directly or through associated microbiota. White clover, for example, increases mycorrhizal colonization and P uptake when integrated as a living mulch in corn silage systems (Deguchi et al., 2007).

The use of mixtures is often advocated due to complementary functions. Combining species from a range of plant families helps ensure a full complement of ecosystem services are supported, as well as buffering against inadequate growth of any unadapted species for any given weather and environmental conditions (Blesh et al., 2019). In addition, combining analogous species in cover crop mixes does offer some functional redundancy and can ensure the provision of the specific functional service (Wortman et al., 2012). Choosing specific combinations of cover crops for a given set of objectives and agroecological niche is facilitated by resources described at the end of this chapter.

3 Cover crop regional assessment

This section explores cover crop use in temperate annual cropping systems, primarily based in North America. We take a regional approach, as the niche for growing a cover crop (time of year, precipitation, available growing degree days, length of the growth window) varies by region (Snapp et al., 2005). Cover crops integrated with annual cash crops are typically established shortly before or just after the harvest of the main crop. Historically this has exploited the winter niche across much of North America, where cash crops are grown over the spring-summer season and into fall in many cases. In southern or western sites, where heat or water limitations make the summer niche unprofitable for cash crops, the potential for cover crop use has grown with interest in leveraging soil services. Through the lens of this regional literature, we consider three sets of related soil health services, as reported for widely grown annual cropping systems within major agricultural regions of North America, here grouped as Northwest and Northeast, the Northern Plains and Midwest, the West Coast and the South (USDA-NRCS, 2006). The soil health services provided by cover crop studies within the considered regions are summarized in association with three main areas: 1) water and soil regulation; 2) biomass, soil building and nutrient supply; and 3) yield and economics.

3.1 Regional niches

Crop rotations in the Northwest and Northeast include areas used for specialty crop production, wheat, corn and forage. These represent the land resource regions A, B, N, R and S outlined by USDA-NRCS (2006). The diverse rotations provide opportunities for winter-seeded cover crops that are adapted to medium or short seasons and are used through relay intercrop and planting in late summer after specialty crops or into cereals in the spring. Species including red clover or other legumes are commonly used before wheat or corn (Ketterings et al., 2015; Schipanski et al., 2014). Other species including cereal rye are grown to increase soil organic matter (SOM), enhance weed control and reduce the risk of N losses (Ketterings et al., 2015). There is considerable interest in cover crops that establish successfully in the fall or spring, and are relatively easy to terminate in the spring before planting a cash crop.

In the Northern Plains typical systems are managed for grain production and are often irrigated. These include land resource regions F and the northern portions of regions G and H (USDA-NRCS, 2006). Examples of rotation sequences are wheat-soybean-sunflower-other pulses. Dryland areas may incorporate a leguminous green manure fallow into wheat-fallow systems. This region has generally a short to medium growing season. Legume and brassicas such as field peas, and lentils, and mustards and canola are typically used as part of the rotation in this region. The Midwest encompasses land resource regions M, N, L and K (USDA-NRCS, 2006). This is a region dominated by summer production of corn-soybean systems in M and N, fruit and vegetables in L and the dairy and forage in region K. Cover crops are grown during the winter in a generally short season. Primary objectives in the Midwest include nutrient retention and erosion prevention, and occasionally nitrogen fixation; however, time to establish a legume in the fall remains a hurdle. Cereal rye and a few other grasses, clovers and hairy vetch are the primary cover crops that can be grown.

The systems in the Western states, particularly in California, are irrigated production systems with specialty crops such as vegetables grown in the winter months or perennials including fruit and nut crops. Often soil moisture is limiting in this region with the Mediterranean climate, and where cover crops are often planted in the summer (which is the offseason in the southern part of the West Coast). Annual species planted include rye/legumes and rye/brassicas mixes and clovers (Brennan et al., 2011; Brennan and Smith, 2005). Irrigated systems provide a unique, and broad, niche for cover crops based on being able to enhance germination and growth through supplemental watering. This has been shown in the Midwest as well as the West Coast, as farmers noted cover crops' performance and specific value is high within irrigated systems in Michigan (Snapp et al., 2005).

In the Southern Plains, the production of cotton and other field crops are represented and include areas from Texas to Georgia (USDA-NRCS, 2006). Cover crops are grown during the summer, in a generally long season. Sorghum Sudangrass is a typical species grown in this region (Roper et al., 2017). Heat, high decomposition and limited moisture to support cover crop growth may limit their benefits in the Southwestern Plains. Nutrient retention and erosion control on the other hand are clear contributions from grass cover crops in the South (Aryal et al., 2018). Common legumes include field pea and hairy vetch.

3.2 Water and soil regulation

Across all North American regions, we found cover crops such as grasses with fibrous root systems to enhance soil structure by promoting aggregation and infiltration and provide indirect evidence that cover crop use helps prevent soil erosion losses (Table 3). The most widely observed service improvement was increased infiltration through the adoption of a wide range of cover crops. This is seen in the Midwest and Northern Plains, where infiltration was enhanced by over two-fold in several winter cover crop systems with corn-soybean (Chalise et al., 2018) and in wheat-sorghum rotational sequences (Blanco-Canqui et al., 2011).

Table 3 The influence of cover crops on water and soil regulation services across North American agricultural regions

Region	System	Cover Crop	Trend	Study
Northwest and northeast				
OR, WA	Small grains-potato	Mustard	• Cover crop did not change dry aggregate stability measurements as influenced by the cover crop	Pi et al. (2019)
PA	Corn-oats	Cereal rye, red clover, canola, radish, oats, pea	• Legume cover crop was less successful at retaining N. • Hairy vetch and red clover when mixed with canola or cereal rye, promoted greater N retention	Finney et al. (2016)
OR	Vegetables	Cereal rye, triticale, rye-wheat mix, vetch-triticale mix	• Long-term experiment • Cover crops reduced NO_3-N in leachate from soils fertilized using recommended practices for vegetable production, but concentrations were still above the EPA recommendations for drinking water	Feaga et al. (2010)
Northern Plains and Midwest				
SD	Corn-soybean	Rye + hairy vetch/vetch	• Infiltration rates were 1.31.8 higher in areas planted with cover crops	Chalise et al. (2018)
CO	Corn-soybean	Rye	• Plots with rye had 2122% more plant-available water. It was typically lower in the spring but available water rebounded by the time of planting	Basche et al. (2016)
MN	Corn silage	Cereal rye	• Soil moisture at corn planting was similar for plots without cereal rye	Krueger et al. (2011)
MT	Wheat-fallow	Lentil	• About 87% of time lentil did not require more water than fallow fields	Allen et al. (2011)
KS	Wheat-sorghum	Vetch, sunhemp, soybean	• Cover crops showed a 3-fold increase in infiltration relative to no-till without cover crops	Blanco-Canqui et al. (2011)
IL	Corn-soybean	Rye-rye, rye-vetch, rye+vetch-rye	• Cover crop mixes improved water holding capacity and plant-available water	Villamil et al. (2006)
IA	Corn-soybean	Rye, oats	• Infiltration and runoff rates only significantly different in one-third of the studied years. • About 85% reduction in rill erosion in areas planted with rye cover crop	Kaspar et al. (2001)
MO	Corn-corn	Rye or wheat	• Cover crops significantly reduced soil loss in no-till systems	Wendt and Burwell (1985)

West				
CA	Vegetables	Mix (mostly triticale rye and vetch)	• Negative water balance • Ground cover achieved, but with significant water usage	Mitchell et al. (2015b)
CA	Vegetables	Rye and rye-legume mixes	• Early ground season cover achieved with monocultures signaling greater potential for erosion control	Brennan et al. (2011)
CA	Tomato-wheat	Oat, hairy vetch	• Systems with cover crop required more water but also exhibited greater water holding capacity and less ET	Colla et al. (2000)
Southern plains				
CO, NE	Proso millet-winter wheat	Mix of rapeseed, flax, oat, pea, lentil, common vetch, berseem clover, barley, safflower	• Cover crop planted prior to winter wheat • No difference in wheat water use efficiency on plots following single or mixed cover crops	Nielsen et al. (2016)
TX	Cotton-cotton	Rye	• Improved infiltration and penetration resistance in systems managed with rye after cotton harvest	DeLaune et al. (2019)
TX	Grazeland	Mix of Austrian winter peas, common vetch, wooly pod vetch, red clover, triticale, winter oats, cereal rye, forage collards, and radish	• Soil moisture was lower in cover crop plots when compared to a bermudagrass control at different times of the year and influenced by location	Teague et al. (2019)
NM	Winter wheat	Pea, oat, canola and mixtures of pea-oat, pea-canola, pea-oat-canola, and pea-oat-canola-hairy vetch-forage radish-barley	• Compared the effect of cover crops on winter wheat and a winter wheat-fallow • At the time of planting winter wheat, soil water content improved by 24% in plots planted with pea-oat-canola, pea-oat-canola-hairy vetch-forage radish-barley. • At the time of cover crop termination, soil water content was lower in plots with cover crops	Ghimire et al. (2019)
GA	Maize or pearl millet-maize or pearl millet-peanuts	Winter rye	• Used cover crops after each cash crop harvest • Carbon gains linked to improved water holding capacity	Strickland et al. (2015)

Soil moisture is generally enhanced, due to increased infiltration of water and storage capacity gains. There is the potential – important in drier regions – that soil moisture will be decreased through cover crop transpiration; however, this can be managed through early termination of cover crops. The empirical evidence in Table 3 illustrates the variability of cover crop effects on soil moisture. Overall, soil moisture is modestly higher or lower depending on the weather, the site and the cover crop species grown. In hot dry areas cover crops are often net users of water, as shown by long-term studies in the West (Colla et al., 2000; Mitchell et al., 2015a,b) and in the Southern Plains (Ghimire et al., 2019; Teague et al., 2019). There is concern over cover crops causing soil moisture competition in these hotter areas; for example, cotton and tomato producers in dry regions are concerned about yield loss associated with the cash crop having insufficient moisture available, particularly in rain-fed production systems.

In cooler areas, cover crop benefits appear much more consistent, both in terms of enhanced water supply and regulation of moisture. There may be initial soil moisture declines, but by crop harvest there were often water holding capacity gains through a cover crop presence (Villamil et al., 2006). A specific benefit of cover crops in more humid regions such as the Midwest and much of Canada is found in field crop production areas with frequent cold, wet spring conditions. Cover crops can take up moisture in the spring, which can markedly improve soil conditions for planting a cash crop. Green fallow systems in the dryland regions like the plains compensate for water use with increased infiltration and water-holding capacity when managed properly (Allen et al., 2011; Mooleki et al., 2016).

3.3 Biomass, soil carbon and nutrient availability

Overall, we found evidence that cover crop soil health interventions help improve soil properties and tighten N cycling (Table 4), but our review of regional studies showed variable ability of cover crops to reduce greenhouse gas emissions and increase soil C sequestration (Kaye and Quemada, 2017; Poeplau and Don, 2015). Cover crops are a reliable means to reduce soil and SOC loss in many environments. This is shown by studies of cover crop mitigation of soil C loss from tillage in sloped soils in the Midwest (Olson et al., 2014), and in a corn-soybean rotation after 12 years, where cover crops increased SOC 10-30% compared to a non-cover crop control, depending on tillage regime (Olson et al., 2014).

Across the regions, many short-term studies have failed to detect statistical differences but other long-term studies provided evidence of increases in SOM concentrations (e.g. Allen et al. (2011). Cover crops need to accrue sufficient biomass above and belowground if they are to improve SOC, and this varies markedly with an average of about 3.5 T biomass/ha typically produced by cereal cover crops in temperate areas (Blanco-Canqui et al., 2015). Humid Northwest and Midwest environments provide a sufficient window of time for cover crops, with moist conditions that support abundant biomass and, generally, SOC accrual (Table 4). However, the cover crop species composition matters as shown in a vegetable cropping system study in the Northeast which found that a cover crop mix of nine species was associated with an 8% increase in SOM, while plots planted with mustard alone exhibited a 9% decrease (Larkin, 2020). Also, in the Northeast, an unfertilized corn-corn system revealed that over 10 years, vetch/rye and rye cover crop treatments both increased SOC by about 30% compared to a bare control (Ding et al., 2006). In the same study, under fertilized conditions, the rye seeded alone increased SOC 22% while the rye/vetch mix failed to alter SOC, potentially due to the more decomposable residue from the vetch tissue and the mineralization effect of N fertilizer addition (Ding et al., 2006). Other long-term studies in the Midwest showed variable results. For example, Snapp and Surapur (2018) reported no SOC changes over time, with the exception of an active soil C fraction in aggregates in a corn–corn–soybean system that used rye as a cover crop in the upper Midwest (Snapp and Surapur, 2018). The long duration of winter in this region and the late cover crop planting times did not allow much biomass accrual by rye. In contrast, a study at a warmer site in the Midwest found that the corn/rye–soybean/vetch and corn/rye–soybean/vetch+rye systems were associated with modest gains in topsoil SOM, as compared to no cover or rye cover crop control systems (Villamil et al., 2006). Under high, arid and cold conditions it also appears possible to increase SOC, as shown in a 15-year no-tillage study of a wheat–sorghum system in the Great Plains where sun hemp and soybean contributed to SOC increases of 30% and 20% respectively compared to no-tillage without cover crop (Blanco-Canqui et al., 2011).

Nitrogen retention is a reliable service provided by cover crops, and it is apparent in a shorter time frame. Cover crops can fix or scavenge N, thus increasing soil health. Nitrogen supply varies with species, but legume cover crops almost always increase the microbial and related active soil N pools (Wyland et al., 1996). In the Northeast, hairy vetch outperformed clovers in N fixation in a cornwheat system (Dou et al., 1994), and in a corn system in North Carolina (Jani et al., 2016). Leguminous cover crop species have been known to supplement plant-available N, but in some cases shorter growing seasons can limit growth and the amount of N fixation (Lawson et al., 2013). In Western tomato systems, grasses, legumes and mixes all were related with increased mineralization, though N supply was somewhat contingent on weather (Mitchell et al., 2017; Wyland et al., 1996). Rye decreased leaching better than legume covers in a corn-corn rotation (White et al., 2017). Although in the plains both grasses, legumes and mixes have shown reduced soil nitrate relative to no-cover

Table 4 The influence of cover crops on biomass production, soil building and nutrient availability across North American regions

Region	System	Cover Crop	Trend	Study
Northwest and Northeast				
MA	Corn–corn	Vetch, vetch + rye fall planting	• An average increase of 22% SOC in fertilized plots seeded with rye • In unfertilized conditions, the combination of vetch/rye increased SOC by about 32% compared to a non-cover crop control • When planted alone, rye increased SOC by 28%	Ding et al. (2006)
ME	Vegetables	Rye, mustard, a mixture of field peas, vetch, crimson clover, rye, ryegrass, oilseed radish, rapeseed and buckwheat, and fallow	• Rye alone promoted a 2% increase in SOM, while plots planted with mustard decreased SOM by 9% • The mix of nine species, increased SOM by 8%	Larkin (2020)
NY	Corn	Vetch with fall planting	• Conventional till plots incorporated the residue and showed increased levels of inorganic N as compared to a no-till counterpart	Sarrantonio and Scott (1988)
PA	Corn–wheat	Red clover and hairy vetch	• Hairy vetch fixed more N than clover • Incorporation of the cover crop promoted higher mineralization	Dou et al. (1994)
PA	Corn–oats	Cereal rye, red clover, canola, radish, oats, pea	• Increasing the proportion of non-legumes promoted immobilization and decreased N supply potential	Finney et al. (2016)
PA	Corn	Cereal rye, red clover, yellow blossom sweet clover, mix of three and four other species	• Reduced NO_3-N leaching from rye-based fields	White et al. (2017)
WA	Sweet corn–potato	Sudangrass, white mustard, winter wheat, rapeseed, rye	• Overwintering cover crops were more effective at taking N and reducing NO_3-N leaching • Sudangrass was unable to establish in a short season region	Weinert et al. (2002)

WA	NA	Barley, rye, hairy vetch	• Legume varieties fixed more N, but not enough to support production of a cash crop	Wayman et al. (2015)
WA	Sweet corn	Rye, hairy vetch, and rye-hairy vetch mix	• Cover crops did not supply significant N to the cash crop • Cover crop growth and biomass accumulation were limited by the short growing season	Lawson et al. (2013)
Northern Plains and Midwest				
MN	Corn silage	Cereal rye	• Available soil NO_3-N was reduced by 35% and 59% after cereal rye was either terminated or harvested, respectively	Krueger et al. (2011)
ND	Wheat-soybean	Cereal rye, winter camelina, winter pea	• Mix, pea, and rye all reduced soil NO_3-N concentration by approximately 66% compared to a no-cover control	Peterson et al. (2019)
KS	Wheat-sorghum	Vetch, sunhemp, soybean	• Sun hemp and soybean contributed to a 30% and 20% greater SOC over no till without cover crop, respectively	Blanco-Canqui et al. (2011)
MI	Corn–corn–soybean, corn–soybean–wheat	Wheat, rye	• Use of cover crops reduced soil mineral N without causing it to become limiting for crop production	McSwiney et al. (2010)
IL	Corn–corn	Cereal rye	• Rye immobilize early season N • When terminated within a week of optimal corn planting in the region and fertilized N supplied at V6, the treatment combination became limiting to crop production	Crandall et al. (2005)
IL	Corn–soybean	Rye, vetch	• Cover crops showed potential to offset SOC losses due to tillage • Concentrations of SOC in NT with cover crop increased 3-fold compared to NT with no-cover crop, or tillage with cover crop • Cover crops increased SOC 1030% depending on the tillage regime	Olson et al. (2014)

(*Continued*)

Table 4 (*Continued*)

Region	System	Cover Crop	Trend	Study
IL	Corn-soy-wheat, or vegetable crops (Brassicaceae and Solanaceae)	Five-way mixes that included grasses, brassicas, legumes and individual monoculture of each species of cover crop (Oat, spring wheat, forage corn, sudangrass, kodiak mustard, purple top turnip, tillage radish, buckwheat (substituted a brassica in parts of the study), faba bean, field pea, cowpea, soybean)	• At cover crop termination, soil NO_3-N in weeded control were 3-fold in spring and 2-fold higher in summer • After incorporation, weeded control had 25% (spring), and 16% (summer) more soil NO_3-N concentrations on average • Plots with winter pea and soybean monoculture at termination showed NO_3-N concentrations similar to weeded control • A five-way mix without radish had 22% lower soil nitrate concentrations compared to the radish monoculture	Holmes et al. (2019)
IL	Corn-soybean	Rye, rye-vetch, rye+vetch-rye	• Corn/rye-soybean/vetch and corn/rye-soybean/vetch+rye increased SOM to 30 cm • Corn/rye-soybean/vetch most effectively reduced soil NO_3-N concentrations, followed by corn/rye-soybean/vetch+rye • Corn/rye-soybean/rye treatment had highest inorganic P	Villamil et al. (2006)
IA	Corn-soybean	Rye	• Rye after soybean had comparable SOC to the control, but rye following corn did increase SOC by 15% • The rye treatment after both cash crops increased potentially mineralizable N by 38%	Moore et al. (2014)
MI	Corn-corn-soybean	Rye	• Rye cover crop had minimal impact on SOC but was an effective N management tool with no corn yield penalty	Snapp and Surapur (2018)
MI	Corn-soybean-wheat	Red clover	• Legume cover crops increased particulate organic matter and maintained bioavailable P over 17 years	Gallaher and Snapp (2015)
MT	Wheat-fallow	Lentil	• Legume treatment promoted slightly higher SOC concentrations • With no external N fertility and after three crop cycles, soil NO_3-N in the legume cover crop matched the fallow treatment and maintained about 26% higher inorganic N concentrations at planting	Allen et al. (2011)

MT	Wheat	Pea	• Rotations with pea had higher potentially mineralizable C and N	O'Dea et al. (2015)
IN	Corn–soybean	Ryegrass or cereal rye	• Cover crops reduced nitrate export by 6990% through tile drainage at field level • Watershed daily nitrate median flow was reduced 1822% when comparing years with and without cover crops	Hanrahan et al. (2018)
West				
CA	Tomato	Medic, triticale+vetch	• Poor nutrient supply and significantly less than fallow	Herrero et al. (2001)
CA	Tomato	Vetch	• System with vetch retained more fertility overtime	Doane et al. (2009)
CA	Tomato–cotton	Triticale rye and vetch	• Over the long term, cover crops promoted more soil C storage	Mitchell et al. (2015a)
CA	Tomato–cotton	Triticale rye and vetch	• Long-term study • Increased N mineralization in treatments with cover crops	Mitchell et al. (2017)
CA	Tomato–maize	Peas and vetch, later bell beans vetch and oats	• Increased N mineralization but supply was dependent on weather	Li et al. (2019)
CA	Vegetables (lettuce, broccoli)	Rye and Phacelia (winter) cover crops	• Reduced nitrate leaching and enhanced microbial N pool • Transient effects	Wyland et al. (1996)
CA	Irrigated vegetables	Rye and cereal legume mixes	• Effective at increasing active soil C pools • Cover crops reduced inorganic N losses	Smukler et al. (2008)
OR	Broccoli	Oat, vetch, phacelia, and two-way mixes	• Vetch supplied N, but not the oats or phacelia treatments	Luna et al. (2018)

(*Continued*)

Table 4 (*Continued*)

Region	System	Cover Crop	Trend	Study
Southern Plains				
TX	Cotton–cotton	Cereal rye	• Long-term study • Reduced soil NO_3-N associated with cereal rye treatments • Increased SOC storage on the surface depths	DeLaune et al. (2019)
TX	Cotton	Long-term rye vs. a mix of rye-hairy vetch, radish and winter pea	• Cover crop treatments increased SOC in surface 15 cm • Cover crops reduced NO_3-N concentrations • No-cover crop effect on the concentrations of inorganic P	Lewis et al. (2018)
NM	Winter wheat	Pea, oat, canola, and mixtures of pea-oat, pea-canola, pea-oat-canola, and pea-oat-canola-hairy vetch-forage radish-barley	• Cover crop biomass was 33% to 142% greater with oat, pean-oat, pea-canola, pea-oat-canola, and pea-oat-canola-hairy vetch-forage radish-barley than pea or canola alone • SOC increased mostly under oat	Ghimire et al. (2019)
GA	Vegetables	Rye, hairy vetch, crimson clover	• SOC was consistently higher under rye than hairy vetch or crimson clover • TN varied among years but was generally higher in the treatments with legume cover crops	Sainju et al. (2002)
GA	Maize or pearl millet-maize or pearl millet-peanuts	Winter rye	• Soil C increased at various rates depending upon slope position and was associated with an increase in TN • Between 72% to 80% of the C accretion was in the top 35 cm	Strickland et al. (2015)
NC	Corn	Pea, clover, vetch	• Vetch supplied larger quantities of inorganic N to the cash crop • Vetch produced more biomass than the other two species	Jani et al. (2016)
MD, NC	Corn	Grass-legume mixes	• Grass legumes mixes can fix N and suppress weeds	Vann et al. (2019)

crop controls (Peterson et al., 2019). In the Northwest, overwintering cover crops were more effective at reducing nitrate leaching (Weinert et al., 2002). In the upper Midwest, rye reduced soil nitrate corresponding to cover crop uptake as modulated by kill date (Krueger et al., 2011). In the corn belt, grass and brassica cover crops are effective at N uptake, reducing N losses via leaching by about one-third, and in some cases by two-thirds or more (Hanrahan et al., 2018).

Even modest amounts of cover crop biomass can reduce N losses through leaching, as shown in Midwest corn-based cropping systems that incorporate small grain winter cover crops in their rotation. This is likely driven by uptake of inorganic N in the fall and by temporary immobilization in the spring (Villamil et al., 2006). High C biomass production, a characteristic of reproductive stage grasses, can immobilize nitrates to the point of limiting cash crop uptake even while gradually increasing potentially mineralizable N (Moore et al., 2014), but with appropriate management, this challenge can be ameliorated (Crandall et al., 2005; Snapp and Surapur, 2018). For example, timely tillage can increase mineralization (Dou et al., 1994; Sarrantonio and Scott, 1988) and N supply for optimal crop production (McSwiney et al., 2010).

Other functions are less understood and warrant further investigation. In the Midwest, continuous use of rye alone as a cover crop in a corn-soybean rotation was reported to increase P supply, in some cases (Villamil et al., 2006). Meanwhile, legume species such as alfalfa and red clover show soil P availability can be enhanced in specific soil types (Gallaher and Snapp, 2015). Overall, there is evidence, including that described in Table 4, that suggests cover crop soil health interventions can help tighten nutrient cycling, which is desired at the farm scale as well as the societal level.

3.4 Crop yield and economics

Overall, we found yield responses to be more apparent in the Northeast region that represents an area that is less limited by available water (Table 5). In contrast, those areas that are hotter and water-limited, such as the West Coast, the Northern and Southern Plains, face yield limitations due in large part to water usage as discussed previously in the section on water and soil regulation. The use of cover crops in some instances, however, also reduces the magnitude of yield penalty due to drought as shown in results from a long-term study in California (Li et al., 2019). This is likely due to increased water holding capacity in soils managed with cover crops. It is also important to note that the cash crop species also matters. Water-demanding crops such as maize may be more negatively impacted by cover crops than others, such as grain sorghum, cotton and tomato.

The evidence in this literature review also suggests that yield benefits depend greatly on the cover crop species. Table 5 shows the effect of cover crops on yield is neutral or inconsistent from year to year when a grass or mixes of grass and legume species are grown as cover crops prior to the cash crop (Basche et al., 2016; Colla et al., 2000; DeLaune et al., 2019; Finney et al., 2016; Herrero et al., 2001; Moore et al., 2014; Strickland et al., 2015). The majority of yield increases are observed in crops following legume cover crops, particularly vetch (Allen et al., 2011; Chalise et al., 2018; Doane et al., 2009; Luna et al., 2018). This newer work is consistent with the earlier foundational research cited in the introduction. Increased nutrient supply under legume cover crops seems to be the main contributor to these differences (Finney et al., 2016).

Factors related to cover crop management reveal an effect on yield. The time of planting and termination time impacts the amount of biomass produced and the quantity and quality of residue return, which in turn, contributes to yield via soil nutrient supply (Marcillo and Miguez, 2017). The amount of biomass produced by the cover crop as a result of early planting is also an important aspect of weed control (Table 5). Legume species, for this purpose, are less effective. Examples from this review highlight the use of cereal rye (Baraibar et al., 2018; Ryan et al., 2011; Wayman et al., 2015) for effective weed control. While biomass production for weed suppression is desirable, cover crops influence pest management by providing habitat to predatory organisms (Rivers et al., 2020; Zhang et al., 2020) or by becoming hosts of important pests (Dunbar et al., 2016). These results emphasize the complexity of management systems that must optimize multiple objectives to achieve the maximum benefits of cover crops.

Farmers have to prioritize cover crop benefits to cash crop yields, and overall, to profitable operation of a farm, in order to remain economically viable (Clark, 2008). Yield benefits are not always clearly attributable to cover crops; although many studies show that cover crops do not reduce yields, they do not necessarily show increases either, particularly in the short term (Snapp and Surapur, 2018). Farmers are interested in all of the above but need to produce sufficient cash crop yields to maintain economic sustainability. Consequently, cover crops, at a minimum, must contribute to yield stability and preferably increased yields. As noted, immediate yield benefits are rarely detected upon the adoption of cover crops (Table 5). In addition to societal level subsidies, farmer adoption in many instances relies on short-term economic incentives through the dual use of cover crops. Examples include cover crops that are grazed by livestock or cover crops that could also produce a harvestable product such as an oil crop (Bergtold et al., 2019). Work by Schomberg et al. (2014) in Georgia provides an example of system-specific analysis that finds increased economic returns associated with grazing of a rye cover crop but at a cost to soil health due to soil compaction.

4 Integration for soil health management by niche

Overall, cover crops provide many services to draw upon in designing farm management for soil health. However, integrating cover crops into cropping sequences and farming systems can be difficult. Primary challenges include socioeconomic and ecological factors such as utilizing the available ecological niche space and managing the water and N balance. The most highly productive niche space is typically occupied by cash crops, which provides an opportunity for cover crop growth during the spring and fall, with the exception of the western region. This results in short growing seasons that challenge sufficient cover crop growth needed for expected functions to occur (Blanco-Canqui et al., 2015). Therefore, cover crop integration can be easier in humid areas with longer growing seasons or in irrigated systems, but there remain practical limitations for the farmer and how to best combine cover cropping and their particular management system. Planting technologies have enabled reduced and no-tillage systems in some locations and integrated with cover crops as part of the rotation, facilitated by the use of herbicides (Williams et al., 1981). Organic farm production systems rely greatly on cover crops to provide services such as nutrient supply, soil health and pest regulation. However, the use of cover crops in organic systems is not managed through herbicides, which poses other challenges, but has propelled research on precision planting and mowing, and improved physical controls such as roller crimpers (Creamer et al., 1995).

Water balance is especially important in the Great Plains and seasonally for other regions. Drier areas need to account for any potential water deficit incurred by the growth of cover crops, as this could reduce subsequent cash crop yields. Conversely, in humid areas the water lost to transpiration can benefit soil drying and field trafficability. Similarly, N balance needs consideration when planting non-legume cash crops. Loading the soil with a heavy C residue can immobilize N and result in yield loss. Legumes and narrow C:N ratio biomass can improve soil N supplying ability by adjusting fixation rates when available N is high (Blesh, 2019).

Expanding cover crop niche space could support greater integration for soil health. As understanding of management broadens, we have seen an increase in cover crop niche exploitation. Historically cover crops were often bound to the winter niche space. Newer options include understory inter-seeding into standing crops, or the spring planting of winter annuals to prevent vernalization and thus reproductive maturity, and dual-use systems such as spring-planted cover crops that are grazed before or after they perform their expected function. In dry areas, leguminous green manures have seen increased adoption in place of tilled or chemical fallow. Increasing evidence suggests

Table 5 The influence of cover crops on yield and economic benefits in cropping systems across North America

Region	System	Cover Crop	Trend	Study
Northwest and Northeast				
NY	Corn	Vetch with fall planting	• No consistent yield increase in vetch-corn in CT vs NT studies	Sarrantonio and Scott (1988)
ME	Vegetables	Rye, mustard, a mixture of field peas, vetch, crimson clover, rye, ryegrass, oilseed radish, rapeseed, and buckwheat and fallow	• Improved yield of green beans, yellow squash and sweet peppers with cover crops relative to fallow	Larkin (2020)
MA	Potato	cereal rye, forage radish, winter pea	• Winter pea supplemented with 75 and 150 kg/ha, outperforms potato yield when compared with forage radish and cereal rye with similar N inputs	Hashemi et al. (2013)
PA	Corn	Cereal rye, red clover, yellow blossom sweet clover, mix of three and four other species	• Mixture cover crops contributed to better yields than monoculture rye	White et al. (2017)
PA	Corn-oats	Cereal rye, red clover, canola, radish, oats, pea	• Residue C:N ratios had a negative relationship with maize yield • Mixtures dominated by cereal rye may prevent the timely release of N for the cash crop	Finney et al. (2016)
PA	Wheat/cover crop-corn-corn	vetch and red clover	• Lower yields in both treatments when comparing CT and NT with synthetic N supply and cover crops • Red clover plots produced higher yields than those with vetch	Dou and Fox (1994)
PA	Corn and wheat	Monoculture and mixtures of cereal rye, oats, forage radish, canola, medium red clover and Austrian winter pea	• Cereal rye followed by mixtures were effective at suppressing weeds • Early planting promoted more biomass production and better weed control • Legume cover crops were the least weed suppressive	Baraibar et al. (2018)
PA, MD	Soybean	Cereal rye	• Increased planting densities of cereal rye reduced weed biomass	Ryan et al. (2011)

WA	Sweet corn–potato	Sudangrass, white mustard, winter wheat, rapeseed, rye	• No difference in potato yield attributed to previous cover crop	Weinert et al. (2002)
WA	NA	Barley, rye, hairy vetch	• Rye was effective at suppressing weeds	Wayman et al. (2015)
Northern Plains and Midwest				
ND	Wheat-soybean	Cereal rye, winter camelina, winter pea	• Reduced subsequent wheat yield if unfertilized • No yield effect on inter seeded companion soybean crop	Peterson et al. (2019)
MT	Wheat-fallow	Lentil	• With no supplemental N fertility yields matched fertilized plots after three crop cycles over 6 years of legume	Allen et al. (2011)
MT	Wheat-fallow	Lentil and winter pea	• The use of cover crops affected subsequent winter wheat yield and protein content relative to continuous wheat • Positive economics relative to continuous wheat	Chen et al. (2012)
SD	Corn-soybean	Rye + hairy vetch/vetch	• Soybean yields were 14% higher in plots with cover crop treatments	Chalise et al. (2018)
CO	Corn-soybean	Rye	• Cover crops had no significant impact on yield	Basche et al. (2016)
IA	Corn-soybean	Rye	• Cover crops had no significant impact on yield	Moore et al. (2014)
MI	Corn-soybean	Rye	• Cover crops had no significant impact on yield	Snapp and Surapur (2018)
West				
CA	Tomato	Medic, triticale+vetch	• Increased tomato yield in one of a 2-year study	Herrero et al. (2001)
CA	Tomato	Vetch	• Tomato yields were initially higher in plots with conservation tillage and cover crops	Doane et al. (2009)
CA	Tomato-cotton	Mix (mostly triticale rye and vetch)	• A 5% tomato yield increase over time resulting from use of cover crops • No consistent cotton yield change	Mitchell et al. (2015a)
CA	Tomato-cotton	Mix (mostly triticale rye and vetch)	• A 5% increase in tomato yield over time due to cover crops • No consistent cotton yield change	Mitchell et al. (2017)

(Continued)

Table 5 (*Continued*)

Region	System	Cover Crop	Trend	Study
CA	Tomato-wheat	Oat, hairy vetch	• No differences in tomato yield	Colla et al. (2000)
CA	Vegetables (lettuce, broccoli)	Rye and Phacelia (winter) cover crops	• Cover crop management effects on economics were really minimal in what is a very high value (high cost/high return) horticulture cropping system	Wyland et al. (1996)
CA	Irrigated vegetables	Legume (Magnus peas, vicia, bell faba) and cereal mix (oats) winter cover crops	• Long-term experiment • Tomato yields were not increased, but they were buffered • Losses in dry years were modest compared to non-cover crop systems • Corn yields are reduced substantially by about 35% when grown with a cover crop	Li et al. (2019)
CA	Irrigated vegetables	Rye and cereal legume mixes	• Effective at suppressing some weeds	Smukler et al. (2008)
CA	Vegetables (lettuce, broccoli)	Rye and Phacelia (winter) cover crops	• No change in insect or disease regulation was detected	Wyland et al. (1996)
CA	Irrigated vegetable systems	Legume/oats, mustard, oats	• Winter cover crops had variable effects on weed suppression • Mustard cover crops that grew rapidly in the spring, with more biomass, were more effective than cereal/legume mixed cover crops	Brennan and Smith, (2005)
Southern Plains				
TX	Cotton	Wheat, and a mix of rye, wheat, pea, vetch	• Not enough evidence to suggest negative influence in crop development and yield due to cover crops	Adams et al. (2020)
LA	Cotton	Winter wheat, hairy vetch, volunteer winter wheat	• Long-term study under irrigation conditions • Cover crops significantly increased yield components over the studied period likely associated with increased N fixation in legume cover crop treatment and no moisture limitation	Boquet et al. (2004)

TX	Cotton	Rye	• Cover crops enhanced lint yields by >10% compared to NT without cover crops	DeLaune et al. (2019)
TX		Long-term rye vs. a mix of rye-hairy vetch, radish, and winter pea	• Cotton lint yield was less with no-tillage and the rye cover when compared with conventional tillage in 2 of 3 years	Lewis et al. (2018)
NM, OK	Wheat-fallow	Pea, oat mix	• None of the cover crop options were economically viable without consideration of erosion control or N benefits	Acharya et al. (2019)
NM	Winter wheat	Pea, oat, canola, and mixtures of pea-oat, pea-canola, pea-oat-canola, and pea-oat-canola-hairy vetch-forage radish-barley	• There was year to year variation in yield of winter wheat, but no yield reductions associated with cover crops	Ghimire et al. (2019)
GA	Maize or pearl millet-maize or pearl millet-peanuts	Winter rye	• Corn yield increased and the magnitude of increase depended on the landscape position of the field. • Soil C content in the 15 cm to 35 cm profile was the soil parameter primarily associated with corn yield	Strickland et al. (2015)
TX	Cotton	Winter wheat	• No detrimental effect of cover crops on lint yield and fiber quality parameters	McDonald et al. (2020)
NC	Soybean	Rye	• Early flowering varieties produced more biomass • Weed control was similar across six different rye cultivars • In one of the 2-year studies, there was a positive relationship between rye biomass from early flowering cultivars and soybean yield	Wells et al. (2016)
AL	Tomato	Rye, crimson clover	• The rye cover crop, consistently produced the highest revenues compared to the use of plastic mulch or crimson clover • Crimson clover produced a higher economic return than plastic mulch in 1 out of 4 years of study	Duzy et al. (2014)

that proper management can optimize both ecological and economic yields from continuous cover, or living mulch systems (Bartel et al., 2017; Deguchi et al., 2007). Additionally, new technologies for planting intercropped cover crops, such as inter-seeding drills, flying on seed through drones and related technologies show potential for expanded niche exploitation by increasing seeding and termination capabilities.

5 Research gaps

It is unclear whether the potential for cover crops is being fully exploited in North American systems (Schipanski et al., 2014). Documentation is urgently needed of barriers to adoption, and the context-specific benefits associated with cover crop adoption, including soil health (Runck et al., 2020). Further, there are many research gaps concerning the biophysical processes associated with cover crops and soil health, many of which were identified in early reviews yet attention in the literature remains scarce (Drinkwater and Snapp, 2007; Reicosky and Forcella, 1998). These include the need for investigation of belowground processes and root system traits, for selection of improved cover crops and their associated microflora, and for better understanding of functional attributes and combinations.

Expanding the library of functional traits available in cover crops could increase their use. If cover crops were bred for tolerance of environmental extremes or susceptibility to chemical termination, then it may be possible to plant a broader range of species (Schipanski et al., 2014). Improved tolerance for growing as an understory is one of the major ways that a temporal niche can be expanded. While this is done in perennial systems (wine and nut crops) in Western states (Baumgartner et al., 2008; Guerra and Steenwerth, 2012) it is relatively untapped in other regions. Red clover is well suited to be grown as a frost-seeded understory with a wheat crop due to tolerance to low light and cold, and dry conditions (Snapp et al., 2005). Another important trait is rapid early growth, to cover soil and maximize biomass above and belowground, as described in a study that evaluated clover cultivars (Mauromicale et al., 2010). Indeed, cover crop species could be selected and managed for complementary traits like these (Blesh et al., 2019). The plasticity of cover crop growth is likely to pose a significant challenge for selection efforts. Few studies have addressed this explicitly but many cite variable response due to spatial, seasonal and cultivar variability.

Managing nutrient access through cover crops and their associated microbiota has shown promise as a means to improve soil fertility. Soil bacteria and fungi play a key role in N scavenging, biological N fixation and enhancing P accessibility from sparingly soluble pools (Drinkwater and Snapp, 2007); however, plant specificity of interaction with microbial communities is not well understood, except in the case of Rhizobia. There is a substantial research gap in terms of breeding for cover crops that facilitate rhizosphere activity for enhanced nutrient supply and other benefits (Tani et al., 2017). Quantification of plant traits for systematic and rapid screening of cover crop genotypes is a first step (Bodner et al., 2010); however, assessment of cover crop and associated microbial species also needs to be considered in a systematic manner.

Control of pests for healthy plants is another understudied research area. This is consistent with the review by Ratnadass et al. (2012) that found limited ties between vegetational diversification and reduced incidence of pests or diseases. In our review of cover crops for soil health in North America, we found few studies that considered the functional mechanism of cover crops in relationship to biodiversity, either above or below ground. There is growing evidence that some diseases are suppressed through diversification of the cropping system and the inclusion of diverse cover crops as part of the rotation. At the field level, cover crops effectively compete with weeds and reduce the need to use herbicides, thus preventing the development of herbicide resistance (Davis and Frisvold, 2017; Weisberger et al., 2019). Similarly, the incorporation of cover crops within a rotation can break the life cycle of pathogens, thus contributing to the balance in the agroecosystem (Chen et al., 2012; Hershman, 1995). Biodiversity is a modestly studied aspect of cover crop interventions in arable cropping systems. Few studies have considered how mixtures influence services including pest regulation and soil health.

To foster soil health the combination of cover crops with other practices needs to be more widely explored, including integration with compost, manure and judicious conservation tillage. There is evidence that manure amendments may be synergistic with cover crops, by stabilizing aggregation processes that protect soil C accrual, and vice versa, cover crops can help prevent nutrient loss from manure use (Mpeketula and Snapp, 2019). Despite interest in the use of mixes, we found relatively few controlled comparisons of combinations or in-depth explorations of niche stratification. The ecological literature can provide important insights, as there has been extensive research on functional diversity and associated enhanced performance through cover cropping (Blesh, 2018). Other pioneering research is suggestive that cover crop diversity is essential if deeper soil C benefits are to be derived from conservation tillage (Ugarte et al., 2014). Overall, data regarding cover crop polycultures remain generally inconclusive and there is a need for clear guidance on pairings, and on cover crop management, to enhance the ability to predict the impact on soil function. A positive relationship between microbial abundance and cover crops is established as shown by a recent meta-analysis (Kim et al., 2020). However, research is limited in relation to the functional roles for enhanced microbiota under cover crop systems. For example, there is a lack of studies linking cover crop types to microbial responses or the microbiome and associated functions like disease suppression (de Corato, 2020) or C sequestration (Frasier et al., 2016).

Greater understanding of farmer decision-making and policies that enhance the adoption of cover crops for soil health are other important research areas. This warrants further research on policy approaches on increasing farmer participation, such as programs for decreased crop insurance premiums for cover cropping adoption (e.g. Illinois Department of Agriculture, 2020; Iowa Department of Agriculture and Land Stewardship, 2020). Policies may need to take into account both societal level and farmer level soil health attributes (Snapp et al., 2005). Dual-use of cover crops for grazing or other economic products needs more attention in the literature, as this could support broader adoption. Indeed, adoption by farmers is often influenced by the ability to generate co-products such as grazing or haylage, and this supports the uptake of the practice (Bartel et al., 2017). Yet there have been few initiatives that support dual use or consider how policy could enhance this farmer-friendly approach to cover crop integration.

6 Conclusion

A cropping system niche framework provides insights into which species and growth types of cover crops can be feasibly grown in a given agroecosystem, to promote soil health. Cover crop benefits depend on sufficient growth being achieved, which requires a temporal niche for growth, sufficiently warm but not too hot temperatures and moisture that is adequate. In such circumstances, cover crop biomass is optimal prior to being terminated and the gains in soil health properties are discernable. Soil health at its foundation depends on SOM, which covers crops that are effective at promoting. First and foremost, this is through conserving topsoil and associated organic matter through erosion prevention and protection of aggregates.

Conservation of soil is an important service that operates at both the farm and society levels. However, erosion prevention is difficult to measure and it occurs over long time periods, thus societal scale documentation and investment in cover crops is important. Finally, cover crops that are integrated in a manner that allows substantial biomass accrual, especially belowground, are associated with soil carbon accrual.

Soil physical structure is enhanced by organic matter, which enhances aeration and soil moisture management. Indeed, infiltration and water holding capacity are key to growing a healthy and productive cash crop, and to overall cropping system resilience in a changing climate. A related soil function that benefits farmers, as well as providing societal benefits, is the regulation of water quality. Cover crops that establish rapidly and produce copious root systems have been shown to be highly effective at reducing nitrate losses, to protect water quality. Further, if managed well, temporary immobilization of inorganic N during cover crop residue can protect soil water quality and the environment, with no or limited impact on crop yield. Finally, cover crops such as legumes that have high biochemical quality residues, provide nutrient-rich and soluble carbon sources for microbial function and are highly successful at building enhanced soil labile nitrogen pools, for reduced reliance on inorganic nitrogen, leading to cost savings and environmental benefits.

Although the potential is high, in the complexity of real-world agricultural systems, soil health varies considerably. Cover crops require adequate growth conditions to fully express their potential, and this is a challenge to achieve within the context of commercial agricultural production. In this chapter we took a regional approach, to consider realistic soil health benefits, whereby cover crops are conditioned by crop species present, and by the environmental context. For example, in hot, dry environments such as the West Coast and the Great Plains water use by cover crops is considered in a much different way than humid areas such as the Upper Midwest where spring rainfall is often associated with inundation. In both cases, cover crops that promote aggregation are valued. However, in dry, hot areas, producers select cover crop species that are parsimonious of soil moisture, whereas, in humid areas, producers prefer cover crops that take up water and reduce soil moisture at times of the year when typical cultural practices take place. Soil water dynamics are complex, as hotter and drier regions might benefit from cover crop biomass shielding the soil to reduce evaporation and conserve soil moisture, yet moisture retention for a subsequent cash crop must also be considered at least in rain-fed cropping systems. Thus, the impact of cover crops can either increase or decrease soil moisture depending on region, species, management and time of year.

On a societal scale, cover crop benefits include climate change mitigation, which is related to soil C sequestration and reduced requirements for fossil-fuel-derived inputs. There is a moderate body of literature that is consistent with greenhouse gas mitigation through soil C sequestration being associated with cover crop adoption. The literature on suppression of nitrous oxide generation is also limited, with cover crops providing variable impacts. A large research gap is the potential of cover crop diversification to reduce requirements for fertilizers and pesticides through gains in soil function, e.g. nutrient supply capacity, and healthy crop growth. Long-term studies are required, including scientific investment in plant breeding and agronomic strategies that support such slow accruing soil health benefits. These areas are difficult to research and may require participatory and on-farm investigation due to the complex interactions with cropping sequences, management, edaphic properties and the environmental context.

Overall, cover crops are fundamental to diversification above and belowground, to enhance living cover, and to healthy function in production agriculture. How to understand and enhance soil health through cover crops is a rapidly growing field, one that will benefit both farmers and society at large.

7 Where to look for further information

7.1 Websites

- https://www.sare.org/resources/cover-crops/.
- https://mccc.msu.edu/covercroptool/.
- http://northeastcovercrops.com/.
- https://westerncovercrops.org/.
- https://southerncovercrops.org/.
- https://mccc.msu.edu/.

7.2 Books and reviews

- Clark, A. ed., 2008. *Managing Cover Crops Profitably*. Diane Publishing. https://www.sare.org/resources/managing-cover-crops-profitably-3rd-edition/.
- Shackelford, G. E., Kelsey, R. and Dicks, L. V., 2019. Effects of cover crops on multiple ecosystem services: ten meta-analyses of data from arable farmland in California and the Mediterranean. *Land Use Policy*, *88*, 104204.
- Snapp, S. S., Swinton, S. M., Labarta, R., Mutch, D. R., Black, J. R., Leep, R., Nyiraneza J. and O'Neil, K., 2005. Evaluating benefits and costs of cover crops for cropping system niches. *Agronomy Journal*, *97*, 322–332.

8 References

Abawi, G. S. and Widmer, T. L. (2000). Impact of soil health management practices on soilborne pathogens, nematodes and root diseases of vegetable crops, *Appl. Soil Ecol.* 15(1), 37–47.

Abdalla, M., Hastings, A., Cheng, K., Yue, Q., Chadwick, D., Espenberg, M., Truu, J., Rees, R. M. and Smith, P. (2019). A critical review of the impacts of cover crops on nitrogen leaching, net greenhouse gas balance and crop productivity, *Glob. Chang. Biol.* 25(8), 2530–2543.

Acharya, R. N., Ghimire, R., Gc, A. and Blayney, D. (2019). Effect of cover crop on farm profitability and risk in the southern high plains, *Sustainability* 11(24), 1–13.

Adams, C., Thapa, S. and DeLaune, P. (2020). Growth and development of irrigated cotton in conservation management systems of the Texas rolling plains, *Agron. J.* 112(3), 1805–1814.

Allen, B. L., Pikul, J. L., Waddell, J. T. and Cochran, V. L. (2011). Long-term lentil green-manure replacement for fallow in the semiarid northern great plains, *Agron. J.* 103(4), 1292–1298.

Aryal, N., Reba, M. L., Straitt, N., Teague, T. G., Bouldin, J. and Dabney, S. (2018). Impact of cover crop and season on nutrients and sediment in runoff water measured at the edge of fields in the Mississippi Delta of Arkansas, *J. Soil Water Conserv.* 73(1), 24–34.

Baraibar, B., Hunter, M. C., Schipanski, M. E., Hamilton, A. and Mortensen, D. A. (2018). Weed suppression in cover crop monocultures and mixtures, *Weed Sci.* 66(1), 121–133.

Bartel, C. A., Banik, C., Lenssen, A. W., Moore, K. J., Laird, D. A., Archontoulis, S. V. v. and Lamkey, K. R. (2017). Establishment of perennial groundcovers for maize-based bioenergy production systems, *Agron. J.* 109(3), 822–835.

Basche, A. D., Kaspar, T. C., Archontoulis, S. V. v., Jaynes, D. B., Sauer, T. J., Parkin, T. B. and Miguez, F. E. (2016). Soil water improvements with the long-term use of a winter rye cover crop, *Agric. Water Manag.* 172, 40–50.

Baumgartner, K., Steenwerth, K. L. and Veilleux, L. (2008). Cover-crop systems affect weed communities in a California Vineyard, *Weed Sci.* 56(4), 596–605.

Bergtold, J. S., Ramsey, S., Maddy, L. and Williams, J. R. (2019). A review of economic considerations for cover crops as a conservation practice, *Renew. Agr. Food Syst.* 34(1), 62–76.

Blanco-Canqui, H., Mikha, M. M., Presley, D. R. and Claassen, M. M. (2011). Addition of cover crops enhances no-till potential for improving soil physical properties, *Soil Sci. Soc. Am. J.* 75(4), 1471–1482.

Blanco-Canqui, H., Shaver, T. M., Lindquist, J. L., Shapiro, C. A., Elmore, R. W., Francis, C. A. and Hergert, G. W. (2015). Cover crops and ecosystem services: insights from studies in temperate soils, *Agron. J.* 107(6), 2449–2474.

Blesh, J. (2018). Functional traits in cover crop mixtures: biological nitrogen fixation and multifunctionality, *J. Appl. Ecol.* 55(1), 38–48.

Blesh, J. (2019). Feedbacks between nitrogen fixation and soil organic matter increase ecosystem functions in diversified agroecosystems, *Ecol. Appl.* 29(8), e01986.

Blesh, J., VanDusen, B. M. and Brainard, D. C. (2019). Managing ecosystem services with cover crop mixtures on organic farms, *Agron. J.* 111(2), 826–840.

Bodner, G., Himmelbauer, M., Loiskandl, W. and Kaul, H.-P. (2010). Improved evaluation of cover crop species by growth and root factors, *Agron. Sustain. Dev.* 30(2), 455–464.

Boquet, D. J., Hutchinson, R. L. and Breitenbeck, G. A. (2004). Long-term tillage, cover crop, and nitrogen rate effects on cotton: plant growth and yield components, *Agron. J.* 96(5), 1443–1452.

Brennan, E. B., Boyd, N. S., Smith, R. F. and Foster, P. (2011). Comparison of rye and legume-rye cover crop mixtures for vegetable production in California, *Agron. J.* 103(2), 449–463.

Brennan, E. B. and Smith, R. F. (2005). Winter cover crop growth and weed suppression on the central coast of California 1, *Weed Technol.* 19(4), 1017–1024.

Chalise, K. S., Singh, S., Wegner, B. R., Kumar, S., Pérez-Gutiérrez, J. D., Osborne, S. L., Nleya, T., Guzman, J. and Rohila, J. S. (2018). Cover crops and returning residue impact on soil organic carbon, bulk density, penetration resistance, water retention, infiltration, and soybean yield, *Agron. J.* 111(1), 99–108.

Chen, C., Neill, K., Burgess, M. and Bekkerman, A. (2012). Agronomic benefit and economic potential of introducing fall-seeded pea and lentil into conventional wheat-based crop rotations, *Agron. J.* 104(2), 215–224.

Clark, A. (2008). *Managing Cover Crops Profitably* (3rd edn.), Diane Publishing, SARE https://www.sare.org/resources/managing-cover-crops-profitably-3rd-edition/.
Colla, G., Mitchell, J. P., Joyce, B. A., Huyck, L. M., Wallender, W. W., Temple, S. R., Hsiao, T. C. and Poudel, D. D. (2000). Soil physical properties and tomato yield and quality in alternative cropping systems, *Agron. J.* 92(5), 924–932.
Couëdel, A., Kirkegaard, J., Alletto, L. and Justes, É. (2019). Crucifer-legume cover crop mixtures for biocontrol: toward a new multi-service paradigm, *Adv. Agron.* 157, 55–139.
Crandall, S. M., Ruffo, M. L. and Bollero, G. A. (2005). Cropping system and nitrogen dynamics under a cereal winter cover crop preceding corn, *Plant Soil* 268(1), 209–219.
Creamer, N. G., Plassman, B., Bennett, M. A., Wood, R. K., Stinner, B. R. and Cardina, J. (1995). A method for mechanically killing cover crops to optimize weed suppression, *Am. J. Altern. Agr.* 10(4), 157–162.
Dabney, S. M., Delgado, J. A. and Reeves, D. W. (2001). Using winter cover crops to improve soil and water quality, *Commun. Soil Sci. Plant Anal.* 32(7–8), 1221–1250.
Daryanto, S., Fu, B., Wang, L., Jacinthe, P. A. and Zhao, W. (2018). Quantitative synthesis on the ecosystem services of cover crops, *Earth Sci. Rev.* 185, 357–373.
Davis, A. S. and Frisvold, G. B. (2017). Are herbicides a once in a century method of weed control?, *Pest Manag. Sci.* 73(11), 2209–2220.
de Corato, U. (2020). Soil microbiota manipulation and its role in suppressing soil-borne plant pathogens in organic farming systems under the light of microbiome-assisted strategies, *Chem. Biol. Technol. Agric.* 7, 1–26.
Deguchi, S., Shimazaki, Y., Uozumi, S., Tawaraya, K., Kawamoto, H. and Tanaka, O. (2007). White clover living mulch increases the yield of silage corn via arbuscular mycorrhizal fungus colonization, *Plant Soil* 291(1–2), 291–299.
DeLaune, P. B., Mubvumba, P., Lewis, K. L. and Keeling, J. W. (2019). Rye cover crop impacts soil properties in a long-term cotton system, *Soil Sci. Soc. Am. J.* 83(5), 1451–1458.
Di, H. J. and Cameron, K. C. (2002). Nitrate leaching in temperate agroecosystems: sources, factors and mitigating strategies, *Nutr. Cycl. Agroecosys* 64(3), 237–256.
Ding, G., Liu, X., Herbert, S., Novak, J., Amarasiriwardena, D. and Xing, B. (2006). Effect of cover crop management on soil organic matter, *Geoderma* 130(3–4), 229–239.
Doane, T. A., Horwath, W. R., Mitchell, J. P., Jackson, J., Miyao, G. and Brittan, K. (2009). Nitrogen supply from fertilizer and legume cover crop in the transition to no-tillage for irrigated row crops, *Nutr. Cycl. Agroecosyst.* 85(3), 253–262.
Dou, Z. and Fox, R. H. (1994). The contribution of nitrogen from legume cover crops double-cropped with winter wheat to tilled and non-tilled maize, *Eur. J. Agron.* 3(2), 93–100.
Dou, Z., Fox, R. H. and Toth, J. D. (1994). Tillage effect on seasonal nitrogen availability in corn supplied with legume green manures, *Plant Soil* 162(2), 203–210.
Drinkwater, L. E. and Snapp, S. S. (2007). Understanding and managing the rhizosphere in agroecosystems. In: Cardon, Z. G. and Whitbecjm, J. L. (Eds), *The Rhizosphere-an Ecological Perspective*, Elsevier, San Diego, CA, pp. 127–154.
Dunbar, M. W., O'Neal, M. E. and Gassmann, A. J. (2016). Increased risk of insect injury to corn following rye cover crop, *J. Econ. Entomol.* 109(4), 1691–1697.
Duniway, M. C., Pfennigwerth, A. A., Fick, S. E., Nauman, T. W., Belnap, J. and Barger, N. N. (2019). Wind erosion and dust from US drylands: a review of causes, consequences, and solutions in a changing world, *Ecosphere* 10(3), e02650. doi: 10.1002/ecs2.2650.
Duzy, L. M., Kornecki, T. S., Balkcom, K. S. and Arriaga, F. J. (2014). Net returns and risk for cover crop use in Alabama tomato production, *Renew. Agr. Food Syst.* 29(4), 334–344.

Fageria, N. K., Baligar, V. C. and Bailey, B. A. (2005). Role of cover crops in improving soil and row crop productivity, *Commun. Soil Sci. Plant Anal.* 36(19-20), 2733-2757.

Feaga, J. B., Selker, J. S., Dick, R. P. and Hemphill, D. D. (2010). Long-term nitrate leaching under vegetable production with cover crops in the Pacific Northwest, *Soil Sci. Soc. Am. J.* 74(1), 186-195.

Finney, D. M., White, C. M. and Kaye, J. P. (2016). Biomass production and carbon/nitrogen ratio influence ecosystem services from cover crop mixtures, *Agron. J.* 108(1), 39-52.

Franco, J. G., King, S. R., Masabni, J. G. and Volder, A. (2015). Plant functional diversity improves short-term yields in a low-input intercropping system, *Agric. Ecosyst. Environ.* 203, 1-10.

Frasier, I., Noellemeyer, E., Figuerola, E., Erijman, L., Permingeat, H. and Quiroga, A. (2016). High quality residues from cover crops favor changes in microbial community and enhance C and N sequestration, *Glob. Ecol. Conserv.* 6, 242-256.

Gabriel, J. L., Quemada, M., Martín-Lammerding, D. and Vanclooster, M. (2019). Assessing the cover crop effect on soil hydraulic properties by inverse modelling in a 10-year field trial, *Agric. Water Manag.* 222, 62-71.

Gallaher, C. and Snapp, S. S. (2015). Organic management and legume presence maintained phosphorus bioavailability in a 17-year field crop experiment, *Renew. Agr. Food Syst.* 30(3), 211-222.

Ghimire, R., Ghimire, B., Mesbah, A. O., Sainju, U. M. and Idowu, O. J. (2019). Soil health response of cover crops in winter wheat-fallow system, *Agron. J.* 111(4), 2108-2115.

Guerra, B. and Steenwerth, K. (2012). Influence of floor management technique on grapevine growth, disease pressure, and juice and wine composition: a review, *Am. J. Enol. Vitic.* 63(2), 149-164.

Hanrahan, B. R., Tank, J. L., Christopher, S. F., Mahl, U. H., Trentman, M. T. and Royer, T. V. v. (2018). Winter cover crops reduce nitrate loss in an agricultural watershed in the central U.S., *Agric. Ecosyst. Environ.* 265, 513-523.

Hartwig, N. L. and Ammon, H. U. (2002). Cover crops and living mulches, *Weed Sci.* 50(6), 688-699.

Hartwig, N. L. and Hoffman, L. D. (1975). Suppression of perennial legume and grass cover crops for no-tillage corn. *Proc. Northeast. Weed Sci. Soc.* 29, 82-88.

Haruna, S. I., Anderson, S. H., Udawatta, R. P., Gantzer, C. J., Phillips, N. C., Cui, S. and Gao, Y. (2020). Improving soil physical properties through the use of cover crops: a review, *Agrosyst. Geosci. Environ.* 3(1), e20105.

Hashemi, M., Farsad, A., Sadeghpour, A., Weis, S. A. and Herbert, S. J. (2013). Cover-crop seeding-date influence on fall nitrogen recovery, *J. Plant Nutr. Soil Sci.* 176(1), 69-75.

Herrero, E. V. v., Mitchell, J. P., Lanini, W. T., Temple, S. R., Miyao, E. M., Morse, R. D. and Campiglia, E. (2001). Use of cover crop mulches in a no-till furrow-irrigated processing tomato production system, *HortTech* 11(1), 43-48.

Hershman, D. E. (1995). Effect of wheat residue and tillage on *Heterodera glycines* and yield of double crop soybean in Kentucky, *Plant Dis.* 79(6), 631-633.

Holmes, A. A., Thompson, A. A., Lovell, S. T., Villamil, M. B., Yannarell, A. C., Dawson, J. O. and Wortman, S. E. (2019). Nitrogen provisioned and recycled by cover crops in monoculture and mixture across two organic farms, *Nutr. Cycl. Agroecosyst.* 115(3), 441-453.

Idowu, O. J., van Es, H. M., Abawi, G. S., Wolfe, D. W., Schindelbeck, R. R., Moebius-Clune, B. N. and Gugino, B. K. (2009). Use of an integrative soil health test for evaluation of soil management impacts, *Renew. Agr. Food Syst.* 24(3), 214–224.

Illinois Department of Agriculture (2020). Cover crop premium discount program returns for 2020, Available at: https://www2.illinois.gov/IISNews/22030-Cover_Crop_Premium_Discount_Program_Returns_for_2020.pdf (accessed 4 December 2020).

Iowa Department of Agriculture and Land Stewardship (2020). Crop insurance discounts available for farmers who plant cover crops, Available at: https://iowaagriculture.gov/news/crop-insurance-discounts-available-farmers-who-plant-cover-crops (accessed 4 December 2020).

Jani, A. D., Grossman, J., Smyth, T. J. and Hu, S. (2016). Winter legume cover-crop root decomposition and N release dynamics under disking and roller-crimping termination approaches, *Renew. Agr. Food Syst.* 31(3), 214–229.

Jarecki, M. K., Parkin, T. B., Chan, A. S. K., Kaspar, T. C., Moorman, T. B., Singer, J. W., Kerr, B. J., Hatfield, J. L. and Jones, R. (2009). Cover crop effects on nitrous oxide emission from a manure-treated mollisol, *Agric. Ecosyst. Environ.* 134(1–2), 29–35.

Kadziene, G., Suproniene, S., Auskalniene, O., Pranaitiene, S., Svegzda, P., Versuliene, A., Ceseviciene, J., Janusauskaite, D. and Feiza, V. (2020). Tillage and cover crop influence on weed pressure and Fusarium infection in spring cereals, *Crop Prot.* 127, 104966.

Karathanasis, A. D., Murdock, L. W., Matocha, C. J., Grove, J. and Thompson, Y. L. (2014). Fragipan horizon fragmentation in slaking experiments with amendment materials and ryegrass root tissue extracts, *Sci. World J.* 2014, 276892. doi: 10.1155/2014/276892.

Kaspar, T. C., Radke, J. K. and Laflen, J. M. (2001). Small grain cover crops and wheel traffic effects on infiltration, runoff, and erosion, *J. Soil Water Conserv.* 56, 160–164.

Kaye, J. P. and Quemada, M. (2017). Using cover crops to mitigate and adapt to climate change: a review, *Agron. Sustain. Dev.* 37(1), 4.

Ketterings, Q. M., Swink, S. N., Duiker, S. W., Czymmek, K. J., Beegle, D. B. and Cox, W. J. (2015). Integrating cover crops for nitrogen management in corn systems on Northeastern U.S. dairies, *Agron. J.* 107(4), 1365–1376.

Kim, N., Zabaloy, M. C., Guan, K. and Villamil, M. B. (2020). Do cover crops benefit soil microbiome? A meta-analysis of current research, *Soil Biol. Biochem.* 142, 107701.

Krueger, E. S., Ochsner, T. E., Porter, P. M. and Baker, J. M. (2011). Winter rye cover crop management influences on soil water, soil nitrate, and corn development, *Agron. J.* 103(2), 316–323.

Kuo, S., Sainju, U. M. and Jellum, E. J. (1997). Winter cover crop effects on soil organic carbon and carbohydrate in soil, *Soil Sci. Soc. Am. J.* 61(1), 145–152.

Langdale, G. W., Blevins, R. L., Karlen, D. L., McCool, D. K., Nearing, M. A., Skidmore, E. L., Thomas, A. W., et al. (1991). Cover crop effects on soil erosion by wind and water, *Wind Water Erosion*, 15–22.

Larkin, R. P. (2020). Effects of selected soil amendments and mulch type on soil properties and productivity in organic vegetable production, *Agronomy* 10, 171–186.

Lawson, A., Fortuna, A. M., Cogger, C., Bary, A. and Stubbs, T. (2013). Nitrogen contribution of rye-hairy vetch cover crop mixtures to organically grown sweet corn, *Renew. Agr. Food Syst.* 28(1), 59–69.

Lewis, K. L., Burke, J. A., Keeling, W. S., McCallister, D. M., Delaune, P. B. and Keeling, J. W. (2018). Soil benefits and yield limitations of cover crop use in Texas high plains cotton, *Agron. J.* 110(4), 1616–1623.

Li, M., Peterson, C. A., Tautges, N. E., Scow, K. M. and Gaudin, A. C. M. (2019). Yields and resilience outcomes of organic, cover crop, and conventional practices in a Mediterranean climate, *Sci. Rep.* 9(1), 12283.

Liebman, M. and Davis, A. S. (2000). Integration of soil, crop and weed management in low-external-input farming systems, *Weed Res.* 40(1), 27–47.

Luna, J. M., Sullivan, D., Garrett, A. M. and Xue, L. (2018). Cover crop nitrogen contribution to organic broccoli production, *Renew. Agr. Food Syst.* 35, 49–58.

Marcillo, G. S. and Miguez, F. E. (2017). Corn yield response to winter cover crops: an updated meta-analysis, *J. Soil Water Conserv.* 72(3), 226–239.

Mauromicale, G., Occhipinti, A. and Mauro, R. P. (2010). Selection of shade-adapted subterranean clover species for cover cropping in orchards, *Agron. Sustain. Dev.* 30(2), 473–480.

McDonald, M. D., Lewis, K. L. and Ritchie, G. L. (2020). Short term cotton lint yield improvement with cover crop and no-tillage implementation, *Agronomy* 10, 1–14.

McSwiney, C. P., Snapp, S. S. and Gentry, L. E. (2010). Use of N immobilization to tighten the N cycle in conventional agroecosystems, *Ecol. Appl.* 20(3), 648–662.

Meisinger, J. J., Hargrove, W. L., Mikkelsen, R. L., Williams, J. R. and Benson, V. (1991). Effects of cover crops on groundwater quality. In: Hargrove, W. L. (Ed.), *Cover Crops for Clean Water*, Soil & Water Conservation Society, Ankeny, pp. 57–68.

Messiga, A. J., Gallant, K. S., Sharifi, M., Hammermeister, A., Fuller, K., Tango, M. and Fillmore, S. (2016). Grape yield and quality response to cover crops and amendments in a vineyard in Nova Scotia, Canada, *Am. J. Enol. Vitic.* 67(1), 77–85.

Mitchell, J. P., Shrestha, A., Horwath, W. R., Southard, R. J., Madden, N., Veenstra, J. and Munk, D. S. (2015a). Tillage and cover cropping affect crop yields and soil carbon in the San Joaquin valley, California, *Agron. J.* 107(2), 588–596.

Mitchell, J. P., Shrestha, A. and Irmak, S. (2015b). Trade-offs between winter cover crop production and soil water depletion in the San Joaquin Valley, California, *J. Soil Water Conserv.* 70(6), 430–440.

Mitchell, J. P., Shrestha, A., Mathesius, K., Scow, K. M., Southard, R. J., Haney, R. L., Schmidt, R., Munk, D. S. and Horwath, W. R. (2017). Cover cropping and no-tillage improve soil health in an arid irrigated cropping system in California's San Joaquin Valley, USA, *Soil Till. Res.* 165, 325–335.

Mooleki, S. P., Gan, Y., Lemke, R. L., Zentner, R. P. and Hamel, C. (2016). Effect of green manure crops, termination method, stubble crops, and fallow on soil water, available N, and exchangeable P, *Can. J. Plant Sci.* 96(5), 867–886.

Moore, E. B., Wiedenhoeft, M. H., Kaspar, T. C. and Cambardella, C. A. (2014). Rye cover crop effects on soil quality in no-till corn silage-soybean cropping systems, *Soil Sci. Soc. Am. J.* 78(3), 968–976.

Mpeketula, P. M. G. and Snapp, S. S. (2019). Structural stability conditions soil carbon gains from compost management and rotational diversity, *Soil Sci. Soc. Am. J.* 83(1), 203–211.

Nielsen, D. C., Lyon, D. J., Higgins, R. K., Hergert, G. W., Holman, J. D. and Vigil, M. F. (2016). Cover crop effect on subsequent wheat yield in the central great plains, *Agron. J.* 108(1), 243–256.

O'Dea, J. K., Jones, C. A., Zabinski, C. A., Miller, P. R. and Keren, I. N. (2015). Legume, cropping intensity, and N-fertilization effects on soil attributes and processes from an eight-year-old semiarid wheat system, *Nutr. Cycl. Agroecosyst.* 102(2), 179–194.

Olson, K., Ebelhar, S. A. and Lang, J. M. (2014). Long-term effects of cover crops on crop yields, soil organic carbon stocks and sequestration, *Open J. Soil Sci.* 04(8), 284–292.

Paustian, K., Larson, E., Kent, J., Marx, E. and Swan, A. (2019). Soil C sequestration as a biological negative emission strategy, *Front. Clim.* 1, 8.

Peterson, A. T., Berti, M. T. and Samarappuli, D. (2019). Intersowing cover crops into standing soybean in the US upper midwest, *Agronomy* 9(5), 1–14.

Pi, H., Huggins, D. R. and Sharratt, B. (2019). Dry aggregate stability of soils influenced by crop rotation, soil amendment, and tillage in the Columbia Plateau, *Aeolian Res.* 40, 65–73.

Poeplau, C. and Don, A. (2015). Carbon sequestration in agricultural soils via cultivation of cover crops – a meta-analysis, *Agr. Ecosyst. Environ.* 200, 33–41.

Ratnadass, A., Fernandes, P., Avelino, J. and Habib, R. (2012). Plant species diversity for sustainable management of crop pests and diseases in agroecosystems: a review, *Agron. Sustain. Dev.* 32(1), 273–303.

Reicosky, D. C. and Forcella, F. (1998). Cover crop and soil quality interactions in agroecosystems, *J. Soil Water Conserv.* 53, 224–229.

Reiss, A., Fomsgaard, I. S., Mathiassen, S. K. and Kudsk, P. (2018). Weed suppressive traits of winter cereals: allelopathy and competition, *Biochem. Syst. Ecol.* 76, 35–41.

Rivers, A., Voortman, C. and Barbercheck, M. (2020). Cover crops support arthropod predator activity with variable effects on crop damage during transition to organic management, *Biol. Control* 151, 104377.

Roper, W. R., Osmond, D. L., Heitman, J. L., Wagger, M. G. and Reberg-Horton, S. C. (2017). Soil health indicators do not differentiate among agronomic management systems in North Carolina soils, *Soil Sci. Soc. Am. J.* 81(4), 828–843.

Runck, B. C., Khoury, C. K., Ewing, P. M. and Kantar, M. (2020). The hidden land use cost of upscaling cover crops, *Commun. Biol.* 3(1), 300.

Rusch, H. L., Coulter, J. A., Grossman, J. M., Johnson, G. A., Porter, P. M. and Garcia Garcia, A. (2020). Towards sustainable maize production in the U.S. upper Midwest with interseeded cover crops, *PLoS ONE* 15(4), e0231032.

Ryan, M. R., Mirsky, S. B., Mortensen, D. A., Teasdale, J. R. and Curran, W. S. (2011). Potential synergistic effects of cereal rye biomass and soybean planting density on weed suppression, *Weed Sci.* 59(2), 238–246.

Sainju, U. M., Singh, B. P. and Whitehead, W. F. (2002). Long-term effects of tillage, cover crops, and nitrogen fertilization on organic carbon and nitrogen concentrations in sandy loam soils in Georgia, USA, *Soil Till. Res.* 63(3–4), 167–179.

Sarrantonio, M. and Scott, T. W. (1988). Tillage effects on availability of nitrogen to corn following a winter green manure crop, *Soil Sci. Soc. Am. J.* 52(6), 1661 –1668.

Schipanski, M. E., Barbercheck, M., Douglas, M. R., Finney, D. M., Haider, K., Kaye, J. P., Kemanian, A. R., Mortensen, D. A., Ryan, M. R., Tooker, J. and White, C. (2014). A framework for evaluating ecosystem services provided by cover crops in agroecosystems, *Agric. Syst.* 125, 12–22.

Schomberg, H. H., Fisher, D. S., Reeves, D. W., Endale, D. M., Raper, R. L., Jayaratne, K. S. U., Gamble, G. R. and Jenkins, M. B. (2014). Grazing winter rye cover crop in a cotton no-till system: yield and economics, *Agron. J.* 106(3), 1041–1050.

Singh, H. P., Batish, D. R. and Kohli, R. K. (2003). Allelopathic interactions and allelochemicals: new possibilities for sustainable weed management, *Crit. Rev. Plant Sci.* 22(3–4), 239–311.

Skidmore, E. L. and Siddoway, F. H. (1978). Crop residue requirements to control wind erosion. In: Oschwald, W. R. (Ed.), *Crop Residue Management Systems*, American Society of Agronomy/Crop Science Society of America/Soil Science Society of America, Madison, WI, pp. 17-33.

Smith, R. G., Gross, K. L. and Robertson, G. P. (2008). Effects of crop diversity on agroecosystem function: crop yield response, *Ecosystems* 11(3), 355-366.

Smukler, S. M., Jackson, L. E., Murphree, L., Yokota, R., Koike, S. T. and Smith, R. F. (2008). Transition to large-scale organic vegetable production in the Salinas Valley, California, *Agr. Ecosyst. Environ.* 126(3-4), 168-188.

Snapp, S. and Surapur, S. (2018). Rye cover crop retains nitrogen and doesn't reduce corn yields, *Soil Till. Res.* 180, 107-115.

Snapp, S. S., Swinton, S. M., Labarta, R., Mutch, D., Black, J. R., Leep, R., Nyiraneza, J. and O'Neil, K. (2005). Evaluating cover crops for benefits, costs and performance within cropping system niches, *Agron. J.* 97(1), 322-332.

Sojka, R. E., Langdale, G. W. and Karlen, D. L. (1984). Vegetative techniques for reducing water erosion of cropland in the southeastern United States, *Adv. Agron.* 37, 155-181.

Strickland, T. C., Scully, B. T., Hubbard, R. K., Sullivan, D. G., Abdo, Z., Savabi, M. R., Lee, R. D., Olson, D. M. and Hawkins, G. L. (2015). Effect of conservation practices on soil carbon and nitrogen accretion and crop yield in a corn production system in the southeastern coastal plain, United States, *J. Soil Water Conserv.* 70(3), 170-181.

Tani, E., Abraham, E., Chachalis, D. and Travlos, I. (2017). Molecular, genetic and agronomic approaches to utilizing pulses as cover crops and green manure into cropping systems, *Int. J. Mol. Sci.* 18(6), 1202.

Teague, R., DeLaune, P. B. and Dowhower, S. L. (2019). Impacts of over-seeding Bermudagrass pasture with multispecies cover crops on soil water availability, microbiology, and nutrient status in North Texas, *Agric. Ecosyst. Environ.* 273, 117-129.

Thapa, R., Mirsky, S. B. and Tully, K. L. (2018). Cover crops reduce nitrate leaching in agroecosystems: a global meta-analysis, *J. Environ. Qual.* 47(6), 1400-1411.

Thorup-Kristensen, K., Magid, J. and Jensen, L. S. (2003). Catch crops and green manures as biological tools in nitrogen management in temperate zones, *Adv. Agron.* 79, 227-302.

Tiemann, L. K., Grandy, A. S., Atkinson, E. E., Marin-Spiotta, E. and McDaniel, M. D. (2015). Crop rotational diversity enhances belowground communities and functions in an agroecosystem, *Ecol. Lett.* 18(8), 761-771.

Turner, B. L., Fuhrer, J., Wuellner, M., Menendez, H. M., Dunn, B. H. and Gates, R. (2018). Scientific case studies in land-use driven soil erosion in the central United States: why soil potential and risk concepts should be included in the principles of soil health, *Int. Soil Water Conserv.* 6, 63-78.

Ugarte, C. M., Kwon, H., Andrews, S. S. and Wander, M. M. (2014). A meta-analysis of soil organic matter response to soil management practices: an approach to evaluate conservation indicators, *J. Soil Water Conserv.* 69(5), 422-430.

Unger, P. W. and Vigil, M. F. (1998). Cover crop effects on soil water relationships, *J. Soil Water Conserv.* 53, 200-207.

USDA-NRCS (2006). *Land Resource Regions and Major Land Resource Areas of the United States, the Caribbean, and the Pacific Basin*. Handbook 296.

Vann, R. A., Reberg-Horton, S. C., Castillo, M. S., McGee, R. J. and Mirsky, S. B. (2019). Winter pea, crimson clover, and hairy vetch planted in mixture with small grains in the southeast United States, *Agron. J.* 111(2), 805-815.

Villamil, M. B., Bollero, G. A., Darmody, R. G., Simmons, F. W. and Bullock, D. G. (2006). No-till corn/soybean systems including winter cover crops: effects on soil properties, *Soil Sci. Soc. Am. J.* 70(6), 1936–1944.

Wanic, M., Żuk-Gołaszewska, K., Wanic, M. and Orzech, K. (2018). Catch crops and the soil environment – a review of the literature, *J. Elem.* 24, 31–45.

Wayman, S., Cogger, C., Benedict, C., Burke, I., Collins, D. and Bary, A. (2015). The influence of cover crop variety, termination timing and termination method on mulch, weed cover and soil nitrate in reduced-tillage organic systems, *Renew. Agr. Food Syst.* 30(5), 450–460.

Weinert, T. L., Pan, W. L., Moneymaker, M. R., Santo, G. S. and Stevens, R. G. (2002). Nitrogen Recycling by Nonleguminous Winter Cover Crops to Reduce Leaching in Potato Rotations, *Agron. J.* 94(2), 365–372.

Weisberger, D., Nichols, V. and Liebman, M. (2019). Does diversifying crop rotations suppress weeds? A meta-analysis, *PLoS ONE* 14(7), e0219847.

Wells, M. S., Brinton, C. M. and Reberg-Horton, S. C. (2016). Weed suppression and soybean yield in a no-till cover-crop mulched system as influenced by six rye cultivars, *Renew. Agr. Food Syst.* 31(5), 429–440.

Wendt, R. C. and Burwell, R. E. (1985). Runoff and soil losses for conventional, reduced, and no-till corn, *J. Soil Water Conserv.* 40, 450–454.

White, C. M., DuPont, S. T., Hautau, M., Hartman, D., Finney, D. M., Bradley, B., LaChance, J. C. and Kaye, J. P. (2017). Managing the trade off between nitrogen supply and retention with cover crop mixtures, *Agr. Ecosyst. Environ.* 237, 121–133.

Williams, J. R., Allmaras, R. R., Renard, K. G., Lyles, L., Moldenhauer, W. C., Langdale, G. W., Meyer, L. D., et al. (1981). Soil erosion effects on soil productivity: a research perspective, *J. Soil Water Conserv.* 36, 82–90.

Wischmeier, W. H. (1960). Cropping-management factor evaluations for a Universal Soil-Loss Equation, *Soil Sci. Soc. Am. J.* 24(4), 322–326.

Woodruff, N. P. and Siddoway, F. H. (1965). A wind erosion equation, *Soil Sci. Soc. Am. J.* 29(5), 602–608.

Wortman, S. E., Francis, C. A., Bernards, M. L., Drijber, R. A. and Lindquist, J. L. (2012). Optimizing cover crop benefits with diverse mixtures and an alternative termination method, *Agron. J.* 104(5), 1425–1435.

Wyland, L. J., Jackson, L. E., Chaney, W. E., Klonsky, K., Koike, S. T. and Kimple, B. (1996). Winter cover crops in a vegetable cropping system: impacts on nitrate leaching, soil water, crop yield, pests and management costs, *Agr. Ecosyst. Environ.* 59(1–2), 1–17.

Zentner, R. P., Campbell, C. A., Biederbeck, V. O., Selles, F., Lemke, R., Jefferson, P. G. and Gan, Y. (2004). Long-term assessment of management of an annual legume green manure crop for fallow replacement in the Brown soil zone, *Can. J. Plant Sci.* 84(1), 11–22.

Zhang, Y., Haan, N. L. and Landis, D. A. (2020). Landscape composition and configuration have scale-dependent effects on agricultural pest suppression, *Agr. Ecosyst. Environ.* 302, 107085.

Chapter 8

Assessing the effects of crop residue retention on soil health

Hero T. Gollany, USDA-ARS, USA

1 Introduction

Plant residues are the largest fraction of organic matter entering a soil system, and crop residues are the most accessible form of plant residues or biomass. Conventional agricultural cropping systems rely on synthetic fertilizer; however, these systems also require regular organic inputs from crop residue, root biomass, and rhizodeposition to maintain soil organic matter (SOM) and associated soil organic carbon (SOC), improve soil nutrient cycling, soil microbial population, and soil fauna. Soil organic matter helps to define soil fertility through plant macro- and micronutrient cycling and retention and enhances soil productivity. Retaining crop residue on soils is important for enhancing soil physical properties, protecting soils from erosion and soil water loss, regulating soil temperature, maintaining SOM, enhancing water and solute transport, and ultimately soil health and land sustainability. Soil health is defined as 'the capacity of soil to function as a living system' (FAO, 2011). Soil health and 'soil quality' are commonly used interchangeably, although they are different. Wander et al. (2019) defined soil quality as the characteristics and dynamic of soil properties, while soil health is a continued capacity of soil to supply a service based on the existing attributes or process and function as a vital ecosystem within land use (Liu et al., 2022). Soil-related attributes and processes that determine healthy soil and agroecosystem sustainability related

http://dx.doi.org/10.19103/AS.2021.0094.12
Published by Burleigh Dodds Science Publishing Limited, 2023.

to crop residue retention can be broadly grouped into chemical, biological, and physical processes. Soil organic carbon has become one of the primary index/indicators of soil health parameters, with SOC stock losses having implications for chemical, biological, and physical properties, attributes of soil health, and long-term sustainability and resiliency of agricultural production systems.

The world's growing population will increase the need for food, feed, fiber, fuel, and shelter, leading to increased demand for crop residue removal. Crop residue removal could affect soil health, accelerate soil degradation, and promote environmental problems. The maintenance of healthy soil is crucial to life. This chapter covers the effect of crop residue retention/removal on SOC attributes of soil health. It will explore the forms of crop residue and their quality effect on soil chemical properties including SOM and SOC stabilization, soil pH, cation exchange capacity (CEC) and nutrient supply (Section 2); crop residue and soil biological processes (Section 3); crop residue and related soil physical properties (Section 4); trends in soil health research (Section 5); and summary and conclusions (Section 6).

2 Crop residue and soil chemical properties

Chemical properties include organic matter (a source of many plant-available nutrients), associated organic carbon, and total soil nutrient content. The rate at which these nutrients are converted to plant-available forms (mineralization, weathering, and mineral dissolution), the ability of soils to retain mineralized nutrients (exchange capacity and adsorption), and pH largely determine the rate at which nutrients become available for plant growth or lost from the site through leaching or volatilization (Fig. 1). Acidity, conductivity, and the exchange properties of organic molecules, inorganic oxides, and clay minerals also play important roles in nutrient cycling processes.

2.1 Soil organic matter

Soil organic matter (SOM) is a heterogeneous organic substance, a reservoir of plant nutrients, and a principal regulator of soil health. Soil organic matter is important in maintaining land productivity and an important indicator of soil health and agricultural sustainability (Lal, 2009a; Powlson et al., 2012). Soil organic matter consists of stable and labile organic matter fractions (Six et al., 2002a; Wander, 2004). Stable organic matter fractions are relatively resistant to decomposition with turnover time ranging from years to centuries, while labile organic matter fractions are a readily available energy source for microorganisms with turnover time ranging from weeks to months. Plant inputs drive decomposition and SOM formation and control the carbon (C) cycle. Agricultural soils play a key role in C cycling and can act as a source or sink

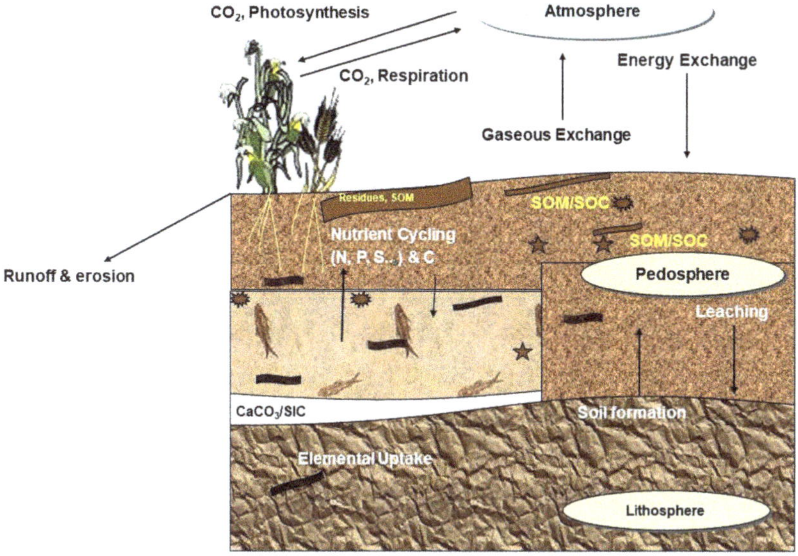

Figure 1 A simplified diagram of crop residue inputs and interaction between soils transformed by soil microorganisms. Crops capture carbon from the atmosphere through photosynthesis and produce plant residue biomass. Residue retained in the soil is then decomposed by microorganisms and soil fauna, releasing nutrients. Decomposition and respiration produce CO_2 and microbial residue which releases nutrients (N, S, P, K ...) and carbon. Nutrients are then taken up by plants and nutrient cycling continues. The organic residue is also physically and chemically attached to soil particles.

of C and therefore impact global C pools. The current worldwide SOC pool within 1-m soil depth is estimated at ~1500 Pg (Eswaran et al., 1993, 1995) and ~2500 Pg within 2-m depth (Wiesmeier et al., 2019). Anthropogenic loss of C from terrestrial ecosystems is estimated to range from 48 to 114 Pg before the industrial revolution and about 25% of this loss is attributed to SOM mineralization (Lal, 2004; Houghton, 2012).

The incorporation of crop residues, such as stover and straw, is very important to maintain or increase SOM levels under cropland production systems (Lugato et al., 2014). Wheat shoots incorporated 69% of labeled ^{14}C aboveground, 14% belowground, and respired 17% (Paul, 2016). Depending on plant age, tracer studies indicate from 20% to 40% of photosynthetic C is translocated underground; of this portion, approximately 50% remains in the roots, 33% is respired, only 2–18% converted to SOM, and a small amount is lost by leaching (Mertens et al., 2007; Jones et al., 2009; Paul, 2016). Crop inputs drive SOM formation and decomposition, controlling the C cycle. However, long-term field experiments have shown that the incorporation of cereal straw is often ineffective at increasing SOM (Heitkamp et al., 2012; Poeplau et al., 2015). Only 10–20% of added cereal straw biomass remains in the soil after 2 years (Buyanovsky et al., 1997). The efficiency of residue incorporation (i.e. the increase of SOM per unit of input) decreases with increasing residue amount added, as shown in a long-term experiment on a silty Luvisol soil (Heitkamp et al., 2012). Reasons for this may be due to (1) low belowground plant biomass because of genetic selection and breeding for high-yielding cereals, which is supposed to be more recalcitrant and have a longer mean residence time in soil (Rasse et al., 2005); (2) priming of SOM by incorporated plant residues (Fontaine et al., 2003; Kuzyakov, 2010); and (3) a finite capacity of soil aggregates, which provide physical protection for SOM against mineralization (Shahbaz et al., 2017).

2.2 Soil organic carbon

Soil organic carbon (SOC) is the most often used indicator of soil health in agroecosystems, receiving considerable attention from growers, environmental advocates, and policymakers. Soil organic carbon provides energy and substrates for microbial metabolism and promotes biological diversity, thus playing an important role in agroecosystem function (Loveland and Webb, 2003). Therefore, SOC change is used as an index of soil health/soil quality evaluation (Haynes, 2005; Karlen et al., 2019; Sparling and Schipper, 2004). However, change in SOC is not easily detected in short-term studies (Bhattacharyya et al., 2011; Gong et al., 2009) because soil C can be present in many pools and protected within the soil matrix. The stable C pools consist of either C associated with silt and clay particles, or C occluded within microaggregates. This C pool is protected from the decomposer community and has a long residence time in the soil. The stabilization of SOC is an important indicator of soil fertility (Lal, 1997; Paustian et al., 1997) and productivity, enhancing soil health and the global C cycle. However, some long-term experiments in tropical and temperate agroecosystems show little or no SOC storage change in response to C input from crop residues, suggesting an upper limit or 'saturation level' for SOC sequestration (Campbell et al., 1991; Six et al., 2002a,b; Nash et al., 2018; Novak et al., 2020). Therefore, it is important to understand that crop residue retention effects can vary across the existing SOC stocks, the factors controlling C stabilization, and SOC storage of new C addition.

2.3 Crop residue and soil organic carbon stabilization

In agroecosystems, soil C levels are governed by the balance between C inputs through crop residues and C losses, primarily through decomposition (i.e. heterotrophic soil respiration) (Paustian et al., 2000). Many factors affect SOC stabilization and storage at different spatial scales. These factors are microorganisms, soil fauna, soil type, parent material, clay mineralogy, specific surface area, metal oxides, Ca and Mg cations, aggregation, texture, natural vegetation, land use and management, topography, and climate (Wiesmeier et al., 2019). Three main mechanisms contribute to the stabilization of SOC. These are as follows: (1) inherent stability against biochemical decomposition (i.e. the chemical recalcitrance); (2) low accessibility for biological degradation caused by occlusion or encapsulation of organic C and hydrophobicity; and (3) interactions with mineral particles which include metal-OC complex formation, ligand exchange, and polyvalent cation bridges (von Lützow et al., 2006). The recalcitrant C pool is resistant to decomposition, and therefore allows C storage in soils through both physical and chemical stabilization (Oades, 1993), while the labile C pool or light fraction (LF, density < 1.6–1.8 g cm^{-3}) is easily decomposed by microorganisms. Soil labile organic C fractions are dominated by newly incorporated plant-derived materials (e.g. plant root exudates) and can be influenced by changes in soil management practice. The labile organic C fractions are easily decomposed by microorganisms (Oades, 1993) and are valuable indicators of changes in total SOC, nutrient cycling, and microbial activity. In a residue incubation study, Ge et al. (2021) found that the straw-residue C in soil was mainly retained as particulate organic carbon (POC). The residue-derived POC was significantly increased, by 3, 5, and 20 times, whereas the residue-derived dissolved organic carbon was significantly decreased by 71%, 57%, and 95% with the straw only, straw with synthetic fertilizer, and synthetic fertilizer with manure treatments, respectively, during straw residue decomposition.

Crop residue addition can affect the mineralization of native SOM (Soon and Arshad, 2002; Chen et al., 2014; Mitchell et al., 2018). The recalcitrant C pool or heavy fraction (HF, density >1.8 g cm^{-3}) is a stable SOC pool existing as an organo-mineral complex and is critical to C sequestration by persisting in soil for many years with less sensitivity to land management practices (Yan et al., 2007). Mitchell et al. (2018) reported that most residue-derived SOM accumulated as particulate organic matter (65%) with small gains in mineral-associated fractions or stable pool. Rapid accumulation of residue-derived SOM in the mineral-associated fractions during initial stages of decomposition coincided with a high loss of labile residue components. They recommended future research should examine whether increased crop residue-derived SOM stabilization is offset by increased decomposition of existing SOM and whether the incorporation of residue results in more residue stabilization over long periods.

Under field conditions, crop residues are usually at various decomposition stages. The abundance of complex compounds or the availability of labile compounds determines SOC storage or losses (Bertrand et al., 2006; Wang et al., 2015). The SOC in the LF is mainly composed of plant-derived remains like cell structure, fungal hyphae, and spores (Gregorich et al., 1995). The LF has a fast turnover time, is readily decomposed by microorganisms, and constitutes 20–30% of SOC (Modak et al., 2020). Microbes can start mining SOM to meet their nutritional and energy demands under a lack of sufficient labile substrate availability, resulting in a positive priming effect (Fontaine et al., 2003). However, it is unknown whether the quality of decaying residues or their decomposition stage is more important for SOM priming when microorganisms experience a pulse of labile C input (e.g. through root exudate).

2.4 Residue management and residue quality effect on soil organic carbon

Residue quality and residue management can have significant effects on SOC storage, SOC fractions, and thus soil health. Residue retention can improve SOC storage by increasing soil HF content. In a semiarid climate in China after 25 years of millet cultivation, the HF of SOC improved as the land converted to grassland (Han et al., 2015). Under temperate and semiarid environments, plant diversity influenced SOC stocks at different spatial scales (Soussana et al., 2004). The crops or cover crop species type and their decomposition rate can impact SOC stock (Poeplau and Don, 2015; Gregorutti and Caviglia, 2019). About 40% of plant primary productions are transferred to soils during rhizodeposition (Lynch and Whipps, 1990), which are comprised of root exudates, mycorrhizal hyphae, sloughed-off root cells, hair, and senescing roots (Lu et al., 2002; Leake et al., 2006). However, roots contain higher amounts of complex compounds (e.g. acid-insoluble fractions and lignin) than aboveground crop biomass and therefore provide less labile C to microorganisms (Gentile et al., 2011; Shahbaz et al., 2017).

Crop residue quality also affects SOC stocks and plant nutrient availability. Increasing crop rotational diversity has been shown to play a major role in increasing SOC storage, driven by enhanced root C input and soil microbial diversity. Plant diversity with synergy between functional traits can be positively correlated with SOC stocks, such as interactions between N-fixing legumes and grasses which enhance N mineralization in species-rich grasslands without legumes (Fornara and Tilman, 2008; Steinbeiss et al., 2008; Cong et al., 2014; Wiesmeier et al., 2019; Zuffo et al., 2022). In a 9-year experiment, SOC in the top 0–5 cm soil increased by ~48%, 42%, and 36%, respectively, with wheat (*Triticum aestivum* L.) + soybean (*Glycine max* L.) residue, soybean residue, and wheat residue addition compared against no residue added (Modak et al., 2020). Furthermore, in a study on canola (*Brassica napus* L.) versus wheat residue-amended soil, the canola residue with lower C-to-nutrient ratios increased native SOM mineralization resulting in a greater release of plant-available nutrients, particularly N and S (Sarker et al., 2018). Also, microbial activity increased across all the tillage treatments following the canola residue input, with a priming effect from added canola residue resulting in SOC losses relative to wheat residue (Sarker et al., 2018). The increased total organic C and native SOC mineralization by canola versus wheat residues were related to differences in the nutrient stoichiometry of the added canola residues. Furthermore, residues from cover crops have been found to enhance nutrient cycling, soil physicochemical and biological properties, weed suppression, and pest management (Koudahe et al., 2022; Teasdale and Daughtry, 1993).

In a microcosm experiment to investigate stover decomposition from two corn (*Zea mays* L.) cultivars (XY and LY) under field conditions, the stover was either field incorporated or surface applied; up to 71.7% of C and 86.9% of N in incorporated stover was decomposed during the study period, and significantly greater than 32.8% of C and 55.3% N from the surface-applied stover for both corn cultivars (Liu et al., 2019a). The LY stover had a higher decomposition rate and greater capacity for SOC sequestration than XY when incorporated into the soil, but when applied on the soil surface had 16.8% higher CO_2 emissions (Liu et al., 2019a). They concluded that incorporating LY stover into the soil rather than surface mulching could enhance SOC sequestration and reduce CO_2 emissions.

During a 6-year experiment on SOC storage under corn production of two regions in Mexico, greater SOC content was found under four conservation tillage treatments when residues were retained at several rates compared to the conventional tillage treatment (plowing and disking) and in conservation tillage with 0% residue retained (Salinas-Garcia et al., 2001). Greater SOC stocks in conservation tillage with stover retained were attributed to less residue contact with soil microorganisms and decreased decomposition rate compared to conventional tillage, where residues were mixed in the soil and in close contact with microorganisms (Reicosky et al., 1997). Corn stover mulching can facilitate the accumulation of more fungal residue than bacterial, accelerate the accumulation of microbial residues C, and enhance the contribution of microbial residues to C sequestration compared to corn stover removal treatment (Liu et al., 2019b). In another study, an 84.5% increase in corn stover incorporation in non-fertilized treatments was reported at the end of 360 days incubation (Xu et al., 2019). The increased SOC from corn stover amendment (1.08 g C kg^{-1}) could not compensate for the loss of native SOC (2.49 g C kg^{-1} soil); however, the addition of 5- and 10-times corn stover rate was sufficient to offset the native SOC loss. Soil type, N fertilizer application, and a threshold amount of corn stover input are needed to drive net SOC sequestration. Inputs of corn stover increase the mineralization of native SOM by influencing microbial biomass and activity via co-metabolism (Blagodatskaya and Kuzyakov, 2008), and through 'nutrient mining' by microbes, especially when C-rich residues are returned to nutrient-poor soils (Kuzyakov, 2010). Furthermore, free particulate organic matter or occluded within macroaggregates when exposed through mixing or tillage would be prone to microbial decomposition more than those associated with microaggregates (Six et al., 1998).

2.5 Effect of crop residue on soil pH

Crop residue retention can affect soil pH and the direction of change depends on crop residue quality, nitrogen fertilization rates, and soil properties. Retention of crop residues and changes in soil pH are related to the residue chemical composition and soil properties (Tang and Yu, 1999; Xu and Coventry, 2003; Gollany et al., 2005, 2006; Butterly et al., 2011). The initial soil pH is important in determining the magnitude and direction of pH change (Tang and Yu, 1999). In a 59-day incubation study in two soils, wheat, canola, and chickpea (*Cicer arietinum* L.) residues increased pH in both soils, and the magnitude of the pH increase was as follows: chickpea > canola > wheat (Butterly et al., 2011). The magnitude of the pH change was related to the concentration of excess cations of the residue (Noble et al., 1996; Tang and Yu, 1999). Legume residues high in ash alkalinity and N content will have a greater effect on soil pH compared to a wheat residue with low N content (Xu and Coventry, 2003). The change in soil pH is highly dependent on the quality of the residue added and soil characteristics (Luo et al., 2016). Changes in soil pH due to crop residue addition are also correlated with the concentration of organic anions in the residue and residue N content (Tang and Yu, 1999; Butterly et al., 2011). Furthermore, soil properties, such as initial soil pH, texture, soil water content, temperature, available N, and SOC content control the effect of residue on soil pH (Jarvis et al., 1996). The decomposition of organic anions contributes to the change in soil pH by two mechanisms: (1) decarboxylation of organic anions that consume H+ cations (Bonn and Fish, 1991), and (2) ligand exchange with hydroxyl groups of aluminum and iron oxides. Mineralization of crop residue N contributes to changes in soil pH by initially consuming H+ during the ammonification process and subsequently releasing H+ during nitrification, resulting in overall soil acidification because of disequilibrium in the N cycle. Therefore, the final soil pH is determined by the balance of these reactions (Xu and Coventry, 2003). Generally, the effect of residue retention on soil pH is restricted to the topsoil. In central Mexico, when corn and wheat residues were retained on the surface for 5 years, soil pH was higher (7.0) in the 0–5 cm surface layer compared to when residues were removed (pH = 6.6) or when residues were incorporated in conventional tillage (pH = 6.7); however, there was no significant effect in the subsoil layers (Govaerts et al., 2007). In contrast, in a 5-year study in the semiarid soils of central Spain, soil pH in the 0–5 cm zone was lower under NT with residue retained and zone tillage with paraplow (where a large amount of residue remains on the surface) compared to conventional tillage and minimum tillage with chisel plow (López-Fando and Pardo, 2009).

Soil acidification can restrict crop growth and favor growth of acid-tolerant plants. Soil acidification from long-term synthetic N fertilizer application to soils with low SOM content can lead to silica (Si) solubilization and movement within the soil profile, forming a layer that impedes root penetration and water movement into the soil profile (Gollany et al., 2005, 2006). The impacts of SOC and pH on the leaching of water-soluble silica from the plow layer in wheat fallow under the moldboard and sweep tillage systems confirmed Si illuviation (Gollany et al., 2005). Pea vine or manure amendments with crop residue retention reduced Si movement and siliceous pan formation that impaired infiltration and internal drainage (Gollany et al., 2006).

2.6 Crop residue retention, cation exchange capacity, and plant nutrient supply

Cation exchange capacity (CEC) is the capacity of soil to retain and exchange cations with the soil solution. It is an indicator of soil fertility, as it indicates the capacity of the soil to retain nutrients (e.g. K^+, NH_4^+, Ca^{2+}, Fe^{2+}, Zn^{2+}) in plant-available form. Since crop residue retention increases SOM content, this leads to an increase in soil pH-dependent CEC. The CEC in the top 5-cm soil depth increased when residues were retained compared with soils when residue was incorporated, but there was no difference in the 5–20 cm soil depth (Govaerts et al., 2007). An increase in SOM in the top 5 cm of the NT resulted in an increase in the CEC and the amount of extractable P, K, and Ca (Duiker and Beegle, 2006). The effect of crop residue on CEC may be limited to the topsoil layer (5-cm soil depth) as with soil pH and SOM.

Crop residues contain macro- and micronutrients, such as N, P, K, Fe, and S which are needed by crops. Degradation of crop residues and mineralization of SOM release organic acids (Guppy et al., 2005; Kumari et al., 2008). These processes facilitate the release of nutrients such as N and P from soil aggregates during dissolution and desorption reactions (Guppy et al., 2005; Keiluweit et al., 2015). Residue retention has been found to increase the concentration of P in the top 5-cm topsoil (Zibilske et al., 2002; Duiker and Beegle, 2006). This can be attributed to the redistribution of P from the lower soil layers by plant uptake (Zibilske et al., 2002). In weathered soils, the addition of residues can increase the availability of P by releasing organic acids during the decomposition of crop residues, which can block Al-oxide adsorption sites and reduce the overall adsorption of P (Haynes and Mokolobate, 2001). This will depend on the type and quality of the residue added. Legumes are generally more effective because of their high N content and low C/N ratio, which increases residue decomposition rates (Nziguheba et al., 1998). Zuffo et al. (2022) used unharvested soybean crop residues as a source of N for the subsequent corn crop in soils of medium fertility with medium acidity level and high fertility with low acidity level. They concluded soybean residues remaining on the soil surface supplied most of the N requirement to off-season corn grown in succession, especially in tropical soils of medium fertility.

In agricultural production systems that target improvement of soil health and ecosystem services recommended management practices include using cover crops, or cropping system diversification, allowing aboveground biomass to remain on the soil surface while root biomass is directly incorporated into the soil. Using a ^{15}N technique to measure residue N turnover in a corn-soybean rotation and winter wheat-red clover rotation, 9.7–13.5% of corn residue-N was recovered in the soybean seed, 26% of soybean residue-N was recovered in the corn grain, and 30% of wheat-red clover residue-N was recovered in the corn grain (Taveira et al., 2020). Recoveries of belowground residue-N ranged from 8 to 18 times more N than aboveground residues. Furthermore, the indigenous soil N or N-fixed contributed to crop N uptake more than the crop residues-N or N fertilizer combined (Taveira et al., 2020). The authors postulated that an accumulated legacy of N fertilizer applications was likely responsible for building a soil N reservoir for the crop and masking the effect that diversification might have on residue-N turnover and subsequent N uptake (Taveira et al., 2020). Diversified crop systems and including legume cover crops contributed to improved N use efficiency (Smith et al., 2008; Gaudin et al., 2015a) due to N contribution from legumes residue decomposition, reduced nitrate leaching, and reduced nitrous oxide emissions in these rotations (Pappa et al., 2011). Diversified cropping systems improved overall crop yields (Smith et al., 2008; Gaudin et al., 2015b; Bowles et al., 2020), and the resiliency of crops to overcome environmental stresses (Gaudin et al., 2015b), which might have contributed to improved soil health and enhanced ecosystem services (Scherr and McNeely, 2008).

3 Crop residue and soil biological processes

Crop residue is an essential food source for soil biota. Soil biota are critical components of soils, and most depend on organic matter as a source of energy, responsible for regulating nutrient transformation, plant growth, and decomposition of organic matter. Biological soil processes help determine decomposition rates of organic matter. Crop residue retention and residue quality determine microbial and fungal populations and their ratios, consequently SOC pools and soil health as well. Because the rhizosphere is the zone of the highest soil microbial activity (Lynch and Whipps, 1990), rhizodeposition can provide an important C source for soil microorganisms (Marx et al., 2007; Lu et al., 2002) and is key to soil C and nutrient cycling in terrestrial ecosystems (Helal and Sauerbeck, 1989). Retention of crop residue in the soil can increase microbial activity and biomass C, N, and P by over 30%

without a measurable increase in SOM concentration (Malhi and Lemke, 2007; Powlson et al., 2011). Residue retention is an important factor in stimulating soil microbial biomass (SMB) and microbial activity. Crop residue provides energy and substrates for microbial metabolism and promotes biodiversity (Loveland and Webb, 2003). Soil fauna and their interactions, both with the biotic and the abiotic environment (i.e. temperature, moisture, and organic matter quantity and quality), are involved in several processes that promote C stabilization or mineralization and subsequently affect soil health. Soil microbial activity and community composition are affected by residue retention and other abiotic factors, such as soil water content and temperature, soil pH, SOC, N, and P levels, and the C/N ratio (Henriksen and Breland, 2002; Geisseler et al., 2011; Kamble and Bååth, 2016; Zhao et al., 2019). The low rates of crop residue retention under synthetic fertilizer application have no significant effects on soil microbial community and enzyme activities. However, a high rate of residue retention increases total phospholipid fatty acids, alters microbial community structure, and stimulates earthworm activities and populations (Zhao et al., 2019). Soil respiration was positively correlated with soil water potential and significantly increased with the residue addition (Geisseler et al., 2011; Jin et al., 2018). The soil community composition regulates SOC dynamics and improves soil health parameters (Fierer et al., 2003, 2009; Wallenstein and Hall, 2012).

Heterotrophic soil microorganisms (bacteria, fungi, and archaea) obtain their nutrient and energy needs for maintenance and growth while mediating the turnover of SOM, and are essential to the mineralization, immobilization, and turnover of SOM. Furthermore, the SMB plays a dual role in decomposition (mineralization) of SOM, and stabilization of SOC processes, therefore influencing soil health. The SMB pool changes more rapidly than total SOC with turnover times of 1–5 years (Wiesmeier et al., 2019). The strongest relationships between SMB-C and SOC have been found in soils with <2.5% SOC (Anderson and Domsch, 1980). A wide continuum of SMB-C to SOC ratios ranging from 0.3% to over 7.0% is reported in the literature (von Lützow et al., 2007) and can be attributed to different soil ecosystems (Anderson and Domsch, 1980), and environmental factors influencing C-input such as microclimate, soil texture, soil N, and soil pH (Insam and Domsch, 1988). Significant increases in SMB-C level with straw retained was attributed to improved C and N contents, increased soil moisture, and decreased soil temperature due to the residue cover (Lou et al., 2011). In many soil ecosystems where soil N is a limiting factor, N rather than soil C may influence immobilization of SOM by the microbial biomass. Crop residues with a high C:N ratio can induce N immobilization and reduce SOC decomposition in the absence of inorganic N. The addition of inorganic N fertilizer can supplement microbial demand for available N, consequently increasing the microbial biomass and activity, and as a result, accelerating the SOC decomposition (Conde et al., 2005; Kochsiek et al., 2013; Li et al.,

2013; Dimassi et al., 2014; Zhao et al., 2019). Furthermore, when microbes are nutrient limited, SOC losses increase because of mineralization despite more crop residue inputs and more C inputs, which indicates how constraints on microbial physiology may influence SOC storage (Allison et al., 2010; Kirkby et al., 2014) and affect soil health.

Fungi are more efficient at decomposing recalcitrant constituents, while bacteria favor more bioavailable substrates (Schneider et al., 2012). Therefore, fungi are considered more effective in crop straw decomposition, and fungal residues dominantly contribute to long-term SOC build-up at the decade scale. In an 8-year field experiment, fungal residues accumulated in the top 10-cm soil and contributed to long-term SOM build-up, whereas bacterial residues were actively involved in SOM turnover (Liu et al., 2019b). Liu et al. (2019b) concluded that corn stover mulching enhances the contribution of microbial residues, especially fungal residues, and not only enlarged the capacity of the microbial residue pool in the soil but also strengthened the ability of the soil to retain microbial residues. Ge et al. (2021) reported that the content of residue-derived microbial biomass carbon (MBC) was higher (40.6 mg kg^{1}) in synthetic fertilizer with manure treatments and synthetic fertilizer only (33.0 mg kg^{1}) compared to the unfertilized control treatment (27.0 mg kg^{1}) at the end of 150-day incubation.

Soil fauna and their interactions with the biotic and the abiotic environment promote C stabilization or mineralization, and therefore affect soil health. They can stimulate or inhibit microbial activity (Hättenschwiler and Bretscher, 2001; Scheu et al., 2002), and change the microbial community composition when selectively grazing (Lavelle et al., 1997; Scheu et al., 2002). Additionally, soil fauna affects SOC storage by affecting soil aggregate stability and soil structure (see Section 6). Residue quality, organic matter quality and distribution (Bossuyt et al., 2006; Coq et al., 2007; Fonte and Six, 2010), agricultural management (Fonte et al., 2007; Pulleman et al., 2005), or land use (van Vliet et al., 1993; Pulleman et al., 2005; Hedde et al., 2013) could increase or decrease extent of soil faunal effects and subsequent soil health. Residue retention of corn, soybean, and wheat in a 21-year mulching management experiment stimulated the accumulation of microbial residues compared with no residue addition (Ding et al., 2013). In a controlled laboratory experiment using soil amended with five litters differing in chemistry found earthworms regulated microbial C and N acquisition as C to N-degrading enzyme activity ratio was nearly doubled in the low lignin litter (clover, stover, and straw), while it was decreased in the high lignin species bagasse and Rumex fiber (Zheng et al., 2018). The C loss induced by earthworms was mainly attributed to their effects on soil fauna and the role of residue chemistry in regulating soil fauna. Earthworm feeding on crop residue has been suggested to induce long-term stabilization of SOC by the formation of microaggregates within macroaggregates in earthworm casts

(Six et al., 2004; Bossuyt et al., 2005; Lubbers et al., 2013) and thus protecting C from microbial attack.

4 Crop residue and related soil physical properties

Soil aggregate stability, texture, soil bulk density, water holding capacity and storage, infiltration, and erosion play key roles in soil function. Physical damage to soils can have long-term impacts on soil health but can be reduced through the maintenance of an adequate crop residue cover, with retained residue promoting C sequestration and other positive effects. Corn residue incorporation or removal induces changes in soil bulk density and total porosity (Lal, 2009b). Crop residue amount and placement have significant and complex interactions with soil water and thermal regimes and important consequences for soil C dynamics (Power and Doran, 1988), biogeochemical cycles (Gollany et al., 2015), solute transport, gas exchange (Malhi et al., 2006), and soil health. Most studies report positive effects of retaining crop residues on soil bulk density (Tormena et al., 2017; Li et al., 2019; Yuan et al., 2019), soil compaction (Tormena et al., 2017; Koudahe et al., 2022), soil aggregate stability (Karlen et al., 1994; Blanco-Canqui et al., 2006; Villamil et al., 2015; Yuan et al., 2019), soil water retention (Liu et al., 2006; Govaerts et al., 2009; Sharma et al., 2018; Li et al., 2019), saturated hydraulic conductivity (Yuan et al., 2019), infiltration rate (Govaerts et al., 2009), soil available water capacity (Li et al., 2019), and soil erosion and runoff. Crop residue retention also reduces raindrop impact on the soil surface and minimizes the intensity of freeze-thaw and wind/water erosion by providing cover (Layton et al., 1993; Miner et al., 2013). Retaining crop residue reduces soil temperature (Sharratt and Campbell, 1994; Akhtar et al., 2020), improves soil water retention, and enhances soil health benefits in most soils. However, reduced soil temperature from residue cover can delay early crop growth and lower crop yields in some cold climate soils (Verhulst et al., 2010). Agronomic practices such as mulching and cover crop improve the soil water content by regulating soil temperature and reducing water evaporation from the soil (Sharma et al., 2018). Surface residue cover protects soil aggregates from raindrop impact. The addition of crop residue improves surface cover and the stability of soil aggregates and thus soil porosity (Pulido-Fernández et al., 2013) that led to a decrease in soil surface runoff and evaporation (Akhtar et al., 2020). Furthermore, crop residue retention increases infiltration rates under NT compared to NT without residue retention (Govaerts et al., 2009), and reduces sediment load in surface runoff during rainfall events. Although crop residue modifies the soil structure and increases total pore volume, the effect on overall pore size classes is limited. Residue decreases mesoporosity (30–75 mm), but its effect on macropores (>75 mm) is in terms of shape: an increase in elongated and irregular pores (Pituello et al., 2016). Therefore, the overall effect of crop residue on soil structure is limited; however, even minor pore changes can affect gas exchange (Malhi et al., 2006; Deurer et al., 2009), water movement, and solute transport (Pituello et al., 2016) that could be beneficial for soil functioning and soil health.

Macropores play a key role in several soil processes such as gas exchange with the atmosphere, and water and biogeochemical cycles (Jarvis, 2007). Water plays a major role in biogeochemical cycling because C, nutrient, and water cycles are intricately connected (Wright et al., 2004; Asbjornsen et al., 2011). Generally, photosynthesis and plant growth are limited by soil water content and nutrient availability. The uptake of nutrients by plants and plant growth are both affected by water availability (Cramer et al., 2009) and hence biogeochemical cycling, crop residue production, and soil health. In a 3-year study, a significant amount of soil water content was observed in the 0–15 and 15–30 cm soil depths of the silty-clay loam soils. Higher wheat grain yield was observed under conservation tillage with the surface mulch of wild sage and eupatorium compared to conventional management practices where crop residue is incorporated in the soil (Acharya et al., 1998). Crop residue retention also affects the least limiting water range (LLWR), which defines a range of soil water content within root growth that is the least limited by water availability (water potential), aeration, and soil resistance to root penetration (Tormena et al., 2017). In a study to quantify the impact of two tillage practices and three levels of corn stover removal on near-surface soil physical quality through the LLWR, soil penetration resistance determined the lower limit of the LLWR regardless of tillage or stover treatment, whereas soil aeration controlled the upper limit only at soil bulk density (ρ_d) > 1.45 and ρ_d > 1.55 Mg m^{-3} for chisel plow and no-tillage, respectively (Tormena et al., 2017). The LLWR was smallest for no-tillage with moderate or high corn stover removal, indicating poor soil physical condition for plant growth; however, the introduction of alfalfa (*Medicago sativa* L.) into an extended rotation improved the LLWR by reducing the soil penetration resistance and potential root growth restriction (Tormena et al., 2017).

Fresh plant and microbial residues may bind clay particles and microaggregates into macroaggregates and megaaggregates, thus preserving partially decaying crop residues and microbial products within the aggregates and consequently improving soil health (Edwards and Bremner, 1967; Six et al., 2002a). However, macroaggregates tend to have low structural stability and are more sensitive to agricultural management practices such as tillage than microaggregates (Six et al., 1998, 2002a). Crop residue retention on the soil surface increases aggregate stability more than when residues are incorporated into the soil (Caesar-TonThat et al., 2010; Hammerbeck et al., 2012). Incorporating residues into the soil increases soil temperature and can lead to higher residue mineralization compared to residue retention on the soil surface as in a no-till system (Campbell and Souster, 1982).

The effectiveness of various types of crop residues and management may also differ. Application of various agricultural straws including fresh straw, decomposed straw, straw-derived biochar to soil alters the soil C pool and soil aggregate stability (Huang et al., 2018). In an in-situ mesocosm experiment conducted under five treatments: (1) control with no straw or no biochar added, (2) straw addition, (3) straw with a straw decay bacterium, (4) biochar added, and (5) a combination of straw and biochar, the application of straw decay bacteria increased the percentages of macroaggregates (i.e. 0.25–2 mm and >2 mm), while biochar increased 0.053–0.25 mm aggregate and fine intra-aggregate particulate organic C (Huang et al., 2018). Straw and straw with biochar had the highest coarse-intra-aggregate particulate organic C in 0.25–2 mm and >2 mm aggregates (Huang et al., 2018). In another study, crop residue retention on the soil resulted in 6.7% fewer erodible aggregates <0.87-mm diameter and 8.6% more less-erodible >38.0-mm diameter aggregates (Malhi and Lemke, 2007). Soil aggregates and soil structure are affected by plant residue C inputs, tillage, and agricultural management practices (Fig. 2).

The findings from these studies suggest the importance of crop residue retention and residue cover in protecting soil aggregates, reducing rainfall

Figure 2 Soil structure and aggregates in a continuous chemical fallow under no-till (CCF/NT) for 16 years without crop residue input and periodic weed spraying with glyphosate, a winter wheat-fallow rotation under a conventional moldboard plowed tillage during the fallow phase of the rotation after plowing (F-W/CT), after 23 years of conversion to continuous winter wheat under no-till (CW/NT), and winter wheat-fallow rotation under conventional moldboard plowed tillage (W-F/CT) in wheat phase.

impact and soil loss due to erosion and surface runoff, improving soil water content, regulating soil temperature, and reducing the penetration resistance with potential for root growth restriction.

5 Crop residue retention and future trends in soil health research

The benefits of residue retention are site specific and depend not only on agroclimatic factors but also on the landscape position, cropping system, and the residue quantity and quality. Although many studies report positive effects of crop residue retention on soil water content and crop yields, in some high rainfall areas, residue retention can lead to excessive soil water content and waterlogging due to reduced runoff (Thierfelder and Wall, 2009, 2012). However, crop yields could increase at the same experimental site during dry years and improve resiliency to climate change. Climate change and other global and regional impacts may generate unforeseen consequences for crop residue production and soil health. Climate change may alter precipitation patterns and result in higher air temperatures, resulting in warmer soil temperatures, improved soil aeration, and higher rates of SOM decomposition (Billings et al., 1984; Oechel and Vourlitis, 1995).

In a meta-analysis of 142 experiments, crop yields increased with increased straw incorporation in 92% (131 sites) of experiments carried out over more than 3 years; in contrast, crop yields decreased only in 8% of these experiments because of straw return (Zhao et al., 2015a,b). The effect of residue retention on various soil properties indicates the importance of considering abiotic factors, such as soil texture, clay minerals, agronomic practices, study duration, sampling methods, and regional climate when evaluating different residue management practices and soil health. Incorporating residue or retaining crop residue on the soil surface is known to have multiple soil health benefits (Blanco-Canqui and Lal, 2008, 2009; Karlen et al., 1994; Wilhelm et al., 1986, 2004, 2010; Gollany et al., 2015).

Plant nutrient scarcity concerns have arisen, especially for P ('peak P'; Cordell et al., 2009). The need for crop residue retention and nutrient conservation may increase the pressure to leave crop residue on-site in the future to retain P and K, because quantities of P and K are finite, based on current technologies (Gollany et al., 2015). Soil pH changes with increased synthetic N fertilizer use may lead to plant nutrients and residue yield limitations. Monitoring the SMB pool size and labile organic C can be used as an early indicator for changes in SOC contents. These may lead to new ways to monitor microbiome, SOC, and in-situ soil health. The development of sensor technology can also improve our ability to manage soils more sustainably, or recover scarce minerals exported from sites for feed, fiber, and fuel.

6 Conclusion

Crop residue is an essential food source for soil biota and soil health. Agricultural management practices that retain crop residue will sustain SOC and healthy soil biotic communities, promote plant nutrient cycling, and improve soil chemical and physical properties. These are priorities for healthy soils and ecosystem functions and services. Agricultural management practices, such as diversified crop rotation, cover cropping, expanding the use of perennials, and stover or straw retention increase or maintain SOM and sustain a healthy microbial community that actively cycles plant nutrients with improved soil health. Increased SOC improves soil aggregate stability, soil carbon storage, and soil physical properties while protecting air and water quality. Residue cover moderates soil temperature and reduces evaporation and soil water losses, intercepts raindrop energy, increases water infiltration, reduces surface runoff, and mitigates water and wind soil erosion.

The growing world population increases the need for food, feed, fiber, fuel, and shelter, which could lead to increased demand for crop residue removal. However, crop residue removal could potentially worsen soil health, accelerate soil degradation, and exacerbate environmental problems. Removing crop residue reduces SOM and SOC, which can change soil fauna population dynamics and microbial community structure and function. The ongoing and impending impacts of biomass removal require management strategies ensuring that soil properties and functions are sustainable for all combinations of crops, soils, sites, climates, and management systems.

7 Where to look for further information

- USDA-NRCS soil health at: https://www.nrcs.usda.gov/wps/portal/nrcs/main/soils/health/.
- Building Soils for Better Crops: Ch 23. How Good Are Your Soils?

8 References

Acharya, C. L., Kapur, O. C. and Dixit, S. P. (1998). Moisture conservation for rainfed wheat production with alternative mulches and conservation tillage in the hills of north-west India. *Soil Till. Res.* 46(3-4), 153-163.

Akhtar, K., Wang, W., Ren, G., Khan, A., Enguang, N., Khan, A., Feng, Y., Yang, G. and Wang, H. (2020). Straw mulching with inorganic nitrogen fertilizer reduces soil CO2 and N2O emissions and improves wheat yield. *Sci. Total Environ.* 741, 140488 doi.org/10.1016/j.scitotenv.2020.140488.

Allison, S. D., Wallenstein, M. D. and Bradford, M. A. (2010). Soil-carbon response to warming dependent on microbial physiology. *Nat. Geosci.* 3(5), 336-340.

Anderson, J. P. E. and Domsch, K. H. (1980). Quantities of plant nutrients in the microbial biomass of selected soils. *Soil Sci.* 130(4), 211-216.

Asbjornsen, H., Goldsmith, G. R., Alvarado-Barrientos, M. S., Rebel, K., Van Osch, F. P., Rietkerk, M., Chen, J., Gotsch, S., Tobon, C., Geissert, D. R., Gomez-Tagle, A., Vache, K. and Dawson, T. E. (2011). Ecohydrological advances and applications in plant water relations research: a review. *J. Plant Ecol.* 4(1-2), 3-22. doi.org/10.1093/jpe/rtr005.

Bertrand, I., Chabbert, B., Kurek, B. and Recous, S. (2006). Can the biochemical features and histology of wheat residues explain their decomposition in soil? *Plant Soil* 281(1-2), 291-307.

Bhattacharyya, R., Kundu, S., Srivastva, A. K., Gupta, H. S., Prakash, V. and Bhatt, J. C. (2011). Long term fertilization effects on soil organic carbon pools in a sandy loam soil of the Indian sub-Himalayas. *Plant Soil* 341(1-2), 109-124.

Billings, W. D., Peterson, K. M., Luken, J. O. and Mortensen, D. A. (1984). Interaction of increasing atmospheric carbon dioxide and soil nitrogen on the carbon balance of tundra ecosystems. *Oecol.* 65, 26-29.

Blagodatskaya, E. and Kuzyakov, Y. (2008). Mechanisms of real and apparent priming effects and their dependence on soil microbial biomass and community structure: critical review. *Biol. Fertil. Soils* 45(2), 115-131.

Blanco-Canqui, H. and Lal, R. (2008). Corn stover removal impacts on micro-scale soil physical properties. *Geoderma* 145(3-4), 335-346. doi.org/10.1016/j.geoderma.2008.03.016.

Blanco-Canqui, H. and Lal, R. (2009). Corn stover removal for expanded uses reduces soil fertility and structural stability. *Soil Sci. Soc. Am. J.* 73(2), 418-426. doi.org/10.2136/sssaj2008.0141.

Blanco-Canqui, H., Lal, R., Post, W. M., Izaurralde, R. C. and Owens, L. B. (2006). Rapid changes in soil carbon and structural properties due to stover removal from notill corn plots. *Soil Sci.* 171(6), 468-482.

Bonn, B. A. and Fish, W. (1991). Variability in the measurement of humic carboxyl content. *Environ. Sci. Technol.* 25(2), 232-240.

Bossuyt, H., Six, J. and Hendrix, P. F. (2005). Protection of soil carbon by microaggregates within earthworm casts. *Soil Biol. Biochem.* 37(2), 251-258.

Bossuyt, H., Six, J. and Hendrix, P. F. (2006). Interactive effects of functionally different earthworm species on aggregation and incorporation and decomposition of newly added residue carbon. *Geoderma* 130(1-2), 14-25.

Bowles, T. M., Mooshammer, M., Socolar, Y., Calderón, F., Cavigelli, M. A., Culman, S. W., Deen, W., Drury, C. F., Garcia y Garcia, A., Gaudin, A. C. M., Harkcom, W. S., Lehman, R. M., Osborne, S. L., Robertson, G. P., Salerno, J., Schmer, M. R., Strock, J. and Grandy, A. S. (2020). Long-term evidence shows that crop-rotation diversification increases agricultural resilience to adverse growing conditions in North America. *One Earth* 2(3), 284-293. doi.org/10.1016/J.ONEEAR.2020.02.007.

Butterly, C. R., Kaudal, B. B., Baldock, J. A. and Tang, C. (2011). Contribution of soluble and insoluble fractions of agricultural residues to short-term pH changes. *Eur. J. Soil Sci.* 62(5), 718-727.

Buyanovsky, G. A., Brown, J. R. and Wagner, G. H. (1997). Sanborn field effect of 100 years of cropping on soil parameter. In: Paul, E. A., et al. (Eds). *Soil Organic Matter in Temperate Agroecosystems. Long Term Experiments in North America*. CRC Press, Boca Raton, FL, pp 205-226.

Caesar-TonThat, T. C., Sainju, U. M., Wright, S. F., Shelver, W. L., Kolberg, R. L. and West, M. (2010). Long-term tillage and cropping effects on microbiological properties associated with aggregation in a semi-arid soil. *Biol. Fertil. Soils* 47(2), 157-165.

Campbell, C. A. and Souster, W. (1982). Loss of organic matter and potentially mineralizable nitrogen from Saskatchewan soils due to cropping. *Can. J. Soil. Sci.* 62(4), 651-656.

Campbell, C. A., Zentner, R. P., Bowren, K. E., Townley-Smith, L. and Schnitzer, M. (1991). Effect of crop rotations and fertilization on soil biochemical properties in a thick Black chernozem. *Can. J. Soil Sci.* 71(3), 377-387.

Chen, R., Senbayram, M., Blagodatsky, S., Myachina, O., Dittert, K., Lin, X., Blagodatskaya, E. and Kuzyakov, Y. (2014). Soil C and N availability determine the priming effect: microbial N mining and stoichiometric decomposition theories. *Glob. Change Biol.* 20(7), 2356-2367.

Conde, E., Cardenas, M., Ponce-Mendoza, A., Lunaguido, M., Cruzmondragon, C. and Dendooven, L. (2005). The impacts of inorganic nitrogen application on mineralization of 14C-labelled maize and glucose, and on priming effect in saline alkaline soil. *Soil Biol. Biochem.* 37(4), 681-691.

Cong, W., van Ruijven, J., Mommer, L., De Deyn, G. B., Berendse, F. and Hoffland, E. (2014) Plant species richness promotes soil carbon and nitrogen stocks in grasslands without legumes. *J. Ecol.* 102(5), 1163-1170. doi.org/10.1111/1365-2745.12280.

Coq, S., Barthes, B. G., Oliver, R., Rabary, B. and Blanchart, E. (2007). Earthworm activity affects soil aggregation and organic matter dynamics according to the quality and localization of crop residues - an experimental study (Madagascar). *Soil Biol. Biochem.* 39(8), 2119-2128. Available at: http://refhub.elsevier.com/S0016-7061(17)31984-5/rf0230.

Cordell, D., Drangert, J.-O. and White, S. (2009). The story of phosphorus: global food security and food for thought. *Glob. Environ.* 19(2), 292-305. doi.org/10.1016/j.gloenvcha.2008.10.009.

Cramer, M. D., Hawkins, H. J. and Verboom, G. A. (2009). The importance of nutritional regulation of plant water flux. *Oecologia* 161(1), 15-24.

Deurer, M., Grinev, D., Young, I., Clothier, B. E. and Müller, K. (2009). The impact of soil carbon management on soil macropore structure: a comparison of two apple orchard systems in New Zealand. *Eur. J. Soil Sci.* 60(6), 945-955. doi.org/10.1111/j.1365-2389.2009.01164.x.

Dimassi, B., Mary, B., Fontaine, S., Perveen, N., Revaillot, S. and Cohan, J. (2014). Effect of nutrients availability and long-term tillage on priming effect and soil C mineralization. *Soil Biol. Biochem.* 78, 332-339.

Ding, X., Han, X. and Zhang, X. (2013). Long-term impacts of manure, straw, and fertilizer on amino sugars in a silty clay loam soil under temperate conditions. *Biol. Fertil. Soils* 49(7), 949-954.

Duiker, S. W. and Beegle, D. B. (2006). Soil fertility distributions in long-term no-till, chisel/disk and moldboard plow/disk systems. *Soil Till. Res.* 88(1-2), 30-41.

Edwards, A. P. and Bremner, J. M. (1967). Microaggregates in soils. *J. Soil Sci.* 18(1), 64-73.

Eswaran, H., Van Den Berg, E. and Reich, P. (1993). Organic carbon in soils of the world. *SSSAJ.* 57, 192-194.

Eswaran, H., Van Den Berg, E. and Reich, P. (1995). Global soil carbon resources. In: Lal, R., Kimble, J., Levine, E. and Stewart, B. A. (Eds), *Soils and Global Change*. CRC/Lewis Publishers, Boca Raton, FL, pp 27-43.

FAO (2011). *Soil Health: Technologies That Save and Grow.* Food and Agriculture Organization, Rome.

Fierer, N., Allen, A. S., Schimel, J. P. and Holden, P. A. (2003). Controls on microbial CO2 production: a comparison of surface and subsurface soil horizons. *Glob. Change Biol.* 9(9), 1322–1332. doi.org/10.1046/j.1365-2486.2003.00663.x.

Fierer, N., Strickland, M. S., Liptzin, D., Bradford, M. A. and Cleveland, C. C. (2009). Global patterns in belowground communities. *Ecol. Lett.* 12(11), 1238–1249. doi.org/10.1111/j.1461-0248.2009.01360.x.

Fontaine, S., Mariotti, A. and Abbadie, L. (2003). The priming effect of organic matter: a question of microbial competition? *Soil Biol. Biochem.* 35(6), 837–843.

Fonte, S. J., Kong, A. Y. Y., van Kessel, C., Hendrix, P. F. and Six, J. (2007). Influence of earthworm activity on aggregate-associated carbon and nitrogen dynamics differs with agroecosystem management. *Soil Biol. Biochem.* 39(5), 1014–1022.

Fonte, S. J. and Six, J. (2010). Earthworms and litter management contributions to ecosystem services in a tropical agroforestry system. *Ecol. Appl.* 20(4), 1061–1073.

Fornara, D. A. and Tilman, D. (2008). Plant functional composition influences rates of soil carbon and nitrogen accumulation. *J. Ecol.* 96(2), 314–322.

Gaudin, A. C. M., Janovicek, K., Deen, B. and Hooker, D. C. (2015a). Wheat improves nitrogen use efficiency of maize and soybean-based cropping systems. *Agric. Ecosyst. Environ.* 210, 1–10. doi.org/10.1016/j.agee.2015.04.034.

Gaudin, A. C. M., Tolhurst, T. N., Ker, A. P., Janovicek, K., Tortora, C., Martin, R. C. and Deen, W. (2015b). Increasing crop diversity mitigates weather variations and improves yield stability. *PLoS ONE* 10(2), e0113261. doi.org/10.1371/journal.pone.0113261.

Ge, Z., Li, S., Bol, R., Zhu, P., Peng, C., An, T., Cheng, N., Liu, X., Li, T., Xu, Z. and Wang, J. (2021). Differential long-term fertilization alters residue-derived labile organic carbon fractions and microbial community during straw residue decomposition. *Soil Till. Res.* 213, 105120. doi.org/10.1016/j.still.2021.105120.

Geisseler, D., Horwath, W. R. and Scow, K. M. (2011). Soil moisture and plant residue addition interact in their effect on extracellular enzyme activity. *Pedobiologia* 54(2), 71–78. doi.org/10.1016/j.pedobi.2010.10.001.

Gentile, R., Vanlauwe, B. and Six, J. (2011). Litter quality impacts short-but not long-term soil carbon dynamics in soil aggregate fractions. *Ecol. Appl.* 21(3), 695–703.

Gollany, H. T., Allmaras, R. R., Copeland, S. M., Albrecht, S. L. and Douglas, C. L. (2005). Tillage and nitrogen fertilizer influence on carbon and silica distribution in a mollisol of the Pacific Northwest. *Soil Sci. Soc. Am. J.* 69(4), 1102–1109.

Gollany, H. T., Allmaras, R. R., Copeland, S. M., Albrecht, S. L. and Douglas, C. L. (2006). Incorporated source carbon and nitrogen fertilization effects on carbon storage and soluble silica in a haploxeroll. *Soil Sci.* 171(8), 585–597.

Gollany, H. T., Titus, B. D., Scott, D. A., Asbjornsen, H., Resh, S. C., Chimner, R. A., Kaczmarek, D. J., Leite, L. F., Ferreira, A. C., Rod, K. A., Hilbert, J., Galdos, M. V. and Cisz, M. E. (2015). Biogeochemical research priorities for sustainable biofuel and bioenergy feedstock production in the Americas. *Environ. Manag.* 56(6), 1330-1355. doi.org/10.1007/s00267-015-0536-7.

Gong, W., Yan, X. Y., Wang, J. Y., Hu, T. X. and Gong, Y. B. (2009). Long-term manure and fertilizer effects on soil organic matter fractions and microbes under a wheat- maize cropping system in northern China. *Geoderma* 149(3–4), 318–324.

Govaerts, B., Sayre, K. D., Lichter, K., Dendooven, L. and Deckers, J. (2007). Influence of permanent raised bed planting and residue management on physical and chemical soil quality in rain-fed maize/wheat systems. *Plant Soil* 291(1–2), 39–54.

Govaerts, B., Sayre, K. D., Goudeseune, B., De Corte, P., Lichter, K., Dendooven, L. and Deckers, J. (2009). Conservation agriculture as a sustainable option for the central Mexican highlands. *Soil Till. Res.* 103(2), 222-230.

Gregorich, E. G., Monreal, C. M. and Ellert, B. H. (1995). Turnover of soil organic matter and storage of corn residue carbon estimated from natural 13C abundance. *Can. J. Soil Sci.* 75(2), 161-167. doi.org/10.4141/cjss95-023.

Gregorutti, V. C. and Caviglia, O. P. (2019). Impact of crop aerial and root biomass inputs on soil nitrifiers and cellulolytic microorganisms. *Soil Till. Res.* 191, 85-97.

Guppy, C. N., Menzies, N. W., Moody, P. W. and Blamey, F. P. C. (2005). Competitive sorption reactions between phosphorus and organic matter in soil: a review. *Soil Res.* 43(2), 189-202.

Hammerbeck, A. L., Stetson, S. J., Osborne, S. L., Schumacher, T. E. and Pikul, J. L. (2012). Corn residue removal impact on soil aggregates in a no-till corn/soybean rotation. *Soil Sci. Soc. Am. J.* 76(4), 1390-1398. doi.org/10.2136/sssaj2011.0421.

Han, H., Li, X., Niu, D., Hall, S. J., Guo, D., Wan, C., Learned, J. K. and Fu, H. (2015). Conventional tillage improves the storage of soil organic carbon in heavy fractions in the Loess Plateau, China. *J. Arid Land* 7(5), 636-643.

Hättenschwiler, S. and Bretscher, D. (2001). Isopod effects on decomposition of litter produced under elevated CO_2, N deposition and different soil types. *Glob. Change Biol.* 7(5), 565-579.

Haynes, R. J. (2005). Labile organic matter fractions as central components of the quality of agricultural soils: an overview. *Adv. Agron.* 85, 221-268.

Haynes, R. J. and Mokolobate, M. S. (2001). Amelioration of Al toxicity and P deficiency in acid soils by additions of organic residues: a critical review of the phenomenon and the mechanisms involved. *Nutr. Cycl. Agroecosyst.* 59(1), 47-63.

Hedde, M., Bureau, F., Delporte, P., Cécillon, L. and Decaëns, T. (2013). The effects of earthworm species on soil behavior depend on land use. *Soil Biol. Biochem.* 65, 264-273.

Heitkamp, F., Wendland, M., Offenberger, K. and Gerold, G. (2012). Implications of input estimation, residue quality and carbon saturation on the predictive power of the Rothamsted Carbon Model. *Geoderma* 170, 168-175. doi.org/10.1016/j.geoderma.2011.11.005.

Helal, H. M. and Sauerbeck, D. (1989). Carbon turnover in the rhizosphere. *Z. Pflanzenernaehr Bodenk.* 152(2), 211-216.

Henriksen, T. M. and Breland, T. A. (2002). Carbon mineralization, fungal and bacterial growth, and enzyme activities as affected by contact between crop residues and soil. *Biol. Fertil. Soils* 35(1), 41-48.

Houghton, R. A. (2012). Historic changes in terrestrial carbon storage. In: Lal, R., Lorenz, K. and Hutt, R. F., Scheid Er, B. U. and von Braun, J. (Eds.). *Recarbonization of the Biosphere*. Springer, Houghton, pp S59-82.

Huang, R. T., Tian, D., Liu, J., Lv, S., He, X. and Gao, M. (2018). Responses of soil carbon pool and soil aggregates associated organic carbon to straw and straw-derived biochar addition in a dryland cropping mesocosm system. *Agric. Ecosyst. Environ.* 265, 576-586. doi.org/10.1016/j.agee.2018.07.013.

Insam, H. and Domsch, K. H. (1988). Relationship between soil organic carbon and microbial biomass on chronosequences of reclamation sites. *Microb. Ecol.* 15(2), 177-188.

Jarvis, N. J. (2007). A review of non-equilibrium water flow and solute transport in soil macropores: principles, controlling factors and consequences for water quality. *Eur. J. Soil Sci.* 58, 523–546. doi.org/10.1111/j.1365-2389.2007.00915.x.

Jarvis, S. C., Stockdale, E. A., Shepherd, M. A. and Powlson, D. S. (1996). Nitrogen mineralization in temperate agricultural soils: processes and measurement. *Adv. Agron.* 57, 187–235.

Jin, X., An, T., Gall, A. R., Li, S., Filley, T. and Wang, J. (2018). Enhanced conversion of newly added maize straw to soil microbial biomass C under plastic film mulching and organic manure management. *Geoderma* 313, 154–162. doi.org/10.1016/j.geoderma.2017.10.036.

Jones, D. L., Nguyen, C. and Finlay, R. D. (2009). Carbon flow in rhizosphere: carbon trading at the soil-root interface. *Plant Soil* 321(1–2), 5–33.

Kamble, P. N. and Bååth, E. (2016). Comparison of fungal and bacterial growth after alleviating induced N-limitation in soil. *Soil Biol. Biochem.* 103, 97–105.

Karlen, D. L., Wollenhaupt, N. C., Erbach, D. C., Berry, E. C., Swan, J. B., Eash, N. S. and Jordahl, J. L. (1994). Crop residue effects on soil quality following 10-years of no-till corn. *Soil Till. Res.* 31(2–3), 149–167. doi.org/10.1016/0167-1987(94)90077-9.

Karlen, D. L., Veum, K. S., Sudduth, K. A., Obrycki, J. F. and Nunes, M. R. (2019). Soil health assessment: past accomplishments, current activities, and future opportunities. *Soil Till. Res.* 195, 104365. doi.org/10.1016/j.still.2019.104365.

Keiluweit, M., Bougoure, J. J., Nico, P. S., Pett-Ridge, J., Weber, P. K. and Kleber, M. (2015). Mineral protection of soil carbon counteracted by root exudates. *Nat. Clim. Change* 5(6), 588–595. doi.org/10.1038/NCLIMATE2580.

Kirkby, C. A., Richardson, A. E., Wade, L. J., Passioura, J. B., Batten, G. D., Blanchard, C. and Kirkegaard, J. A. (2014). Nutrient availability limits carbon sequestration in arable soils. *Soil Biol. Biochem.* 68 (Suppl. C), 402–409. doi.org/10.1016/j.soilbio.2013.09.032.

Kochsiek, A., Knops, J. and Zhang, W. (2013). Effects of nitrogen availability on the fate of litter-carbon and soil organic matter decomposition. *Br. J. Environ. Clim. Change* 3, 24–43.

Koudahe, K., Allen, S. C. and Djaman, K. (2022). Critical review of the impact of cover crops on soil properties. *International Soil and Water Conservation Research* 10(3), 343–354. doi.org/10.1016/j.iswcr.2022.03.003.

Kumari, A., Kapoor, K. K., Kundu, B. S. and Kumari Mehta, R. (2008). Identification of organic acids produced during rice straw decomposition and their role in rock phosphate solubilisation. *Plant Soil Environ.* 54(2), 72–77.

Kuzyakov, Y. (2010). Priming effects: interactions between living and dead organic matter. *Soil Biol. Biochem.* 42(9), 1363–1371.

Lal, R. (1997). Long-term tillage and maize monoculture effects on a tropical alfisol in western Nigeria. II: soil chemical properties. *Soil Till. Res.* 42(3), 161–174.

Lal, R. (2004). Soil carbon sequestration impacts on global climate change and food security. *Science* 304(5677), 1623–1627. doi.org/10.1126/science.1097396.

Lal, R. (2009a). Soil quality impacts of residue removal for bioethanol production. *Soil Till. Res.* 102(2), 233–241. doi.org/10.1016/j.still.2008.07.003.

Lal, R. (2009b). Challenges and opportunities in soil organic matter research. *Eur. J. Soil Sci.* 60(2), 158–169.

Lavelle, P., Bignell, D., Lepage, M., Woltes, V., Roger, P., Ineson, P., Heal, O. W. and Dhillion, S. (1997). Soil function in a changing world: the role of invertebrate ecosystem engineers. *Eur. J. Soil Biol.* 33(4), 159–193.

Layton, J. B., Skidmore, E. L. and Thompson, C. A. (1993). Winter-associated changes in dry-soil aggregation as influenced by management. *Soil Sci. Soc. Am. J.* 57(6), 1568–1572.

Leake, J. R., Ostle, N. J., Rangel-Castro, J. I. and Johnson, D. (2006). Carbon fluxes from plants through soil organisms determined by field 13CO2 pulse-labelling in an upland grassland. *Appl. Soil Ecol.* 33(2), 152–175.

Li, L., Han, X., You, M., Yuan, Y., Ding, X. and Qiao, Y. (2013). Carbon and nitrogen mineralization patterns of two contrasting crop residues in a mollisol: effects of residue type and placement in soils. *Eur. J. Soil Biol.* 54, 1–6.

Li, Y., Li, Z., Cui, S., Jagadamma, S. and Zhang, Q. (2019). Residue retention and minimum tillage improve physical environment of the soil in croplands: a global meta-analysis. *Soil Till. Res.*194104292. doi.org/10.1016/j.still.2019.06.009.

Liu, C., Plaza-Bonill, D., Coulter, J. A., Kutcher, H. R., Beckie, H. J., Wang, L., Floc'h, J., Hamel, C., Siddique, K. H. M., Li, L. and Gan, Y. (2022). Diversifying crop rotations enhances agroecosystem services and resilience. *Adv. Agron.* 173, 299–335. doi.org /10.1016/bs.agron.2022.02.007.

Liu, X. H., Herbert, S. J., Hashemi, A. M., Zhang, X. and Ding, G. (2006). Effects of agricultural management on soil organic matter and carbon transformation – a review. *Plant Soil Environ.* 52, 531–543. doi.doi.org/10.17221/3544-PSE.

Liu, S., Fan, R., Yang, X., Zhang, Z., Zhang, X. and Liang, A. (2019a). Decomposition of maize stover varies with maize type and stover management strategies: a microcosm study on a Black soil (mollisol) in northeast China. *J. Environ. Manag.* 234, 226-236. doi.org/10.1016/j.jenvman.2019.01.008.

Liu, X., Zhoua, F., Hu, G., Shao, S., He, H., Zhang, W., Zhang, X. and Li, L. (2019b). Dynamic contribution of microbial residues to soil organic matter accumulation influenced by maize straw mulching. *Geoderma* 333, 35–42. doi.org/10.1016/j.geoderma.2018.0 7.017.

López-Fando, C. and Pardo, M. T. (2009). Changes in soil chemical characteristics with different tillage practices in a semi-arid environment. *Soil Till. Res.* 104(2), 278–284.

Lou, Y., Liang, W., Xu, M., He, X., Wang, Y. and Zhao, K. (2011). Straw coverage alleviates seasonal variability of the topsoil microbial biomass and activity. *CATENA* 86(2), 117–120.

Loveland, P. and Webb, J. (2003). Is there a critical level of organic matter in the agricultural soils of temperate regions: a review. *Soil Till. Res.* 70(1), 1–18.

Lu, Y., Watanabe, A. and Kimura, M. (2002). Contribution of plant-derived carbon to soil microbial biomass dynamics in a paddy rice microcosm. *Biol. Fertil. Soils* 36(2), 136–142.

Lubbers, I. M., van Groenigen, K. J., Fonte, S. J., Six, J., Brussaard, L. and van Groenigen, J. W. (2013). Greenhouse-gas emissions from soils increased by earthworms. *Nat. Climate Chang.* 3(3), 187–194. doi: 10.1038/NCLIMATE1692.

Lugato, E., Panagos, P., Bampa, F., Jones, A. and Montanarella, L. (2014). A new baseline of organic carbon stock in European agricultural soils using a modeling approach. *Glob. Change Biol.* 20(1), 313-326.

Luo, Z., Wang, E. and Sun, O. J. (2016). A meta-analysis of the temporal dynamics of priming soil carbon decomposition by fresh carbon inputs across ecosystems. *Soil Biol. Biochem.* 101, 96–103. doi.org/10.1016/j.soilbio.2016.07.011.

Lynch, J. M. and Whipps, J. M. (1990). Substrate flow in the rhizosphere. *Plant Soil* 129(1), 1–10.

Malhi, S. S., Lemke, R., Wang, Z. H. and Chhabra, B. S. (2006). Tillage, nitrogen and crop residue effects on crop yield, nutrient uptake, soil quality and greenhouse gas emissions. *Soil Till. Res.* 90(1-2), 171-183. doi.org/10.1016/j.still.2005.09.001.

Malhi, S. S. and Lemke, R. (2007). Tillage, crop residue and N fertilizer effects on crop yield, nutrient uptake, soil quality and nitrous oxide gas emissions in a second 4-yr rotation cycle. *Soil Till. Res.* 96(1-2), 269-283. doi.org/10.1016/j.still.2007.06.011.

Marx, M., Buegger, F., Gattinger, A., Marschner, B., Zsolnay, Á and Munch, J. C. (2007). Determination of the fate of 13C labelled maize and wheat rhizodeposit-C in two agricultural soils in a greenhouse experiment under 13C-CO2-enriched atmosphere. *Soil Biol. Biochem.* 39(12), 3043-3055.

Mertens, J., Vanderborght, J., Kasteel, R., Pütz, T., Merckx, R., Feyen, J. and Smolders, E. (2007). Dissolved organic carbon fluxes under bare soil. *J. Environ. Qual.* 36(2), 597-606.

Miner, G. L., Hansen, N. C., Inman, D., Sherrod, L. A. and Peterson, G. A. (2013). Constraints of no-till dryland agroecosystems as bioenergy production systems. *Agron. J.* 105(2), 364-376.

Mitchell, E., Scheer, C., Rowlings, D., Conant, R. T., Cotrufo, M. F. and Grace, P. (2018). Amount and incorporation of plant residue inputs modify residue stabilisation dynamics in soil organic matter fractions. *Agric. Ecosyst. Environ.* 256, 82-91 doi.org /10.1016/j.agee.2017.12.006.

Modak, K., Biswas, D. R., Ghosh, A., Pramanik, P., Das, T. K., Das, S., Kumar, S., Krishnan, P. and Bhattacharyya, R. (2020). Zero tillage and residue retention impact on soil aggregation and carbon stabilization within aggregates in subtropical India. *Soil Till. Res.* 202, 104649. doi.org/10.1016/j.still.2020.104649.

Nash, P. R., Gollany, H. T., Novak, J. M., Bauer, P. J., Hunt, P. G. and Karlen, D. L. (2018). Simulated soil organic carbon response to tillage, yield, and climate change in the southeastern Coastal Plains. *J. Environ. Qual.* 47(4), 663-673. doi.org/10.2134/jeq2017.05.0190.

Noble, A. D., Zenneck, I. and Randall, P. J. (1996). Leaf litter ash alkalinity and neutralisation of soil acidity. *Plant Soil* 179(2), 293-302.

Novak, J. M., Watts, D. W., Bauer, P. J., Karlen, D. L., Hunt, P. G. and Mishra, U. (2020). Loamy sand soil approaches organic carbon saturation after 37 years of conservation tillage. *Agron. J.* 112(4), 3152-3162. doi.org/10.1002/agj2.20184.

Nziguheba, G. P., Palm, C. A., Buresh, R. J. and Smithson, P. C. (1998). Soil phosphorus fractions and adsorption as affected by organic and inorganic sources. *Plant and Soi1.* 1198, 159-168.

Oades, J. M. (1993). The role of biology in the formation, stabilization and degradation of soil structure. *Geoderma* 56(1-4), 377-400. doi.org/10.1016/0016-7061(93)90123-3.

Oechel, W. C. and Vourlitis, G. L. (1995). Effects of global change on carbon storage in cold soils. In: Lal, R., et al. (Eds) *Soils Global Change*. CRC Press, Boca Raton, FL, pp 177-187.

Pappa, V. A., Rees, R. M., Walker, R. L., Baddeley, J. A. and Watson, C. A. (2011). Nitrous oxide emissions and nitrate leaching in an arable rotation resulting from the presence of an intercrop. *Agric. Ecosyst. Environ.* 141(1-2), 153-161. doi.org/10.1016/j.agee.2011.02.025.

Paul, E. A. (2016). The nature and dynamics of soil organic matter: plant inputs, microbial transformations, and organic matter stabilization. *Soil Biol. Biochem.* 98, 109-126.

Paustian, K., Collins, H. P. and Paul, E. A. (1997). Management controls on soil carbon. In: Paul, E. A., Paustian, K., Elliott, E. T. and Cole, C. V. (Eds). CRC Press, Boca Raton, FL, pp 15-49.

Paustian, K., Six, J., Elliott, E. T. and Hunt, H. W. (2000). Management options for reducing CO2 emissions from agricultural soils. *Biogeochemistry* 48(1), 147-163.

Pituello, C., Dal Ferro, N., Simonetti, G., Berti, A. and Morari, F. (2016). Nano to macro pore structure changes induced by long-term residue management in three different soils. *Agric. Ecosyst. Environ.* 217, 49-58. doi.org/10.1016/j.agee.2015.10.029.

Poeplau, C. and Don, A. (2015). Carbon sequestration in agricultural soils via cultivation of cover crops: a metanalysis. *Agric. Ecosyst. Environ.* 200, 33-41. doi.org/10.1016/j.agee.2014.10.024.

Poeplau, C., Kätterer, T., Bolinder, M. A., Börjesson, G., Berti, A. and Lugato, E. (2015). Low stabilization of aboveground crop residue carbon in sandy soils of Swedish long-term experiments. *Geoderma* 237-238, 246-255. doi.org/10.1016/j.geoderma.2014.09.010.

Pulido-Fernández, M., Schnabel, S., Lavado-Contador, J. F., Miralles Mellado, I. and Ortega Pérez, R. (2013). Soil organic matter of Iberian open woodland rangelands as influenced by vegetation cover and land management. *CATENA* 109, 13-24.

Pulleman, M. M., Six, J., Uyl, A., Marinissen, J. C. Y. and Jongmans, A. G. (2005). Earthworms and management affect organic matter incorporation and microaggregate formation in agricultural soils. *Appl. Soil Ecol.* 29(1), 1-15.

Power, J. F. and Doran, J. W. (1988). Role of crop residue management in nitrogen cycling and use. In: Hargrove, W. L., et al. (Eds). *Cropping Strategies for Efficient Use of Water and Nitrogen*, Special Publication No. 51. CSSA, ASA-SSSA. Madison Book Company, Madison, WI, pp 101-113.

Powlson, D. S., Glendining, M. J., Coleman, K. and Whitmore, A. P. (2011). Implications for soil properties of removing cereal straw: results from long-term studies 1. *Agron. J.* 103(1), 279-287. doi.org/10.2134/agronj2010.0146s.

Powlson, D. S., Bhogal, A., Chambers, B. J., Coleman, K., Macdonald, A. J., Goulding, K. W. T. and Whitmore, A. P. (2012). The potential to increase soil carbon stocks through reduced tillage or organic material additions in England and Wales: a case study. *Agric. Ecosyst. Environ.* 146(1), 23-33.

Rasse, D. P., Rumpel, C. and Dignac, M. F. (2005). Is soil carbon mostly root carbon? Mechanisms for a specific stabilisation. *Plant Soil* 269(1-2), 341-356.

Reicosky, D. C., Dugas, W. A. and Torbert, H. A. (1997). Tillage-induced soil carbon dioxide loss from different cropping systems. *Soil Till. Res.* 41(1-2), 105-118. doi.org/10.1016/S0167-1987(96)01080-X.

Salinas-Garcia, J. R., Baez-Gonzalez, A. D., Tiscareno-Lopez, M. and Rosales-Robles, E. (2001). Residue removal and tillage interaction effects on soil properties under rain-fed corn production in central Mexico. *Soil Till. Res.* 59(1-2), 67-79.

Sarker, J. R., Singh, B. P. and Cowiea, A. L. Fang, Y., Collins, D., Dougherty, W. J. and Singh, B. K. (2018). Carbon and nutrient mineralisation dynamics in aggregate-size classes from different tillage systems after input of canola and wheat residues. *Biol. Biochem.* 116, 22-38. doi.org/10.1016/j.soilbio.2017.09.030.

Scherr, S. J. and McNeely, J. A. (2008). Biodiversity conservation and agricultural sustainability: towards a new paradigm of "ecoagriculture" landscapes. *Philos. Trans. R. Soc. Lond. B Biol. Sci.* 363(1491), 477-494. doi.org/10.1098/rstb.2007.2165.

Scheu, S., Schlitt, N., Tiunov, A. V., Newington, J. E. and Jones, H. T. (2002). Effects of the presence and community composition of earthworms on microbial community functioning. *Oecologia* 133(2), 254–260.

Schneider, T., Keiblinger, K. M., Schmid, E., Sterflinger-Gleixner, K., Ellersdorfer, G., Roschitzki, B., Richter, A., Eberl, L., Zechmeister-Boltenstern, S. and Riedel, K. (2012). Who is who in litter decomposition? Metaproteomics reveals major microbial players and their biogeochemical functions. *ISME J.* 6(9), 1749–1762.

Shahbaz, M., Kuzyakov, Y. and Heitkamp, F. (2017). Decrease of soil organic matter stabilization with increasing inputs: mechanisms and controls. *Geoderma* 304, 76–82. doi.org/10.1016/j.geoderma.2016.05.019.

Sharratt, B. S. and Campbell, G. S. (1994). Radiation balance of a soil-straw surface modified by straw color. *Agron. J.* 86(1), 200–203.

Sharma, P., Singh, A., Kahlon, C. S., Brar, A. S., Grover, K. K., Dia, M. and Steiner, R. L. (2018). The role of cover crops towards sustainable soil health and agriculture: a review paper. *Am. J. Plant Sci.* 09(9), 1935–1951. doi.org/10.4236/ajps.2018.99140.

Six, J., Elliott, E. T., Paustian, K. and Doran, J. W. (1998). Aggregation and soil organic matter accumulation in cultivated and native grassland soils. *Soil Sci. Soc. Am. J.* 62(5), 1367–1377.

Six, J., Feller, C., Denef, K., Ogle, S. M., de Moraes, J. C. and Albrecht, A. (2002a). Soil organic matter, biota and aggregation in temperate and tropical soils - effects of no-tillage. *Agronomie* 22(7–8), 755–775.

Six, J., Conant, R. T., Paul, E. A. and Paustian, K. (2002b). Stabilization mechanisms of soil organic matter: implications for C-saturation of soils. *Plant Soil* 241(2), 155–176. doi.org/10.1023/A:1016125726789.

Six, J., Bossuyt, H., Degryze, S. and Denef, K. (2004). A history of research on the link between (micro)aggregates, soil biota, and soil organic matter dynamics. *Soil Till. Res.* 79(1), 7–31. doi.org/10.1016/j.still.2004.03.008.

Smith, R. G., Gross, K. L. and Robertson, G. P. (2008). Effects of crop diversity on agroecosystem function: crop yield response. *Ecosystems* 11(3), 355–366. doi.org/10.1007/s10021-008-9124-5.

Soon, Y. K. and Arshad, M. A. (2002). Comparison of the decomposition and N and P mineralization of canola, pea and wheat residues. *Biol. Fertil. Soils* 36(1), 10–17.

Soussana, J.-F., Loiseau, P., Vuichard, N., Ceschia, E., Balesdent, J., Chevallier, T. and Arrouays, D. (2004). Carbon cycling and sequestration opportunities in temperate grasslands. *Soil Use Manag.* 20(2), 219–230. doi: 10.1079/SUM2003234.

Sparling, G. and Schipper, L. (2004). Soil quality monitoring in New Zealand: trends and issues arising from a broad-scale survey agriculture. *Agriculture Ecosystems & Environment* 104(3), 545–552. doi.org/10.1016/j.agee.2003.11.014.

Steinbeiss, S., Bebler, H., Engles, C., Temperton, V. M., Buchmann, N., Roscher, C., Kreutziger, Y., Baade, J., Habekost, M. and Gleixner, G. (2008). Plant diversity positively affects short-term soil carbon storage in experimental grasslands. *Glob. Change Biol.* 14(12), 2937–2949. doi.org/10.1111/j.1365-2486.2008.01697.x.

Tang, C. and Yu, Q. (1999). Impact of chemical composition of legume residues and initial soil pH on pH change of a soil after residue incorporation. *Plant Soil* 215(1), 29–38.

Taveira, C. J., Farrell, R. E., Wagner-Riddle, C., Machado, P. V. F., Deen, B. and Congreves, K. A. (2020). Tracing crop residue N into subsequent crops: insight from long-term crop rotations that vary in diversity. *Field Crops Res.* 255, 107904. doi.org/10.1016/j.fcr.2020.107904.

Teasdale, J. R. and Daughtry, C. S. T. (1993). Weed suppression by live and desiccated hairy vetch (Vicia villosa). *Weed Sci.* 41(2), 207-212.

Thierfelder, C. and Wall, P. C. (2009). Effects of conservation agriculture techniques on infiltration and soil water content in Zambia and Zimbabwe. *Soil Till. Res.* 105(2), 217-227.

Thierfelder, C. and Wall, P. C. (2012). Effects of conservation agriculture on soil quality and productivity in contrasting agro-ecological environments of Zimbabwe. *Soil Use Manag.* 28(2), 209-220.

Tormena, C. A., Karlen, D. L., Logsdon, S. and Cherubin, M. R. (2017). Corn stover harvest and tillage impacts on near-surface soil physical quality. *Soil Till. Res.* 166, 122-130 doi.org/10.1016/j.still.2016.09.015.

van Vliet, P. C. J., West, L. T., Hendrix, P. F. and Coleman, D. C. (1993). The influence of Enchytraeidae (Oligochaeta) on the soil porosity of small microcosms. *Geoderma* 56(1-4), 287-299. Available at: http://refhub.elsevier.com/S0016-7061(17)31984-5/rf1295.

Verhulst, N., Govaerts, B., Verachtert, E., Mezzalama, M., Wall, P. C., Chocobar, A., Deckers, J. and Sayre, K. D. (2010). Conservation agriculture, improving soil quality for sustainable production systems? In: Lal, R. and Stewart, B. A. (Eds). Boca Raton, FL, pp 137-208.

Villamil, M. B., Little, J. and Nafziger, E. D. (2015). Corn residue, tillage, and nitrogen rate effects on soil properties. *Soil Till. Res.* 151, 61-66. doi.org/10.1016/j.still.2015.03.005.

von Lützow, M., Kögel-Knabner, I., Ekschmitt, K., Flessa, H., Guggenberger, G., Matzner, E. and Marschner, B. (2007). SOM fractionation methods: relevance to functional pools and to stabilization mechanisms. *Soil Biol. Biochem.* 39(9), 2183-2207.

von Lützow, M., Kogel-Knabner, I., Ludwig, B., Matzner, E., Flessa, H., Ekschmitt, K., Guggenberger, G., Marschner, B. and Kalbitz, K. (2006). Stabilization mechanisms of organic matter in four temperate soils: development and application of a conceptual model. *J. Plant Nutr. Soil Sci.* 171(1), 111-124.

Wallenstein, M. D. and Hall, E. K. (2012). A trait-based framework for predicting when and where microbial adaptation to climate change will affect ecosystem functioning. *Biogeochemistry* 109(1-3), 35-47. doi.org/10.1007/s10533-011-9641-8.

Wander, M. (2004). Soil organic matter fractions and their relevance to soil function. In: Magdoff, F. R. and Weil, R. R. (Eds). *Soil Organic Matter in Sustainable Agriculture*. CRC Press, Boca Raton, pp 67-102.

Wander, M. M., Cihacek, L. J., Coyne, M., Drijber, R. A., Grossman, J. M., Gutknecht, J. L. M., Horwath, W. R., Jagadamma, S., Olk, D. C., Ruark, M., Snapp, S. S., Tiemann, L. K., Weil, R. and Turco, R. F. (2019). Developments in soil quality and health: reflections by the research committee on soil organic matter management. *Front. Environ. Sci.* 7, 1-9. doi.org/10.3389/fenvs.2019.00109.

Wang, H., Boutton, T. W., Xu, W., Hu, G., Jiang, P. and Bai, E. (2015). Quality of fresh organic matter affects priming of soil organic matter and substrate utilization patterns of microbes. *Sci. Rep.* 5, 10102.

Wiesmeier, M., Urbanski, L., Hobley, E., Lang, B., von Lützow, M., Marin-Spiotta, E., van Wesemael, B., Rabot, E., Ließ, M., Garcia-Franco, N., Wollschläger, U., Vogel, H. and

Kögel-Knabner, I. (2019). Soil organic carbon storage as a key function of soils - a review of drivers and indicators at various scales. *Geoderma* 333, 149–162. doi.org/10.1016/j.geoderma.2018.07.026.

Wilhelm, W. W., Doran, J. W. and Power, J. F. (1986). Corn and soybean yield response to crop residue management under no-tillage production systems 1. *Agron. J.* 78(1), 184–189.

Wilhelm, W. W., Johnson, J. M. F., Hatfield, J. L., Voorhees, W. B. and Linden, D. R. (2004). Crop and soil productivity response to corn residue removal: a literature review. *Agron. J.* 96(1), 1–17.

Wilhelm, W. W., Hess, J. R., Karlen, D. L., Johnson, J. M. F., Muth, D. J., Baker, J. M., Gollany, H. T., Novak, J. M., Stott, D. E. and Varvel, G. E. (2010). Review: balancing limiting factors and economic drivers for sustainable Midwestern US agricultural residue feedstock supplies. *Ind. Biotechnol.* 6(5), 271–287.

Wright, I. J., Reich, P. B., Westoby, M., Ackerly, D. D., Baruch, Z., Bongers, F., Cavender-Bares, J., Chapin, T., Cornelissen, J. H., Diemer, M., Flexas, J., Garnier, E., Groom, P. K., Gulias, J., Hikosaka, K., Lamont, B. B., Lee, T., Lee, W., Lusk, C., Midgley, J. J., Navas, M. L., Niinemets, U., Oleksyn, J., Osada, N., Poorter, H., Poot, P., Prior, L., Pyankov, V. I., Roumet, C., Thomas, S. C., Tjoelker, M. G., Veneklaas, E. J. and Villar, R. (2004). The worldwide leaf economics spectrum. *Nature* 428(6985), 821–827.

Xu, R. K. and Coventry, D. R. (2003). Soil pH changes associated with lupin and wheat plant materials incorporated in a red-brown earth soil. *Plant Soil* 250(1), 113–119.

Xu, X., An, T., Zhang, J., Sun, Z., Schaeffer, S. and Wang, J. (2019). Transformation and stabilization of straw residue carbon in soil affected by soil types, maize straw addition and fertilized levels of soil. *Geoderma* 337, 622–629. doi.org/10.1016/j.geoderma.2018.08.018.

Yan, D., Wang, D. and Yang, L. (2007). Long-term effect of chemical fertilizer, straw, and manure on labile organic matter fractions in a paddy soil. *Biol. Fertil. Soils* 44(1), 93–101.

Yuan, L., Zhou, L., Song, C., Sindhu, J. and Qingping, Z. (2019). Residue retention and minimum tillage improve physical environment of the soil in croplands: A global meta-analysis. *Soil Tillage Res.* 194, 104292. doi.org/10.1016/j.still.2019.06.009.

Zhao, H. S., Lu, B. F., Lu, F., Zhang, G., Wang, X. and Ouyang, Z. (2015a). Straw incorporation strategy on cereal crop yield in China. *Crop Sci.* 55(4), 1773–1781.

Zhao, J., Yang, X., Dai, S., Lv, S. and Wang, J. (2015b). Increased utilization of lengthening growing season and warming temperatures by adjusting sowing dates and cultivar selection for spring maize in Northeast China. *Eur. J. Agron.* 67, 12–19.

Zhao, S., Qiu, S., Xu, X., Ciampitti, I. A., Zhang, S. and He, P. (2019). Change in straw decomposition rate and soil microbial community. *Appl. Soil Ecol.* 138, 123–133. doi.org/10.1016/j.apsoil.2019.02.018.

Zheng, Y., Wang, S., Bonkowski, M., Chen, X., Griffiths, B., Hu, F., and Liu, M. (2018). Litter chemistry influences earthworm effects on soil carbon loss and microbial carbon acquisition. *Soil Biol. Biochem.* 123, 105–114. doi.org/10.1016/j.soilbio.2018.05.012.

Zibilske, L. M., Bradford, J. M. and Smart, J. R. (2002). Conservation tillage induced changes in organic carbon, total nitrogen and available phosphorus in a semi-arid alkaline subtropical soil. *Soil Till. Res.* 66, 153–163, doi.org/10.1016/j.apsoil.2019.02.018.Zibilske.

Published by Burleigh Dodds Science Publishing Limited, 2023.

Zuffo, A. M., Ratke, R. F., Okla, M. K., Al-Hashimi, A., González Aguilera, J., Trento, A. C. S., Pereira da Silva, N., de Souza, E. D., Nogueira, B. K. A., Coutinho, J. H., Steiner, F., de Alcântara Neto, F., da Silva Júnior, G. B., Dos Santos Silva, F. C., Sobrinho, R. L. and AbdElgawad, H. (2022). Understanding the contribution of soybean crop residues inoculated with *Bradyrhizobium* spp. and not harvested on nitrogen supply in off-season corn cultivars. *PLoS ONE* 17(6), e0269799. doi.org/10.1371/journal.pone.0269799.

Part 3

The role of organic and other soil amendments

Chapter 9

Assessing the effects of compost on soil health

Cristina Lazcano, University of California-Davis, USA; Charlotte Decock, California Polytechnic State University, USA; Connie T. F. Wong, University of California-Davis, USA; and Kamille Garcia-Brucher, California Polytechnic State University, USA

1 Introduction

Returning organic waste materials to the soil in the form of compost or other organic amendments has been identified as one of the key strategies to promote soil health and slow down or reverse rapid degradation of agricultural lands (Karlen et al., 2019). Organic amendments or organic fertilizers are naturally occurring organic materials from animal or plant origin that can be found in different degrees of decomposition or biochemical stability and support plant growth by improving soil physical properties and/or providing essential plant nutrients. Examples of organic amendments include animal manures, compost, vermicompost, biochar, or crop residues, among others.

http://dx.doi.org/10.19103/AS.2021.0094.06

These amendments provide diverse food sources for soil biota and part of the organic material becomes incorporated into soil organic matter (SOM) upon decomposition (Lehmann and Kleber, 2015). Buildup of SOM, changes in the soil microbiome, and beneficial effects of decomposition by-products such as plant nutrients and organic compounds ultimately determine the outcome of organic amendments for soil health (Diacono and Montemurro, 2010; Mehta et al., 2014). Nevertheless, given the diversity of organic amendments, both in terms of chemical composition and in biochemical stability, the benefits for soil health are also highly variable in magnitude. Important properties such as the total plant macro- and micronutrient content, carbon to nitrogen ratio (C:N), salinity, and presence of pathogenic microorganisms or heavy metals can significantly affect the outcomes for soil health. Factors such as the initial feedstock, composting process, or age can significantly affect the physicochemical and biological properties of the finished compost (Hargreaves et al., 2008).

When management strategies to improve soil health such as compost application are adopted, it is therefore pertinent to track management-induced improvements in soil health over time, taking into account the specific properties of the compost used. Standard soil testing for agronomic management is heavily focused on chemical soil health indicators such as soil nutrient concentrations and pH, greatly ignoring biological soil health and the role of soil physical properties in promoting a suitable habitat for soil biota and root growth (Lazcano et al., 2021a). The concept of soil health, on the other hand, not only incorporates soil chemical parameters but integrates them with physical and biological soil parameters, emphasizing the functioning of soils as ecosystems (Pawlett et al., 2021). Various indicators to assess soil health have been proposed over the past few decades. These include measurable physical, chemical, and biological soil properties that are indicative of the functions and ecosystem services provided by healthy soils (Fig. 1). Nevertheless, poor standardization and the lack of robust interpretation schemes limit the capacity of soil health indicators for predicting and assessing soil threats, soil functioning, or provisioning of ecosystem services (Bünemann et al., 2018). In the light of an increasing need to assess and quantify the impact of management practices that promote soil health, recent studies have aimed to standardize the measurement and interpretability of soil health indicators (Moebius-Clune, 2016; Norris et al., 2020; Nunes et al., 2020; Rinot et al., 2019). Given the multifunctional nature of soil health, no single soil health indicator would accurately reflect soil health. Additionally, narrowing down soil health assessments to a set of indicators that comprehensively capture and accurately rate the many functions healthy soils provide, from suppressing plant disease and supplying plant nutrients to filtering water and sequestering carbon, is a challenging task (Wade et al., 2022).

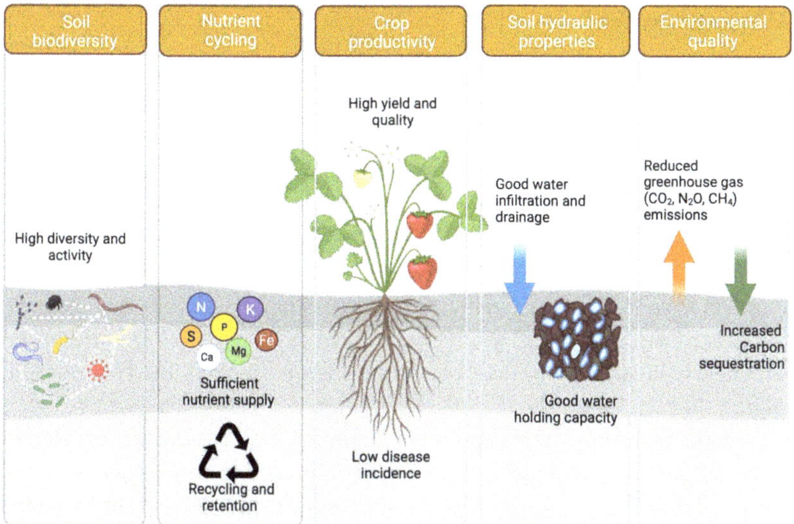

Figure 1 Overview of the functions performed by a healthy soil. Figure created with Biorender.com.

A great wealth of field and incubation experiments carried out in the last decades has shown that the addition of compost to soils has well-known short- and long-term, direct, and indirect benefits to soil biological, chemical, and physical properties that support ecosystem functioning and soil health (Diacono and Montemurro, 2010; Martinez-Blanco et al., 2013a,b). Nevertheless, these studies show substantial variability in the observed effects. This lack of clarity may result in the decrease of grower adoption rates and make the precision management of this soil organic amendment very difficult. As growers around the world move towards more sustainable soil management practices and embrace the principles of the circular economy, it is expected that the use of compost will increase and become more widespread across production systems. Thus, there is a need for a critical evaluation of the potential of compost to improve soil health, what are the best indicators to evaluate these benefits, and what are the expected ranges of variability.

This chapter provides an overview of the current literature evaluating the effects of compost on soil health in the short-, medium-, and long-term. By looking at a wide range of soil health indicators and crops, we aim to identify those indicators that are consistently responsive to compost application, as well as to characterize potential sources of variability. Finally, we identify potential risks of the use of compost for environmental quality, pointing at future research needs. Two case studies illustrate examples of current research evaluating the potential tradeoffs between benefit and risks in the use of compost in perennial and annual crops.

2 Why compost?

Approximately 44% of the waste generated in the world is organic in nature (Kaza et al., 2018). Organic waste materials constitute a valuable source of plant nutrients that, if used as fertilizers, can contribute to closing the loop of nutrients in agriculture and increasing soil organic matter with benefits for soil physical, chemical, and biological properties (Diacono and Montemurro, 2010). Furthermore, increasing soil organic matter contributes to increasing soil C and therefore is an appealing pathway for soil C sequestration and climate change mitigation. Finally, recycling of these materials eliminates the emissions of pollutants (i.e. greenhouse gasses, heavy metals) and nutrients associated with landfilling, thus having additional benefits for environmental health.

Organic materials such as animal manure or crop residues are traditional sources of plant nutrients which are commonly used in small-scale and crop-livestock integrated systems. These materials are also important in organic production systems where crop nutrient inputs rely heavily on organic fertilizers (Ramos et al., 2019). Production of animal manures has experienced a steady increase worldwide due to the increase in the consumption of livestock and poultry (Bai et al., 2018; Chadwick et al., 2015), which could help satisfy the needs for fertilizer products in agricultural systems. For example, only in 2011, 120 Tg of manure N were excreted, an amount which was similar or larger than the total synthetic fertilizer N requirements worldwide (Liu et al., 2017). Nevertheless, the reintegration of animal manures into cropping systems is challenging due to the presence of enteric microorganisms in animal manures such as *Escherichia coli* and *Salmonella* sp. (Tran et al., 2020). The spread of animal manures is also a source of antibiotic resistance in agroecosystems (Wind et al., 2021). Furthermore, mismanagement of animal manures can increase the contents of salts in the soil and the availability of labile forms of N and C that can induce nitrification and denitrification processes, leading to the loss of large amounts of N (Charles et al., 2017; Lazcano et al., 2021b). Similar concerns are associated with the agricultural use of certain plant-based organic waste materials. For example, grape pomace is a nutrient-rich solid waste material produced in large amounts in wine-growing regions of the world, which consists of the seeds, skins, and residual stalks of pressed grapes (Oliveira and Duarte, 2016). Due to its high phenolic content, low pH, high C:N, and large concentration of salts, grape pomace is highly toxic for plant growth. Similarly, the application of sewage sludge or municipal solid waste in agriculture poses several environmental and agronomic challenges due to the presence of human pathogens, plastics, and contaminants of different nature including heavy metals or solvents (Langdon et al., 2019; McBride, 1995; Petrie et al., 2015).

Composting involves the accelerated aerobic decomposition of organic waste materials by microorganisms, under controlled conditions of temperature and humidity (Debertoldi et al., 1983). The breakdown of labile organic compounds allows for the biological stabilization of the organic waste material decreasing the amount of easily available C and N and therefore the risks of environmental pollution after application to the soil compared to uncomposted materials (Charles et al., 2017; Li et al., 2016; Oudart et al., 2012). During composting, the organic waste typically undergoes a characteristic thermophilic stage, resulting from the intense microbial activity, which allows for the elimination of pathogenic microorganisms (Mc Carthy et al., 2011). Subsequently, during the maturation stage, decomposition rates decrease, together with compost temperature. Composting has demonstrated to be a robust, low-cost technology to efficiently stabilize and valorize several types of organic waste materials including animal manures (Bernal et al., 2009; Gómez-Brandón et al., 2008), sewage sludge (Lü et al., 2021), municipal solid waste (Kumar, 2011), anaerobic digestate (Bustamante et al., 2012); grape pomace (Ruggieri et al., 2009), paper pulp (Jain et al., 2018; Sesay et al., 1997), and other agricultural and food processing waste materials. The highest process efficiency and compost quality is achieved through co-composting which consists in the mix of different feedstocks with different N contents and particle sizes, to obtain a material with optimal physicochemical properties that support microbial activity and decomposition (Bustamante et al., 2008; Hazarika and Khwairakpam, 2018). Thus, most commercial composts are produced with a mix of plant and animal organic waste materials (Bernal et al., 2017).

Composting of organic waste and subsequent application to agricultural land is arguably one of the most impactful strategies to restore and maintain soil health due to the accessibility and low cost of composting technologies, particularly at the farm scale (Colvero et al., 2020; Pergola et al., 2018). Different composting systems exist, ranging from closed, automated in-vessel reactor systems to open static piles (Bernal et al., 2017). Closed in-vessel systems provide forced aeration and controlled ventilation, therefore accelerating decomposition and reducing the losses of N through leaching of nitrates or through the release of gases (NH_3 or N_2O). On the other hand, static piles typically involve minimal external intervention and rely on the presence of a critical mass that will result in the self-heating of the waste. Windrow composting involves a higher manipulation of the pile using specialized tractors to mix and aerate the waste material. The output of gases and leachates in windrow and static pile composting systems is not controlled, which results in larger losses of N compared to in-vessel reactors (Liu et al., 2020; Makan and Fadili, 2020). While closed composting systems are certainly more efficient and environmentally friendly, they are also less affordable. Thus, static piles or windrow composting may be more feasible and accessible for small commercial or on-farm operations. The simplicity, robustness and affordability of composting makes it an attractive strategy for management of organic waste at different scales, from home composting to integration of several waste streams at the regional level. Pergola et al. (2018) determined that on-farm composting was the most environmentally friendly strategy for the conversion of organic solid waste into compost in southern Italy. In China, about 76% of the animal manure produced in intensive livestock farms is composted on-site, while 9% is composted off-farm (Chadwick et al., 2015). In urban regions, home composting is an effective strategy in diverting organic waste away from landfilling, as shown by Pai et al. (2019) in a study carried out in Chicago (USA).

Compost, the final product of composting, is a stabilized and decomposed organic material which is compatible and beneficial to plant growth (Bernal et al., 2009; Insam and de Bertoldi, 2007). Given the broad definition of compost and the large range of feedstock materials that can be used in composting, there is also large variability in compost quality and physicochemical properties (Epelde et al., 2018). Most regions have established quality standards for commercially produced compost that regulate parameters such as odors, microbial stability, nutrient contents, contaminants (e.g. heavy metals, PAHs), percentage of impurities (plastic, glass, and metal), pH, soluble salts, moisture content, organic matter content, pathogens (*Salmonella* and fecal coliforms), and maturity (determined in seed germination assays) (Bernal et al., 2017; Bläsing and Amelung, 2018). The quality of the initial feedstock has a great impact on the quality of the final compost and, while many quality parameters can be regulated through adequate management of the composting process (Lannan et al., 2013), some others such as impurities or contaminants are more challenging. Typically, compost obtained from the organic fraction of municipal solid waste tends to be of lower quality than compost obtained from manure and/or green waste, due to a higher percentage of salts (contained in MSW and further increased through mineralization), impurities, and contaminants which are not eliminated through composting (de Araújo et al., 2010; Hargreaves et al., 2008). Thus, many countries such as Spain have strict quality criteria that regulates the end use of municipal solid waste compost defining limits for heavy metal concentration and pathogens (Bernal et al., 2017; Siles-Castellano et al., 2020). Nevertheless, other countries have looser requirements for contaminants and pathogens, and several researchers are calling for a better harmonization of criteria at the international level (Bernal et al., 2017). For this review, we will assume that compost meets the highest quality standards to be used as a soil amendment or fertilizer in agricultural land.

3 Effects of compost on soil nutrient cycling

3.1 Soil macro- and micronutrient content

Compost, as stabilized organic matter, is generally a nutrient-rich material which contains all essential plant macro- and micronutrients (Diacono and Montemurro, 2010; Duong et al., 2013; Martínez-Blanco et al., 2013b). Nevertheless, both the total nutrient content and nutrient availability depend largely on the compost feedstock(s) and composting process. The content of available or potentially plant available N, the most limiting nutrient for plant growth, can range from 4 g kg^{-1} dry matter to 120 g kg^{-1} dry matter, depending on the compost feedstock (Bernal et al., 2017). Total P contents range from 0.4 g kg^{-1} dry matter to 23 g kg^{-1} dry matter, and K contents range from 0.7 g kg^{-1} to 12 g kg^{-1} dry matter (Stoffella et al., 2001). Generally, N, P, and K contents tend to be lower in plant-based or compost produced from certain industrial feedstocks such as paper pulp, while compost produced from animal manures, on the other hand, typically have higher N and P concentrations (Chen et al., 2018; Habteweld et al., 2018). The concentration of micronutrients in compost is also highly variable; for example, Stoffella et al. (2001) and Siles-Castellano et al. (2020) reported that in municipal solid waste, compost samples collected in the US and Europe Fe content ranged between 9300 mg kg^{-1} dw and 16400 mg kg^{-1} dw, B between 3 mg kg^{-1} dw and 60 mg kg^{-1} dw, Cu between 31 mg kg^{-1} dw and 1052 mg kg^{-1} dw, Mn between 400 mg kg^{-1} dw and 600 mg kg^{-1} dw, Mo 7.2 mg kg^{-1} dw, and Zn between 15 mg kg^{-1} dw and 1650 mg kg^{-1} dw. The composting process also decreases nutrient availability and therefore potential release to the soil as compared to raw organic waste materials (Chalk et al., 2013).

Compost can be used to replace synthetic fertilizers in agricultural systems partially or completely (Al-Suhaibani et al., 2020; Lazcano et al., 2013; Wu and Ma, 2015). Interestingly, due to the complex chemical composition of composts as compared to the most common NPK synthetic fertilizers, compost constitutes an excellent way of replenishing both plant macro- and micronutrients in soils. Ample evidence in the literature shows that compost application increases soil fertility (Diacono and Montemurro, 2010; Martínez-Blanco et al., 2013a,b), defined as the total soil content of plant macro and micronutrients (Havlin, 2013), compared to soils without compost application. Nevertheless, the magnitude of increase depends on compost properties (i.e. maturity, nutrient content, particle size), rate, number of applications, and time of soil sampling (Duong et al., 2013; Liu et al., 2019). Solaiman et al. (2019) studied the short-term change in total soil nutrient content after the application of six different types of compost to a potting soil compared to an unamended control. It was observed that N contents increased between 0% and 40%, total soil P increased between 7% and 13%, total K increased between 35% and 297%, S increased between 7% and 186% and Zn increased between 12.5% and 290% depending on the compost. Remarkably, a study in Mediterranean grasslands by Ryals et al. (2014) showed that the effects of a single compost application on soil fertility can be long lasting and persist in the soil for three years, due to the recycling of plant nutrients and stabilization of the added nutrients.

In cropping systems, due to the generally higher plant nutrient uptake rates and low nutrient recycling rates (i.e. high nutrient exports in harvested products), compost needs to be reapplied regularly to maintain nutrient fertility. After 2 years of compost application to a sugar beet-wheat rotation, Crecchio et al. (2001) reported increases of 2.1% and 7.8% in total soil N, compared to an unamended control, after application of municipal solid waste compost at 120 kg ha^{-1} and 240 kg ha^{-1} respectively (Table 1). Hartl et al. (2003) reported that soil K increased on average by 26% as compared to a control, after 5 years of compost additions. Application of four different types of compost at a rate of 40 Mg ha^{-1} to a sandy soil resulted in increases in soil Olsen-P between 25% and 525% compared to an unamended control (Castán et al., 2016). Eghball and Power (1999) reported increases in soil P of 57%-159% with two different rates of composted manure applied over 4 years as compared to an unamended control.

Long-term (>10 years) application of composted organic waste leads to the buildup of soil organic matter and plant macro- and micronutrients in soil (Table 1) (Chen et al., 2018) to levels sometimes higher than the use of synthetic fertilizers (Lehtinen et al., 2017). It should be highlighted that compost is typically applied at a rate to satisfy the crop N demands which results in overapplication of P. High concentrations of P can be concerning due to possible runoff and leaching and subsequent contamination of the aquifers (Antille et al., 2014; Maltais-Landry et al., 2015). Thus, compost application rates and soil levels of available P need to be closely monitored to avoid losses and subsequent environmental damage through eutrophication of aquatic ecosystems (Chen et al., 2018; Maltais-Landry et al., 2015). Similarly, certain types of compost such as those produced from municipal solid waste can increase micronutrient content to a point where it can cause environmental pollution; in these cases, application rates need to be controlled to avoid environmental pollution (Fagnano et al., 2011).

3.2 Mineralization rates

Most of the nutrients supplied with compost are in organic form and therefore not immediately plant available. Thus, besides the total concentration of nutrients provided with compost, the rates at which these nutrients locked up in organic compounds become available for plants once in the soil through mineralization, is one of the most important components of soil fertility.

Table 1 Examples of changes in common soil health indicators reported in the literature with short- (<1 year), medium- (1–10 year), and long- term (>10 years) compost application, relative to non-amended soils

Soil function	Goal	Indicator	Short-term	Medium-term	Long-term	References
Nutrient cycling	Supply	Available soil N	0% to 40%	2.1% to 7.8%	NA	Crecchio et al. (2001), Eghball and Power (1999), Hartl et al. (2003), Solaiman et al. (2019)
		Soil P	7% to 13%	25% to 525%	NA	
		Soil K	35% to 297%	26%	NA	
		Soil S	7% to 186%	NA	NA	
		Soil Zn	12.5% to 290%	NA	NA	
		CEC	NA	NA	NA	
		pH	NA	NA	NA	
	Recycling and retention	N mineralization	5% to 30%	40% to 50%	50% to 60%	Boldrin et al. (2009), Hartz et al. (2000), Lazicki et al. (2020), Martínez-Blanco et al. (2013a)
		P mineralization	2% to 50%	90% to 100%	90% to 100%	
		K mineralization	75% to 80%	20% to 100%	NA	
Soil hydraulic properties	Water retention/ infiltration	Aggregate stability	18% to 41%	0% to 63%	29% to 238%	Celik et al. (2010), Daynes et al. (2013), Leroy et al. (2008), Mangalassery et al. (2019), Martínez-Blanco et al. (2013a), Tejada et al. (2008), Tejada and Gonzalez, (2006a), Xin et al. (2016)
	Water infiltration	Infiltration rate	0% to 300%	24% to 396%	339%	Kranz et al. (2020), Leelamanie and Manawardana (2019), Logsdon et al. (2017), Martínez-Blanco et al. (2013a), Mohammadshirazi et al. (2017), Rivers et al. (2021), Weindorf et al. (2006), Xin et al. (2016)
	Compaction	Bulk density	−2.5% to −55%	0.7% to −40%	−20.6% to −40%	Aggelides and Londra (2000), Courtney and Mullen (2008), Diana et al. (2008), Kranz et al. (2020), Martínez-Blanco et al. (2013a), Mohammadshirazi et al. (2017), Mylavarapu and Zinati (2009), Sax et al. (2017), Stamatiadis et al. (1999)

(*Continued*)

Table 1 (*Continued*)

Soil function	Goal	Indicator	Short-term	Medium-term	Long-term	References
	Water retention	Water holding capacity	0% to 50%	0% to 2.6%	−0.4% to 1.7%	Arthur et al. (2011), Eden et al. (2017), Leroy et al. (2008), Mamo et al. (2000), Martínez-Blanco et al. (2013a), Miller et al. (2015)
Crop productivity	High crop yield	Crop yield	−48% to 106%	0% to 71%	10% to 27%	Arthur et al. (2012), Chen et al. (2018), Choi et al. (2001), Erhart et al. (2005), Giannakis et al. (2014), Hartl et al. (2003), Iglesias-Jimenez and Alvarez (1993), Mamo et al. (1999), Tejada and Gonzalez, (2006a)
	High crop quality	Vitamin content	24%	NA	NA	Chan et al. (2010), Coria-Cayupán et al. (2009), Martínez-Blanco et al. (2011), Morlat and Symoneaux (2008), Mugnai et al. (2012), Ramzani et al. (2017), Rubio et al. (2013), Saha et al. (2007), Wang and Lin (2003)
		Phytochemicals (phenols, sinapic acid, anthocyanins)	−33% to 76%	NA	−2%	
		Micronutrient content (Zn, Fe)	+ 12% to 26%	NA	NA	
	Crop health	Disease incidence	−83% to 53%	NA	NA	Avilés and Borrero (2017), Kanaan et al. (2018), Tubeileh and Stephenson (2020)
Soil biodiversity	High microbial abundance	Microbial biomass	0% to 106%	10% to 97%	10% to 400%	Bastida et al. (2007), (2008), Franco-Otero et al. (2012), García-Gil et al. (2000), Lazcano et al. (2013), Li et al. (2015), Nair and Ngouajio (2012), Pezzolla et al. (2013), Ros et al. (2006), Zhen et al. (2014)
	High diversity	Microbial diversity	18% to 34%	NA	3% to 4%	Bastida et al. (2008), Li et al. (2015), Martínez-Blanco et al. (2013a), Zhen et al. (2014)
	High activity	Soil basal respiration	133% to 500%	29%	0% to 43%	García-Gil et al. (2000), Jindo et al. (2016), Li et al. (2015), Pérez-Piqueres et al. (2006), Ros et al. (2003), Tian et al. (2015)
		Enzyme activity	0% to 500%	−18% to 730%	−29% to 300%	
Environmental quality	Mitigate climate change	Soil C sequestration (% of applied C)	40% to 53%	30%	2% to 16%	(Martínez-Blanco et al. (2013a)

Mineralization of organic compounds is carried out by soil microorganisms and strongly dependent on environmental conditions that regulate microbial activity (i.e. soil temperature, moisture, and pH) (Carvalhais et al., 2014; Frey et al., 2013). These environmental conditions depend on large-scale factors such as the predominant climate and seasonality, but also on small-scale factors such as soil texture. It has been recently suggested that fine-textured soils support higher mineralization rates than coarse-textured soils through mediation of soil water content and its distribution in the soil matrix (Li et al., 2020).

Furthermore, mineralization of organic compounds is strongly dependent on the quality of the organic matter being decomposed. The relative amount and recalcitrance of the organic compounds (i.e. molecular complexity and strength of the chemical bonds) as well as stoichiometric relationships between C and plant nutrients (mostly N and P) can substantially affect the amount of nutrients being released from compost. High recalcitrance and C to N ratio (between 20 and 30) result in the slow release of nutrients and will even decrease the amount of plant-available nutrients in the soil if the C:N ratio of the added compost is above 30 (Lazicki et al., 2020). Immature composts can have large C:N ratios and high amount of labile C compounds that results in a net immobilization of nutrients such as P and N rather than a net increase (Lazicki et al., 2020). The C:N ratio is also strongly associated with the predominant feedstock used in the composting process; typically high C:N composts, such as those produced from yard trimmings, release up to 5% of the N within the first three months after application whereas low C:N composts produced from animal manures can release up to 15-30% of the applied N over the same amount of time (Hartz et al., 2000; Lazicki et al., 2020).

Besides the chemical properties, it has also been suggested that the physical properties of compost and precisely the particle size will influence the total amount of nutrients being released over a certain time. In particular, it has been observed that finer particle sizes typically result in higher mineralization rates than larger particle sizes (Duong et al., 2013). Higher mineralization rates with smaller particle sizes could be attributed to the larger surface area available for microbial attack, lower C:N ratio, and an increase in ash and nutrient content (Haynes et al., 2015).

Finally, nutrient mineralization from compost also depends on the timeframe considered; it has been observed that 5%–22% of the N in compost is mineralized the first year, and 40–50% in 3–5 years, and 50–60% beyond 5 years (Table 1) (Boldrin et al., 2009; Martínez-Blanco et al., 2013a). For P, between 2 and 50% can be mineralized within 1 year and up to 100% within 5 years (Martínez-Blanco et al., 2013a). The existing literature showing release rates of K from compost is scarce; some studies have suggested that release rates reach up to 75-80% in the first year and 20-100% in three to five years of application.

In summary, understanding compost mineralization is essential to predict nutrient availability in soils and synchronize nutrient release with crop nutrient uptake. For this, the quality of the compost amendment, the environmental conditions of temperature and moisture, and the time frame of interest need to be considered to match plant demands and avoid losses of nutrients.

3.3 Soil cation exchange capacity

Cation exchange capacity (CEC) is an important component of soil fertility which defines the capacity of the soil to supply and retain nutrients. CEC depends on the number of negative charges in a specific soil that would be able to adsorb cations, including many essential plant nutrients. These charges are mostly associated with clay minerals and soil organic matter. Thus, increases in soil organic matter content with compost application are associated with higher CEC as compared to unamended soils (Miller et al., 2015; Weber et al., 2007). Previous research shows variable effects of compost on CEC, which may depend on the rate of compost application, time since application and characteristics of the compost such as compost maturity and feedstock (Harada and Inoko, 1980; Shiralipour et al., 1992). For example, increasing application rates from 20 to 80 Mg ha^{-1} y^{-1} can increase soil CEC of a soil between 7% and 18% respectively compared to unamended soil (Table 1) (Miller et al., 2015). However higher increases in soil CEC can be expected with aged or mature compost compared to fresh compost (Liu et al., 2019).

3.4 Soil pH

Soil pH drives the solubility of elements including plant nutrients, being a very reliable indicator of plant nutrient availability in soils. Plant macronutrients are mostly soluble at pH 6.5-7.5, becoming unavailable at higher or lower pH through complexation with other elements in the soil solution. For instance, phosphate (HPO_4^{-2} or $H_2PO_4^{-}$), the plant available form of P, co-precipitates with calcium (Ca^{2+}) at pH higher than 7 and with aluminum (Al^{3+}) and iron (Fe^{3+}) when pH is below 5 (Brady and Weil, 2016). Therefore, slightly acidic conditions are desired to avoid nutrient deficiencies. The literature shows inconsistent effects of compost on soil pH. For example, depending on compost stability, short-term decreases in soil pH can be observed due to organic acid production during the decomposition of labile organic molecules and oxidation of N and S compounds (Franco-Otero et al., 2012). However, other studies reported slight increases in soil pH shortly after compost addition (Tian et al., 2015) and in fact, compost has been used successfully to buffer soil acidity in degraded, soils in tropical (Latifah et al., 2018; Ouédraogo, 2001) and temperate regions of the world (Domínguez et al., 2019). Reports on the long-term effects of compost application on soil pH are also inconsistent showing increases (Butler and Muir, 2006; García-Gil et al., 2004), decreases (Bastida et al., 2008; Meng et al., 2005), or no differences after 13 years of continuous application of compost (Calleja-Cervantes et al., 2015). Besides compost stability, other factors that seem to be important in driving the effects of compost on soil pH are the pH of the compost and application rates (Diacono and Montemurro, 2010).

4 Effects of compost on soil hydraulic properties

4.1 Soil aggregate stability

Soil structure regulates the balance between macro and micropores and therefore has a strong influence in soil hydraulic properties, allowing for the balancing of a soil's water holding capacity and infiltration, reducing soil erosion. Furthermore, a good soil structure facilitates plant root growth, creates habitat for soil biota (Erktan et al., 2020), and leads to C sequestration (Six et al., 2000). The formation of aggregates through the binding of soil particles by organic compounds is one of the major processes contributing to the creation of structure in the topsoil (Bronick and Lal, 2005). Compost can increase the formation of water-stable aggregates directly by supplying organic compounds (i.e. polysaccharides, proteins, lipids) that will increase particle inter-cohesion and aggregate hydrophobicity or indirectly through the increase in microbial biomass and the subsequent generation of microbial residues, known to be excellent binding agents. Several studies show that compost application can increase aggregate stability which depends on the quality of the compost, the number of consecutive applications, and length of the experiment (Table 1) (Diacono and Montemurro, 2010). For instance, application of compost with high content of easily decomposable organic compounds such as glucose and other monosaccharides leads to intense and transient increases in aggregate stability, whereas composts with high contents of recalcitrant compounds such as lignin lead to lower but longer lasting effects (Abiven et al., 2009). In an incubation experiment, Temiz et al (2021) observed that aggregate stability was increased up to 18% compared to a control only 60 days after compost application (Table 1). Nevertheless, this depended on the compost feedstock and application rates, with compost produced from animal manures and higher compost application rate leading to the largest increases. In a review of the literature published between 1990 and 2012, Martinez-Blanco et al. (2013a,b) pointed out higher increases in aggregate stability (29–41%) in the short-term (<1 year after compost application). The short-term formation of stable aggregates after compost application is further accelerated by the presence of plant roots and arbuscular mycorrhizal fungi, which suggests potential synergies and highlights the importance of soil agroecological management (Celik et al., 2010; Daynes et al., 2013). In the medium term, previous studies showed either no changes in aggregate stability (Tejada and Gonzalez, 2006a) or significant increases of 10.5%, 37.8% or 63% compared to a control (Leroy et al., 2008; Mangalassery et al., 2019; Tejada et al., 2008). In the long term, it was observed that repeated compost application for 13 and 23 years increased the proportion of water stable aggregates by 29% and 238%, respectively (Celik et al., 2010; Xin et al., 2016).

4.2 Water infiltration rate

A good soil structure and porosity are essential to reduce ponding and facilitate the movement of water into the soil profile (i.e. infiltration). By improving aggregation and infiltration rates, compost application facilitates the recharge of aquifers and reduces soil erosion rates (Martínez-Blanco et al., 2013b). The published literature shows variable effects of composts on soil infiltration rates in the short term (<1 year after compost application), with either no observed effects (Leelamanie and Manawardana, 2019; Weindorf et al., 2006) or increases of 300% compared to a control soil (Table 1) (Rivers et al., 2021). In the mid-term, between 1 and 10 years after compost application, studies report increases between 24% and 396% compared to a control (Kranz et al., 2020; Logsdon et al., 2017; Mohammadshirazi et al., 2017). In the long term, Xin et al. (2016) reported increases of 339% in infiltration rates compared after 23 years of compost application compared to a control soil without compost.

4.3 Compaction

Soil compaction can be assessed through the measurement of soil bulk density or penetration resistance in the field. By promoting aggregate formation, and increasing soil organic matter content, compost can increase soil porosity which is associated with lower bulk density and penetration resistance (Diacono and Montemurro, 2010; Ruehlmann and Körschens, 2009). In the short term, i.e. <1 year, the application of different types of compost decreased soil bulk density between 2.5% and 55% as compared to soil without compost (Table 1) (Aggelides and Londra, 2000; Courtney and Mullen, 2008; Diana et al., 2008; Kranz et al., 2020; Mohammadshirazi et al., 2017; Mylavarapu and Zinati, 2009; Stamatiadis et al., 1999), while in the mid-term (1–10 years after compost application), the decrease may be as much as 40%, with a minimum decrease of 0.7% (Kranz et al., 2020; Martínez-Blanco et al., 2013b). Long-term (> 10 years) compost application can decrease soil bulk density between 20.6% and 40% (Kranz et al., 2020; Martínez-Blanco et al., 2013b; Sax et al., 2017). Compost application seems to be more effective in the short term (Weber et al., 2007) and with increasing application rates (Hemmat et al., 2010), as larger doses produce larger decreases in bulk density.

4.4 Water holding capacity

Increased drought severity due to climate change is increasing the interest in improving water use efficiency in crop production. Besides the use of water-efficient crops and irrigation systems, improving soil water holding capacity is considered as a key strategy to adapt to current and future water shortages. Soil water holding capacity (WHC) is linked to porosity and therefore aggregate formation and soil structure are critical in soil water retention. Given the important role of soil organic matter in driving aggregate formation and soil porosity, it has been observed that increasing soil organic matter leads to the increase in soil WHC (Rawls et al., 2003). In a meta-analysis of the published literature, Minasny and McBratney (2018) observed that a 1% increase in soil organic C was associated with significant increases in soil plant-available water-holding capacity, yet these increases were highly dependent on soil texture. Consequently, the effects of compost on soil WHC observed in the literature are also highly variable (Table 1). Martínez-Blanco et al. (2013a,b) showed that, in the short term, compost application resulted in 0–50% increases in soil WHC. In the medium term, consecutive applications of 90 tn ha^{-1} yr^{-1} compost to corn over 3 years have shown to produce no changes in soil WHC (Mamo et al., 2000), whereas Ramos (2017) observed that compost application for 3 years resulted in increases between 1.2% and 2.6% of soil WHC in a Mediterranean vineyard. In the long term, subsequent applications of compost for more than 10 years have shown to either decrease soil WHC by 0.4% (Leroy et al., 2008) or increase soil WHC by between 0.7% and 1.7% (Arthur et al., 2011; Eden et al., 2017; Miller et al., 2015). This large variability in the observed results is due not only to the soil texture but also to the application rate and particle size of the compost applied. Higher rates and finer particle sizes result in larger increases in soil WHC (Głąb et al., 2020; Şeker and Manirakiza, 2020).

5 Effect of compost on crop productivity

5.1 Crop yield

Crop yield, the mass of marketable crop produced per unit of surface area, is often the single most important variable for agricultural producers. The effects of compost application on crop yield depend largely on the crop, timeframe considered, compost properties, compost application rates, baseline soil fertility, characteristics of the control treatment, and environmental conditions (moisture and temperature). In a meta-analysis of 53 studies, Wortman et al. (2017) reported that larger yield increases are observed in leafy greens and annual crops in the first season of compost application compared to root/tuber/bulb crops, especially when the compost had relatively low C:N ratio, and was applied to soils with initial low fertility. This is due to the fact that leafy greens and herbs are often more responsive to increasing fertility (either organic or synthetic) than fruiting, grain, or root crops (Wortman, 2015).

In one single season, compost application has shown to increase the yield of ryegrass between 30% and 106% compared to a control, depending on the application rate (Table 1) (Iglesias-Jimenez and Alvarez, 1993). Nevertheless yield decreases between 35% and 48% have been observed in crops such as corn, tomatoes, and lettuce, due to the immobilization of N when compost with high C:N ratio is used (Choi et al., 2001; Giannakis et al., 2014). In the medium term, subsequent applications of compost for 3 years increased the yield of a rice crop between 5% and 7% (Tejada and Gonzalez, 2006b) and corn by 71% compared to a control (Mamo et al., 1999). Nevertheless, Hartl et al., (2003) saw no effects on the yield of a rye crop after 5 years of compost application.

Long-term studies also show variable effects of compost on crop yields. For example, Arthur et al. (2012) observed an increase in tomato yields of 27% after 10 years of subsequent compost additions (7.5 ton ha^{-1} yr^{-1}) as compared to a control. Erhart et al. (2005) observed 10% yield increases after 10 years of compost amendment to cereals and potatoes. In a recent meta-analysis, Chen et al. (2018) pointed out that the effects of long-term application of compost for crop yield depends on compost-, soil-, and climate-dependent factors. Soils with low initial fertility, near neutral pH, low soil organic matter, finer texture, and in tropical climates show the highest increases in crop yield after long-term application of organic fertilizers (Chen et al., 2018). Consistently, several studies show that compost use is a great alternative in low-input tropical cropping systems such as sorghum in West Africa, or maize, wheat, potato, and faba bean smallholder farms in Ethiopia, where access to synthetic fertilizers is limited (Bedada et al., 2014; Ouédraogo, 2001). Finally, it should be highlighted that the effects of compost on crop yield can be smaller than those of equivalent amounts of synthetic N, P, and K inputs due to the slower mineralization and nutrient release in compost (Martínez-Blanco et al., 2013a). Similar yields can be achieved when compost is supplemented with inorganic fertilizers. Furthermore, some studies have reported larger increases in crop yield and nutritional quality with compost as compared with equal amounts of synthetic fertilizers, which could be attributed to the supply of secondary macronutrients or micronutrients, as well as improvement of soil biological and physical properties (Celestina et al., 2019; Lazcano et al., 2013).

5.2 Crop quality

Scientific evidence shows that the importance of a healthy diet for health and wellbeing has increased consumer and farmer interest in the nutritional quality of crops (Willett et al., 2019). High concentrations of macro- and micronutrients, sugars, phytochemicals, and secondary metabolites in agricultural products are generally desired, although the parameters that define crop quality are crop dependent (Brandt et al., 2011). Because of this, it is extremely difficult to summarize and compare current literature studying the effects of compost on crop quality. Generally, it has been observed in large-scale studies that continued application of high-analysis simple synthetic fertilizers and export of harvest products can lead to the gradual depletion of micronutrients such as Zn and Fe in soils (Chen et al., 2017). Even if this does not negatively affect crop growth and yield, it can decrease the micronutrient concentration in the harvested parts of the plant and therefore the crop nutritional value (Chen et al., 2017). This decrease in crop nutritional value, called 'hidden hunger,' is further exacerbated by increasing crop yields due to a 'dilution effect' (Liu et al., 2014). On the other hand, higher crop nutritional quality has been observed in several organic systems that use organic fertilizers, an effect which is attributed to the supply of both macro and micronutrients, and the lower crop yields that avoid the abovementioned 'dilution effect' (Helfenstein et al., 2016; Lairon, 2011). Several studies have evaluated the effects of compost on crop nutritional quality (Table 1). In the short term (< 1 year after compost application), Ramzani et al. (2017) showed that application of small amounts of acidified compost to a saline soil could increase Zn and Fe concentrations in grain by 26% and 12.5% respectively, compared to a control. Furthermore, compost increased grain protein content by 3%, vitamins by 24%, although polyphenol concentration decreased by 33% compared to the control. In rice, Saha et al. (2007) observed that compost application increased grain Zn and Fe contents by 14 and 12% respectively, compared to inorganic fertilizers. In strawberries, Wang and Lin (2003) reported significant increases in antioxidant activity and flavonoid contents in plants grown in pots with 50% soil and 50% compost as compared to plants grown in pots with 100% soil. In lettuce, Coria-Cayupán et al. (2009) reported higher contents of chlorophylls and carotenoids, but lower contents or non-significant differences for phenol compounds and antiradical activity. In cauliflower, Martínez-Blanco et al. (2011) observed up to 76% and 24% higher content of sinapic acids and phenols, respectively, in plants grown with compost compared to inorganic fertilizers. In winegrapes, a single application of compost produced from exhausted grape marc, citrus juice waste, and cattle manure resulted in 9% increase in tartaric acid and 16% increase in Fe in fruit, while decreasing total anthocyanins, polyphenols, and Zn by 1%, 9%, and 7% respectively as compared to an unamended control (Rubio et al., 2013).

Consecutive annual applications of compost to perennial crops have also shown inconsistent effects on crop quality. For example, application of green waste compost to Chardonnay winegrapes for 8 years resulted in no change in soluble solids (measured as °Brix), titratable acidity, malic and tartaric acid concentrations compared to inorganic fertilizers, parameters than are known to influence grape fermentation and wine quality (Mugnai et al., 2012). Chan et al. (2010) observed that application of compost to winegrapes for three consecutive years resulted in only slight increases (0.4%) in grape pH. In the long term, application of compost to winegrapes for 28 consecutive years reduced the content of total soluble solids (°Brix) by 4%, anthocyanins by 2%, and tannin content by 9%, and increased pH by 1.7% compared to an unamended control (Morlat and Symoneaux, 2008). Thus, there is evidence that compost use can change crop quality, although general conclusions cannot be made, given the large variability in the observed results and diversity of quality criteria used in each crop. Furthermore, there is a shortage of studies evaluating the long-term effects of compost on crop quality.

5.3 Crop health

Pests and diseases can greatly decrease the yield and quality of crops and are therefore one of the main concerns of agricultural producers. Conventional agriculture relies in heavy applications of pesticides which in many cases have shown to be harmful for soil, environmental, and human health. Furthermore, repeated applications of these pesticides lead to the generation of resistance in the targeted pathogens and a decrease in the efficiency of the fumigants over time (Lucas et al., 2015). The capacity of compost to suppress or reduce the incidence of plant diseases has been demonstrated in numerous studies and is extensively reviewed in (Bonanomi et al., 2007; De Corato, 2020; Martínez-Blanco et al., 2013a; Noble and Coventry, 2005). Compost can support plant health indirectly, by supplying the required nutrients and providing favorable conditions for plant growth. In addition, compost can directly suppress the presence of several soil-borne plant pathogens and increase the resistance of the plants by inducing systemic resistance (Noble and Coventry, 2005). Direct suppression is known to be well correlated to changes in soil microbial community structure and function with compost application and with the presence of microbes that directly compete with the soil-borne pathogens or that produce antibiotic substances (Bonanomi et al., 2010; De Corato, 2020; Mehta et al., 2014; Weller et al., 2002). Given the large differences in microbial and physicochemical properties of different types of compost, it is expected that suppressiveness is also highly variable (Table 1). For instance, in a pot experiment, Kanaan et al. (2018) reported a decrease of 83% in disease severity of *Verticillium dahliae* in eggplants grown in compost produced from tomato waste, but increase of 53% when plants were grown with compost produced from olive waste, as compared to a non-amended peat potting media. In a different pot experiment comparing four types of olive waste compost, Avilés and Borrero (2017) saw both increases and decreases in *V. dahliae* disease severity in cotton seedlings, depending on the type of compost. Finally, compost suppressiveness is often pathogen specific. For instance, compost has shown higher suppression rates in pathogens with a limited saprophytic capacity such as *Thielaviopsis basicola* and *V. dahliae* but low suppression rates or even increases in the abundance of the soil-borne fungi *Phytophthora* spp., *Rhizoctonia solani*, and *Pythium* spp. (Bonanomi et al., 2010). In a field study with bell peppers, Tubeileh and Stephenson (2020) observed that plant-based composts reduced the incidence of *V. dahliae* by 60%, while compost produced from dairy manure was not suppressive. Furthermore, disease suppression was transient and disappeared 2 weeks after compost application. In summary, while the use of compost to suppress soil-borne diseases seems like a promising strategy, much research is still needed to determine the specific mechanisms of suppression, and general management recommendations cannot be made.

6 Effects of compost on soil biodiversity

Soil biota encompasses a wide diversity of organisms, represented across all life kingdoms, and ranging in size from microorganisms such as bacteria and fungi to megafauna such as moles or ground squirrels. The abundance of soil biota can be extremely high, particularly in the case of microorganisms, with billions of microbial cells per gram of soil (Bardgett and van der Putten, 2014). By supporting the major processes of decomposition, nutrient cycling, and soil formation, soil biota is crucial for the maintenance of soil health and crop productivity (Bardgett and van der Putten, 2014; de Vries et al., 2013; Lehman et al., 2015). Thus, a mechanistic understanding of the effects of compost on soil health necessarily involves a thorough study of how it affects soil biota. For this, the development of reliable, robust, and informative soil biological indicators is of high importance for the scientific community. Soil biological communities can be studied from the perspective of '*who is there*' or community structure, and '*what are they doing*' or community function. Community structure is described as both the number of different species that form a specific community, and their biomass. Measurements of function of soil communities include microbial respiration (Degens and Harris, 1997), stable isotope probing (Neufeld et al., 2007), reverse transcription of functional genes coding for specific enzymes (Sharma et al., 2005), or potential activity of hydrolytic enzymes (Fierer et al., 2021; Nannipieri et al., 1980; Tabatabai, 1994).

6.1 Microbial biomass

Microbial biomass constitutes only a small percentage (1–5%) of the total soil organic matter, yet it is extremely important for soil functioning supporting not only nutrient cycling, but also acting as a reservoir of nutrients and a vehicle for C sequestration (Zhang et al., 2017; Zhu et al., 2020). Soil microbial biomass can be estimated as the sum of the concentration of phospholipid fatty acids (PLFAs) for bacteria and fungi, the concentration of the biomarker ergosterol for fungi, through substrate-induced respiration, adenosine triphosphate (ATP) concentration, or as total microbial biomass - C (MBC), N (MBN), measured through fumigation-extraction (Frostegård et al., 1993; Horwath and Paul, 1994; Vance et al., 1987; Young, 1995). Even though it has been observed that a compost with poor quality (i.e. with heavy metals or high salinity) can decrease microbial biomass (de Araújo et al., 2010), typically, the organic compounds found in compost are a source of energy and nutrients for soil microorganisms. Therefore, compost application can boost microbial growth in the short term (Table 1). In a 45-day incubation study, Pezzolla et al. (2013) observed that application of municipal solid waste compost increased soil MBC up to 55% relative to an unamended control. In a pot experiment, Zhen et al. (2014) observed that manure compost increased soil microbial biomass by 58% in just one season. In a field experiment, Bastida et al. (2008) showed that composted sewage sludge could increase MBC by 106% in only one season. Nevertheless, Franco-Otero et al. (2012) did not see increases in MBC after application of municipal solid waste or sewage sludge compost in a semi-arid Mediterranean soil. Similarly, Lazcano et al. (2013) reported no increases in total PLFAs after adding vermicompost to a corn crop. These differences in the short-term effects of compost application observed in the literature have been attributed to the relative proportion of labile and recalcitrant C compounds in the compost (Calbrix et al., 2007; Lazcano et al., 2013).

Studies with repeated applications of compost over several years show more consistent positive effects on microbial biomass (Table 1). For example, two years of compost application to a Mediterranean soil resulted in a 97% increase in MBC compared to an unamended control (Bastida et al., 2008). Nair and Ngouajio (2012) reported a 50% increase in MBC after 3 years of dairy compost application at 25 t ha^{-1}. Similarly, García-Gil et al. (2000) observed that consecutive annual applications of 20 t ha^{-1} or 80 t ha^{-1} of municipal solid waste compost for 9 years increased MBC by 10% and 46%, respectively. In the long term, it was observed that 17 years of annual applications of compost to degraded semi-arid soil resulted in 84–400% increases in MBC, depending on the application rate (65 t ha^{-1} and 260 t ha^{-1} respectively) (Bastida et al., 2007). Ros et al. (2006) observed increases of up to 10% in MBC after 12 years of green waste compost application at 17 t ha^{-1} yr^{-1}. Li et al. (2015) reported increases of 83% to 138% in MBC after 25 years of continuous application of composted cattle manure at 375 kg N $h^{-1}yr^{-1}$ and 750 kg N $h^{-1}yr^{-1}$, compared to an unamended control in a corn–wheat rotation.

6.2 Microbial diversity

Microbial diversity is an important component of soil health since it is responsible for the provision of ecosystem service and resilience to external disturbances (Wagg et al., 2014). Changes in microbial diversity induced with soil management have been associated with the overall decline in the function of the microbial community (Tsiafouli et al., 2015). Soil microbial diversity can be assessed with different molecular techniques. Fingerprinting DNA-based techniques such as terminal restriction fragment length polymorphism (T-RFLP) and denaturing gradient electrophoresis (DGGE) were commonly used in the 1990–2000s. Recently, high-throughput sequencing of gene markers, such as 16S rRNA (for prokaryotes) and ITS (for fungi), is commonly used in assessments of soil health (Fierer et al., 2021; Lindahl et al., 2013; Ward et al., 1990). The diversity of soil microbial communities is frequently calculated through indices that take into account species richness and the relative abundance of each species (i.e. Shannon diversity index, Simpson, ACE, Chao 1) (Kim et al., 2017).

Compost has short-, medium-, and long-term effects on soil microbial diversity (Table 1) (Martínez-Blanco et al., 2013b). Using DGGE, Zhen et al. (2014) reported that, in one single season, compost addition could increase bacterial and fungal diversity by 34% and 18% respectively compared to an unamended control. Studies show that, generally, in the short term, gram-negative and gram-positive bacteria are stimulated by the addition of labile C compounds with compost (Lazcano et al., 2013; Willekens et al., 2014) whereas fungi are not so strongly affected (Lazcano et al., 2013). This selection of soil microorganisms results in differences in microbial community structure between compost amended and unamended soils observed within only one season (Farrell et al., 2010; Pérez-Piqueres et al., 2006; Watts et al., 2010). The changes exerted on the composition of the microbial community by a single application of compost can be long-lasting and visible 24 years after as shown by Torres et al. (2015).

Repeated applications over more than 1 year can also shift the composition of the microbial community. Two years after applying sewage sludge compost to degraded soil, Bastida et al. (2008) reported significant changes in the structure of the soil microbial community as assessed by PLFAs, a change that was associated with higher microbial biomass activity and soil C. Nevertheless, other studies show no differences in bacterial community structure between compost-amended soils and controls even after 5 years of subsequent compost applications (Cherif et al., 2009). In the long term, 25 years of continued applications of composted cattle manure at 375 kg N $ha^{-1}yr^{-1}$ and 750 kg N $ha^{-1}yr^{-1}$ to a corn-wheat rotation increased bacterial diversity by 3% and 4% respectively, compared to an unamended control in a corn-wheat rotation (Li et al., 2015). The contrasting results observed in the effects of compost on soil microbial diversity have been attributed to various factors such as the application rate (higher application rates result in stronger changes), and the proportion of mineralizable carbon and lignin in the applied compost (Saison et al., 2006).

6.3 Microbial activity

Measurements of microbial activity are often preferred to measurements of microbial diversity or community structure since they are directly correlated to soil processes such as mineralization of organic compounds and release of plant-available nutrients (Kibblewhite et al., 2008). Microbial activity can be estimated, among other methods, by measuring soil respiration or potential activity of microbially produced enzymes involved in the catabolism of organic compounds (Degens and Harris, 1997; Nannipieri et al., 1980; Tabatabai, 1994). Both microbial respiration and enzyme activity are sensitive to changes in soil management and have shown to change with compost application (Martínez-Blanco et al., 2013b). High concentration of heavy metals and other pollutants in compost can decrease soil microbial activity; however, if compost quality is good, typically an increase in microbial activity is expected after compost application (Knapp et al., 2010). In addition to estimating overall microbial activity, many studies assess the diversity of metabolic functions supported by the soil microbial community (i.e. functional diversity) using community-level physiological profiles or CLPPs. The use of BIOLOG Ecoplates® allows for the simultaneous analysis of 31 C sources using a tetrazolium redox dye as an indicator of microbial metabolism (Garland and Mills, 1991). This method stems from the early work of Degens and Harris (1997) who developed a method to estimate substrate-induced respiration using an array of C sources of different complexity.

In a short-term incubation experiment with two soils, Pérez-Piqueres et al. (2006) observed large increases (between 133% and 500%) in soil basal respiration after application of compost as compared to an unamended control (Table 1). Furthermore, this same study reported significant changes in the CLPPs in amended soils as compared to unamended soils only 10 days after the start of the experiment, indicating a rapid shift in the capacity of the bacterial community to degrade C compounds. In a field experiment, Ros et al. (2003) observed that right after addition of 30 kg compost m^{-2} to a degraded soil, microbial respiration increased by 500%, dehydrogenase activity by 300%, protease activity by 200%, but did not change glucosidase activity compared to an unamended control.

In the medium term (1–10 years of compost application), Jindo et al. (2016) reported that 5 years of sheep manure compost application to plum trees at 20 t ha^{-1} yr^{-1} resulted in a 29% increase in microbial respiration but

18% decrease in dehydrogenase enzyme activity, as compared to the control. Application of composted manure for 9 consecutive years increased the activity of dehydrogenase and catalase by 730% and 200% (García-Gil et al., 2000). Tian et al. (2015) reported that microbial activity measured as hydrolysis of fluorescein diacetate increased by 250% after 3 years of application of 300 kg ha^{-1} yr^{-1} of composted cattle manure.

In a long-term experiment, Ros et al. (2006) reported significant increases in microbial activity with annual application of compost for 12 years, although this depended on the type of compost. Sewage sludge compost increased basal respiration by 43% whereas municipal solid waste and green waste compost did not have any effects. Li et al. (2015) reported that annual application of composted manure for 25 years increased urease activity by 300%, alkaline phosphatase by 175%, α-Glucosidase by 65%, β-Glucosidase by 175%, β-Xylosidase by 225%, β-Cellobiosidase by 280%, and L-Leucine Aminopeptidase by 25%, compared to non-amended soil. This increased enzyme activity was attributed to higher soil organic C and microbial biomass in the plots amended with composted manure. Nonetheless, two other enzymes, peroxidase and phenol oxidase, were decreased by 12.5% and 29% compared to the non-amended soil, a decrease that could be attributed to higher N inputs and lower fungal abundance compared to the non-amended control (Li et al., 2015).

7 Effects of compost on environmental quality

7.1 Carbon sequestration

The global biophysical potential to decrease greenhouse gas emissions by building soil C was estimated at an upper limit of 4–5 Gt CO_2 per year, if soil conservation practices were extensively adopted worldwide (Paustian et al., 2019). Compost application to agricultural lands is among the soil conservation practices well known to sequester C. In fact, several studies that synthesized data on the impact of conservation practices showed greater soil carbon sequestration rates associated with the addition of organic amendments, including compost, compared to other practices such as cover cropping or reduced tillage intensity (Payen et al., 2021; Tiefenbacher et al., 2021). Compost application directly impacts soil C by adding organic matter to the system, as approximately half of the organic matter in compost is C. Compost can also indirectly impact soil C inputs by stimulating plant growth. Not all carbon added as compost to the soil is sequestered. Upon application, compost undergoes decomposition, during which part of the C is respired and lost as CO_2. A culmination of research findings in recent decades supports the concept that soil organic matter is a continuum of progressively decomposing organic compounds, during which C can be stabilized in pools with relatively long turnover times, for example inside aggregates or adsorbed onto soil mineral surfaces (Lehmann and Kleber, 2015). Thus, the amount of C sequestered following compost application depends on the amount of C that is respired versus assimilated into biomass, and the amount of C that can be stabilized in pools with slower turnover times. These processes, in turn, are affected by soil physical, chemical, and biological properties as well as the quality and quantity of the organic matter inputs. Therefore, it is not surprising to see a wide range in soil C sequestration rates following compost application (Table 1). For example, a global meta-analysis of 92 studies in which organic amendments had been added to arid, semiarid, or Mediterranean rangelands showed a net increase in soil organic carbon following compost application (Gravuer et al., 2019). However, the magnitude of gains in soil organic carbon was affected by complex interactions between climate, the compost nitrogen concentration, and the number of years following compost application. Given the limited supply of compost relative to the acreage of agricultural lands, it is pertinent to carefully assess under which conditions compost application can lead to the greatest soil organic carbon gains.

7.2 Nitrogen and phosphorus losses

Nitrogen and phosphorus are essential nutrients for plant growth but cause major environmental problems when lost from agricultural fields. Given that compost contains N and P, compost can be used as fertilizer source, but also runs the risk of causing environmental pollution when excess N or P are applied. When compost is used as a fertilizer, it is typically applied at a rate to meet the crop's N requirement. This is particularly prominent in organic cropping systems, where compost can be the dominant nutrient source. For example, several long-term experiments comparing organic and conventional practices have used compost as dominant nutrient source in treatments under organic management (Krause et al., 2020; Tautges and Scow, 2020). In such experiments, unbalanced soil N:P ratios have been observed after long-term compost application (Maltais-Landry et al., 2016). The dominant form of N in compost is organic N, which needs to be mineralized before N becomes plant available. As such, N applied in the form of compost is typically more slowly available to the plant compared to synthetic fertilizer N. In comparison to soils that receive no N-containing amendments at all, compost application can enhance N leaching (Li et al., 1997). However, when compost is used instead of mineral fertilizer N, reduced N leaching losses are typically observed (Siedt et al., 2021). The amount of N leached appears to be positively correlated to the C:N ratio of the compost (Leclerc et al., 1995; Li et al., 1997). This is in line with close correlations observed between the C:N ratio of organic amendments and associated N mineralization rates (Lazicki et al., 2020). To minimize N leaching after compost application, it is recommended to closely match potentially mineralizable compost N to crop N requirements, considering potential legacy effects of previous years' compost inputs.

As opposed to N, compost P can be highly soluble and more easily lost from the soil compared to its inorganic counterpart (Bernal et al., 2017). This is compounded with nutrient imbalances, where P application often exceeds crop demand by matching compost application rates to meet the crop's N requirement (Eghball, 2003). In a column experiment using three soils and seven livestock manure composts, P leaching was positively correlated with the total P concentration of the compost as well as the soil, and negatively correlated with the P sorption capacity of the soil (Kim et al., 2011). In another study, P leaching increased with increased simulated rainfall and losses were greater following application of dairy compost compared to raw dairy manure (McDowell and Sharpley, 2004). In general, less research has been conducted on the effect of compost application on P loss compared to N loss. However, it is recommended that compost application rates are carefully selected to prevent excessive P inputs, especially in soils with high soil P content or low P sorption capacity.

7.3 Greenhouse gas emissions

The effect of compost application on the net greenhouse gas (GHG) budget does not only depend on soil carbon sequestration, but also on emissions of nitrous oxide (N_2O) from soil, as well as potential upstream emissions (CO_2, N_2O, and methane (CH_4)) associated with compost production and distribution. Nitrous oxide is a GHG with a potency 265 times that of CO_2 on a 100-year horizon (IPCC, 2014), predominately emitted from agricultural soils through the microbially mediated processes of nitrification and denitrification (Butterbach-Bahl et al., 2013). The strong correlation between N_2O and N application has led to the development of emission factors to quantify N_2O emissions based on land application of N inputs for national inventory purposes (Solomon et al., 2007). Extensive literature review settled on a default emission factor of 1%, meaning that for each kilogram of N applied, 0.01 kg of N_2O-N is estimated to be emitted (Solomon et al., 2007). Following the release of the 1% default emission factors, various efforts were geared towards developing more context-specific emission factors, including emission factors that differentiate emissions associated with organic versus synthetic fertilizer N inputs. In general, N_2O emissions appear to be lower following application of compost compared to synthetic N fertilizer or unprocessed organic fertilizers such as raw manure (Lazcano et al., 2021b; Walling and Vaneeckhaute, 2020). However, between-study variability is large. For example, a recent literature synthesis showed N_2O

emission factors for compost ranged between 0.06% and 5.60% across studies (Walling and Vaneeckhaute, 2020). Clear patterns to explain variability between studies are yet to emerge, but the variability is likely attributed to variation in soil type, climate, compost quality, and mode of compost application. Compost is often applied as a soil amendment, meant to improve soil health rather than meeting the full crop nutrient requirement. In these situations, compost is often complemented with synthetic fertilizer N inputs, to meet both goals of building soil health and ensuring immediate crop N requirements. The combined use of compost and synthetic fertilizer N and lack of a consistent control treatment across studies further complicate unraveling the effect of compost on N_2O emissions (Graham et al., 2017). The large variability in effects of compost on soil C sequestration as well as N_2O emissions underlines the importance of further research into potential tradeoffs between C sequestration and N_2O emissions at the field scale (Graham et al., 2017; Walling and Vaneeckhaute, 2020). To assess the effect of compost application on GHG emissions beyond the field scale, upstream emissions associated with compost production and distribution need to be accounted for, as well as avoided emissions associated with alternative management of the organic waste feedstock. Moreover, the net effect on GHG emissions will also depend on whether compost is used as substitute for synthetic fertilizer, replaces raw manure inputs, or is simply added to the standard nutrient management plan. Greenhouse gas emissions during compost production vary widely but can be reduced through strategic selection of the composting method (Ba et al., 2020; Walling and Vaneeckhaute, 2020). For example, silo composting, adding sawdust or straw, and good aeration can reduce GHG emissions during the composting process. Given that typical compost application rates range between 4 Mg ha^{-1} and 40 Mg ha^{-1}, GHGs associated with transportation from the composting facility to the field site add up quickly (Tiefenbacher et al., 2021; Walling and Vaneeckhaute, 2020). Therefore, it is recommended that compost is locally sourced.

8 The use of compost to improve soil health in annual crops: a case study with strawberries

Cultivated strawberry (*Fragaria x ananassa* Duchesne ex Rozier) is a fruit crop highly appreciated throughout the world and of major economic importance in countries such as China, USA, Mexico, Egypt, Turkey, and Spain. One of the biggest challenges in strawberry cultivation is the management of soilborne diseases. Since the 1960s, methyl bromide has been almost universally used to control soilborne diseases in conventional strawberry (Chellemi, 2002), but this fumigant has been phased out of production and use as of 2016 due to its negative impacts on public health and the environment. To combat the increased disease pressure in strawberry production, researchers are proposing an integrated approach including breeding for increased disease resistance, development of novel pesticide chemistries, optimization of cultivation practices, and improvement of the health and disease-suppressiveness of the soil (Lloyd, 2016). To date, strawberry cultivars have been identified that are tolerant to soil-borne pathogens like *Verticillium dahliae* (Shaw et al., 2010, 1997) and *Fusarium oxysporum* f. sp. *fragariae*. Additionally, the strawberry industry is looking for cultivars that are tolerant to *Macrophomina phaseolina*, an emerging disease that causes crown rot in strawberries and is seriously threatening the strawberry industry in California (Koike, 2008). The mechanisms of disease tolerance are still unclear, but recent research points out at possible interactions between resistant cultivars and beneficial microorganisms colonizing in the rhizosphere (Lazcano et al., 2021a).

Another major challenge in strawberry production where soils play a critical role is the management of nitrogen (N). California strawberry growers commonly apply N fertilizer in two forms: a pelletized, coated pre-plant fertilizer or control release fertilizer (CRF) and in-season liquid N fertilizer via drip irrigation, a method called fertigation. California strawberry fields typically receive between 168 kg ha^{-1} and 336 kg ha^{-1} fertilizer N over a production season while strawberry crop N uptake has been reported to range from 201 kg ha^{-1} to 246 kg ha^{-1}, but N fertilization rates and timing vary widely among growers. Nitrate pollution of freshwater resources associated with leaching of fertilizer nitrogen from agricultural lands is a major challenge in many areas of the world where strawberries are grown (Carle et al., 2006). The strawberry industry has focused on 'fine-tuning' fertility and water management for more efficient resource use (Bottoms et al., 2013a,b). Several studies reported a high potential for leaching of N fertilizer during the early stages of plant development, when plant uptake is low (Bottoms et al., 2013b). Thus, there is a strong need to find slow-release fertilizers that supply enough nutrients to support crop growth while reducing the risks of N loss.

By improving soil health, compost application presents one potential strategy to address these two major challenges in strawberry production. Long-term management of soils with organic amendments increases soil organic matter, microbial biomass, and activity compared with fertilized or unamended soils (Diacono and Montemurro, 2010). By increasing the size and activity of the soil microbial community, compost could be conducive to the establishment of beneficial plant-microbial interactions, resulting in improved plant health and nutritional quality. In fact, compost has been shown to suppress soilborne disease in several published studies (Noble and Coventry, 2005). Lodha et al. (2002) found that amendment of soil with compost prepared from crop or weed residues reduced the severity of dry root rot in cluster beans caused by *M. phaseolina*. However, an increase in some diseases due to compost usage has also been observed, since compost is a product that varies considerably in chemical, physical, and biotic composition, and, consequently, also in its ability to suppress soil-borne diseases (Pugliese et al., 2015).

While compost application adds nutrients to the soil, most of them are in organic form and need to be mineralized before it becomes plant available. Thus, the gradual release of N from compost may reduce the likelihood of nitrate leaching. Lloyd et al. (2016) found that vermicompost and mushroom compost are good sources of early season nitrate for strawberry plants. If compost can supply sufficient nitrogen to the crop during the early season without triggering nitrate leaching, compost may be an effective alternative to synthetic slow-release fertilizers. Here we evaluated the effects of pre-plant compost application on N dynamics and suppression of *M. phaseolina* in strawberry cultivars with different levels of resistance to this soil-borne fungal pathogen.

8.1 Experimental design

A 2-year field trial was conducted to assess the effect of strawberry cultivar and preplant fertilizer management on nitrogen dynamics and disease incidence in a strawberry field infested with *M. phaseolina*. The experimental site was located at the California Polytechnic State University in San Luis Obispo, California (USA) on a silty clay loam soil. The experiment was laid as a randomized complete block design with four replications. Pre-plant fertilizers represented the main plots and included three treatments: a control with no pre-plant fertilizer applied (control), a treatment with AgRx brand 18-8-13 urea coated custom blended strawberry preplant controlled release fertilizer (CRF), and a treatment with preplant compost application (compost). CRF was applied at a rate equivalent to 112 kg N ha^{-1} representing typical grower practice. Dairy manure-based compost was incorporated through hand-tilling in the top 15 cm of soil and applied at a rate equivalent to 100 kg N ha^{-1}, equating to 10–12 t ha^{-1} accounting for year-to-year variation in compost moisture content. This rate also matched the total N input in the CRF treatment. This application rate equated to 1.2 t C ha^{-1} per dry weight basis (Fig. 2). It should be noted that only a fraction of this N is expected to be mineralized in the first year following compost application. All treatments received the same in-season fertilizer applications with conventional liquid fertilizer commonly used in CA strawberry production through the drip irrigation system. Each main plot was split into 4 subplots, each of which was planted with one of four strawberry cultivars: Albion, Monterey, San Andreas, and a proprietary variety. Cultivars were selected based on preliminary evidence suggesting differences in N requirements. Yield data was collected by hand-picking treatment plots twice per week during the harvesting season between April and July each year, representative of grower practice. Disease incidence was expressed as area under disease progress curve (AUDPC), a metric that

Figure 2 From left to right: Rhizon pore water sampler with porous ceramic tip installed to 30 cm depth (below root zone); preparation of strawberry fields includes shaping soil into raised beds with a tractor implement, a similar implement is used to apply synthetic pre-plant fertilizer; strawberry plants photographed in May 2020 showing symptoms of infection by *Macrophomina phaseolina* (brown plants).

considers the plant mortality rate (determined visually) as well as the timing of disease incidence within the growing season and is expressed as %-days (Jeger and Viljanen-Rollinson, 2001; Madden et al., 2017). Soil pore water samples were collected every two weeks (from April until July) from rhizon samplers (Rhizosphere Research Products (RRP), the Netherlands) installed at 15 cm and 30 cm depth to assess the concentration of nitrate in and below the root zone, respectively (Fig. 2). Soil pore water NO_3^--N concentration was converted to soil NO_3^--N concentration based on soil moisture content. To determine soil NO_3^--N exposure, the summed concentrations were plotted against time and trapezoidal- integration used to represent cumulative soil NO_3^--N availability across the entire analysis period. Linear-mixed effects analysis (ANOVA) was performed in R to assess the effects of factors on response variables. The Shapiro-Wilk test and Levene's test were performed to confirm that our data met the assumptions of normality and homogeneity of variance.

8.2 Results and discussion

There was a significant effect of pre-plant fertilizer treatment in *M. phaseolina* disease incidence in the 2019 and 2020 growing seasons ($P = 0.017$ and $P < .001$, respectively). Nevertheless, opposite to what we had initially hypothesized, compost application did not suppress disease incidence in this case study, but it increased disease incidence compared to the control treatment where no pre-plant fertilizer was applied in one of the growing seasons (Fig. 3). In 2019, disease incidence in the control treatment (169.8%-days) was significantly less than in the compost and CRF treatments (249.9%-days and 256.3%-days, respectively), and in 2020, disease incidence in the control and compost treatments (343.7%-days and 451.2%-days, respectively) were significantly less than in the CRF treatment (9863%-days). While various mechanisms by which

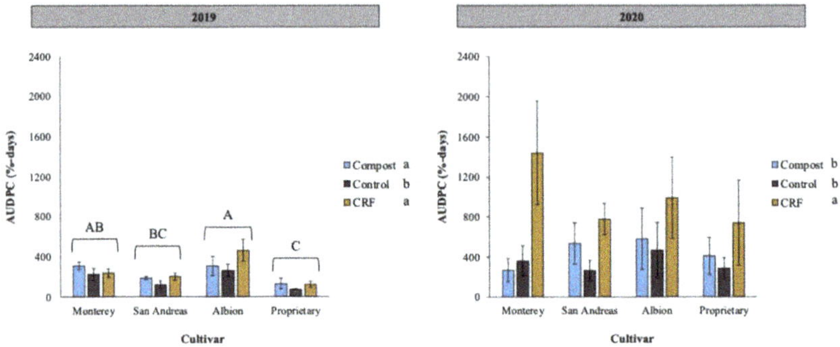

Figure 3 Disease incidence determined using the area under the disease progress curve (AUDPC) from observation data collected from May 10, 2019, to July 19, 2019 (year 1) and May 1, 2020, to July 24, 2020 (year 2). Error bars represent standard errors of the mean for each treatment (n = 4). Uppercase letters indicate significant differences in disease incidence between cultivar in 2019 measurements at $P < 0.05$. Lowercase letters indicate significant differences in disease incidence between pre-plant fertilizer treatments in 2019 and 2020 measurements at $P < 0.05$.

compost contributes to disease suppressiveness have been demonstrated, mixed effects of compost application on disease incidence have been observed (Mehta et al., 2014). Variations in effects of compost application on disease suppressiveness across studies have been attributed to variations in the cropping system, the compost application rate, method of application, frequency of compost amendments, compost feedstock and quality, and the compost microbiome (De Corato, 2020). For example, Tubeileh and Stephenson (2020) observed a reduction in the soil load of *Verticillium dahliae* in soils applied with 25 t ha^{-1} of composts derived from grape pomace, olive pomace-dairy manure mix, or mixed crop residue, but compost made of dairy and horse manure alone was much less effective at suppressing the pathogen. If compost application is aimed at suppressing disease, identifying a suitable compost source, rate, and application method is essential.

There was a significant effect of cultivar on total yield in both 2019 and 2020 ($P < 0.001$) (Fig. 4). There was no significant effect of pre-plant fertilizer on total yield in 2019, but there was a significant effect of pre-plant fertilizer on total yield in 2020 ($P = 0.041$) (Fig. 4). The compost and control treatment had a greater mean total yield in 2020 compared with the CRF treatment. Prior to planting, soil tests indicated a residual soil nitrate concentration of 75 mg kg^{-1} in 2019 and 21 mg kg^{-1} in 2020. Despite the stark differences in N applications between treatments, no effects of preplant fertilizer treatment on nitrate exposure in or below the root zone were observed during the 2019 or 2020 growing seasons (Fig. 5). Potentially, the release of nitrogen from both the compost and the CRF treatment was slow enough for plant uptake, nitrate immobilization, and denitrification to reduce the concentration of

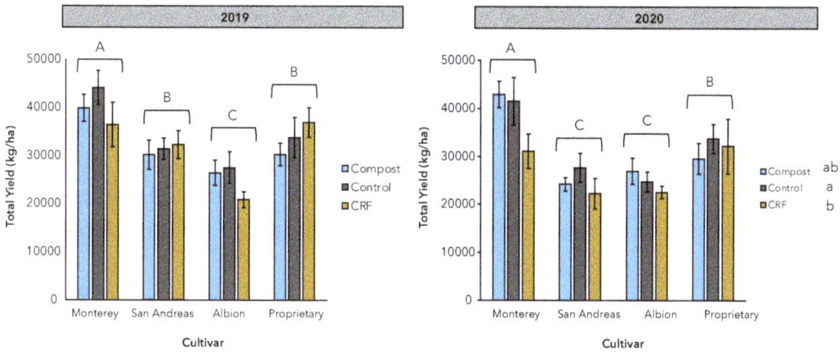

Figure 4 Total yield (kg ha^{-1}) from strawberry harvests collected from April 1, 2019, to July 19, 2019 (year 1) and April 4, 2020, to July 24, 2020 (year 2). Error bars represent standard errors of the mean for each treatment (n = 4). Uppercase letters indicate significant differences in total yield between cultivars in 2019 and 2020 measurements at $P < 0.05$. Lowercase letters indicate significant differences in total yield between pre-plant fertilizer treatments in 2020 measurements at $P < 0.05$.

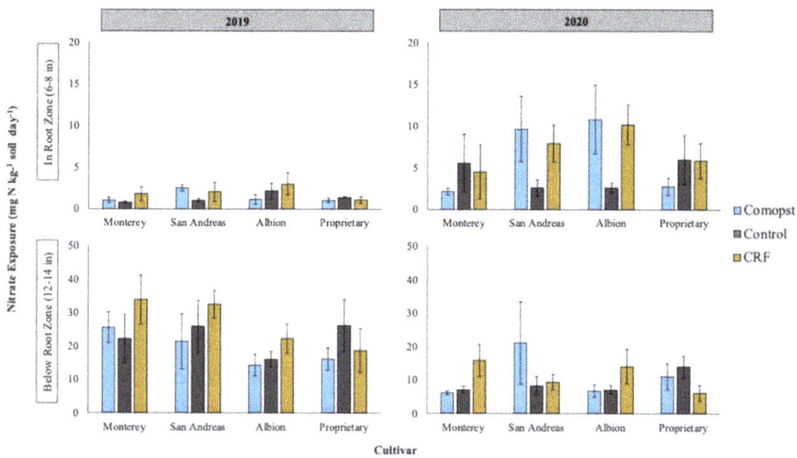

Figure 5 Soil nitrate exposure from January to July 2019 and November 2019 to April 2020 for the 2019 and 2020 growing seasons, respectively. Error bars represent standard errors of the mean for each treatment (n = 4).

NO_3^- in the soil. Tightening of N cycling by more heavily relying on microbial turnover of N between the soil organic matter pool and plant available forms has been identified as a key strategy to reduce N leaching from agricultural systems (Drinkwater and Snapp, 2007; Schimel and Bennett, 2004). Given that organic agriculture typically relies on compost amendments, comparison of organic versus conventional systems can be used as a proxy to assess the effect of compost on N loss. In a recent review, Seufert and Ramankutty (2017) established that there is great uncertainty in the impact of organic management on water quality but N leaching per unit of surface area tended to be lower in organic agriculture compared to conventional agriculture.

In summary, despite the lack of short-term benefits observed in this experiment, compost use did not reduce yields and should therefore be considered as a sustainable alternative to control released fertilizers, given the long-term benefits for soil health shown in the literature. More research is needed to better understand the impact of compost source, rate, timing, and placement on crop nutrition and N leaching.

9 The use of compost to improve soil health, sequester carbon and reduce greenhouse gas emissions in perennial crops: a case study in a Mediterranean vineyard

Soil health in vineyards around the world is threatened by intensive agricultural practices such as tillage and herbicide use, particularly in semi-arid Mediterranean regions (Salomé et al., 2016). Sustainable soil management practices have gradually been adopted by the wine industry in the past few decades to minimize soil disturbance and improve soil health. The primary focus on healthy soil management practices is usually on increasing soil organic matter content, because of its fundamental role in supporting many biological, physical, and chemical dynamic aspects of soil health (Liu et al., 2016; Rawls et al., 2003). Practices such as the use of cover crops, reduced till, and the application of compost have the potential to improve soil health through the increase in soil organic matter. Furthermore, by increasing soil organic matter, these practices lead to sequestration of C in soils, having a large potential for mitigation of climate change. Compost application provides a direct input of C and nutrients to the soil (Lal, 2004) which increases microbial biomass and promotes aggregate formation, facilitating C stabilization and sequestration as biomass or within soil aggregates (Six et al., 2004). Macroaggregates provide a preferential site for microaggregate formation and the associated C stabilization (Six et al., 2000), and long-term compost input was found to increase the fraction of large macroaggregate (Liu et al., 2021). Furthermore, compost can indirectly increase soil C storage by supplying nutrients and increasing above and belowground vine biomass and root exudation, a mechanism which has been recently found to contribute to most of the sequestered or stable C in soils. As woody perennials, grapevines have great potential for C sequestration and climate change mitigation under the right management practices (Wolff et al., 2018). On the other hand, the mineralization of labile organic C and N from compost application can lead to the subsequent release of carbon dioxide (CO_2) and nitrous oxide (N_2O) through increased microbial respiration, nitrification, and denitrification (Bustamante et al., 2011; Lazcano et al., 2013;

Verhoeven et al., 2019). This release of CO_2 and N_2O can offset the climate change mitigation benefits of compost, particularly in a situation where soil C sequestration potential has reached its plateau (Guenet et al., 2020; Lugato et al., 2018). Compost application rate is one of the main factors affecting the potential benefits for soil health (Martínez-Blanco et al., 2013b). While high application rates can increase nutrient and C inputs to the soil therefore having stronger benefits for soil health, they could also trigger large emissions of greenhouse gases, offsetting the benefits of compost for climate change mitigation. Here, we evaluated these potential tradeoffs by measuring the effects of increasing doses of compost application on soil C, CO_2, and N_2O emissions in a Mediterranean vineyard.

9.1 Experimental design and analyses

A 3-year study investigating the effects of compost application rate on soil C sequestration and greenhouse gas (GHG) emissions was conducted in a commercial wine grape vineyard (*Vitis vinifera*, var. Cabernet Sauvignon) located at J. Lohr Vineyards and Wines in the Paso Robles American Viticultural Area (AVA) near Paso Robles, California (USA) between November 2018 and November 2020. The experimental setup was a randomized completed block design with four treatment levels (compost application rates) replicated four times: 0 (control), 4.5, 9.0, and 13.5 Mg ha^{-1} $year^{-1}$ (fresh weight, 0, 0.45, 0.91, and 1.36 Mg C ha^{-1} equivalent). Individual plots included four rows of 30 vines and three interrow areas (tractor rows), while the two middle vine rows and the middle tractor rows were the sampling areas, within which two functional locations (vine and tractor row) were designated for soil and GHG sampling. Compost (certified organic through California Department of Food and Agriculture's Organic Input Materials Program) made of processed livestock manure and green waste was broadcasted annually over the entire orchard floor after harvest when the rain (wet) season started in the region. During the experimental period, soil samples to 60 cm depth were collected in spring each year from the two functional locations. The effects of compost application rate on soil C sequestration were measured by soil health indicators including total soil C (%), permanganate oxidizable carbon (POXC), and aggregate size distribution. Total soil C was determined by combustion of approximately 1 g of air-dried, ground soil in a Vario Max CNS elemental analyzer (Elementar, Langenselbold, Hesse, Germany) at 900°C. POXC was colorimetrically determined by reacting air-dried soil samples with 2 M potassium permanganate solution using the revised protocol of Weil et al. (2003). Aggregate size distribution was done by wet-sieving based on the methodology described in Six et al. (2000), and four aggregate-size fractions were separated from this procedure: silt and clay (S+C; <53 μm), microaggregate (m; 53–250 μm), small macroaggregate (SM;

250-2000 μm), and large macroaggregate (LM; >2000 μm). GHG samples were also collected from both functional locations to measure cumulative CO_2 and N_2O emissions using the static chamber technique (Hutchinson and Mosier, 1981; Parkin and Venterea, 2010) and USDA GraceNet guidelines for static chamber design and GHG sampling procedures (Fig. 6). The seasons were arbitrarily defined: dry season started on the first day of cover crop mowing in early to mid-April, and the wet season began after harvest when the associated N_2O fluxes had subsided to background level. Linear mixed effects analysis (ANOVA) was performed in R to assess the effects of factors on response variables. The Shapiro-Wilk test and Levene's test were performed to confirm that our data met the assumptions of normality and homogeneity of variance.

9.2 Results and discussion

After 2 years of compost application, total soil C content did not increase significantly when compost input increased (Table 2). This is likely due to the slow C turnover rates in soil. However, POXC, which represents the active C pool that is sensitive to soil management and tends to be stabilized in soil (Hurisso et al., 2016), was significantly increased by the rate of compost application (Fig. 7). Higher compost application rates resulted in higher concentrations of POXC to a depth of 60 cm ($P < 0.001$). This suggests that annually applying compost up to the rate of 13.5 Mg ha^{-1} in a vineyard can increase the tendency of C stabilization and potential sequestration in soil. In addition, compost application rate also had a significant interactive effect on the distribution of the large macroaggregate (LM) fraction of soil (Fig. 8). Significantly more LM was found in the tractor row topsoil than the subsoil and both depth increments under the vine ($P < 0.001$). The LM distribution within the tractor row topsoil was marginally and significantly greater at compost application rates of 9.0 and 13.5 Mg ha^{-1} than the control treatment ($P = 0.0755$ at 90% confidence interval), suggesting an early sign towards increased macroaggregate

Figure 6 From left to right: broadcasted of compost at the vineyard and placement of the static chambers used to sample GHG at the two functional locations within each experimental plot.

Table 2 C content (Mg C ha^{-1}) and the respective standard errors of the means (n = 4) by compost application rate, functional location, and depth in April 2020, after two annual compost applications.

	0–15 cm		15–30 cm		30–60 cm	
Rate (t ha^{-1})	Tractor row	Vine row	Tractor row	Vine row	Tractor row	Vine row
0	15.3 ± 0.48	10.1 ± 0.61	6.3 ± 0.23	6.0 ± 0.23	10.5 ± 0.54	10.3 ± 0.43
4.5	16.9 ± 1.46	7.2 ± 0.41	6.9 ± 0.50	6.0 ± 0.24	9.7 ± 1.14	8.3 ± 0.99
9.0	17.5 ± 1.05	8.4 ± 0.34	6.7 ± 0.58	6.4 ± 0.53	9.4 ± 0.70	9.2 ± 0.37
13.5	18.9 ± 1.61	9.4 ± 1.35	6.2 ± 0.13	6.7 ± 0.19	9.2 ± 0.18	9.8 ± 0.57
Location by depth interaction						
Tractor row	17.2 ± 1.29a		6.5 ± 0.39d		9.7 ± 0.69c	
Vine row	9.3 ± 0.76b		6.3 ± 0.33d		9.4 ± 0.68cd	

Lower case letters indicate significant differences between locations and depths at $P < 0.05$.

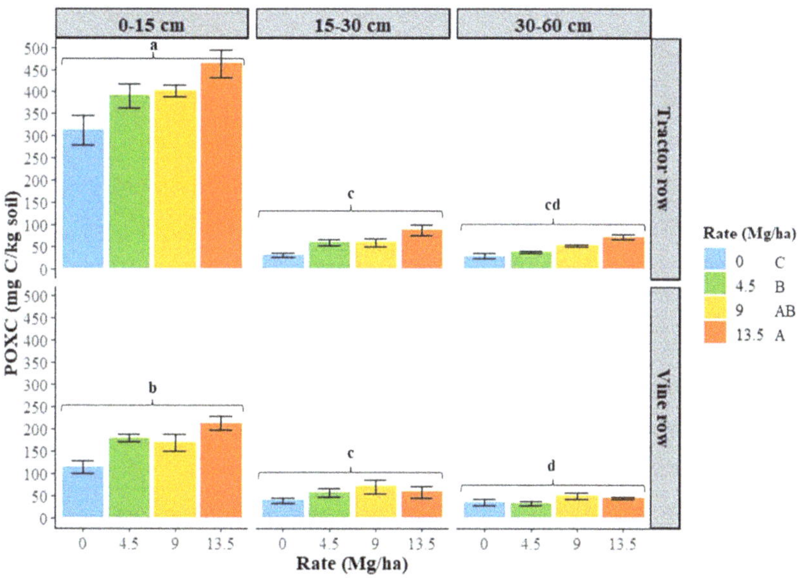

Figure 7 POXC (mg C kg^{-1} soil) by depth and functional location measured in April 2020, after two annual compost applications. Error bars represent standard errors of the means (n = 4). Lowercase letters indicate significant differences between depths and locations at $P < 0.05$. Upper case letters indicate significant differences between compost application rates at $P < 0.05$.

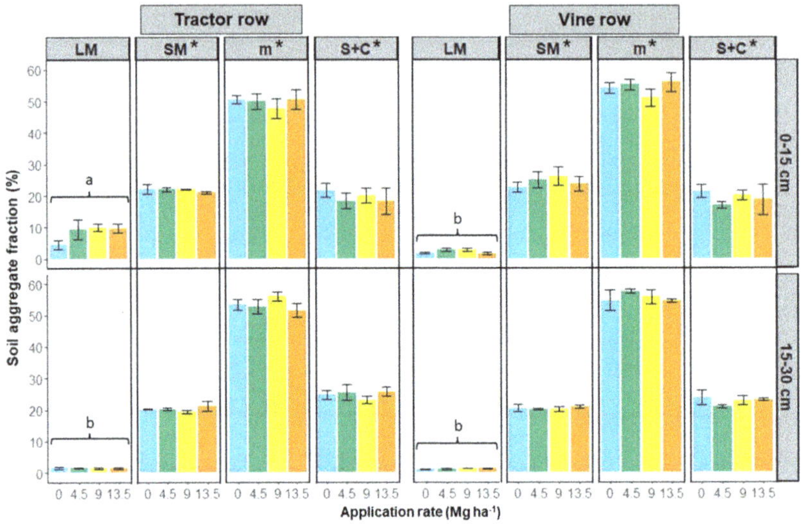

Figure 8 Fractions (%) of soil aggregates by aggregate size and functional location at the depths of 0–15 and 15–30 cm in April 2020. LM, SM, m, and S+C represent the four aggregate size categories: large macroaggregate, small macroaggregate, microaggregate, and silt and clay, respectively. Error bars are standard errors of the means (n = 4). Lowercase letters indicate significant differences between locations and depths, while asterisks (*) indicate significant differences between depths at $P < 0.05$ within each aggregate size category.

formation through biological activity boosted by the addition of compost, and further C stabilization within these macroaggregates (Liu et al., 2021; Six et al., 2004). When it comes to GHG emissions, both cumulative CO_2 (Fig. 9) and N_2O (Fig. 10) emissions at both functional locations measured in two wet (winter-spring) and two dry (summer-fall) seasons were not significantly affected by compost application. This lack of significant effect of compost addition and the application rates was a welcomed finding that suggests the environmental tradeoff in the form of CO_2 and N_2O emissions from the use of compost up to 6 tons/acre may be minimal in vineyards. Similar to GHG emissions, grape yield was also not significantly affected by compost application in this study (Table 3).

In conclusion, although no significant changes in total soil C content were observed, applying compost annually up to 6 tons/acre was shown to have positive effects on early signs of C sequestration, as indicated by increased POXC and large macroaggregate formation. Long-term commitment to this practice will be required to show the desired increase in total soil C content. Since compost application did not significantly increase cumulative GHG emissions in this study, nor have detrimental effects on grape yield, it should be recommended to wine growers for the purpose of conserving soil organic matter and maintaining soil health.

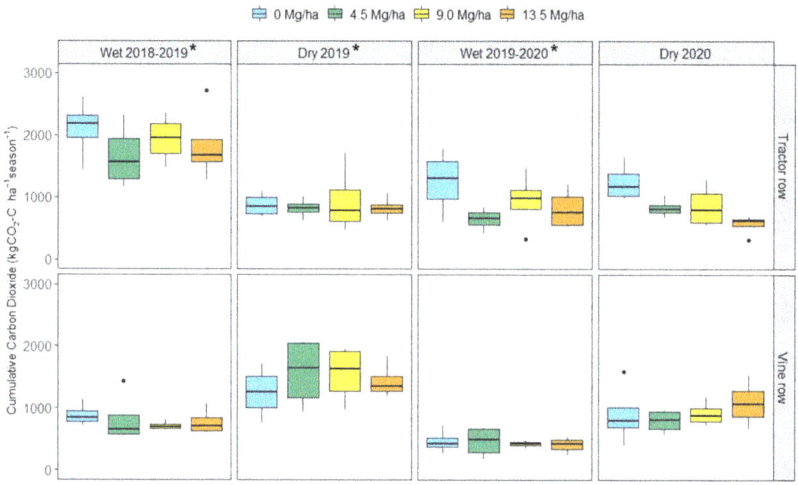

Figure 9 Cumulative CO_2 emissions measured in soil by functional location under compost application treatments at rates of 0 Mg ha^{-1}, 4.5 Mg ha^{-1}, 9.0 Mg ha^{-1}, and 13.5 Mg ha^{-1} during the two wet and two dry seasons from 2018 to 2020. Error bars represent standard errors of the mean (n = 4). Asterisks (*) indicate significant differences between locations within the same season and application rate at $P < 0.05$.

Figure 10 Cumulative N_2O emissions measured in soil by functional location under compost application treatments at rates of 0 Mg ha^{-1}, 4.5 Mg ha^{-1}, 9.0 Mg ha^{-1}, and 13.5 Mg ha^{-1} during the two wet and two dry seasons from 2018 to 2020. Error bars represent standard errors of the mean (n = 4). Asterisk (*) indicate significant differences between locations at $P < 0.05$ within the dry season.

Table 3 Grape yield components and standard errors of the means (n = 4) by compost application rate in J. Lohr Vineyards and Wines in 2020

Rate (Mg ha^{-1})	Clusters vine^{-1}	Cluster weight (g)	Berry mass (g)	Yield (t ha^{-1})
0	52 ± 1.5	69.8 ± 2.60	0.69 ± 0.03	7.27 ± 1.63
4.5	50 ± 4.1	64.2 ± 1.41	0.75 ± 0.02	6.56 ± 1.85
9.0	58 ± 5.6	68.7 ± 2.34	0.70 ± 0.03	7.96 ± 1.58
13.5	54 ± 3.5	67.3 ± 3.25	0.68 ± 0.02	7.23 ± 2.01

10 Conclusion

Compost application has short-, medium-, and long-term benefits for soil health. There is strong experimental evidence that shows that compost constitutes a valuable source of plant macro and micronutrients. Furthermore, these nutrients are released gradually which can help prevent losses to the environment. The supply of organic matter boosts microbial biomass and activity, shifting the composition of the microbial community. Increases in microbial biomass and organic matter inputs lead to sequestration of C which can help mitigate climate change. Nevertheless, the observed effects were extremely variable. Furthermore, evidence showing the effects of compost on crop yield, quality, and health is weaker, ranging from positive to negative effects.

Evidence shows that compost can impede plant growth due to the supply of excessive amounts of salts or phytotoxic compounds. There are also environmental risks associated with the use of compost such as the excess application of easily available nitrogen and P which can lead to eutrophication of water courses when transported through leaching and runoff, or to the emission of greenhouse gasses such as nitrous oxide. Certain composts can carry excess amounts of contaminants such as heavy metals or impurities such as microplastics which are known to have clear negative consequences for the structure and function of the soil ecosystem. Finally, the potential presence of enteric pathogens such as *Salmonella* or *E. coli* poses a serious risk to human health, particularly when compost is applied to crops that will be consumed fresh.

Thus, even though there are clear benefits to the use of compost, there are also certain risks which can be avoided by understanding the factors that drive this large variability in agronomic, environmental, and health outcomes. Most risks can be avoided through the adequate management of the composting process. Particularly, the thermophilic stage is critical for the elimination of phytotoxic substances and enteric pathogens and therefore for compost stabilization. Selecting high-quality feedstocks, free of impurities and contaminants, contributes to increasing compost quality and therefore can maximize the benefits of compost to soil health while decreasing environmental risks.

Variability in compost agronomic effects can be attributed to different crop requirements, compost management, compost quality, soil properties, and environmental conditions. In terms of management, the number and rate of applications strongly influence the outcomes for soil health. Repeated and large application rates lead to stronger beneficial effects in the soil health indicators evaluated but can also lead to the buildup of easily available N and P and therefore the associated risk for environmental pollution. Compost C:N ratio or particle size also drives the magnitude of the effects by regulating the rate of nutrient release. Soil properties such as texture have a strong effect on organic matter mineralization rates and C sequestration after compost application. In terms of environmental properties, seasonality, average temperatures, and annual precipitation rates were revealed to be important in driving the magnitude of the effects of compost on soil health.

Altogether, this evidences the need for site-specific knowledge and management in order to maximize compost benefits and reduce risks. Selecting high-quality feedstocks, managing the composting process to allow for the stabilization, sanitization, and curing of the organic material, understanding the properties of the finished compost, adjusting compost application rates and times to match crop needs are all required steps to maximize the benefits of compost for soil health. Furthermore, the potential benefits or risks of compost need to be considered within the context of other management strategies that may affect its performance, such as irrigation, the use of cover crops, or soil tillage, among others. Finally, evidence showing long-term effects of compost is particularly scarce; there is a strong need for standardized and reproducible long-term research that disentangles the interactions between compost, soil biota, and plants, across different soils and environmental conditions.

11 Where to look for further information

The journal 'Compost Science and Utilization' is a quarterly peer-reviewed journal that reports on advancements and breakthroughs in the science and engineering of compost production, compost product quality, and the utilization of composted materials.

The Test Method for the Examination of Composting and Compost (TMECC) of the US Composting Council, provides detailed protocols for the composting industry to verify the physical, chemical, and biological condition of composting feedstocks, material in process and compost products at the point of sale.

The European Compost Network, provides information on European policies on the use of compost and protection of soil health. https://www.compostnetwork.info.

Current Approaches and Future Trends in Compost Quality Criteria for Agronomic, Environmental and Human Health Benefits by Maria Pilar Bernal, Sven G. Sommer, Dave Chadwick, Chen Qing, Li Guoxue and Frederick C. Michel Jr. 2017, Advances in Agronomy.

12 References

Abiven, S., Menasseri, S. and Chenu, C. 2009. The effects of organic inputs over time on soil aggregate stability – a literature analysis. *Soil Biology and Biochemistry* 41(1), 1–12. https://doi.org/10.1016/j.soilbio.2008.09.015.

Aggelides, S. M. and Londra, P. A. 2000. Effects of compost produced from town wastes and sewage sludge on the physical properties of a loamy and a clay soil. *Bioresource Technology* 71(3), 253–259. https://doi.org/10.1016/S0960-8524(99)00074-7.

Al-Suhaibani, N., Selim, M., Alderfasi, A. and El-Hendawy, S. 2020. Comparative performance of integrated nutrient management between composted agricultural wastes, chemical fertilizers, and biofertilizers in improving soil quantitative and qualitative properties and crop yields under arid conditions. *Agronomy* 10(10), 1503. https://doi.org/10.3390/agronomy10101503.

Antille, D. L., Sakrabani, R. and Godwin, R. J. 2014. Phosphorus release characteristics from biosolids-derived organomineral fertilizers. *Communications in Soil Science and Plant Analysis* 45(19), 2565–2576. https://doi.org/10.1080/00103624.2014.912300.

Arthur, E., Cornelis, W. and Razzaghi, F. 2012. Compost amendment to sandy soil affects soil properties and greenhouse tomato productivity. *Compost Science and Utilization* 20(4), 215–221. https://doi.org/10.1080/1065657X.2012.10737051.

Arthur, E., Cornelis, W. M., Vermang, J. and De Rocker, E. 2011. Amending a loamy sand with three compost types: impact on soil quality: compost effect on the soil quality of loamy sand. *Soil Use and Management* 27(1), 116–123. https://doi.org/10.1111/j.1475-2743.2010.00319.x.

Avilés, M. and Borrero, C. 2017. Identifying characteristics of Verticillium Wilt suppressiveness in olive mill composts. *Plant Disease* 101(9), 1568–1577. https://doi.org/10.1094/PDIS-08-16-1172-RE.

Ba, S., Qu, Q., Zhang, K. and Groot, J. C. J. 2020. Meta-analysis of greenhouse gas and ammonia emissions from dairy manure composting. *Biosystems Engineering* 193, 126–137. https://doi.org/10.1016/j.biosystemseng.2020.02.015.

Bai, Z., Ma, W., Ma, L., Velthof, G. L., Wei, Z., Havlík, P., Oenema, O., Lee, M. R. F. and Zhang, F. 2018. China's livestock transition: driving forces, impacts, and consequences. *Science Advances* 4(7), eaar8534. https://doi.org/10.1126/sciadv.aar8534.

Bardgett, R. D. and van der Putten, W. H. 2014. Belowground biodiversity and ecosystem functioning. *Nature* 515(7528), 505–511. https://doi.org/10.1038/nature13855.

Bastida, F., Kandeler, E., Moreno, J. L., Ros, M., García, C. and Hernández, T. 2008. Application of fresh and composted organic wastes modifies structure, size and activity of soil microbial community under semiarid climate. *Applied Soil Ecology* 40(2), 318-329. https://doi.org/10.1016/j.apsoil.2008.05.007.

Bastida, F., Moreno, J. L., García, C. and Hernández, T. 2007. Addition of urban waste to semiarid degraded soil: long-term effect. *Pedosphere* 17(5), 557-567. https://doi.org/10.1016/S1002-0160(07)60066-6.

Bedada, W., Karltun, E., Lemenih, M. and Tolera, M. 2014. Long-term addition of compost and NP fertilizer increases crop yield and improves soil quality in experiments on smallholder farms. *Agriculture, Ecosystems and Environment* 195, 193-201. https://doi.org/10.1016/j.agee.2014.06.017.

Bernal, M. P., Alburquerque, J. A. and Moral, R. 2009. Composting of animal manures and chemical criteria for compost maturity assessment. A review. *Bioresource Technology* 100(22), 5444-5453. https://doi.org/10.1016/j.biortech.2008.11.027.

Bernal, M. P., Sommer, S. G., Chadwick, D., Qing, C., Guoxue, L. and Michel, F. C. 2017. Current approaches and future trends in compost quality criteria for agronomic, environmental, and human health benefits. In: *Advances in Agronomy*. Elsevier, pp. 143-233. https://doi.org/10.1016/bs.agron.2017.03.002.

Bläsing, M. and Amelung, W. 2018. Plastics in soil: analytical methods and possible sources. *Science of the Total Environment* 612, 422-435. https://doi.org/10.1016/j.scitotenv.2017.08.086.

Boldrin, A., Andersen, J. K., Møller, J., Christensen, T. H. and Favoino, E. 2009. Composting and compost utilization: accounting of greenhouse gases and global warming contributions. *Waste Management and Research* 27(8), 800-812. https://doi.org/10.1177/0734242X09345275.

Bonanomi, G., Antignani, V., Capodilupo, M. and Scala, F. 2010. Identifying the characteristics of organic soil amendments that suppress soilborne plant diseases. *Soil Biology and Biochemistry* 42(2), 136-144. https://doi.org/10.1016/j.soilbio.2009.10.012.

Bonanomi, G., Antignani, V., Pane, C. and Scala, F. 2007. Suppression of soilborne fungal diseases with organic amendments. *Journal of Plant Pathology* 89, 311-324.

Bottoms, T. G., Bolda, M. P., Gaskell, M. L. and Hartz, T. K. 2013a. Determination of strawberry nutrient optimum ranges through diagnosis and recommendation integrated system analysis. *HortTechnology* 23(3), 312-318. https://doi.org/10.21273/HORTTECH.23.3.312.

Bottoms, T. G., Hartz, T. K., Cahn, M. D. and Farrara, B. F. 2013b. Crop and soil nitrogen dynamics in annual strawberry production in California. *HortScience* 48(8), 1034-1039. https://doi.org/10.21273/HORTSCI.48.8.1034.

Brady, N. C. and Weil, R. R. 2016. *The Nature and Properties of Soils* (15th ed.). Pearson, Columbus.

Brandt, K., Leifert, C., Sanderson, R. and Seal, C. J. 2011. Agroecosystem management and nutritional quality of plant foods: the case of organic fruits and vegetables. *Critical Reviews in Plant Sciences* 30(1-2), 177-197. https://doi.org/10.1080/07352689.2011.554417.

Bronick, C. J. and Lal, R. 2005. Soil structure and management: a review. *Geoderma* 124(1-2), 3-22. https://doi.org/10.1016/j.geoderma.2004.03.005.

Bünemann, E. K., Bongiorno, G., Bai, Z., Creamer, R. E., De Deyn, G., de Goede, R., Fleskens, L., Geissen, V., Kuyper, T. W., Mäder, P., Pulleman, M., Sukkel, W., van Groenigen, J. W. and Brussaard, L. 2018. Soil quality - a critical review. *Soil Biology and Biochemistry* 120, 105-125. https://doi.org/10.1016/j.soilbio.2018.01.030.

Bustamante, M. A., Alburquerque, J. A., Restrepo, A. P., de la Fuente, C., Paredes, C., Moral, R. and Bernal, M. P. 2012. Co-composting of the solid fraction of anaerobic digestates, to obtain added-value materials for use in agriculture. *Biomass and Bioenergy* 43, 26–35. https://doi.org/10.1016/j.biombioe.2012.04.010.

Bustamante, M. A., Paredes, C., Marhuenda-Egea, F. C., Pérez-Espinosa, A., Bernal, M. P. and Moral, R. 2008. Co-composting of distillery wastes with animal manures: carbon and nitrogen transformations in the evaluation of compost stability. *Chemosphere* 72(4), 551–557. https://doi.org/10.1016/j.chemosphere.2008.03.030.

Bustamante, M. A., Said-Pullicino, D., Agulló, E., Andreu, J., Paredes, C. and Moral, R. 2011. Application of winery and distillery waste composts to a Jumilla (SE Spain) vineyard: effects on the characteristics of a calcareous sandy-loam soil. *Agriculture, Ecosystems and Environment* 140(1–2), 80–87. https://doi.org/10.1016/j.agee.2010.11.014.

Butler, T. J. and Muir, J. P. 2006. Dairy manure compost improves soil and increases tall wheatgrass yield. *Agronomy Journal* 98(4), 1090–1096. https://doi.org/10.2134/agronj2005.0348.

Butterbach-Bahl, K., Baggs, E. M., Dannenmann, M., Kiese, R. and Zechmeister-Boltenstern, S. 2013. Nitrous oxide emissions from soils: how well do we understand the processes and their controls? *Philosophical Transactions of the Royal Society of London. Series B, Biological Sciences* 368(1621), 20130122. https://doi.org/10.1098/rstb.2013.0122.

Calbrix, R., Barray, S., Chabrerie, O., Fourrie, L. and Laval, K. 2007. Impact of organic amendments on the dynamics of soil microbial biomass and bacterial communities in cultivated land. *Applied Soil Ecology* 35(3), 511–522. https://doi.org/10.1016/j.apsoil.2006.10.007.

Calleja-Cervantes, M. E., Fernández-González, A. J., Irigoyen, I., Fernández-López, M., Aparicio-Tejo, P. M. and Menéndez, S. 2015. Thirteen years of continued application of composted organic wastes in a vineyard modify soil quality characteristics. *Soil Biology and Biochemistry* 90, 241–254. https://doi.org/10.1016/j.soilbio.2015.07.002.

Carle, S. F., Esser, B. K. and Moran, J. E. 2006. High-resolution simulation of basin-scale nitrate transport considering aquifer system heterogeneity. *Geosphere* 2(4), 195. https://doi.org/10.1130/GES00032.1.

Carvalhais, N., Forkel, M., Khomik, M., Bellarby, J., Jung, M., Migliavacca, M., Mu, M., Saatchi, S., Santoro, M., Thurner, M., Weber, U., Ahrens, B., Beer, C., Cescatti, A., Randerson, J. T. and Reichstein, M. 2014. Global covariation of carbon turnover times with climate in terrestrial ecosystems. *Nature* 514, 213–217. https://doi.org/10.1038/nature13731.

Castán, E., Satti, P., González-Polo, M., Iglesias, M. C. and Mazzarino, M. J. 2016. Managing the value of composts as organic amendments and fertilizers in sandy soils. *Agriculture, Ecosystems and Environment* 224, 29–38. https://doi.org/10.1016/j.agee.2016.03.016.

Celestina, C., Hunt, J. R., Sale, P. W. G. and Franks, A. E. 2019. Attribution of crop yield responses to application of organic amendments: a critical review. *Soil and Tillage Research* 186, 135–145. https://doi.org/10.1016/j.still.2018.10.002.

Celik, I., Gunal, H., Budak, M. and Akpinar, C. 2010. Effects of long-term organic and mineral fertilizers on bulk density and penetration resistance in semi-arid Mediterranean soil conditions. *Geoderma* 160(2), 236–243. https://doi.org/10.1016/j.geoderma.2010.09.028.

Chadwick, D., Wei, J., Yan'an, T., Guanghui, Y., Qirong, S. and Qing, C. 2015. Improving manure nutrient management towards sustainable agricultural intensification in China. *Agriculture, Ecosystems and Environment* 209, 34–46. https://doi.org/10.1016/j.agee.2015.03.025.

Chalk, P. M., Magalhães, A. M. T. and Inácio, C. T. 2013. Towards an understanding of the dynamics of compost N in the soil-plant-atmosphere system using 15N tracer. *Plant and Soil* 362(1–2), 373–388. https://doi.org/10.1007/s11104-012-1358-5.

Chan, K. Y., Fahey, D. J., Newell, M. and Barchia, I. 2010. Using composted mulch in vineyards–effects on grape yield and quality. *International Journal of Fruit Science* 10(4), 441–453. https://doi.org/10.1080/15538362.2010.530135.

Charles, A., Rochette, P., Whalen, J. K., Angers, D. A., Chantigny, M. H. and Bertrand, N. 2017. Global nitrous oxide emission factors from agricultural soils after addition of organic amendments: a meta-analysis. *Agriculture, Ecosystems and Environment* 236, 88–98. https://doi.org/10.1016/j.agee.2016.11.021.

Chellemi, D. O. 2002. Nonchemical management of soilborne pests in fresh market vegetable production systems. *Phytopathology* 92(12), 1367–1372. https://doi.org/10.1094/PHYTO.2002.92.12.1367.

Chen, X. P., Zhang, Y. Q., Tong, Y. P., Xue, Y. F., Liu, D. Y., Zhang, W., Deng, Y., Meng, Q. F., Yue, S. C., Yan, P., Cui, Z. L., Shi, X. J., Guo, S. W., Sun, Y. X., Ye, Y. L., Wang, Z. H., Jia, L. L., Ma, W. Q., He, M. R., Zhang, X. Y., Kou, C. L., Li, Y. T., Tan, D. S., Cakmak, I., Zhang, F. S. and Zou, C. Q. 2017. Harvesting more grain zinc of wheat for human health. *Scientific Reports* 7(1), 7016. https://doi.org/10.1038/s41598-017-07484-2.

Chen, Y., Camps-Arbestain, M., Shen, Q., Singh, B. and Cayuela, M. L. 2018. The long-term role of organic amendments in building soil nutrient fertility: a meta-analysis and review. *Nutrient Cycling in Agroecosystems* 111(2–3), 103–125. https://doi.org/10.1007/s10705-017-9903-5.

Cherif, H., Ayari, F., Ouzari, H., Marzorati, M., Brusetti, L., Jedidi, N., Hassen, A. and Daffonchio, D. 2009. Effects of municipal solid waste compost, farmyard manure and chemical fertilizers on wheat growth, soil composition and soil bacterial characteristics under Tunisian arid climate. *European Journal of Soil Biology* 45(2), 138–145. https://doi.org/10.1016/j.ejsobi.2008.11.003.

Choi, W.-J., Jin, S.-A., Lee, S.-M., Ro, H.-M. and Yoo, S.-H. 2001. Corn uptake and microbial immobilization of 15N-labeled urea-N in soil as affected by composted pig manure. *Plant and Soil* 235(1), 1–9. https://doi.org/10.1023/A:1011896912888.

Colvero, D. A., Ramalho, J., Gomes, A. P. D., Matos, M. A. A. and Tarelho, L. A. D. C. 2020. Economic analysis of a shared municipal solid waste management facility in a metropolitan region. *Waste Management* 102, 823–837. https://doi.org/10.1016/j.wasman.2019.11.033.

Coria-Cayupán, Y. S., Sánchez de Pinto, M. I. and Nazareno, M. A. 2009. Variations in bioactive substance contents and crop yields of lettuce (Lactuca sativa L.) cultivated in soils with different fertilization treatments. *Journal of Agricultural and Food Chemistry* 57(21), 10122–10129. https://doi.org/10.1021/jf903019d.

Courtney, R. G. and Mullen, G. J. 2008. Soil quality and barley growth as influenced by the land application of two compost types. *Bioresource Technology* 99(8), 2913–2918. https://doi.org/10.1016/j.biortech.2007.06.034.

Crecchio, C., Curci, M., Mininni, R., Ricciuti, P. and Ruggiero, P. 2001. Short-term effects of municipal solid waste compost amendments on soil carbon and nitrogen content, some enzyme activities and genetic diversity. *Biology and Fertility of Soils* 34(5), 311-318. https://doi.org/10.1007/s003740100413.

Daynes, C. N., Field, D. J., Saleeba, J. A., Cole, M. A. and McGee, P. A. 2013. Development and stabilisation of soil structure via interactions between organic matter, arbuscular mycorrhizal fungi and plant roots. *Soil Biology and Biochemistry* 57, 683-694. https://doi.org/10.1016/j.soilbio.2012.09.020.

de Araújo, A. S. F., de Melo, W. J. and Singh, R. P. 2010. Municipal solid waste compost amendment in agricultural soil: changes in soil microbial biomass. *Reviews in Environmental Science and Bio/Technology* 9(1), 41-49. https://doi.org/10.1007/s11157-009-9179-6.

De Corato, U. 2020. Agricultural waste recycling in horticultural intensive farming systems by on-farm composting and compost-based tea application improves soil quality and plant health: a review under the perspective of a circular economy. *Science of the Total Environment* 738, 139840. https://doi.org/10.1016/j.scitotenv.2020.139840.

de Vries, F. T., Thebault, E., Liiri, M., Birkhofer, K., Tsiafouli, M. A., Bjornlund, L., Bracht Jorgensen, H., Brady, M. V., Christensen, S., de Ruiter, P. C., d'Hertefeldt, T., Frouz, J., Hedlund, K., Hemerik, L., Hol, W. H. G., Hotes, S., Mortimer, S. R., Setala, H., Sgardelis, S. P., Uteseny, K., van der Putten, W. H., Wolters, V. and Bardgett, R. D. 2013. Soil food web properties explain ecosystem services across European land use systems. *Proceedings of the National Academy of Sciences of the United States of America* 110(35), 14296-14301. https://doi.org/10.1073/pnas.1305198110.

Debertoldi, M., Vallini, G. and Pera, A. 1983. The biology of composting: a review. *Waste Management and Research* 1(2), 157-176. https://doi.org/10.1016/0734-242X(83)90055-1.

Degens, B. P. and Harris, J. A. 1997. Development of a physiological approach to measuring the catabolic diversity of soil microbial communities. *Soil Biology and Biochemistry* 29(9-10), 1309-1320. https://doi.org/10.1016/S0038-0717(97)00076-X.

Diacono, M. and Montemurro, F. 2010. Long-term effects of organic amendments on soil fertility. A review. *Agronomy for Sustainable Development* 30(2), 401-422. https://doi.org/10.1051/agro/2009040.

Diana, G., Beni, C. and Marconi, S. 2008. Organic and mineral fertilization: effects on physical characteristics and boron dynamic in an agricultural soil. *Communications in Soil Science and Plant Analysis* 39(9-10), 1332-1351. https://doi.org/10.1080/00103620802004037.

Domínguez, M., Paradelo Núñez, R., Piñeiro, J. and Barral, M. T. 2019. Physicochemical and biochemical properties of an acid soil under potato culture amended with municipal solid waste compost. *International Journal of Recycling of Organic Waste in Agriculture* 8(2), 171-178. https://doi.org/10.1007/s40093-019-0246-x.

Drinkwater, L. E. and Snapp, S. S. 2007. Nutrients in agroecosystems: rethinking the management paradigm. In: *Advances in Agronomy*. Elsevier, pp. 163-186. https://doi.org/10.1016/S0065-2113(04)92003-2.

Duong, T. T. T., Verma, S. L., Penfold, C. and Marschner, P. 2013. Nutrient release from composts into the surrounding soil. *Geoderma* 195-196, 42-47. https://doi.org/10.1016/j.geoderma.2012.11.010.

Eden, M., Gerke, H. H. and Houot, S. 2017. Organic waste recycling in agriculture and related effects on soil water retention and plant available water: a review. *Agronomy* for *Sustainable Development* 37(2), 11. https://doi.org/10.1007/s13593-017-0419-9.

Eghball, B. 2003. Leaching of phosphorus fractions following manure or compost application. *Communications in Soil Science and Plant Analysis* 34(19-20), 2803-2815. https://doi.org/10.1081/CSS-120025207.

Eghball, B. and Power, J. F. 1999. Phosphorus- and nitrogen-based manure and compost applications corn production and soil phosphorus. *Soil Science Society of America Journal* 63(4), 895-901. https://doi.org/10.2136/sssaj1999.634895x.

Epelde, L., Jauregi, L., Urra, J., Ibarretxe, L., Romo, J., Goikoetxea, I. and Garbisu, C. 2018. Characterization of composted organic amendments for agricultural use. *Frontiers in Sustainable Food Systems* 2, 44. https://doi.org/10.3389/fsufs.2018.00044.

Erhart, E., Hartl, W. and Putz, B. 2005. Biowaste compost affects yield, nitrogen supply during the vegetation period and crop quality of agricultural crops. *European Journal of Agronomy* 23(3), 305-314. https://doi.org/10.1016/j.eja.2005.01.002.

Erktan, A., Or, D. and Scheu, S. 2020. The physical structure of soil: determinant and consequence of trophic interactions. *Soil Biology and Biochemistry* 148, 107876. https://doi.org/10.1016/j.soilbio.2020.107876.

Fagnano, M., Adamo, P., Zampella, M. and Fiorentino, N. 2011. Environmental and agronomic impact of fertilization with composted organic fraction from municipal solid waste: a case study in the region of Naples, Italy. *Agriculture, Ecosystems and Environment* 141(1-2), 100-107. https://doi.org/10.1016/j.agee.2011.02.019.

Farrell, M., Griffith, G. W., Hobbs, P. J., Perkins, W. T. and Jones, D. L. 2010. Microbial diversity and activity are increased by compost amendment of metal-contaminated soil. *FEMS Microbiology Ecology* 71(1), 94-105. https://doi.org/10.1111/j.1574-6941.2009.00793.x.

Fierer, N., Wood, S. A. and Bueno de Mesquita, C. P. 2021. How microbes can, and cannot, be used to assess soil health. *Soil Biology and Biochemistry* 153, 108111. https://doi.org/10.1016/j.soilbio.2020.108111.

Franco-Otero, V. G., Soler-Rovira, P., Hernández, D., López-de-Sá, E. G. and Plaza, C. 2012. Short-term effects of organic municipal wastes on wheat yield, microbial biomass, microbial activity, and chemical properties of soil. *Biology and Fertility of Soils* 48(2), 205-216. https://doi.org/10.1007/s00374-011-0620-y.

Frey, S. D., Lee, J., Melillo, J. M. and Six, J. 2013. The temperature response of soil microbial efficiency and its feedback to climate. *Nature Climate Change* 3(4), 395-398. https://doi.org/10.1038/nclimate1796.

Frostegård, Å., Bååth, E. and Tunlio, A. 1993. Shifts in the structure of soil microbial communities in limed forests as revealed by phospholipid fatty acid analysis. *Soil Biology and Biochemistry* 25(6), 723-730. https://doi.org/10.1016/0038-0717(93)90113-P.

García-Gil, J. C., Ceppi, S. B., Velasco, M. I., Polo, A. and Senesi, N. 2004. Long-term effects of amendment with municipal solid waste compost on the elemental and acidic functional group composition and pH-buffer capacity of soil humic acids. *Geoderma* 121(1-2), 135-142. https://doi.org/10.1016/j.geoderma.2003.11.004.

García-Gil, J. C., Plaza, C., Soler-Rovira, P. and Polo, A. 2000. Long-term effects of municipal solid waste compost application on soil enzyme activities and microbial biomass. *Soil Biology and Biochemistry* 32(13), 1907-1913. https://doi.org/10.1016/S0038-0717(00)00165-6.

Garland, J. L. and Mills, A. L. 1991. Classification and characterization of heterotrophic microbial communities on the basis of patterns of community-level sole-carbon-source utilization. *Applied and Environmental Microbiology* 57(8), 2351–2359. https://doi.org/10.1128/aem.57.8.2351-2359.1991.

Giannakis, G. V., Kourgialas, N. N., Paranychianakis, N. V., Nikolaidis, N. P. and Kalogerakis, N. 2014. Effects of municipal solid waste compost on soil properties and vegetables growth. *Compost Science and Utilization* 22(3), 116–131. https://doi.org/10.1080/1065657X.2014.899938.

Głąb, T., Żabiński, A., Sadowska, U., Gondek, K., Kopeć, M., Mierzwa-Hersztek, M., Tabor, S. and Stanek-Tarkowska, J. 2020. Fertilization effects of compost produced from maize, sewage sludge and biochar on soil water retention and chemical properties. *Soil and Tillage Research* 197, 104493. https://doi.org/10.1016/j.still.2019.104493.

Gómez-Brandón, M., Lazcano, C. and Domínguez, J. 2008. The evaluation of stability and maturity during the composting of cattle manure. *Chemosphere* 70(3), 436–444. https://doi.org/10.1016/j.chemosphere.2007.06.065.

Graham, R., Wortman, S. and Pittelkow, C. 2017. Comparison of organic and integrated nutrient management strategies for reducing soil N2O emissions. *Sustainability* 9(4), 510. https://doi.org/10.3390/su9040510.

Gravuer, K., Gennet, S. and Throop, H. L. 2019. Organic amendment additions to rangelands: a meta-analysis of multiple ecosystem outcomes. *Global Change Biology* 25(3), 1152–1170. https://doi.org/10.1111/gcb.14535.

Guenet, B., Gabrielle, B., Chenu, C., Arrouays, D., Balesdent, J., Bernoux, M., Bruni, E., Caliman, J., Cardinael, R., Chen, S., Ciais, P., Desbois, D., Fouche, J., Frank, S., Henault, C., Lugato, E., Naipal, V., Nesme, T., Obersteiner, M., Pellerin, S., Powlson, D. S., Rasse, D. P., Rees, F., Soussana, J., Su, Y., Tian, H., Valin, H. and Zhou, F. 2020. Can N2O emissions offset the benefits from soil organic carbon storage? *Global Change Biology* 27(2), 237–256. https://doi.org/10.1111/gcb.15342.

Habteweld, A. W., Brainard, D., Kravchenko, A., Grewal, P. S. and Melakeberhan, H. 2018. Effects of plant and animal waste-based compost amendments on the soil food web, soil properties, and yield and quality of fresh market and processing carrot cultivars. *Nematology* 20(2), 147–168. https://doi.org/10.1163/15685411-00003130.

Harada, Y. and Inoko, A. 1980. Relationship between cation-exchange capacity and degree of maturity of city refuse composts. *Soil Science and Plant Nutrition* 26(3), 353-362. https://doi.org/10.1080/00380768.1980.10431220.

Hargreaves, J., Adl, M. and Warman, P. 2008. A review of the use of composted municipal solid waste in agriculture. *Agriculture, Ecosystems and Environment* 123(1–3), 1–14. https://doi.org/10.1016/j.agee.2007.07.004.

Hartl, W., Putz, B. and Erhart, E. 2003. Influence of rates and timing of biowaste compost application on rye yield and soil nitrate levels. *European Journal of Soil Biology* 39(3), 129–139. https://doi.org/10.1016/S1164-5563(03)00028-1.

Hartz, T. K., Mitchell, J. P. and Giannini, C. 2000. Nitrogen and carbon mineralization dynamics of manures and composts. *HortScience* 35(2), 209–212. https://doi.org/10.21273/HORTSCI.35.2.209.

Havlin, J. (Ed.) 2013. *Soil Fertility and Fertilizers: An Introduction to Nutrient Management* (8th ed.). Pearson, Upper Saddle River, NJ.

Haynes, R. J., Belyaeva, O. N. and Zhou, Y. F. 2015. Particle size fractionation as a method for characterizing the nutrient content of municipal green waste used for composting. *Waste Management* 35, 48–54.

Hazarika, J. and Khwairakpam, M. 2018. Evaluation of biodegradation feasibility through rotary drum composting recalcitrant primary paper mill sludge. *Waste Management* 76, 275–283. https://doi.org/10.1016/j.wasman.2018.03.044.

Helfenstein, J., Müller, I., Grüter, R., Bhullar, G., Mandloi, L., Papritz, A., Siegrist, M., Schulin, R. and Frossard, E. 2016. Organic wheat farming improves grain zinc concentration. *PLoS ONE* 11(8), e0160729. https://doi.org/10.1371/journal.pone.0160729.

Hemmat, A., Aghilinategh, N., Rezainejad, Y. and Sadeghi, M. 2010. Long-term impacts of municipal solid waste compost, sewage sludge and farmyard manure application on organic carbon, bulk density and consistency limits of a calcareous soil in central Iran. *Soil and Tillage Research* 108(1–2), 43–50. https://doi.org/10.1016/j.still.2010.03.007.

Horwath, W. R. and Paul, E. A. 1994. Microbial biomass. In: Weaver, R. W., Angle, S., Bottomley, P., Bezdicek, D., Smith, S., Tabatabai, A. and Wollum, A. (Eds), *SSSA Book Series*. Soil Science Society of America, Madison, WI, pp. 753-773. https://doi.org/10.2136/sssabookser5.2.c36.

Hurisso, T. T., Culman, S. W., Horwath, W. R., Wade, J., Cass, D., Beniston, J. W., Bowles, T. M., Grandy, A. S., Franzluebbers, A. J., Schipanski, M. E., Lucas, S. T. and Ugarte, C. M. 2016. Comparison of permanganate-oxidizable carbon and mineralizable carbon for assessment of organic matter stabilization and mineralization. *Soil Science Society of America Journal* 80(5), 1352–1364. https://doi.org/10.2136/sssaj2016.04.0106.

Hutchinson, G. L. and Mosier, A. R. 1981. Improved soil cover method for field measurement of nitrous oxide fluxes. *Soil Science Society of America Journal* 45(2), 311–316. https://doi.org/10.2136/sssaj1981.03615995004500020017x.

Iglesias-Jimenez, E. and Alvarez, C. E. 1993. Apparent availability of nitrogen in composted municipal refuse. *Biology and Fertility of Soils* 16(4), 313–318. https://doi.org/10.1007/BF00369312.

Insam, H. and de Bertoldi, M. 2007. Chapter 3 Microbiology of the composting process. In: *Waste Management Series*. Elsevier, pp. 25–48. https://doi.org/10.1016/S1478-7482(07)80006-6.

IPCC 2014. *2013 Supplement to the 2006 IPCC Guidelines for National Greenhouse Gas Inventories: Wetlands*. The Intergovernmental Panel on Climate Change (IPCC), Switzerland.

Jain, M. S., Daga, M. and Kalamdhad, A. S. 2018. Composting physics: a degradation process-determining tool for industrial sludge. *Ecological Engineering* 116, 14–20. https://doi.org/10.1016/j.ecoleng.2018.02.015.

Jeger, M. J. and Viljanen-Rollinson, S. L. H. 2001. The use of the area under the disease-progress curve (AUDPC) to assess quantitative disease resistance in crop cultivars. *Theoretical and Applied Genetics* 102(1), 32–40. https://doi.org/10.1007/s001220051615.

Jindo, K., Chocano, C., Melgares de Aguilar, J., González, D., Hernandez, T. and García, C. 2016. Impact of compost application during 5 years on crop production, soil microbial activity, carbon fraction, and humification process. *Communications in Soil Science and Plant Analysis*. https://doi.org/10.1080/00103624.2016.1206922.

Kanaan, H., Hadar, Y., Medina, S., Krasnovsky, A., Mordechai-Lebiush, S., Tsror, L., Katan, J. and Raviv, M. 2018. Effect of compost properties on progress rate of Verticillium dahliae attack on eggplant (Solanum melongena L.). *Compost Science and Utilization* 26(2), 71-78. https://doi.org/10.1080/1065657X.2017.1366375.

Karlen, D. L., Veum, K. S., Sudduth, K. A., Obrycki, J. F. and Nunes, M. R. 2019. Soil health assessment: past accomplishments, current activities, and future opportunities. *Soil and Tillage Research* 195, 104365. https://doi.org/10.1016/j.still.2019.104365.

Kaza, S., Yao, L., Bhada-Tata, P. and Van Woerden, F. 2018. *What a Waste 2.0: A Global Snapshot of Solid Waste Management to 2050*. The World Bank. https://doi.org/10.1596/978-1-4648-1329-0.

Kibblewhite, M. G., Ritz, K. and Swift, M. J. 2008. Soil health in agricultural systems. *Philosophical Transactions of the Royal Society of London. Series B, Biological Sciences* 363(1492), 685-701. https://doi.org/10.1098/rstb.2007.2178.

Kim, B. R., Shin, J., Guevarra, R. B., Lee, J. H., Kim, D. W., Seol, K. H., Lee, J. H., Kim, H. B. and Isaacson, R. E. 2017. Deciphering diversity indices for a better understanding of microbial communities. *Journal of Microbiology and Biotechnology* 27(12), 2089-2093. https://doi.org/10.4014/jmb.1709.09027.

Kim, H.-Y., Lim, S.-S., Kwak, J.-H., Lee, S.-I., Lee, D.-S., Hao, X., Yoon, K.-S. and Choi, W.-J. 2011. Soil and compost type affect phosphorus leaching from inceptisol, ultisol, and andisol in a column experiment. *Communications in Soil Science and Plant Analysis* 42(18), 2188-2199. https://doi.org/10.1080/00103624.2011.602450.

Knapp, B. A., Ros, M. and Insam, H. 2010. Do composts affect the soil microbial community? In: Insam, H., Franke-Whittle, I. and Goberna, M. (Eds), *Microbes at Work: From Wastes to Resources*. Springer, Berlin, Heidelberg, pp. 271-291. https://doi.org/10.1007/978-3-642-04043-6_14.

Koike, S. T. 2008. Crown rot of strawberry caused by Macrophomina phaseolina in California. *Plant Disease* 92(8), 1253. https://doi.org/10.1094/PDIS-92-8-1253B.

Kranz, C. N., McLaughlin, R. A., Johnson, A., Miller, G. and Heitman, J. L. 2020. The effects of compost incorporation on soil physical properties in urban soils - a concise review. *Journal of Environmental Management* 261, 110209. https://doi.org/10.1016/j.jenvman.2020.110209.

Krause, H.-M., Fliessbach, A., Mayer, J. and Mäder, P. 2020. Implementation and management of the DOK long-term system comparison trial. In: *Long-Term Farming Systems Research*. Elsevier, pp. 37-51. https://doi.org/10.1016/B978-0-12-818186-7.00003-5.

Kumar, S. 2011. Composting of municipal solid waste. *Critical Reviews in Biotechnology* 31(2), 112-136. https://doi.org/10.3109/07388551.2010.492207.

Lairon, D. 2011. Nutritional quality and safety of organic food. In: Lichtfouse, E., Hamelin, M., Navarrete, M. and Debaeke, P. (Eds), *Sustainable Agriculture* (vol. 2). Springer Netherlands, Dordrecht, pp. 99-110. https://doi.org/10.1007/978-94-007-0394-0_7.

Lal, R. 2004. Soil carbon sequestration to mitigate climate change. *Geoderma* 123(1-2), 1-22. https://doi.org/10.1016/j.geoderma.2004.01.032.

Langdon, K. A., Chandra, A., Bowles, K., Symons, A., Pablo, F. and Osborne, K. 2019. A preliminary ecological and human health risk assessment for organic contaminants in composted municipal solid waste generated in New South Wales, Australia. *Waste Management* 100, 199-207. https://doi.org/10.1016/j.wasman.2019.09.001.

Lannan, A. P., Erich, M. S. and Ohno, T. 2013. Compost feedstock and maturity level affect soil response to amendment. *Biology and Fertility of Soils* 49(3), 273–285. https://doi.org/10.1007/s00374-012-0715-0.

Latifah, O., Ahmed, O. H. and Majid, N. M. A. 2018. Soil pH buffering capacity and nitrogen availability following compost application in a tropical acid soil. *Compost Science and Utilization* 26(1), 1–15. https://doi.org/10.1080/1065657X.2017.1329039.

Lazcano, C., Zhu-Barker, X. and Decock, C. 2021a. Effects of organic fertilizers on the soil microorganisms responsible for N2O emissions: a review. *Microorganisms* 9(5), 983. https://doi.org/10.3390/microorganisms9050983.

Lazcano, C., Boyd, E., Holmes, G., Hewavitharana, S., Pasulka, A. and Ivors, K. 2021b. The rhizosphere microbiome plays a role in the resistance to soil-borne pathogens and nutrient uptake of strawberry cultivars under field conditions. *Scientific Reports* 11(1), 3188. https://doi.org/10.1038/s41598-021-82768-2.

Lazcano, C., Gómez-Brandón, M., Revilla, P. and Domínguez, J. 2013. Short-term effects of organic and inorganic fertilizers on soil microbial community structure and function: A field study with sweet corn. *Biology and Fertility of Soils* 49(6), 723–733. https://doi.org/10.1007/s00374-012-0761-7.

Lazicki, P., Geisseler, D. and Lloyd, M. 2020. Nitrogen mineralization from organic amendments is variable but predictable. *Journal of Environmental Quality* 49(2), 483–495. https://doi.org/10.1002/jeq2.20030.

Leclerc, B., Georges, P., Cauwel, B. and Lairon, D. 1995. A five year study on nitrate leaching under crops fertilised with mineral and organic fertilisers in lysimeters. *Biological Agriculture and Horticulture* 11(1–4), 301–308. https://doi.org/10.1080/01448765.1995.9754714.

Leelamanie, D. A. L. and Manawardana, C. U. 2019. Soil hydrophysical properties as affected by solid waste compost amendments: seasonal and short-term effects in an ultisol. *Journal of Hydrology and Hydromechanics* 67(3), 232–239. https://doi.org/10.2478/johh-2019-0007.

Lehman, R. M., Acosta-Martinez, V., Buyer, J. S., Cambardella, C. A., Collins, H. P., Ducey, T. F., Halvorson, J. J., Jin, V. L., Johnson, J. M. F., Kremer, R. J., Lundgren, J. G., Manter, D. K., Maul, J. E., Smith, J. L. and Stott, D. E. 2015. Soil biology for resilient, healthy soil. *Journal of Soil and Water Conservation* 70(1), 12A–18A. https://doi.org/10.2489/jswc.70.1.12A.

Lehmann, J. and Kleber, M. 2015. The contentious nature of soil organic matter. *Nature* 528(7580), 60–68. https://doi.org/10.1038/nature16069.

Lehtinen, T., Dersch, G., Söllinger, J., Baumgarten, A., Schlatter, N., Aichberger, K. and Spiegel, H. 2017. Long-term amendment of four different compost types on a loamy silt cambisol: impact on soil organic matter, nutrients and yields. *Archives of Agronomy and Soil Science* 63(5), 663–673. https://doi.org/10.1080/03650340.2016.1235264.

Leroy, B. L. M., Herath, M. S. K., De Neve, S., Gabriels, D., Bommele, L., Reheul, D. and Moens, M. 2008. Effect of vegetable, fruit and garden (VFG) waste compost on soil physical properties. *Compost Science and Utilization* 16(1), 43–51. https://doi.org/10.1080/1065657X.2008.10702354.

Li, H., Van den Bulcke, J., Wang, X., Gebremikael, M. T., Hagan, J., De Neve, S. and Sleutel, S. 2020. Soil texture strongly controls exogenous organic matter mineralization indirectly via moisture upon progressive drying – evidence from incubation experiments. *Soil Biology and Biochemistry* 151, 108051. https://doi.org/10.1016/j.soilbio.2020.108051.

Li, J., Cooper, J. M., Lin, Z., Li, Y., Yang, X. and Zhao, B. 2015. Soil microbial community structure and function are significantly affected by long-term organic and mineral fertilization regimes in the North China Plain. *Applied Soil Ecology* 96, 75–87. https://doi.org/10.1016/j.apsoil.2015.07.001.

Li, P., Lang, M., Li, C. and Hao, X. 2016. Nitrous oxide and carbon dioxide emissions from soils amended with compost and manure from cattle fed diets containing wheat dried distillers' grains with solubles. *Canadian Journal of Soil Science*, CJSS-2016-0068. https://doi.org/10.1139/CJSS-2016-0068.

Li, Y. C., Stoffella, P. J., Alva, A. K., Calvert, D. V. and Graetz, D. A. 1997. Leaching of nitrate, ammonium, and phosphate From compost amended soil columns. *Compost Science and Utilization* 5(2), 63–67. https://doi.org/10.1080/1065657X.1997.10701875.

Lindahl, B. D., Nilsson, R. H., Tedersoo, L., Abarenkov, K., Carlsen, T., Kjøller, R., Kõljalg, U., Pennanen, T., Rosendahl, S., Stenlid, J. and Kauserud, H. 2013. Fungal community analysis by high-throughput sequencing of amplified markers – a user's guide. *New Phytologist* 199(1), 288–299. https://doi.org/10.1111/nph.12243.

Liu, H., Wang, Z. H., Li, F., Li, K., Yang, N., Yang, Y., Huang, D., Liang, D., Zhao, H., Mao, H., Liu, J. and Qiu, W. 2014. Grain iron and zinc concentrations of wheat and their relationships to yield in major wheat production areas in China. *Field Crops Research* 156, 151–160. https://doi.org/10.1016/j.fcr.2013.11.011.

Liu, J., Calderón, F. J. and Fonte, S. J. 2021. Compost inputs, cropping system, and rotation phase drive aggregate-associated carbon. *Soil Science Society of America Journal* 85(3), 829–846. https://doi.org/10.1002/saj2.20252.

Liu, L., Wang, S., Guo, X. and Wang, H. 2019. Comparison of the effects of different maturity composts on soil nutrient, plant growth and heavy metal mobility in the contaminated soil. *Journal of Environmental Management* 250, 109525. https://doi.org/10.1016/j.jenvman.2019.109525.

Liu, Q., Wang, J., Bai, Z., Ma, L. and Oenema, O. 2017. Global animal production and nitrogen and phosphorus flows. *Soil Research* 55(6), 451. https://doi.org/10.1071/SR17031.

Liu, T., Chen, X., Hu, F., Ran, W., Shen, Q., Li, H. and Whalen, J. K. 2016. Carbon-rich organic fertilizers to increase soil biodiversity: evidence from a meta-analysis of nematode communities. *Agriculture, Ecosystems and Environment* 232, 199–207. https://doi.org/10.1016/j.agee.2016.07.015.

Liu, Z., Wang, X., Wang, F., Bai, Z., Chadwick, D., Misselbrook, T. and Ma, L. 2020. The progress of composting technologies from static heap to intelligent reactor: benefits and limitations. *Journal of Cleaner Production* 270, 122328. https://doi.org/10.1016/j.jclepro.2020.122328.

Lloyd, M. 2016. Growing for the future: collective action, land stewardship and soilborne pathogens in California strawberry production. *California Agriculture* 70(3), 101–103. https://doi.org/10.3733/ca.2016a0009.

Lloyd, M., Kluepfel, D. and Gordon, T. 2016. Evaluation of four commercial composts on strawberry plant productivity and soil characteristics in California. *International Journal of Fruit Science* 16(sup1), 84–107. https://doi.org/10.1080/15538362.2016.1239562.

Lodha, S., Sharma, S. K. and Aggarwal, R. K. 2002. Inactivation of Macrophomina phaseolina propagules during composting and effect of composts on dry root rot severity and on seed yield of clusterbean. *European Journal of Plant Pathology* 108(3), 253-261. https://doi.org/10.1023/A:1015103315068.

Logsdon, S. D., Sauer, P. A. and Shipitalo, M. J. 2017. Compost improves urban soil and water quality. *Journal of Water Resource and Protection* 09(4), 345-357. https://doi.org/10.4236/jwarp.2017.94023.

Lü, H., Chen, X. H., Mo, C. H., Huang, Y. H., He, M. Y., Li, Y. W., Feng, N. X., Katsoyiannis, A. and Cai, Q. Y. 2021. Occurrence and dissipation mechanism of organic pollutants during the composting of sewage sludge: a critical review. *Bioresource Technology* 328, 124847. https://doi.org/10.1016/j.biortech.2021.124847.

Lucas, J. A., Hawkins, N. J. and Fraaije, B. A. 2015. The evolution of fungicide resistance. In: *Advances in Applied Microbiology*. Elsevier, pp. 29-92. https://doi.org/10.1016/bs.aambs.2014.09.001.

Lugato, E., Leip, A. and Jones, A. 2018. Mitigation potential of soil carbon management overestimated by neglecting N2O emissions. *Nature Climate Change* 8(3), 219-223. https://doi.org/10.1038/s41558-018-0087-z.

Madden, L. V., Hughes, G. and van den Bosch, F. 2017. *The Study of Plant Disease Epidemics*. The American Phytopathological Society. https://doi.org/10.1094/9780890545058.

Makan, A. and Fadili, A. 2020. Sustainability assessment of large-scale composting technologies using PROMETHEE method. *Journal of Cleaner Production* 261, 121244. https://doi.org/10.1016/j.jclepro.2020.121244.

Maltais-Landry, G., Scow, K., Brennan, E., Torbert, E. and Vitousek, P. 2016. Higher flexibility in input N:P ratios results in more balanced phosphorus budgets in two long-term experimental agroecosystems. *Agriculture, Ecosystems and Environment* 223, 197-210. https://doi.org/10.1016/j.agee.2016.03.007.

Maltais-Landry, G., Scow, K., Brennan, E. and Vitousek, P. 2015. Long-term effects of compost and cover crops on soil phosphorus in two California agroecosystems. *Soil Science Society of America Journal* 79(2), 688-697. https://doi.org/10.2136/sssaj2014.09.0369.

Mamo, M., Moncrief, J. F., Rosen, C. J. and Halbach, T. R. 2000. The effect of municipal solid waste compost application on soil water and water stress in irrigated corn. *Compost Science and Utilization* 8(3), 236-246. https://doi.org/10.1080/1065657X.2000.10701996.

Mamo, M., Rosen, C. J. and Halbach, T. R. 1999. Nitrogen availability and leaching from soil amended with municipal solid waste compost. *Journal of Environmental Quality* 28(4), 1074-1082. https://doi.org/10.2134/jeq1999.00472425002800040003x.

Mangalassery, S., Kalaivanan, D. and Philip, P. S. 2019. Effect of inorganic fertilisers and organic amendments on soil aggregation and biochemical characteristics in a weathered tropical soil. *Soil and Tillage Research* 187, 144-151. https://doi.org/10.1016/j.still.2018.12.008.

Martínez-Blanco, J., Antón, A., Rieradevall, J., Castellari, M. and Muñoz, P. 2011. Comparing nutritional value and yield as functional units in the environmental assessment of horticultural production with organic or mineral fertilization: the case of Mediterranean cauliflower production. *International Journal of Life Cycle Assessment* 16(1), 12-26. https://doi.org/10.1007/s11367-010-0238-6.

Martínez-Blanco, J., Lazcano, C., Boldrin, A., Muñoz, P., Rieradevall, J., Møller, J., Antón, A. and Christensen, T. H. 2013a. Assessing the environmental benefits of compost use-on-land through an LCA perspective. In: Lichtfouse, E. (Ed.), *Sustainable Agriculture*

Reviews, Sustainable Agriculture Reviews. Springer Netherlands, Dordrecht, pp. 255–318. https://doi.org/10.1007/978-94-007-5961-9_9.

Martínez-Blanco, J., Lazcano, C., Christensen, T. H., Muñoz, P., Rieradevall, J., Møller, J., Antón, A. and Boldrin, A. 2013b. Compost benefits for agriculture evaluated by life cycle assessment. A review. *Agronomy for Sustainable Development* 33(4), 721–732. https://doi.org/10.1007/s13593-013-0148-7.

McCarthy, G., Lawlor, P. G., Coffey, L., Nolan, T., Gutierrez, M. and Gardiner, G. E. 2011. An assessment of pathogen removal during composting of the separated solid fraction of pig manure. *Bioresource Technology* 102(19), 9059–9067. https://doi.org/10.1016/j.biortech.2011.07.021.

McBride, M. B. 1995. Toxic metal accumulation from agricultural use of sludge: are USEPA regulations protective? *Journal of Environmental Quality* 24(1), 5–18. https://doi.org/10.2134/jeq1995.00472425002400010002x.

McDowell, R. W. and Sharpley, A. N. 2004. Variation of phosphorus leached from Pennsylvanian soils amended with manures, composts or inorganic fertilizer. *Agriculture, Ecosystems and Environment* 102(1), 17–27. https://doi.org/10.1016/j.agee.2003.07.002.

Mehta, C. M., Palni, U., Franke-Whittle, I. H. and Sharma, A. K. 2014. Compost: its role, mechanism and impact on reducing soil-borne plant diseases. *Waste Management* 34(3), 607–622. https://doi.org/10.1016/j.wasman.2013.11.012.

Meng, L., Ding, W. and Cai, Z. 2005. Long-term application of organic manure and nitrogen fertilizer on N2O emissions, soil quality and crop production in a sandy loam soil. *Soil Biology and Biochemistry* 37(11), 2037–2045. https://doi.org/10.1016/j.soilbio.2005.03.007.

Miller, J. J., Beasley, B. W., Drury, C. F., Larney, F. J. and Hao, X. 2015. Influence of long-term (9 yr) composted and stockpiled feedlot manure application on selected soil physical properties of a clay loam soil in Southern Alberta. *Compost Science and Utilization* 23(1), 1–10. https://doi.org/10.1080/1065657X.2014.963741.

Minasny, B. and McBratney, A. B. 2018. Limited effect of organic matter on soil available water capacity. *European Journal of Soil Science* 69(1), 39–47. https://doi.org/10.1111/ejss.12475.

Moebius-Clune, B. N. 2016. Comprehensive assessment of soil health: the Cornell framework manual. Cornell University, Ithaca, New York. Available at: http://www.css.cornell.edu/extension/soil-health/manual.pdf

Mohammadshirazi, F., McLaughlin, R. A., Heitman, J. L. and Brown, V. K. 2017. A multi-year study of tillage and amendment effects on compacted soils. *Journal of Environmental Management* 203(1), 533–541. https://doi.org/10.1016/j.jenvman.2017.07.031.

Morlat, R. and Symoneaux, R. 2008. Long-term additions of organic amendments in a Loire Valley vineyard on a calcareous sandy soil. III. Effects on fruit composition and chemical and sensory characteristics of Cabernet Franc wine. *American Journal of Enology and Viticulture* 59, 375–386.

Mugnai, S., Masi, E., Azzarello, E. and Mancuso, S. 2012. Influence of long-term application of green waste compost on soil characteristics and growth, yield and quality of grape (*Vitis vinifera* L.). *Compost Science and Utilization* 20(1), 29–33. https://doi.org/10.1080/1065657X.2012.10737019.

Mylavarapu, R. S. and Zinati, G. M. 2009. Improvement of soil properties using compost for optimum parsley production in sandy soils. *Scientia Horticulturae* 120(3), 426–430. https://doi.org/10.1016/j.scienta.2008.11.038.

Nair, A. and Ngouajio, M. 2012. Soil microbial biomass, functional microbial diversity, and nematode community structure as affected by cover crops and compost in an organic vegetable production system. *Applied Soil Ecology* 58, 45–55. https://doi.org/10.1016/j.apsoil.2012.03.008.

Nannipieri, P., Ceccanti, B., Cervelli, S. and Matarese, E. 1980. Extraction of phosphatase, urease, proteases, organic carbon, and nitrogen from soil. *Soil Science Society of America Journal* 44(5), 1011–1016. https://doi.org/10.2136/sssaj1980.03615995004400050028x.

Neufeld, J. D., Dumont, M. G., Vohra, J. and Murrell, J. C. 2007. Methodological considerations for the use of stable isotope probing in microbial ecology. *Microbial Ecology* 53(3), 435–442. https://doi.org/10.1007/s00248-006-9125-x.

Noble, R. and Coventry, E. 2005. Suppression of soil-borne plant diseases with composts: a review. *Biocontrol Science and Technology* 15(1), 3–20. https://doi.org/10.1080/09583150400015904.

Norris, C. E., Bean, G. M., Cappellazzi, S. B., Cope, M., Greub, K. L. H., Liptzin, D., Rieke, E. L., Tracy, P. W., Morgan, C. L. S. and Honeycutt, C. W. 2020. Introducing the North American project to evaluate soil health measurements. *Agronomy Journal* 112, 3195–3215. https://doi.org/10.1002/agj2.20234.

Nunes, M. R., Karlen, D. L., Veum, K. S., Moorman, T. B. and Cambardella, C. A. 2020. Biological soil health indicators respond to tillage intensity: a US meta-analysis. *Geoderma* 369, 114335. https://doi.org/10.1016/j.geoderma.2020.114335.

Oliveira, M. and Duarte, E. 2016. Integrated approach to winery waste: waste generation and data consolidation. *Frontiers of Environmental Science and Engineering* 10(1), 168–176. https://doi.org/10.1007/s11783-014-0693-6.

Oudart, D., Paul, E., Robin, P. and Paillat, J. M. 2012. Modeling organic matter stabilization during windrow composting of livestock effluents. *Environmental Technology* 33(19–21), 2235–2243. https://doi.org/10.1080/09593330.2012.728736.

Ouédraogo, E. 2001. Use of compost to improve soil properties and crop productivity under low input agricultural system in West Africa. *Agriculture, Ecosystems and Environment* 84(3), 259–266. https://doi.org/10.1016/S0167-8809(00)00246-2.

Pai, S., Ai, N. and Zheng, J. 2019. Decentralized community composting feasibility analysis for residential food waste: a Chicago case study. *Sustainable Cities and Society* 50, 101683. https://doi.org/10.1016/j.scs.2019.101683.

Parkin, T. B. and Venterea, R. T. 2010. Chapter 3. Chamber-based trace gas flux measurements. In: R. F. Follett (Ed.), Sampling Protocols. USDA-ARS, pp. 1–39. Available at: (https://www.ars.usda.gov/natural-resources-and-sustainable-agricultural-systems/soil-and-air/docs/gracenet-sampling-protocols/

Paustian, K., Larson, E., Kent, J., Marx, E. and Swan, A. 2019. Soil C sequestration as a biological negative emission strategy. *Frontiers in Climate* 1, 8. https://doi.org/10.3389/fclim.2019.00008.

Pawlett, M., Hannam, J. A. and Knox, J. W. 2021. Redefining soil health. *Microbiology* 167(1). https://doi.org/10.1099/mic.0.001030.

Payen, F. T., Sykes, A., Aitkenhead, M., Alexander, P., Moran, D. and MacLeod, M. 2021. Predicting the abatement rates of soil organic carbon sequestration management in Western European vineyards using random forest regression. *Cleaner Environmental Systems* 2, 100024. https://doi.org/10.1016/j.cesys.2021.100024.

Pérez-Piqueres, A., Edel-Hermann, V., Alabouvette, C. and Steinberg, C. 2006. Response of soil microbial communities to compost amendments. *Soil Biology and Biochemistry* 38(3), 460–470. https://doi.org/10.1016/j.soilbio.2005.05.025.

Pergola, M., Persiani, A., Palese, A. M., Di Meo, V., Pastore, V., D'Adamo, C. and Celano, G. 2018. Composting: the way for a sustainable agriculture. *Applied Soil Ecology* 123, 744–750. https://doi.org/10.1016/j.apsoil.2017.10.016.

Petrie, B., Barden, R. and Kasprzyk-Hordern, B. 2015. A review on emerging contaminants in wastewaters and the environment: current knowledge, understudied areas and recommendations for future monitoring. *Water Research* 72, 3–27. https://doi.org/10.1016/j.watres.2014.08.053.

Pezzolla, D., Said-Pullicino, D., Raggi, L., Albertini, E. and Gigliotti, G. 2013. Short-term variations in labile organic C and microbial biomass activity and structure After organic amendment of arable soils. *Soil Science* 178(9), 474–485. https://doi.org/10.1097/SS.0000000000000012.

Pugliese, M., Gilardi, G., Garibaldi, A. and Gullino, M. L. 2015. Organic amendments and soil suppressiveness: results with vegetable and ornamental crops. In: Meghvansi, M. K. and Varma, A. (Eds), *Organic Amendments and Soil Suppressiveness in Plant Disease Management, Soil Biology*. Springer International Publishing, Cham, pp. 495–509. https://doi.org/10.1007/978-3-319-23075-7_24.

Ramos, M. C. 2017. Effects of compost amendment on the available soil water and grape yield in vineyards planted after land levelling. *Agricultural Water Management* 191, 67–76. https://doi.org/10.1016/j.agwat.2017.05.013.

Ramos, T. M., Jay-Russell, M. T., Millner, P. D., Shade, J., Misiewicz, T., Sorge, U. S., Hutchinson, M., Lilley, J. and Pires, A. F. A. 2019. Assessment of biological soil amendments of animal origin use, research needs, and extension opportunities in organic production. *Frontiers in Sustainable Food Systems* 3, 73. https://doi.org/10.3389/fsufs.2019.00073.

Ramzani, P. M. A., Shan, L., Anjum, S., Khan, W. U., Ronggui, H., Iqbal, M., Virk, Z. A. and Kausar, S. 2017. Improved quinoa growth, physiological response, and seed nutritional quality in three soils having different stresses by the application of acidified biochar and compost. *Plant Physiology and Biochemistry* 116, 127–138. https://doi.org/10.1016/j.plaphy.2017.05.003.

Rawls, W. J., Pachepsky, Y. A., Ritchie, J. C., Sobecki, T. M. and Bloodworth, H. 2003. Effect of soil organic carbon on soil water retention. *Geoderma* 116(1–2), 61–76. https://doi.org/10.1016/S0016-7061(03)00094-6.

Rinot, O., Levy, G. J., Steinberger, Y., Svoray, T. and Eshel, G. 2019. Soil health assessment: a critical review of current methodologies and a proposed new approach. *Science of the Total Environment* 648, 1484–1491. https://doi.org/10.1016/j.scitotenv.2018.08.259.

Rivers, E. N., Heitman, J. L., McLaughlin, R. A. and Howard, A. M. 2021. Reducing roadside runoff: tillage and compost improve stormwater mitigation in urban soils. *Journal of Environmental Management* 280, 111732. https://doi.org/10.1016/j.jenvman.2020.111732.

Ros, M., Hernandez, M. T. and Garcia, C. 2003. Soil microbial activity after restoration of a semiarid soil by organic amendments. *Soil Biology and Biochemistry* 35(3), 463–469. https://doi.org/10.1016/S0038-0717(02)00298-5.

Ros, M., Klammer, S., Knapp, B., Aichberger, K. and Insam, H. 2006. Long-term effects of compost amendment of soil on functional and structural diversity and microbial activity. *Soil Use and Management* 22(2), 209–218. https://doi.org/10.1111/j.1475-2743.2006.00027.x.

Rubio, R., Pérez-Murcia, M. D., Agulló, E., Bustamante, M. A., Sánchez, C., Paredes, C. and Moral, R. 2013. Recycling of agro-food wastes into vineyards by composting: agronomic validation in field conditions. *Communications in Soil Science and Plant Analysis* 44(1–4), 502–516. https://doi.org/10.1080/00103624.2013.744152.

Ruehlmann, J. and Körschens, M. 2009. Calculating the effect of soil organic matter concentration on soil bulk density. *Soil Science Society of America Journal* 73(3), 876–885. https://doi.org/10.2136/sssaj2007.0149.

Ruggieri, L., Cadena, E., Martínez-Blanco, J., Gasol, C. M., Rieradevall, J., Gabarrell, X., Gea, T., Sort, X. and Sánchez, A. 2009. Recovery of organic wastes in the Spanish wine industry. Technical, economic and environmental analyses of the composting process. *Journal of Cleaner Production* 17(9), 830–838. https://doi.org/10.1016/j.jclepro.2008.12.005.

Ryals, R., Kaiser, M., Torn, M. S., Berhe, A. A. and Silver, W. L. 2014. Impacts of organic matter amendments on carbon and nitrogen dynamics in grassland soils. *Soil Biology and Biochemistry* 68, 52–61. https://doi.org/10.1016/j.soilbio.2013.09.011.

Saha, S., Pandey, A. K., Gopinath, K. A., Bhattacharaya, R., Kundu, S. and Gupta, H. S. 2007. Nutritional quality of organic rice grown on organic composts. *Agronomy for Sustainable Development* 27(3), 223–229. https://doi.org/10.1051/agro:2007002.

Saison, C., Degrange, V., Oliver, R., Millard, P., Commeaux, C., Montange, D. and Le Roux, X. 2006. Alteration and resilience of the soil microbial community following compost amendment: effects of compost level and compost-borne microbial community. *Environmental Microbiology* 8(2), 247–257. https://doi.org/10.1111/j.1462-2920.2005.00892.x.

Salomé, C., Coll, P., Lardo, E., Metay, A., Villenave, C., Marsden, C., Blanchart, E., Hinsinger, P. and Le Cadre, E. 2016. The soil quality concept as a framework to assess management practices in vulnerable agroecosystems: a case study in Mediterranean vineyards. *Ecological Indicators* 61, 456–465. https://doi.org/10.1016/j.ecolind.2015.09.047.

Sax, M. S., Bassuk, N., van Es, H. and Rakow, D. 2017. Long-term remediation of compacted urban soils by physical fracturing and incorporation of compost. *Urban Forestry and Urban Greening* 24, 149–156. https://doi.org/10.1016/j.ufug.2017.03.023.

Schimel, J. P. and Bennett, J. 2004. Nitrogen mineralization: challenges of a changing paradigm. *Ecology* 85(3), 591-602. https://doi.org/10.1890/03-8002.

Şeker, C. and Manirakiza, N. 2020. Effectiveness of compost and biochar in improving water retention characteristics and aggregation of a sandy clay loam soil under wind erosion. *Carpathian Journal of Earth and Environmental Sciences* 15(1), 5–18. https://doi.org/10.26471/cjees/2020/015/103.

Sesay, A. A., Lasaridi, K., Stentiford, E. and Budd, T. 1997. Controlled composting of paper pulp sludge using the aerated static pile method. *Compost Science and Utilization* 5(1), 82–96. https://doi.org/10.1080/1065657X.1997.10701866.

Seufert, V. and Ramankutty, N. 2017. Many shades of gray–the context-dependent performance of organic agriculture. *Science Advances* 3(3), e1602638. https://doi.org/10.1126/sciadv.1602638.

Sharma, S., Aneja, M. K., Mayer, J., Munch, J. C. and Schloter, M. 2005. Diversity of transcripts of nitrite reductase genes (*nirK* and *nirS*) in rhizospheres of grain legumes.

Applied and Environmental Microbiology 71(4), 2001–2007. https://doi.org/10.1128/AEM.71.4.2001-2007.2005.

Shaw, D. V., Gordon, T. R., Hansen, J. and Kirkpatrick, S. C. 2010. Relationship between the extent of colonization by *Verticillium dahliae* and symptom expression in strawberry (*Fragaria × ananassa*) genotypes resistant to Verticillium wilt. *Plant Pathology* 59(2), 376–381. https://doi.org/10.1111/j.1365-3059.2009.02203.x.

Shaw, D. V., Gubler, W. D. and Hansen, J. 1997. Field resistance of California strawberries to Verticilium dahliae at three conidial innoculum concentrations. *HortScience* 32(4), 711–713.

Shiralipour, A., McConnell, D. B. and Smith, W. H. 1992. Physical and chemical properties of soils as affected by municipal solid waste compost application. *Biomass and Bioenergy* 3(3–4), 261–266. https://doi.org/10.1016/0961-9534(92)90030-T.

Siedt, M., Schäffer, A., Smith, K. E. C., Nabel, M., Roß-Nickoll, M. and van Dongen, J. T. 2021. Comparing straw, compost, and biochar regarding their suitability as agricultural soil amendments to affect soil structure, nutrient leaching, microbial communities, and the fate of pesticides. *Science of the Total Environment* 751, 141607. https://doi.org/10.1016/j.scitotenv.2020.141607.

Siles-Castellano, A. B., López, M. J., López-González, J. A., Suárez-Estrella, F., Jurado, M. M., Estrella-González, M. J. and Moreno, J. 2020. Comparative analysis of phytotoxicity and compost quality in industrial composting facilities processing different organic wastes. *Journal of Cleaner Production* 252, 119820. https://doi.org/10.1016/j.jclepro.2019.119820.

Six, J., Bossuyt, H., Degryze, S. and Denef, K. 2004. A history of research on the link between (micro)aggregates, soil biota, and soil organic matter dynamics. *Soil and Tillage Research* 79(1), 7–31. https://doi.org/10.1016/j.still.2004.03.008.

Six, J., Elliott, E. T. and Paustian, K. 2000. Soil macroaggregate turnover and microaggregate formation: a mechanism for C sequestration under no-tillage agriculture. *Soil Biology and Biochemistry* 32(14), 2099–2103. https://doi.org/10.1016/S0038-0717(00)00179-6.

Solaiman, Z. M., Yang, H., Archdeacon, D., Tippett, O., Tibi, M. and Whiteley, A. S. 2019. Humus-rich compost increases lettuce growth, nutrient uptake, mycorrhizal colonisation, and soil fertility. *Pedosphere* 29(2), 170–179. https://doi.org/10.1016/S1002-0160(19)60794-0.

Solomon, S., Intergovernmental Panel on Climate Change and Intergovernmental Panel on Climate Change (Eds) 2007. Climate Change *2007: The Physical Science Basis: Contribution of Working Group I to the Fourth Assessment Report of the Intergovernmental Panel on Climate Change*. Cambridge University Press, Cambridge. New York.

Stamatiadis, S., Werner, M. and Buchanan, M. 1999. Field assessment of soil quality as affected by compost and fertilizer application in a broccoli field (San Benito County, California). *Applied Soil Ecology* 12(3), 217–225. https://doi.org/10.1016/S0929-1393(99)00013-X.

Stoffella, P., He, Z., Kahn, B., Yang, X. and Calvert, D. 2001. Plant nutrition benefits of phosphorus, potassium, calcium, magnesium, and micronutrients from compost utilization. In: Stoffella, P. and Kahn, B. (Eds), *Compost Utilization in Horticultural Cropping Systems*. CRC Press. https://doi.org/10.1201/9781420026221.ch15.

Tabatabai, M. A. 1994. Soil enzymes. In: Weaver, R. W., Angle, S., Bottomley, P., Bezdicek, D., Smith, S., Tabatabai, A. and Wollum, A. (Eds), *SSSA Book Series*. Soil Science Society of America, Madison, WI, pp. 775-833. https://doi.org/10.2136/sssabookser5.2.c37.

Tautges, N. and Scow, K. 2020. Pursuing agroecosystem resilience in a long-term Mediterranean agricultural experiment. In: *Long-Term Farming Systems Research*. Elsevier, pp. 53-69. https://doi.org/10.1016/B978-0-12-818186-7.00004-7.

Tejada, M. and Gonzalez, J. L. 2006a. The relationships between erodibility and erosion in a soil treated with two organic amendments. *Soil and Tillage Research* 91(1-2), 186-198. https://doi.org/10.1016/j.still.2005.12.003.

Tejada, M. and Gonzalez, J. L. 2006b. Crushed cotton gin compost on soil biological properties and rice yield. *European Journal of Agronomy* 25(1), 22-29. https://doi.org/10.1016/j.eja.2006.01.007.

Tejada, M., Gonzalez, J. L., García-Martínez, A. M. and Parrado, J. 2008. Application of a green manure and green manure composted with beet vinasse on soil restoration: effects on soil properties. *Bioresource Technology* 99(11), 4949-4957. https://doi.org/10.1016/j.biortech.2007.09.026.

Temiz, C., Akca, M. O., Cayci, G. and Baran, A. 2021. Assessment of the effect of different compost materials on aggregation and mechanical properties in an entisol. *Communications in Soil Science and Plant Analysis* 52(8), 871-885. https://doi.org/10.1080/00103624.2020.1869770.

Tian, W., Wang, L., Li, Y., Zhuang, K., Li, G., Zhang, J., Xiao, X. and Xi, Y. 2015. Responses of microbial activity, abundance, and community in wheat soil after three years of heavy fertilization with manure-based compost and inorganic nitrogen. *Agriculture, Ecosystems and Environment* 213, 219-227. https://doi.org/10.1016/j.agee.2015.08.009.

Tiefenbacher, A., Sandén, T., Haslmayr, H.-P., Miloczki, J., Wenzel, W. and Spiegel, H. 2021. Optimizing carbon sequestration in croplands: a synthesis. *Agronomy* 11(5), 882. https://doi.org/10.3390/agronomy11050882.

Torres, I. F., Bastida, F., Hernández, T., Albaladejo, J. and García, C. 2015. Enzyme activity, microbial biomass and community structure in a long-term restored soil under semi-arid conditions. *Soil Research* 53(5), 553. https://doi.org/10.1071/SR14297.

Tran, D. T. Q., Bradbury, M. I., Ogtrop, F. F. V., Bozkurt, H., Jones, B. J. and McConchie, R. 2020. Environmental drivers for persistence of Escherichia coli and Salmonella in manure-amended soils: a meta-analysis. *Journal of Food Protection* 83(7), 1268-1277. https://doi.org/10.4315/0362-028X.JFP-19-460.

Tsiafouli, M. A., Thébault, E., Sgardelis, S. P., de Ruiter, P. C., van der Putten, W. H., Birkhofer, K., Hemerik, L., de Vries, F. T., Bardgett, R. D., Brady, M. V., Bjornlund, L., Jørgensen, H. B., Christensen, S., Hertefeldt, T. D., Hotes, S., Gera Hol, W. H., Frouz, J., Liiri, M., Mortimer, S. R., Setälä, H., Tzanopoulos, J., Uteseny, K., Pižl, V., Stary, J., Wolters, V. and Hedlund, K. 2015. Intensive agriculture reduces soil biodiversity across Europe. *Global Change Biology* 21(2), 973-985. https://doi.org/10.1111/gcb.12752.

Tubeileh, A. M. and Stephenson, G. T. 2020. Soil amendment by composted plant wastes reduces the Verticillium dahliae abundance and changes soil chemical properties in a bell pepper cropping system. *Current Plant Biology* 22, 100148. https://doi.org/10.1016/j.cpb.2020.100148.

Vance, E. D., Brookes, P. C. and Jenkinson, D. S. 1987. An extraction method for measuring soil microbial biomass C. Soil Biology. *Soil Biology and Biochemistry* 19(6), 703-707. https://doi.org/10.1016/0038-0717(87)90052-6.

Verhoeven, E., Decock, C., Garland, G. and Lazcano, C. 2019. Vineyard nitrous oxide (N2O) emissions. Wine Business Monthly 10. Available at: (https://www.winebusiness.com/wbm/?go=getArticleSignIn&dataId=207657).

Wade, J., Hamilton, M. A., Beetstra, M. L., Culman, S. W. and Margenot, A. J. 2022. Soil health conceptualization differs across key stakeholder groups across the Midwest. *Journal of Soil and Water Conservation* 76(6), 1-7.

Wagg, C., Bender, S. F., Widmer, F. and van der Heijden, M. G. A. 2014. Soil biodiversity and soil community composition determine ecosystem multifunctionality. *Proceedings of the National Academy of Sciences of the United States of America* 111(14), 5266-5270. https://doi.org/10.1073/pnas.1320054111.

Walling, E. and Vaneeckhaute, C. 2020. Greenhouse gas emissions from inorganic and organic fertilizer production and use: a review of emission factors and their variability. *Journal of Environmental Management* 276, 111211. https://doi.org/10.1016/j.jenvman.2020.111211.

Wang, S. Y. and Lin, H. S. 2003. Compost as a soil supplement increases the level of antioxidant compounds and oxygen radical absorbance capacity in strawberries. *Journal of Agricultural and Food Chemistry* 51(23), 6844-6850. https://doi.org/10.1021/jf030196x.

Ward, D. M., Weller, R. and Bateson, M. M. 1990. 16S rRNA sequences reveal numerous uncultured microorganisms in a natural community. *Nature* 345(6270), 63-65. https://doi.org/10.1038/345063a0.

Watts, D. B., Torbert, H. A., Feng, Y. and Prior, S. A. 2010. Soil microbial community dynamics as influenced by composted dairy manure, soil properties, and landscape position. *Soil Science* 175(10), 474-486. https://doi.org/10.1097/SS.0b013e3181f7964f.

Weber, J., Karczewska, A., Drozd, J., Licznar, M., Licznar, S., Jamroz, E. and Kocowicz, A. 2007. Agricultural and ecological aspects of a sandy soil as affected by the application of municipal solid waste composts. *Soil Biology and Biochemistry* 39(6), 1294-1302. https://doi.org/10.1016/j.soilbio.2006.12.005.

Weil, R. R., Islam, K. R., Stine, M. A., Gruver, J. B. and Samson-Liebig, S. E. 2003. Estimating active carbon for soil quality assessment. A simplified method for laboratory and field use. *American Journal of Alternative Agriculture* 18(1), 3-17.

Weindorf, D. C., Zartman, R. E. and Allen, B. L. 2006. Effect of compost on soil properties in Dallas, Texas. *Compost Science and Utilization* 14(1), 59-67. https://doi.org/10.1080/1065657X.2006.10702264.

Weller, D. M., Raaijmakers, J. M., Gardener, B. B. M. and Thomashow, L. S. 2002. Microbial populations responsible for specific soil suppressiveness to plant pathogens. *Annual Review of Phytopathology* 40, 309-348. https://doi.org/10.1146/annurev.phyto.40.030402.110010.

Willekens, K., Vandecasteele, B., Buchan, D. and De Neve, S. 2014. Soil quality is positively affected by reduced tillage and compost in an intensive vegetable cropping system. *Applied Soil Ecology* 82, 61-71. https://doi.org/10.1016/j.apsoil.2014.05.009.

Willett, W., Rockström, J., Loken, B., Springmann, M., Lang, T., Vermeulen, S., Garnett, T., Tilman, D., DeClerck, F., Wood, A., Jonell, M., Clark, M., Gordon, L. J., Fanzo, J., Hawkes, C., Zurayk, R., Rivera, J. A., De Vries, W., Majele Sibanda, L., Afshin, A., Chaudhary, A., Herrero, M., Agustina, R., Branca, F., Lartey, A., Fan, S., Crona, B., Fox, E., Bignet, V., Troell, M., Lindahl, T., Singh, S., Cornell, S. E., Srinath Reddy, K., Narain, S., Nishtar, S. and Murray, C. J. L. 2019. Food in the Anthropocene: the EAT-Lancet commission on healthy diets from sustainable food systems. *The Lancet* 393(10170), 447–492. https://doi.org/10.1016/S0140-6736(18)31788-4.

Wind, L., Krometis, L. A., Hession, W. C. and Pruden, A. 2021. Cross-comparison of methods for quantifying antibiotic resistance in agricultural soils amended with dairy manure and compost. *Science of the Total Environment* 766, 144321. https://doi.org/10.1016/j.scitotenv.2020.144321.

Wolff, M. W., Alsina, M. M., Stockert, C. M., Khalsa, S. D. S. and Smart, D. R. 2018. Minimum tillage of a cover crop lowers net GWP and sequesters soil carbon in a California vineyard. *Soil and Tillage Research* 175, 244–254. https://doi.org/10.1016/j.still.2017.06.003.

Wortman, S. E. 2015. Crop physiological response to nutrient solution electrical conductivity and pH in an ebb-and-flow hydroponic system. *Scientia Horticulturae* 194, 34–42. https://doi.org/10.1016/j.scienta.2015.07.045.

Wortman, S. E., Holmes, A. A., Miernicki, E., Knoche, K. and Pittelkow, C. M. 2017. First-season crop yield response to organic soil amendments: a meta-analysis. *Agronomy Journal* 109(4), 1210–1217. https://doi.org/10.2134/agronj2016.10.0627.

Wu, W. and Ma, B. 2015. Integrated nutrient management (INM) for sustaining crop productivity and reducing environmental impact: a review. *Science of the Total Environment* 512–513, 415–427. https://doi.org/10.1016/j.scitotenv.2014.12.101.

Xin, X., Zhang, J., Zhu, A. and Zhang, C. 2016. Effects of long-term (23 years) mineral fertilizer and compost application on physical properties of fluvo-aquic soil in the North China Plain. *Soil and Tillage Research* 156, 166–172. https://doi.org/10.1016/j.still.2015.10.012.

Young, J. C. 1995. Microwave-assisted extraction of the fungal metabolite ergosterol and total fatty acids. *Journal of Agricultural and Food Chemistry* 43(11), 2904–2910. https://doi.org/10.1021/jf00059a025.

Zhang, Q., Miao, F., Wang, Z., Shen, Y. and Wang, G. 2017. Effects of long-term fertilization management practices on soil microbial biomass in China's cropland: a meta-analysis. *Agronomy Journal* 109(4), 1183–1195. https://doi.org/10.2134/agronj2016.09.0553.

Zhen, Z., Liu, H., Wang, N., Guo, L., Meng, J., Ding, N., Wu, G. and Jiang, G. 2014. Effects of manure compost application on soil microbial community diversity and soil microenvironments in a temperate cropland in China. *PLoS ONE* 9(10), e108555. https://doi.org/10.1371/journal.pone.0108555.

Zhu, X., Jackson, R. D., DeLucia, E. H., Tiedje, J. M. and Liang, C. 2020. The soil microbial carbon pump: from conceptual insights to empirical assessments. *Global Change Biology* 26(11), 6032–6039. https://doi.org/10.1111/gcb.15319.

Chapter 10

Assessing the effects of using animal manure on soil health

Ashraf M. Tubeileh, California Polytechnic State University, USA; and Michael J. Goss, University of Guelph, Canada

1 Introduction

The use of animal manure in agriculture has been recorded for more than 2000 years, and both the Greeks and Romans varied the application rate according to the crop and soil type (Beaton, 2009). By late medieval times, there was a clear understanding that animal manure was being applied to replace materials removed in the growing of crops (Wild, 1988). Even by the middle of the nineteenth century, the only external sources of nitrogen (N) available for agriculture, other than farmyard manure, were guano and sodium nitrate (Beaton, 2009). However, superphosphate was developed in 1842 (Dyke, 1993). It was a further century before N fertilizers became readily available in Europe and North America. Animal producers and farmers in resource-poor regions have continued spreading manure on the land (Tubeileh and Thomas, 2022). Although recycling of plant nutrients back to the soil may still be a key reason why manure is applied to land, the recognition that long-term cultivation of land causes a decline in soil organic matter (SOM) (Jenkinson,

http://dx.doi.org/10.19103/AS.2021.0094.14

1991) provides another. The reduction in the organic matter fraction leads to a fall in soil productivity and crop yield (Bauer and Black, 1994). With the importance of carbon (C) sequestration in combating climate change, there are good reasons to take a holistic view of the use of manure in the context of sustainable food production. Here we review the impacts associated with the application of livestock and poultry manure on the ability of soil to deliver a range of ecosystem services that sustain plants, animals, people and the environment in which they live (Table 1).

We first (Section 2) describe how different collection and storage practices alter the form and composition of the various types of manure applied to the soil. Section 3 reports on what happens to manure after its application to soil, in terms of the availability to crops of the plant nutrients it contains. Section 4 discusses manure treatment approaches to reduce risks from manure nutrients to the environment. Section 5 presents the effects of manure application on soil chemical and physical health. In Section 6, we consider the consequences for the biological properties of soil. How all these impacts affect crop yield is discussed in Section 7, and the challenges to human and ecosystem health are considered in Section 8.

Table 1 Benefits and concerns associated with the application of manure in crop production

Benefit	Concerns
Source of essential major nutrients: N, P, K and S for plant growth	Balance of nutrients applied to soil differs from those required for plant growth; potential for losses of P and N compounds in runoff to surface water and leaching to groundwater resources; additional greenhouse gases released; contributions to acid rain and particulate air pollution
Supply of micronutrients	Toxic metals can buildup in soil
Addition of organic carbon to soil	Antibiotics and endocrine-disrupting compounds released into the environment
Enhanced activity of soil microbes	Zoonotic pathogens released to the environment
Soil aggregation enhanced	
Soil bulk density reduced and more resilient to compaction	There is a risk of soil compaction by large heavily laden tankers and spreaders applying manure under wet conditions
Greater soil permeability	Risk of contaminants reaching water resources by preferential flow
Root exploration of soil enhanced	
Soil water holding capacity improved	
Soil more resilient to wind and water erosion	

2 Types and forms of manure applied to soils

Based on the average of total animal numbers in 2017 and 2019, cattle (beef and dairy) operations together contribute the greatest mass of manure worldwide, followed by pig (pork and breeding) production, with poultry production systems (broilers for meat and layers for eggs) making a smaller contribution to annual manure production (FAOSTAT, 2022). Beusen et al. (2008) calculated that of the total manure produced, 42% was excreted by confined animals and mostly spread on cropland but 7% was applied to grassland. Grazing animals deposit some 44% of total manure production on fields. The remainder (14%) was either excreted outside the agricultural system, used as animal feed, or as a fuel for heating or cooking.

Manure from confined livestock and poultry is commonly collected and stored before sufficient material is available or soils and crops are suitable for application to land. Feces are the main source of solids, other than those present in any bedding or uneaten feed material, and consist of undigested food and any feed additives, such as metals and antibiotics, used prophylactically to prevent disease or enhance growth. Urine from livestock, in addition to organic forms of N (urea and creatinine), contains calcium, sodium, magnesium and potassium (K) as well as phosphates, hormones, organic acids and antibiotics. Poultry excreta is a combination of materials from the digestive tract and the kidneys, but most of the N voided is in the form of uric acid.

2.1 Solid and semi-solid manure

Manures are classified based on the percentage of solids. As such, manures are classified into four categories: liquid (<4% solids), slurries (4–10% solids), semi-solids (10–20% solids) and solids (>20% solids) (Lorimor et al., 2004). Some semi-solids and all solids are stackable. Farmyard manure refers to semi-solid or solid manure from beef or dairy operations, although it may refer to manure from mixed animal sources in diverse operations (Pagliari et al., 2019). When the material collected is a combination of excreta and bedding (such as straw), it may be stacked and turned to undergo some level of composting. If solid manure is turned frequently and not tightly packed, the breakdown of organic matter takes place under aerobic conditions. The rate of the process is faster under aerobic than anaerobic conditions and more C is lost as CO_2. In part, this is because aerobic breakdown releases more energy, which in turn can support about 5.5 times more microbial biomass per unit of organic substrate consumed than can anaerobic fermentation.

2.2 Liquid manure or slurry

In barns with bedding-free or sand stalls, the mixture of feces and urine is moved into a holding tank, with or without wash water. The liquid or slurry manure collected can then be applied to the land. Manure is subject to change from the time that excreta are released. The changes in the manure occur throughout the period of temporary storage in the barn, and for the whole time, they are in longer-term storage prior to application to land. As the enzyme 'urease' is both widespread and persistent, the urea content of urine begins rapidly to break down after excretion even if it is not mixed with feces. The essential products are ammonium (NH_4^+), bicarbonate (HCO_3^-) and hydroxide (OH^-). The latter two ions increase the pH, which changes the balance between NH_4^+ and ammonia (NH_3) in solution, leading to the release of 35–70% of the total ammoniacal N into the atmosphere (Meisinger and Jokela, 2000). Although the same processes occur with poultry excreta, there tends to be less loss of NH_3 to the atmosphere because the material is much drier, unless water is added. The loss ranged between 28% and 46% when poultry litter was surface applied to fescue pastures in the southeastern United States (Marshall et al., 1998).

When liquid or slurry manure gets put into longer-term storage, the settling of materials can result in the formation of several layers, the properties of which vary considerably in space and time (Patni and Jui, 1987). Aerobic decomposition of manure organic fraction results in CO_2 and organic C compounds that are more resistant to breakdown. However, one consequence of stratification is that anaerobic conditions develop within the stored manure. When free oxygen is not present, organic matter is converted to C compounds of small molecular weight, mainly volatile organic acids, and CH_4 gas is released. In addition, the breakdown of proteins can lead to the formation of H_2S. The formation of volatile fatty acids, a readily available C source for microorganisms, can cause the manure to become less basic.

3 The nutrient value of manures

3.1 C:N ratio and nutrient availability

Soil microorganisms drive the availability of nutrient elements from organic molecules to plants. An important parameter that provides an index of the rate of mineralization of organic materials is the C:N ratio (Janssen, 1996; Lazicki et al., 2020). Manure C:N ratios are extremely variable depending on factors, such as species, animal feed, manure management and manure storage. Poultry and pig manures tend to have smaller C:N ratio values (Table 2). Solid manures tend to have higher C:N ratios than liquid manures. Values of 30:1 or less for the C:N ratio allow N to be released into the soil. For larger ratios, some N already present in the soil is immobilized before there is a net release (Scarsbrook,

Table 2 C:N ratio of main animal manure types as presented in different reports

	C:N ratio			
Manure type	Rynk et al. (1992)	Chadwick et al. (2000)	Qian and Schoenau (2002)	Ashworth et al. (2019)
Beef cattle, liquid		11.9		
Beef cattle, solid	19	12.3	11.3-21.7	
Dairy cattle, liquid		7.9		
Dairy cattle, solid	16	13.2		
Pig, liquid		3.0		
Pig, solid	9-19	9.7	11.5	
Poultry (broiler), solid	12-15	12.0	7.6	13.8
Poultry (layer), solid	3-10	15.7		9.9-10.5
Sheep, solid	13-20			
Turkey (broiler), solid	16			14.6

1965). Consequently, the application of solid and semi-solid manure results temporarily in the immobilization of mineral N in the soil, irrespective of the rate of application (Mohanty et al., 2011). In addition, the mineralization rate seems to be optimal at 24–25°C compared with higher or lower temperatures (Sims, 1986; Thangarajan et al., 2015). Soil acidity can also be a factor in governing the rate of N transformations. Research using 17 different soils in Australia has shown slower nitrification occurring when the amended soil is acidic due to smaller numbers of NH_3 oxidizing bacteria (Thangarajan et al., 2015).

3.2 Nutrient content of manure

At physiological maturity, crop plants need to have accumulated N and K to an amount ranging between 1% and 4% of their shoot dry weight. Phosphorus (P) contents range from 0.2% to 0.4% of shoot dry weight. Leguminous grain or forage crops tend to contain more N and P than do cereals (Table 3). Manure is commonly applied to the soil to meet the mineral requirements. However, the content of N, P and K in manure is present in different ratios to those needed by crops (Table 4). Although K is always present in manure as an inorganic cation, both N and P are usually in both organic and inorganic forms, with the proportions varying widely (Table 5). More detailed information on nutrient contents in different manures can be found in the study by He et al. (2016). NH_4^+ is formed under both aerobic and anaerobic breakdown of N-containing plant and animal constituents (mineralization), while NO_3^- only forms under aerobic conditions by the oxidation of NH_4^+ via NO_2^- in nitrification. The proportion of the total N present in NH_3 and NH_4^+-N forms is a key characteristic of manure

Table 3 Approximate content of N, P and K for some agricultural crops at harvest, expressed in g kg^{-1} dry matter, together with their ratio based on P content

	Nutrient content (g kg^{-1})			Nutrient ratio
Crop	N	P	K	N:P:K
Maize (*Zea mays* L.)	13.5	1.8	14.1	7.5:1:7.8
Rice (*Oryza sativa* L.)	9.3	1.6	10.3	5.5:1:6.4
Wheat (*Triticum aestivum* L.)	12.4	1.3	13.8	9.5:1:10.6
Oat (*Avena sativa* L.)	9.0	1.8	16.9	5.0:1:9.4
Soybean (*Glycine max* L. Merr.)	31.7	3.8	15.4	8.4:1:4.1
Lentil (*Lens culinaris* Medik.)	15.9	2.9	13.7	5.5:1:4.7
Alfalfa (*Medicago sativa* L.)	39.9	3.3	29.3	12.1:1:8.9

Source: Based on Goss et al. (2013).

Table 4 Approximate content of N, P and K in various livestock and poultry manures, expressed in g kg^{-1} dry matter and their ratio based on P content

	Nutrient content (g kg^{-1})			Nutrient ratio
Source	N	P	K	N:P:K
Liquid manure, dairy cattle	3.9	0.9	2.5	4.3:1:2.7
Solid manure, beef cattle	9.2	3.3	6.6	2.8:1:2.0
Liquid manure, pig	3.9	1.2	1.9	3.3:1:1.6
Solid manure, poultry (layer)	20.7	10.0	9.8	2.1:1:1.0
Solid manure, poultry (broiler)	31.2	14.1	17.9	2.2:1:1.3
Solid manure, turkey (broiler)	33.5	12.1	14.2	2.8:1:1.2
Solid manure, turkey (breeder)	21.6	13.0	13.5	1.7:1:1.0

Source: Based on OMAFRA (2013, 2018).

and other organic amendments. These forms are considered to be as available as fertilizer N (Beauchamp and Paul, 1989) or nearly so (Chantigny et al., 2004; Paul and Beauchamp, 1994) because they can be lost as NH_3. A good overview of NH_3 losses as a result of animal manure application is given in the study by Søgaard et al. (2002). In contrast to inorganic N, the organic N fraction may be only marginally available in the year of application (Beauchamp, 1986).

There are numerous forms of organic P in manure, including glucose phosphates, phospholipids, nucleotides, DNA, simple phosphate esters and phytate (Turner and Leytem, 2004). With the exception of phytate, these forms are only weakly adsorbed onto soil and are therefore mobile, both if entrained in runoff from the soil surface or in water draining through the profile. In contrast,

Table 5 Distribution of N and P forms in some animal manures expressed in g kg^{-1} dry matter

Manure type	Total N (g kg^{-1})	Mineral N: NH_4^+-N (g kg^{-1})	Mineral N: NO_3^--N (g kg^{-1})	Organic N (g kg^{-1})	Dry matter (%)
Liquid manure, dairy cattle	2.5	1.2	0.002	1.2	4.4
Composted manure, dairy cattle	5.9	0.07	0.25	5.6	29.9
Beef cattle manure	27.0	9.0	NR	18.0	17.1
	Total P (g kg^{-1})	**Mineral P[a] (g kg^{-1})**		**Organic P[b] (g kg^{-1})**	
Beef cattle manure	4.1	2.9		1.2	14.0
Dairy cattle manure	8.6	4.8		1.0	17.1
Pig manure	13.7	12.3		1.4	25.0
Poultry (broiler) litter	15.8	5.6		10.2	84.0
Poultry (mixed litter)	21.5	12.1		8.7	66.0

Source: Based on data from Goss et al. (1995), OMAFRA (2018), Pagliari and Laboski (2012), and Turner and Leytem (2004).[a]Mineral P consisted mainly of phosphate but polyphosphate was found in trace amounts in the poultry manure, some in the swine manure but cattle manure contained the greatest proportion.

[b]Organic P has been investigated using different extractants and analytical techniques, sometimes together with various enzymatic treatments. Compounds that are water soluble or readily become suspended in water can be a problem in allowing P to enter watercourses in runoff or transport contaminants into subsurface soil layers. Organic forms of P, such as phytate, that sorb strongly to clays or react with metals to form insoluble precipitates, are less problematic for the environment but are difficult for microorganisms to access once in the soil and hence the nutrient is not available to plants.

phytate is strongly adsorbed on clays or reacts with metals to form insoluble precipitates. In either case, phytate is not readily accessible to microorganisms for breakdown in soil.

The concentration of P in manure can vary widely according to the dietary intake (Barnett, 1994), in part because of the very different amounts of mineral P that can be provided in feed (Pagliari et al., 2019; Van Faassen and Van Dijk, 1987). However, there is evidence from manure sampling programs that in pig manure the content of P has reduced in the past decade or so (Zhang et al., 2020). One change has been the inclusion of microbially sourced phytase in the diet, which allows the animals to digest the P in grain.

Many fields that have received regular applications of manure or composted manure contain substantial amounts of P, even if the concentrations appear to stabilize over time (Komiyama and Ito, 2019). The maximum amount of P that is present in the soil solution is related to the natural level of labile P and how much additional P has been applied in organic and inorganic forms (Raven and Hossner, 1993; Sharpley and Smith, 1989). Still, the percentage of P from cattle manure that is recovered by crops (32%) is less than half its counterpart from equivalent inorganic fertilizer applications (77–86%) (Parham et al., 2002).

Another aspect that affects the perceived nutrient value of manure is the development of heterogeneity in the distribution of nutrients that occurs during storage prior to land application. For example, in solid manure, traces of NO_3^--N can be produced in the surface layer, where the gaseous loss of NH_4^+-N takes place preferentially, leading to considerable variability in mineral-N concentration with depth in the pile. It is difficult to mix solid manure homogeneously, so variation in application occurs during spreading, leading to considerable spatial variability in the distribution of nutrients, particularly N, applied to the land. In contrast, even though variability in the content of nutrients develops between strata in stored liquid manure, mixing before spreading can be relatively easy in manure lagoons and separate storage units; this pre-mixed material applied to the land can be reasonably uniform in its fertilizer value. However, where manure is stored in under-barn units, mixing likely releases noxious gases. Consequently, the mixing of manure cannot take place and there can be considerable heterogeneity in samples taken in preparation for land application (Dou et al., 2001).

The very different proportions of N and P in manure relative to what is needed for plant growth generally mean that if the crop requirements for N are met, the levels of P as well as K (see Tables 3 and 4) and Mg can be excessive (Cela et al., 2010; Edmeades, 2003). If P is used as the basis for the calculation, insufficient N will be provided to the crop unless supplemented with mineral fertilizer.

In addition to macronutrients, manure also supplies trace elements, such as zinc, copper, cadmium (Table 6) as well as molybdenum and boron. As many

Table 6 Range of metal content in different manure types. Means are given in parentheses.

Manure type	Copper	Zinc	Cadmium
	(mg kg^{-1} - dry matter basis)		
Liquid dairy cattle	1.0-352 (55)	<5-727 (266)	<0.08-3.2 (0.31)
Liquid beef cattle	12-267 (49)	88-938 (277)	<0.08-0.80 (0.32)
Solid dairy/beef	2.5-80 (35)	40-448 (181)	0.04-3.1 (0.31)
Liquid pig	<1-1386 (714)	<5-5832 (1427)	0.06-1.3 (0.65)
Solid pig	134-780 (447)	206-1220 (678)	0.19-0.53 (0.34)
Solid sheep	13	104	ND
Solid chicken (layers deep litter)	17-486 (72)	237-789 (515)	0.09-2.04 (0.53)
Solid broiler	8.4-173 (72)	52-473 (355)	0.11-1.16 (0.27)

Source: Adapted from Goss et al. (2013).

soils, especially those in the humid tropics, are deficient in these micronutrients, the use of manure can be important in crop production (Gupta et al., 2008). There is evidence that the level of Cu, often included in the diets of non-ruminant species to reduce disease, has declined in the manure from chickens, which could reduce the buildup of the element in soils receiving such material for crop production (Zhang et al., 2020).

The timing and rate of release of the nutrients applied in organic form are dependent on the nature of the materials themselves as well as the climate, soil and land management practices. There may also be a residual effect from previous applications, which have not been fully mineralized in preceding years (Schröder, 2005). This makes it difficult to be precise about the amount required and when it should be applied to make the availability synchronize with the needs of the plants. Consequently, there is a considerably greater risk that nutrients will be lost to the wider environment from manure than from mineral fertilizers.

4 Manure treatments

Many large animal production enterprises enhance manure utilization by separating manure solids from liquids or composting; others store slurries in large lagoons and capture NH_3 or use anaerobic digestion to produce biogas - especially from cattle manure, given its high calorific value (Font-Palma, 2019). Szogi et al. (2015) provide a review of the treatment methods available to reduce the nutrient content of manure prior to soil application. Methods that allow nutrient recovery include chemical amendments, thermochemical conversion, biological conversion to non-reactive N, NH_3 emission capture inside animal operations, liquid NH_3 capturing using gas-permeable membranes, chemical P precipitation and wet P extraction.

4.1 Composting

Composts are discussed elsewhere in this book in more detail, and so in this chapter, we will only compare untreated and composted manure in terms of their effect on soil health. Composting is one treatment used to stabilize the N in solid and semi-solid manures and improve their handling characteristics, particularly by reducing the volume of organic materials that are to be land-applied (Dolliver et al., 2008).

The main benefits of composting animal manures are the reduction of pathogens and some organic contaminants, such as antibiotics (Dolliver et al., 2008), together with stabilization of the organic and mineral fractions. Nevertheless, composting results in a loss of total N and C content, although concentrations of other elements increase in the finished product (Larney et al., 2008).

Kirchmann and Lundvall (1993) concluded that aerobic composting should not be used to treat amendments containing high levels of NH_4^+-N as it resulted in too great a loss by NH_3 volatilization. Ramaswamy et al. (2010) reported a 36% loss of total Kjeldahl N together with a 10% loss of C from loose-piled poultry manure and hay (initial C:N ratio of 25:1). In contrast, Kirchmann (1985) reported losses of N ranging between 2% and 10% of nitrate (NO_3^-) N under windrowing, although NO_3^--N concentrations generally did not exceed 0.05% of dry matter. During composting, about 50% of the organic matter, 40% of K and 20–30% of N in manure can be lost if manure is windrowed without covering (Lampkin, 1990; Vogtmann and Besson, 1978). Mass reduction and C loss were significantly greater in turned than in unturned windrows (Tiquia et al., 2002). The net result of composting is a reduction in readily mineralizable organic N compounds.

In addition, the concentration of heavy metals increases during composting due to mineralization, decomposition of organic material and loss of C (Ben Achiba et al., 2010). However, the use of additives during composting can promote the transformation and speciation of heavy metals making them less bioavailable (Yousif Abdallah et al., 2022).

4.2 Anaerobic digestion

In anaerobic digestion, organic wastes are broken down in the absence of dissolved oxygen or its precursors to produce biogas (a mixture of methane, CO_2 and traces of hydrogen sulfide, NH_3 and water vapor). The resulting solid material (digestate or biogas slurry in the case of manure) can be used as a fertilizer. Compared to liquid manure, the percentages of total N and total P on dry matter basis increase with the degree of decomposition in digestate. In addition, anaerobic digestion increases the proportion of NH_4^+ in total N and reduces C:N ratio and the percentage of water-soluble P. Consequently, the application of digestate seems to increase crop N uptake compared to undigested animal slurry by about 10–25% (Möller and Müller, 2012).

4.3 Pelleting

Pelleting manure is a relatively new practice that is growing in popularity, especially in areas with large-scale poultry production operations. The organic crop production industry is a main driver for this type of manure treatment with chicken-manure-based fertilizers such as 4-4-4 being among the most popular organic fertilizers. This treatment improves storage stability, transportability and application logistics and reduces the generated dust by 90% compared to ordinary compost (Hara, 2001). Further details on this treatment are given in Hao and He (2020).

5 Impact on soil chemical and physical health

The application of manure to agricultural land provides macro- and micronutrients to crops and improves soil physical and chemical conditions for plant production. When mineral fertilizers are scarce, the use of organic amendments, such as manure, may be the only means of providing nutrients in sufficient quantity to achieve more than subsistence production.

As stated earlier, an important reason for applying manure to cropland is to offset the natural decline in SOM and enhance the physical and chemical properties of arable soils. Based on an analysis of the long-term experiments at Rothamsted Experimental Station (UK), Johnston et al. (2009) concluded that the SOM content was a function of the input of organic material from crops and organic amendments, such as manure, its rate of decomposition, the rate at which existing SOM mineralized, soil texture and climate. Under a given cropping system, SOM content tended toward an equilibrium level, which was greater for clayey than for sandy soils. Application of organic amendments can improve the C and N content in the top 5 cm of the soil (Peacock et al., 2001), although the magnitude of the effect depends on their dry matter content. Consequently, effects can be small with liquid manure, such as that from pigs (Angers et al., 2010). The increase in C is associated with greater nutrient retention capacity.

In terms of physical properties, the increase in soil C leads to better resilience of soil aggregates to raindrop impact (Bresson et al., 2001), which in turn can improve infiltration and decrease runoff. Adding organic matter, such as manure, to mineral soils decreases bulk density (Celik et al., 2010; Rayne and Aula, 2020; Sommerfeldt and Chang, 1987) and increases porosity and water holding capacity (Wesseling et al., 2009). The reduction in soil bulk density is proportional to the amount of manure applied and the associated increase in SOM content. The changes in porosity can modify soil permeability. For example, Munyankusi et al. (1994) compared the macroporosity of soil that had a history of mineral fertilizer application with one regularly receiving liquid manure from dairy cattle. The greatest macroporosity was found in the top 10 cm of the fertilizer treatment but extended to 20 cm where manure was applied. The continuity of macropores also extended to greater depths in the manured plot, so that preferential flow was somewhat greater.

The impact on water holding capacity is not always so clear because changes in water content at both the upper and lower limits can result in no net change in the water held between the two (Khaleel et al., 1981). Sommerfeldt and Chang (1987) showed that the plant-available water content decreased with increasing cattle manure applications, mostly due to a greater increase at the permanent wilting point compared to the increase in field capacity. Differences in permeability, as measured by hydraulic conductivity at saturation, can also be difficult to establish, although some large increases have been recorded (Khaleel et al., 1981).

Changes in the basic chemical properties of soil can result from applying manure. Improvement in CEC following the application of manure was reported by Bernal and Kirchmann (1992) and Bulluck et al. (2002). The effect was observed only on a soil with less than 10% clay: a soil with nearly 20% clay showed no significant difference in organic C content or CEC. In the soil with less clay, the CEC was directly related to the organic C content (Bernal et al., 1992). Soil pH showed a small but significant rise after application, followed by a decline at the end of 2 years. In the longer-term, continued application of manure from cattle or pigs resulted in a decrease in soil pH, with bigger changes resulting from the application of pig manure (Bernal et al., 1992; Chang et al., 1991). Another benefit of manure application resides in its ability to prevent soil acidification caused by mineral N application (Cai et al., 2019). On the negative side, long-term application of manure (low, medium, or high rates corresponding to double the crop N needs) results in higher levels of electrical conductivity throughout the top 30 cm soil profile compared to unamended control or applications of inorganic fertilizer (Ozlu and Kumar, 2018).

Lack of organic C can reduce the surface charges in soil that are required for cation exchange and retaining basic elements, such as calcium, K and magnesium. This can result in increased soil acidity. Manure applications can reverse this process. For example, Simonetti et al. (2012) described a long-term experiment on clay soil in northern Italy, which compared the effects of an annual application of 40 Mg ha^{-1} of cattle farmyard manure over 44 years with those from mineral fertilizers or no additions. Manure application increased the concentration of C in humic compounds to ~26 g C kg^{-1}, while where minerals or no fertilizer was added, the value was ~18 g C kg^{-1}. This application increased the amount of well-humified material in soil aggregate fractions, indicative of enhanced breakdown of readily catabolized organic matter.

The structural improvements brought about by manure help in reclaiming degraded soils. The addition of poultry manure (10 Mg ha^{-1}) to a heavily degraded tropical sandy soil significantly increased SOM content, thereby reducing bulk density and enhancing total porosity, water infiltration and hydraulic conductivity, which allowed mineral fertilizer to be better managed and retained for the crops (Obi and Ebo, 1995). Similarly, the application of poultry manure compost increased plant cover to 88%, compared with only 5% for the control (Tejada et al., 2006).

6 Impact on soil biological health

Continuous manure application to soil influences the formation, storage and turnover of SOC, soil organic nitrogen, and enzyme dynamics, as well as soil microecology (Ozlu et al., 2019). A significant enhancement of the soil microbial biomass has frequently been reported following manure additions (e.g. Peacock et al., 2001). A recent meta-analysis has shown that activities of enzymes acquiring C, N, P, sulfur and energy were 1.3–3.3 times greater in manured soil compared with soil without manure, over a range of study durations. Manure application increased or did not change enzyme activities regardless of soil, manure or climatic conditions (Liu et al., 2020).

In addition to increasing the total microbial population, manure applications can also change the community structure with a significant shift from Gram-positive organisms in soil under crops given mineral fertilizer to one with many Gram-negative bacteria and more protozoa compared to mineral fertilizer applications or untreated controls (Peacock et al., 2001; Zhang et al., 2012). Likewise, the application of manure may enrich both r- and K-strategists when compared with an appropriate control soil (Parham et al., 2003). Continuous manure application has been shown to stimulate arbuscular mycorrhiza fungi (Celik et al., 2010; Ozlu et al., 2019) but decreases actinomycete populations (Ozlu et al., 2019; Zhang et al., 2012). The application of solid cattle manure increased the total number of bacteria, mesophilic spore-forming bacteria, nitrifying bacteria and cellulose decomposers (Edesi et al., 2012). The effects reported by Edesi et al. (2012) were more pronounced in soils managed using organic rather than conventional farming practices.

Changes in microbial communities resulting from manure application can have other beneficial effects. For example, manure can be an effective tool in the management of soil-borne pests. Nahar et al. (2006) reported that untreated and composted manure increased the numbers of non-plant parasitic nematodes, lowering the ratio of plant-parasitic to non-plant-parasitic nematodes, possibly due to antagonistic effects of soil microbial communities. Untreated cattle manure reduced plant-parasitic nematode populations relative to controls, suggesting that manure may have produced nematicidal substances during decomposition. A possible mechanism (Rodríguez-Kábana, 1986) by which manure reduces the levels of plant-parasitic nematodes is the accumulation of toxic compounds, such as CH_4, NH_3, N_2O and NO_2, during manure decomposition (Chadwick et al., 2000). The application of composted manure is less efficient than plant-based composts in suppressing soil-borne pathogens. Field and *in vitro* research has shown that manure compost and compost extract do not reduce *Verticillium dahliae* counts in the soil or growth in Petri dishes (Tubeileh and Stephenson, 2020).

7 Impact on plants and crop yields

Numerous studies have investigated the effects of manure on crop yields. Crop yields have tended to reflect the organic matter content of the soil. In part, this resulted from a larger annual release of N, together with improved availability of P, when the amount of SOM was enhanced. However, the loss of organic C was also greater under these conditions (Johnston et al., 2009). Long-term research in China, in which manure was compared with synthetic fertilizer treatments, has shown that pig solid manure resulted in greater wheat (*Triticum aestivum* L.) and maize (*Zea mays* L.) yields than did mineral fertilizer, mainly as the result of improved soil fertility (Cai et al., 2019). Long-term research at two sites in Germany showed that cattle manure systems (30 Mg ha^{-1} every 2 years) can produce similar or slightly smaller yields than mineral NPK fertilizer treatments (Ludwig et al., 2011). However, continuous manure application can result in environmentally harmful soil P levels (discussed in Section 8.2).

Compared with unamended soils, those soils that receive manure often produce higher yield levels as a result of improved soil physical properties and the provision of macro- and micronutrients. Long-term trends in crop yields have demonstrated that as the potential has increased, production has often been greater on soils with more organic matter compared to those on soils with less (Johnston et al., 2009). According to Lal (2006), crop yields can be increased by 20–70 kg ha^{-1} for wheat, 10–50 kg ha^{-1} for rice (*Oryza sativa* L.), and 30–300 kg ha^{-1} for maize with every 1 Mg ha^{-1} increase in soil organic C pool in the root zone. Organic amendments, such as manure, might be beneficial for crops even when they are not nutrient rich. In the long-term experiments at Rothamsted (UK), yields of spring crops [barley (*Hordeum vulgare* L. and potato (*Solanum tuberosum* L.)] increased following applications of nutrient-poor peat with or without additional mineral N, whereas winter cereals showed no benefit when mineral N was also applied (Johnston et al., 2009). The application of farmyard manure for 15 years produced almost double the yields of wheat and maize compared to unfertilized control. However, the application of similar amounts of inorganic fertilizer resulted in the same level of production (Liang et al., 2012). Edmeades (2003) reviewed long-term studies comparing yields at comparable N inputs and found no difference between organic amendments and fertilizers with regard to crop yield or pasture production. He concluded that large amounts of organic matter inputs over long periods of time are required to achieve larger yields than comparable fertilizer inputs. Similarly, a study in Pakistani Kashmir showed that it is possible to produce the same yield of wheat if 25% of urea N was replaced by poultry or cattle manure that contributed an equivalent amount of N (Abbasi and Tahir, 2012). The effects of N, improvements in soil P availability and other factors are discussed in detail in the study by Johnston et al. (2009).

8 Challenges to the concepts of soil health from manure application

8.1 Risk to human health

Deaths occur sporadically in relation to liquid manure structures, commonly because of inadequate ventilation or personal protective equipment when carrying out maintenance, repairs or agitating prior to land application (Meinen, 2016; Park et al., 2016). Asphyxiation by H_2S is the typical cause of death. Failure of such structures would pose a very significant risk to life because of the large volume of manure that could be released (approximately 2.6×10^6 liters for 180-day storage and 120 milking cows).

Considerable risks to human health are associated with the land application of manure. Deaths do result from workers climbing into partly filled liquid manure or slurry tankers without adequate breathing apparatus (Reuters, 2000). Movement in runoff or leaching from soil to surface or groundwater can contaminate water resources, which has led to disease outbreaks associated with *Escherichia coli* O157:H7 (e.g. Ackman et al., 1997; Goss and Richards, 2008). Potential pathways of exposure to contaminants from manure include their transport in air or water moving from the site of application. The most direct pathway is the ingestion of contaminated fresh fruit or vegetables grown in a soil receiving manure (Salaheen et al., 2016). Formation of bioaerosols and fine particulate material ($PM_{2.5}$), following application and during agitation of liquid storage, is an important aspect of the airborne risk to humans downwind (Jahne et al., 2015; Sommer and Knudsen, 2021). It appears that incorporation of liquid manure by disking forms bioaerosols with a greater emission rate than does the application process (Paez-Rubio et al., 2007). The injection of liquid manure causes less bioaerosol formation than surface spreading. The spreading of dry poultry litter can also create considerable amounts of particulate material and bioaerosols (Millner, 2009).

The risks arising from manure application to human health can be reduced using different agronomic practices. Examples of such practices include manure incorporation using a moldboard plow, solarization, mulch layers and avoiding plant injury to prevent pathogen entry (Köpke et al., 2007).

8.2 Risk of environmental pollution

In addition to pathogens, manure contains other pollutants that can contribute to pollution of water, air and soil such as heavy metals, antibiotics, hormones and some nutrients. A thorough review of these contaminants and their potential transport to pose a risk to human health is presented in the study by Goss et al. (2013).

8.2.1 Nutrients

Among nutrients, N and P levels represent a major concern from an environmental standpoint. Organic N compounds are slowly mineralized to NH_4^+-N, which can be lost by volatilization or oxidized by soil microbes to NO_2^--N and NO_3^--N. Soil chemistry greatly limits the amount of NH_4^+-N that is present in solution compared with NO_3^--N, which is water soluble and is easily leached beyond the root zone with excess water. Adding organic materials can help reduce nitrate leaching. For example, application of coconut husk biochar with anaerobic digestate greatly reduced the abundance of nitrifying archaea and bacteria in soil by 71–83% in the top 4 cm soil layer and 66–80% in the deeper soil layer one month after the digestate application (Plaimart et al., 2021). Plants can utilize NO_3^--N and NH_4^+-N, so the amount available to be lost depends greatly on the instantaneous demand of the crop and the rate at which water can carry the nutrients out of the root zone.

Nitrogen, P and organic compounds from manure represent concerns for surface water quality if they are removed in surface runoff. The N in any organic material left on the soil surface or associated with fine particles can be lost to runoff beginning the process of water erosion. In consequence, the contaminant can move from fields to surface watercourses. The release of NH_4^+-N into the aquatic environment can also have negative impacts on invertebrates (Zhang et al., 2021), whereas some amphibians are susceptible to elevated levels of NO_3^- (Ortiz-Santaliestra et al., 2011).

Most of the P in the manure applied to the soil becomes associated with clay minerals, sparingly soluble calcium phosphates or is in organic form and the concentrations in the soil solution are limited. When P enters freshwater bodies, it makes them more able to support the growth of aquatic plants, particularly algae, potentially resulting in eutrophication. It is the death and breakdown by microorganisms of blooms of these algae that depletes the oxygen concentration in water, limiting its availability to animals: in extreme conditions, fish can be asphyxiated (Chambers et al., 2001). Reducing the agricultural contribution of P to eutrophication requires a variety of socio-economic tools in addition to agronomic and pasture management (Kleinman et al., 2011).

8.2.2 Antibiotics

Antibiotics present in manure may impact plant growth and crop yield, even though they tend to be strongly sorbed onto organic matter and swelling clays.

Reduced growth and seedling death have been observed, consistent with their uptake by plants (Jjemba, 2002). Exposure of adult humans to a number of such veterinary antibiotics was predicted from their concentration in leafy tissue or roots. As the calculated daily intake represented only about 10% of the acceptable value, the risk to humans from this pathway was small but would be increased for compounds with a very small acceptable daily intake (Boxall et al., 2006). There is little evidence for significant antibiotic movement in surface runoff water or in deep transport to groundwater, although tile drains may be a significant pathway to water resources for some.

An important issue is the impact that the application of antibiotics might have on the soil microbial communities and hence on ecosystem services. The biogeochemistry of N in soil and possibilities for increasing the levels of antibiotic resistance have received considerable attention. Results of experiments on nitrification have not provided a clear picture, although changes in microbial communities, for example replacing bacteria NH_4^+ oxidizers with archaea NH_4^+ oxidizers, could maintain or even increase the rates of NH_4^+ oxidation. Denitrification is inhibited by antibiotics but effects at concentrations typically found in soil have not been adequately assessed (DeVries and Zhang, 2016).

The use of antibiotics for therapeutic or growth enhancement purposes can increase the presence of microbes carrying resistance genes. However, there is little evidence that regular manure applications have enhanced microbial resistance to multiple antibiotic drugs in bacteria from soils following the use of antibiotics to improve growth (Chee-Sanford et al., 2009). In part, increases in antibiotic resistance genes (ARG) detectable in the soil after manure application may result from microbes added with the manure. The activities of the indigenous microbes are important in controlling the rate of dissipation of ARG (Pérez-Valera et al., 2019). Although genetic evidence shows that identical ARG have been found in soil and in clinical pathogens (Forsberg et al., 2012), direct impact on the survival of enteric pathogenic microbes in soil from the use of antibiotics in livestock production is still needed.

8.2.3 Natural hormones and endocrine-disrupting substances

Animal feces commonly contain natural hormones that can threaten the reproductive health of other organisms on the land and in water resources (Goss et al., 2013). Once incorporated into the soil, hormones in manure do not appear to persist but are readily broken down by soil microbes (Lucas and Jones, 2006; Xuan et al., 2008), enhanced by the components of manure (Lucas and Jones, 2006). In contrast, the presence of urea may impair the microbial degradation of natural hormones, increasing their environmental persistence (Lucas and Jones, 2009). Kinney et al. (2008) demonstrated that endocrine-disrupting compounds accumulate in earthworms, indicating they can be persistent in soil. Furthermore, Stumpe and Marschner (2010) suggested that hormones adsorbed onto dissolved organic C compounds in manure can exchange between these and soil-based dissolved organic C. The association with organic C makes these endocrine-disrupting compounds less readily subject to microbial degradation but more readily transported in surface runoff to open waters. Similar to antibiotics, estrogens can be taken up into plants through their roots and accumulate in the tissues.

9 Conclusion

Application of manure from livestock and poultry to arable land can provide organic matter as well as macro- and micronutrients, thereby helping to replenish soil resources used in crop production. In addition to the nutrients, manure contains enzymes, such as urease, that can also contribute to the breakdown of organic forms of important nutrients. The organic component can help build SOM and contribute to enhancing soil structure as well as nutrient chemistry. The increase in SOM can also support soil organisms, including those added with manure. These consequences of manure addition are fully consistent with improving soil quality and resilience. However, the larger content of P relative to N in manure requires the nutrient supply in cropland to be augmented if imbalances are to be avoided. The release of bacteria, archaea and protozoa from farm animals into the environment via manure has implications for the spread of pathogens. The introduction and development of methods for handling manure as liquids and slurry has markedly increased the microbial contamination of groundwater, increasing the risk to human health from drinking-water supplies (Goss et al., 1998). Reverting to spreading solid manure can overcome this problem as well as increasing the organic matter being applied. Another benefit derived from solid manure systems is the reduction in NH_4^+-N that is applied, which in turn reduces the loss of N by volatilization and the potential to form fine particulates that are detrimental to human respiratory and cardiovascular functioning (Lovarelli et al., 2020). Other components of manure, including natural and administered hormones, antibiotics and toxic metals, can also adversely affect soil organisms, wildlife as well as humans. All these latter aspects of manure application to land ostensibly challenge the possibility that soil can deliver the range of ecosystem services that allow plants, animals and people to live healthily.

10 Where to look for further information

Research on manure and soil health is still primarily carried out by departments of agriculture around the world. The possibility for transfer of soil-borne zoonotic diseases to humans is still a real threat that limits safe widespread manure use in agriculture. Our understanding of prions and antibiotic-resistant genes from animal production operations is still limited and needs further research attention.

11 References

Abbasi, M. K. and Tahir, M. M. (2012). Economizing nitrogen fertilizer in wheat through combinations with organic manures in Kashmir, Pakistan. *Agronomy Journal* 104(1): 169–177. DOI: 10.2134/agronj2011.0264.

Ackman, D., Marks, S., Mack, P., Caldwell, M., Root, T. and Birkhead, G. (1997). Swimming-associated haemorrhagic colitis due to Escherichia coli O157:H7 infection: evidence of prolonged contamination of a fresh water lake. *Epidemiology and Infection* 119(1): 1–8. DOI: 10.1017/S095026889700770X.

Angers, D. A., Chantigny, M. H., MacDonald, J. D., Rochette, P. and Côté, D. (2010). Differential retention of carbon, nitrogen and phosphorus in grassland soil profiles with long-term manure application. *Nutrient Cycling in Agroecosystems* 86(2): 225–229. DOI: 10.1007/s10705-009-9286-3.

Ashworth, A. J., Chastain, J. P. and Moore, P. A. J. (2019). Nutrient characteristics of poultry manure and litter. In: *ASA Special Publication 67* Waldrip, H. M., Pagliari, H. and He, Z. (Eds). ASA and SSSA, Madison, pp. 63–87. DOI: 10.2134/asaspecpub67.c5.

Barnett, G. M. (1994). Phosphorus forms in animal manure. *Bioresource Technology* 49(2): 139–147. DOI: 10.1016/0960-8524(94)90077-9.

Bauer, A. and Black, A. L. (1994). Quantification of the effect of soil organic matter content on soil productivity. *Soil Science Society of America Journal* 58(1): 185–193. DOI: 10.2136/sssaj1994.03615995005800010027x.

Beaton, J. D. (2009). History. In: *Efficient Fertilizer Use Manual*. Mosaic, Plymouth, MN, USA. https://academic.uprm.edu/dsotomayor/agro6505/History_of_Fertilizer_Beaton.pdf.

Beauchamp, E. G. (1986). Availability of nitrogen from three manures to corn in the field. *Canadian Journal of Soil Science*. NRC Research Press 66(4): 713–720. DOI: 10.4141/cjss86-070.

Beauchamp, E. G. and Paul, J. W. (1989). A simple model to predict manure nitrogen availability to crops. In: *Nitrogen in Organic Wastes Applied to Soils* Hansen, J. A. and Henriksen, K. (Eds). Academic Press, London, UK, pp. 140–149.

Ben Achiba, W., Lakhdar, A., Gabteni, N., DuLaing, G., Verloo, M., Boeckx, P., Van Cleemput, O., Jedidi, N., Gallali, T. (2010). Accumulation and fractionation of trace metals in a Tunisian calcareous soil amended with farmyard manure and municipal solid waste compost. *Journal of Hazardous Materials* 176 (1–3): 99–108.

Bernal, M. P. and Kirchmann, H. (1992). Carbon and nitrogen mineralization and ammonia volatilization from fresh, aerobically and anaerobically treated pig manure during incubation with soil. *Biology and Fertility of Soils* 13(3): 135–141. DOI: 10.1007/BF00336268.

Bernal, M. P., Roig, A., Lax, A. and Navarro, A. F. (1992). Effects of the application of pig slurry on some physico-chemical and physical properties of calcareous soils. *Bioresource Technology* 42(3): 233–239. DOI: 10.1016/0960-8524(92)90026-T.

Beusen, A. H. W., Bouwman, A. F., Heuberger, P. S. C., Van Drecht, G. and Van Der Hoek, K. W. (2008). Bottom-up uncertainty estimates of global ammonia emissions from global agricultural production systems. *Atmospheric Environment* 42(24): 6067-6077. DOI: 10.1016/j.atmosenv.2008.03.044.

Boxall, A. B. A., Johnson, P., Smith, E. J., Sinclair, C. J., Stutt, E. and Levy, L. S. (2006). Uptake of veterinary medicines from soils into plants. *Journal of Agricultural and Food Chemistry* 54(6): 2288-2297. DOI: 10.1021/jf053041t.

Bresson, L. M., Koch, C., Le Bissonnais, Y., Barriuso, E. and Lecomte, V. (2001). Soil surface structure stabilization by municipal waste compost application. *Soil Science Society of America Journal* 65(6): 1804-1811. DOI: 10.2136/sssaj2001.1804.

Bulluck, L. R., Brosius, M., Evanylo, G. K. and Ristaino, J. B. (2002). Organic and synthetic fertility amendments influence soil microbial, physical and chemical properties on organic and conventional farms. *Applied Soil Ecology* 19(2): 147-160. DOI: 10.1016/S0929-1393(01)00187-1.

Cai, A., Xu, M., Wang, B., Zhang, W., Liang, G., Hou, E. and Luo, Y. (2019). Manure acts as a better fertilizer for increasing crop yields than synthetic fertilizer does by improving soil fertility. *Soil and Tillage Research* 189: 168-175. DOI: 10.1016/j.still.2018.12.022.

Cela, S., Berenguer, P., Santiveri, F. and Lloveras, J. (2010). Potential phosphorus, potassium, and magnesium surpluses in an irrigated maize monoculture fertilized with pig slurry. *Agronomy Journal* 102(1): 96-102. DOI: 10.2134/agronj2009.0139.

Celik, I., Gunal, H., Budak, M. and Akpinar, C. (2010). Effects of long-term organic and mineral fertilizers on bulk density and penetration resistance in semi-arid Mediterranean soil conditions. *Geoderma* 160(2): 236-243. DOI: 10.1016/j.geoderma.2010.09.028.

Chadwick, D. R., Pain, B. F. and Brookman, S. K. E. (2000). Nitrous oxide and methane emissions following application of animal manures to grassland. *Journal of Environmental Quality* 29(1): 277-287. DOI: 10.2134/jeq2000.00472425002900010035x.

Chambers, P. K., Dekimpe, C., Foster, N., et al. (2001). Chapter 13. Agricultural and forestry land use impacts. In: *Threats to Sources of Drinking Water and Aquatic Ecosystem Health in Canada*. NWRI Scientific Assessment Report Series 1. Burlington, Ontario, Canada: Environment Canada, National Water Research Institute, pp. 57-62. Available at: http://www.cciw.c/nwri/.

Chang, C., Sommerfeldt, T. G. and Entz, T. (1991). Soil chemistry after eleven annual applications of cattle feedlot manure. *Journal of Environmental Quality* 20(2): 475-480. DOI: 10.2134/jeq1991.00472425002000020022x.

Chantigny, M. H., Angers, D. A., Morvan, T. and Pomar, C. (2004). Dynamics of pig slurry nitrogen in soil and plant as determined with 15N. *Soil Science Society of America Journal* 68(2): 637-643. DOI: 10.2136/sssaj2004.6370.

Chee-Sanford, J. C., Mackie, R. I., Koike, S., Krapac, I. G., Lin, Y. F., Yannarell, A. C., Maxwell, S. and Aminov, R. I. (2009). Fate and transport of antibiotic residues and antibiotic resistance genes following land application of manure waste. *Journal of Environmental Quality* 38(3): 1086-1108. DOI: 10.2134/jeq2008.0128.

DeVries, S. L. and Zhang, P. (2016). Antibiotics and the terrestrial nitrogen cycle: a review. *Current Pollution Reports* 2(1): 51-67. DOI: 10.1007/s40726-016-0027-3.

Dolliver, H., Gupta, S. and Noll, S. (2008). Antibiotic degradation during manure composting. *Journal of Environmental Quality* 37(3): 1245–1253. DOI: 10.2134/jeq2007.0399.

Dou, Z., Galligan, D. T., Allshouse, R. D., Toth, J. D., Ramberg, C. F. and Ferguson, J. D. (2001). Manure sampling for nutrient analysis: variability and sampling efficacy. *Journal of Environmental Quality* 30(4): 1432–1437. DOI: 10.2134/jeq2001.3041432x.

Dyke, G. V. (1993). *John Lawes of Rothamsted: Pioneer of Science, Farming and Industry*. Harpenden, UK: Hoos Press.

Edesi, L., Järvan, M., Noormets, M., Lauringson, E., Adamson, A. and Akk, E. (2012). The importance of solid cattle manure application on soil microorganisms in organic and conventional cultivation. *Acta Agriculturae Scandinavica, Section B - Soil and Plant Science* 62(7): 583–594. DOI: 10.1080/09064710.2012.678380.

Edmeades, D. C. (2003). The long-term effects of manures and fertilisers on soil productivity and quality: a review. *Nutrient Cycling in Agroecosystems* 66(2): 165–180. DOI: 10.1023/A:1023999816690.

FAOSTAT (2022). Crops and livestock products: producing nimals/slaughtered. Available at: https://www.fao.org/faostat/en/#data/QCL (accessed 18 March 2022).

Font-Palma, C. (2019). Methods for the treatment of cattle manure–a review. *C* 5(2): C 5. DOI: 10.3390/c5020027.

Forsberg, K. J., Reyes, A., Wang, B., Selleck, E. M., Sommer, M. O. and Dantas, G. (2012). The shared antibiotic resistome of soil bacteria and human pathogens. *Science* 337(6098): 1107–1111. DOI: 10.1126/science.1220761.

Goss, M. J., Barry, D. A. J. and Rudolph, D. L. (1998). Contamination in Ontario farmstead domestic wells and its association with agriculture: 1. Results from drinking water wells. *Journal of Contaminant Hydrology* 32(3–4): 267–293. DOI: 10.1016/S0169-7722(98)00054-0.

Goss, M. J., Curnoe, W. E., Beauchamp, E. G., Smith, P. S., Nunn, B. D. C., and Barry, D. A. J. (1995). *An Investigation into the Management of Manure-Nitrogen to Safeguard the Quality of Groundwater*. COESA Report No.: LMAP (vols. 013/95). London, Ontario, Canada: Research Branch Agriculture and Agri-Food Canada, Pest Management Research Centre.

Goss, M. J. and Richards, C. (2008). Development of a risk-based index for source water protection planning, which supports the reduction of pathogens from agricultural activity entering water resources. *Journal of Environmental Management* 87(4): 623–632.

Goss, M. J., Tubeileh, A. and Goorahoo, D. (2013). A review of the use of organic amendments and the risk to human health. *Advances in Agronomy* 120: 275–379.

Gupta, U. C., Wu, K. and Liang, S. (2008). Micronutrients in soils, crops, and livestock. *Earth Science Frontiers* 15(5): 110–125. DOI: 10.1016/S1872-5791(09)60003-8.

Hao, X. and He, Z. (2020). Pelletizing animal manures for on- and off-farm use. In: *ASA Special Publication 67*. *ASA Special Publications* Waldrip, H. M., Pagliari, H. and He, Z. (Eds). American Society of Agronomy, Soil Science Society of America, Madison, WI, pp. 323–344.

Hara, M. (2001). Fertilizer pellets made from composted livestock manure. Agriculture Research Division, Mie Prefectural Science and Technology Promotion Center, Mie, Japan.

He, Z., Pagliari, P. H. and Waldrip, H. M. (2016). Applied and environmental chemistry of animal manure: a review. *Pedosphere* 26(6): 779–816. DOI: 10.1016/S1002-0160(15)60087-X.

Jahne, M. A., Rogers, S. W., Holsen, T. M., Grimberg, S. J. and Ramler, I. P. (2015). Emission and dispersion of bioaerosols from dairy manure application sites: human health risk assessment. *Environmental Science and Technology*. American Chemical Society 49(16): 9842–9849. DOI: 10.1021/acs.est.5b01981.

Janssen, B. H. (1996). Nitrogen mineralization in relation to C:N ratio and decomposability of organic materials. *Plant and Soil* 181(1): 39–45. DOI: 10.1007/978-94-011-5450-5_13.

Jenkinson, D. S. (1991). The Rothamsted long-term experiments: are they still of use? *Agronomy Journal* 83(1): 2–10. DOI: 10.2134/agronj1991.00021962008300010008x.

Jjemba, P. K. (2002). The effect of chloroquine, quinacrine, and metronidazole on both soybean plants and soil microbiota. *Chemosphere* 46(7): 1019–1025. DOI: 10.1016/S0045-6535(01)00139-4.

Johnston, A. E., Poulton, P. R. and Coleman, K. (2009). Soil organic matter: its importance in sustainable agriculture and carbon dioxide fluxes. *Advances in Agronomy Sparks DL* 101: 1–57. DOI: 10.1016/S0065-2113(08)00801-8.

Khaleel, R., Reddy, K. R. and Overcash, M. R. (1981). Changes in soil physical properties due to organic waste applications: a review. *Journal of Environmental Quality* 10(2): 133–141. DOI: 10.2134/jeq1981.00472425001000020002x.

Kinney, C. A., Furlong, E. T., Kolpin, D. W., Burkhardt, M. R., Zaugg, S. D., Werner, S. L., Bossio, J. P. and Benotti, M. J. (2008). Bioaccumulation of pharmaceuticals and other anthropogenic waste indicators in earthworms from agricultural soil amended with biosolid or swine manure. *Environmental Science and Technology*. American Chemical Society 42(6): 1863–1870. DOI: 10.1021/es702304c.

Kirchmann, H. (1985). Losses, plant uptake and utilisation of manure nitrogen during a production cycle [cattle manure, poultry manure, NH3 volatilisation, denitrification, nonsymbiotic N2 fixation]. *Acta Agriculturae Scandinavica* 24: 1–77.

Kirchmann, H. and Lundvall, A. (1993). Relationship between N immobilization and volatile fatty acids in soil after application of pig and cattle slurry. *Biology and Fertility of Soils* 15(3): 161–164. DOI: 10.1007/BF00361605.

Kleinman, P. J. A., Sharpley, A. N., McDowell, R. W., Flaten, D. N., Buda, A. R., Tao, L., Bergstrom, L. and Zhu, Q. (2011). Managing agricultural phosphorus for water quality protection: principles for progress. *Plant and Soil* 349(1–2): 169–182. DOI: 10.1007/s11104-011-0832-9.

Komiyama, T. and Ito, T. (2019). The characteristics of phosphorus in animal manure composts. *Soil Science and Plant Nutrition*. Taylor & Francis 65(3): 281–288. DOI: 10.1080/00380768.2019.1615384.

Köpke, U., Krämer, J. and Leifert, C. (2007). Pre-harvest strategies to ensure the microbiological safety of fruit and vegetables from manure-based production systems. In: *Handbook of Organic Food Safety and Quality* (1st edn.). Cambridge, England: Woodhead Publishing, pp. 413–429. Available at: https://www.sciencedirect.com/science/article/pii/B9781845690106500196?via%3Dihub (accessed 31 January 2022).

Lal, R. (2006). Enhancing crop yields in the developing countries through restoration of the soil organic carbon pool in agricultural lands. *Land Degradation and Development* 17(2): 197–209. DOI: 10.1002/ldr.696.

Lampkin, N. (1990). *Organic farming*. Ipswich, UK: Farming Press Books. Available at: https://www.amazon.com/Organic-Farming-Nicolas-Lampkin/dp/0852361912 (accessed 25 July 2021).

Larney, F. J., Olson, A. F., DeMaere, P. R., Handerek, B. P. and Tovell, B. C. (2008). Nutrient and trace element changes during manure composting at four southern Alberta feedlots. *Canadian Journal of Soil Science*. 88(1): 45-59. DOI: 10.4141/CJSS07044.

Lazicki, P., Geisseler, D. and Lloyd, M. (2020). Nitrogen mineralization from organic amendments is variable but predictable. *Journal of Environmental Quality* 49(2): 483–495. DOI: 10.1002/jeq2.20030.

Liang, Q., Chen, H., Gong, Y., Fan, M., Yang, H., Lal, R. and Kuzyakov, Y. (2012). Effects of 15 years of manure and inorganic fertilizers on soil organic carbon fractions in a wheat-maize system in the North China Plain. *Nutrient Cycling in Agroecosystems* 92(1): 21–33. DOI: 10.1007/s10705-011-9469-6.

Liu, T., Wang, M., Awasthi, M. K., Chen, H., Awasthi, S. K., Duan, Y. and Zhang, Z. (2020). Measurement of cow manure compost toxicity and maturity based on weed seed germination. *Journal of Cleaner Production* 245: 118894. DOI: 10.1016/j.jclepro.2019.118894.

Lorimor, J., Powers, W. and Sutton, A. (2004). *Manure Characteristics* (2nd edn.). Manure Management Systems 1. Ames, IA: Iowa State University. Available at: https://store.extension.iastate.edu/product/Manure-Characteristics-Manure-Management-System-Series (accessed 23 February 2022).

Lovarelli, D., Conti, C., Finzi, A., Bacenetti, J. and Guarino, M. (2020). Describing the trend of ammonia, particulate matter and nitrogen oxides: the role of livestock activities in northern Italy during Covid-19 quarantine. *Environmental Research* 191: 110048. DOI: 10.1016/j.envres.2020.110048.

Lucas, S. D. and Jones, D. L. (2006). Biodegradation of estrone and 17 β-estradiol in grassland soils amended with animal wastes. *Soil Biology and Biochemistry* 38(9): 2803–2815. DOI: 10.1016/j.soilbio.2006.04.033.

Lucas, S. D. and Jones, D. L. (2009). Urine enhances the leaching and persistence of estrogens in soils. *Soil Biology and Biochemistry* 41(2): 236–242. DOI: 10.1016/j.soilbio.2008.10.022.

Ludwig, B., Geisseler, D., Michel, K., Joergensen, R. G., Schulz, E., Merbach, I., Raupp, J., Rauber, R., Hu, K., Niu, L. and Liu, X. (2011). Effects of fertilization and soil management on crop yields and carbon stabilization in soils. A review. *Agronomy for Sustainable Development* 31(2): 361–372. DOI: 10.1051/agro/2010030.

Marshall, S. B., Wood, C. W., Braun, L. C., Cabrera, M. L., Mullen, M. D. and Guertal, E. A. (1998). Ammonia volatilization from tall fescue pastures fertilized with broiler litter. *Journal of Environmental Quality* 27(5): 1125–1129. DOI: 10.2134/jeq1998.00472425002700050018x.

Meinen, R. (2016). Reoccurring themes of manure gas deaths. Available at: https://extension.psu.edu/reoccurring-themes-of-manure-gas-deaths (accessed 18 March 2022).

Meisinger, J. J. and Jokela, W. E. (2000). Ammonia volatilization from dairy and poultry manure. In: *Managing Nutrients and Pathogens From Animal Agriculture* Extension Educators Conference for Nutrient Management Consultants (ed.). Natural Resource, Agriculture, and Engineering Service, Camp Hill, PA, p. 21.

Millner, P. D. (2009). Bioaerosols associated with animal production operations. *Bioresource Technology* 100(22): 5379–5385. DOI: 10.1016/j.biortech.2009.03.026.

Mohanty, M., Reddy, K. S., Probert, M. E., Dalal, R. C., Rao, A. S. and Menzies, N. W. (2011). Modelling N mineralization from green manure and farmyard manure from a laboratory incubation study. *Ecological Modelling* 222(3): 719–726. DOI: 10.1016/j.ecolmodel.2010.10.027.

Möller, K. and Müller, T. (2012). Effects of anaerobic digestion on digestate nutrient availability and crop growth: a review. *Engineering in Life Sciences* 12(3): 242–257. DOI: 10.1002/elsc.201100085.

Munyankusi, E., Gupta, S. C., Moncrief, J. F. and Berry, E. C. (1994). Earthworm macropores and preferential transport in a long-term manure applied Typic Hapludalf. *Journal of Environmental Quality* 23(4): 773–784. DOI: 10.2134/jeq1994.00472425002300040023x.

Nahar, M. S., Grewal, P. S., Miller, S. A., Stinner, D., Stinner, B. R., Kleinhenz, M. D., Wszelaki, A. and Doohan, D. (2006). Differential effects of raw and composted manure on nematode community, and its indicative value for soil microbial, physical and chemical properties. *Applied Soil Ecology* 34(2–3): 140–151. DOI: 10.1016/j.apsoil.2006.03.011.

Obi, M. E. and Ebo, P. O. (1995). The effects of organic and inorganic amendments on soil physical properties and maize production in a severely degraded sandy soil in southern Nigeria. *Bioresource Technology* 51(2–3): 117–123. DOI: 10.1016/0960-8524(94)00103-8.

OMAFRA (2013). Available nutrients and value for manure from various livestock types. *Extension Flyer*. Toronto, Ontario, Canada: Ontario Ministry of Agriculture and Food and the Ministry of Rural Affairs, pp. 13–043. Available at: http://www.omafra.gov.on.ca/english/crops/facts/13-043.htm.

OMAFRA (2018). *Soil Fertility Handbook* (3rd edn.). OMAFRA Publication 611. Toronto, Ontario, Canada: Ontario Ministry of Agriculture, Food and Rural Affairs (OMAFRA). Available at: http://www.omafra.gov.on.ca/english/crops/pub611/pub611.pdf

Ortiz-Santaliestra, M. E., Marco, A. and Lizana, M. (2011). Realistic levels of a fertilizer impair Iberian newt embryonic development. *Herpetologica* 67(1): 1–9. DOI: 10.1655/HERPETOLOGICA-D-10-0001.1.

Ozlu, E. and Kumar, S. (2018). Response of soil organic carbon, pH, electrical conductivity, and water stable aggregates to long-term annual manure and inorganic fertilizer. *Soil Science Society of America Journal* 82(5): 1243–1251. DOI: 10.2136/sssaj2018.02.0082.

Ozlu, E., Sandhu, S. S., Kumar, S. and Arriaga, F. J. (2019). Soil health indicators impacted by long-term cattle manure and inorganic fertilizer application in a corn-soybean rotation of South Dakota. *Scientific Reports*. Nature Publishing Group 9(1): 11776. DOI: 10.1038/s41598-019-48207-z.

Paez-Rubio, T., Ramarui, A., Sommer, J., Xin, H., Anderson, J. and Peccia, J. (2007). Emission rates and characterization of aerosols produced during the spreading of dewatered class B biosolids. *Environmental Science and Technology* 41(10): 3537–3544. DOI: 10.1021/es061786p.

Pagliari, P. H. and Laboski, C. A. M. (2012). Investigation of the inorganic and organic phosphorus forms in animal manure. *Journal of Environmental Quality* 41(3): 901–910. DOI: 10.2134/jeq2011.0451.

Pagliari, P. H., Wilson, M., Waldrip, H. M. and He, Z. (2019). Nitrogen and phosphorus characteristics of beef and dairy manure. In: *ASA Special Publication 67. ASA*

Special Publications Waldrip, H. M., Pagliari, H. and He, Z. (Eds). American Society of Agronomy, Soil Science Society of America, Madison, WI, pp. 45–62.

Parham, J., Deng, S., Raun, W. R. and Johnson, G. (2002). Long-term cattle manure application in soil. I. *Biology and Fertility of Soils* 35(5): 328–337. DOI: 10.1007/s00374-002-0476-2.

Parham, J. A., Deng, S. P., Da, H. N., Sun, H. Y. and Raun, W. R. (2003). Long-term cattle manure application in soil. II. Effect on soil microbial populations and community structure. *Biology and Fertility of Soils* 38(4): 209–215. DOI: 10.1007/s00374-003-0657-7.

Park, J., Kang, T., Jin, S., Heo, Y., Kim, K., Lee, K., Tsai, P. and Yoon, C. (2016). Asphyxiation incidents by hydrogen sulfide at manure storage facilities of swine livestock farms in Korea. *Journal of Agromedicine* 21(2): 144–148. DOI: 10.1080/1059924X.2016.1141735.

Patni, N. K. and Jui, P. Y. (1987). Changes in solids and carbon content of dairy-cattle slurry in farm tanks. *Biological Wastes* 20(1): 11–34. DOI: 10.1016/0269-7483(87)90081-4.

Paul, J. W. and Beauchamp, E. G. (1994). Short-term nitrogen dynamics in soil amended with fresh and composted cattle manures. *Canadian Journal of Soil Science* 74(2): 147–155. DOI: 10.4141/cjss94-022.

Peacock, A. D., Mullen, M. D., Ringelberg, D. B., Tyler, D. D., Hedrick, D. B., Gale, P. M. and White, D. C. (2001). Soil microbial community responses to dairy manure or ammonium nitrate applications. *Soil Biology and Biochemistry* 33(7–8): 1011–1019. DOI: 10.1016/S0038-0717(01)00004-9.

Pérez-Valera, E., Kyselková, M., Ahmed, E., Sladecek, F. X. J., Goberna, M. and Elhottová, D. (2019). Native soil microorganisms hinder the soil enrichment with antibiotic resistance genes following manure applications. *Scientific Reports* 9(1): 6760. DOI: 10.1038/s41598-019-42734-5.

Plaimart, J., Acharya, K., Mrozik, W., Davenport, R. J., Vinitnantharat, S. and Werner, D. (2021). Coconut husk biochar amendment enhances nutrient retention by suppressing nitrification in agricultural soil following anaerobic digestate application. *Environmental Pollution* 268(A): 115684. DOI: 10.1016/j.envpol.2020.115684.

Qian, P. and Schoenau, J. J. (2002). Availability of nitrogen in solid manure amendments with different C:N ratios. *Canadian Journal of Soil Science* 82(2): 219–225. DOI: 10.4141/S01-018.

Ramaswamy, J., Prasher, S. O., Patel, R. M., Hussain, S. A. and Barrington, S. F. (2010). The effect of composting on the degradation of a veterinary pharmaceutical. *Bioresource Technology* 101(7): 2294–2299. DOI: 10.1016/j.biortech.2009.10.089.

Raven, K. P. and Hossner, L. R. (1993). Phosphorus desorption quantity-intensity relationships in soils. *Soil Science Society of America Journal* 57(6): 1501–1508. DOI: 10.2136/sssaj1993.03615995005700060018x.

Rayne, N. and Aula, L. (2020). Livestock manure and the impacts on soil health: a review. *Soil Systems*. Multidisciplinary Digital Publishing Institute 4(4): 64. DOI: 10.3390/soilsystems4040064.

Reuters (2000). Liquid manure tank claims three lives. Available at: https://www.iol.co.za/news/world/liquid-manure-tank-claims-three-lives-47492 (accessed 18 March 2022).

Rodríguez-Kábana, R. (1986). Organic and inorganic nitrogen amendments to soil as nematode suppressants. *Journal of Nematology* 18(2): 129–134.

Rynk, R., van de Kamp, M., Willson, G. B., et al. (1992). On-farm composting handbook. *NRAES 54*. Ithaca, NY: Northeast Regional Agricultural Engineering Service (NRAES). Available at: https://ecommons.cornell.edu/handle/1813/67142 (accessed 17 March 2022)

Salaheen, S., Peng, M. and Biswas, D. (2016). Ecological dynamics of Campylobacter in integrated mixed crop-livestock farms and its prevalence and survival ability in post-harvest products. *Zoonoses and Public Health* 63(8): 641–650. DOI: 10.1111/zph.12279.

Scarsbrook, C. E. (1965). Nitrogen availability. In: *Agronomy Monographs 10*. Madison, WI: American Society of Agronomy, pp. 481–502. DOI: 10.2134/agronmonogr10.c13.

Schröder, J. (2005). Revisiting the agronomic benefits of manure: a correct assessment and exploitation of its fertilizer value spares the environment. *Bioresource Technology*. The 10th International Conference on Recycling of Agricultural, Municipal and Industrial Residues in Agriculture 96(2): 253–261. DOI: 10.1016/j.biortech.2004.05.015.

Sharpley, A. N. and Smith, S. J. (1989). Mineralization and leaching of phosphorus from soil incubated with surface-applied and incorporated crop residue. *Journal of Environmental Quality* 18(1): 101–105. DOI: 10.2134/jeq1989.00472425001800010018x.

Simonetti, G., Francioso, O., Nardi, S., Berti, A., Brugnoli, E. and Francesco Morari, E. L. (2012). Characterization of humic carbon in soil aggregates in a long-term experiment with manure and mineral fertilization. *Soil Science Society of America Journal* 76(3): 880–890. DOI: 10.2136/sssaj2011.0243.

Sims, J. T. (1986). Nitrogen transformations in a poultry manure amended soil: temperature and moisture effects. *Journal of Environmental Quality* 15(1): 59–63. DOI: 10.2134/jeq1986.00472425001500010014x.

Søgaard, H. T., Sommer, S. G., Hutchings, N. J., Huijsmans, J. F. M., Bussink, D. W. and Nicholson, F. (2002). Ammonia volatilization from field-applied animal slurry–the ALFAM model. *Atmospheric Environment* 36(20): 3309–3319. DOI: 10.1016/S1352-2310(02)00300-X

Sommer, S. G. and Knudsen, L. (2021). Impact of Danish livestock and manure management regulations on nitrogen pollution, crop production, and economy. *Frontiers in Sustainability* 2: 658231. DOI: 10.3389/frsus.2021.658231.

Sommerfeldt, T. G. and Chang, C. (1987). Soil-water properties as affected by twelve annual applications of cattle feedlot manure. *Soil Science Society of America Journal* 51(1): 7–9. DOI: 10.2136/sssaj1987.03615995005100010002x.

Stumpe, B. and Marschner, B. (2010). Dissolved organic carbon from sewage sludge and manure can affect estrogen sorption and mineralization in soils. *Environmental Pollution* 158(1): 148–154. DOI: 10.1016/j.envpol.2009.07.027.

Szogi, A. A., Vanotti, M. B. and Ro, K. S. (2015). Methods for treatment of animal manures to reduce nutrient pollution prior to soil application. *Current Pollution Reports* 1(1): 47–56. DOI: 10.1007/s40726-015-0005-1.

Tejada, M., Hernandez, M. T. and Garcia, C. (2006). Application of two organic amendments on soil restoration: effects on the soil biological properties. *Journal of Environmental Quality* 35(4): 1010–1017. DOI: 10.2134/jeq2005.0460.

Thangarajan, R., Bolan, N. S., Naidu, R. and Surapaneni, A. (2015). Effects of temperature and amendments on nitrogen mineralization in selected Australian soils.

Environmental Science and Pollution Research International 22(12): 8843–8854. DOI: 10.1007/s11356-013-2191-y.

Tiquia, S. M., Richard, T. L. and Honeyman, M. S. (2002). Carbon, nutrient, and mass loss during composting. *Nutrient Cycling in Agroecosystems* 62(1): 15–24. DOI: 10.1023/A:1015137922816.

Tubeileh, A. M. and Stephenson, G. T. (2020). Soil amendment by composted plant wastes reduces the *Verticillium dahliae* abundance and changes soil chemical properties in a bell pepper cropping system. *Current Plant Biology* 22: 100148. DOI: 10.1016/j.cpb.2020.100148.

Tubeileh, A. M. and Thomas, R. J. (2022). The management of soil systems in arid and semi-arid climates. In: *Biological Approaches to Sustainable Soil Systems* (2nd edn.) Uphoff, N. (Ed.). CRC Press, Boca Raton, FL. In press.

Turner, B. L. and Leytem, A. B. (2004). Phosphorus compounds in sequential extracts of animal manures: chemical speciation and a novel fractionation procedure. *Environmental Science and Technology* 38(22): 6101–6108.

Van Faassen, H. G. and Van Dijk, H. (1987). Manure as a source of nitrogen and phosphorus in soils. In: *Animal Manure on Grassland and Fodder Crops. Fertilizer or Waste? Developments in Plant and Soil Sciences*. Dordrecht, Netherlands: Springer, pp. 27–45. Available at: https://link.springer.com/chapter/10.1007/978-94-009-3659-1_3 (accessed 26 July 2021).

Vogtmann, H. and Besson, J. M. (1978). European composting methods: treatment and use of farm yard manure and slurry. *Compost Science/Land Util.* 19: 1–23.

Wesseling, J. G., Stoof, C. R., Ritsema, C. J., Oostindie, K. and Dekker, L. W. (2009). The effect of soil texture and organic amendment on the hydrological behaviour of coarse-textured soils. *Soil Use and Management* 25(3): 274–283. DOI: 10.1111/j.1475-2743.2009.00224.x.

Wild, A. (1988). Historical. In: *Russell's Soil Conditions and Plant Growth* (11th edn.). Harlow, Essex, UK: Longman Scientific & Technical, pp. 1–30.

Xuan, R., Blassengale, A. A. and Wang, Q. (2008). Degradation of estrogenic hormones in a silt loam soil. *Journal of Agricultural and Food Chemistry* 56(19): 9152–9158. DOI: 10.1021/jf8016942.

Yousif Abdallah, A. Y., Shi, Z. J., Luo, Y. S., Hou, W. T., Yang, X., and Wang, R. L. (2022). Effects of different additives and aerobic composting factors on heavy metal bioavailability reduction and compost parameters: A meta-analysis. *Environmental Pollution* 307:119549.

Zhang, H., Vocasek, F., Antonangelo, J. and Gillespie, C. (2020). Temporal changes of manure chemical compositions and environmental awareness in the southern great plains. In: *ASA Special Publication 67. ASA Special Publications* Waldrip, H.M., Pagliari, H. and He, Z. (Eds). American Society of Agronomy, Soil Science Society of America, Madison, WI, pp. 15–26.

Zhang, Q.-C., Shamsi, I. H., Xu, D.-T., Wang, G., Lin, X., Jilani, G., Hussain, N. and Chaudhry, A. N. (2012). Chemical fertilizer and organic manure inputs in soil exhibit a vice versa pattern of microbial community structure. *Applied Soil Ecology* 57: 1–8. DOI: 10.1016/j.apsoil.2012.02.012.

Zhang, Y., Leung, J. Y. S., Zhang, Y., Cai, Y., Zhang, Z. and Li, K. (2021). Agricultural activities compromise ecosystem health and functioning of rivers: insights from multivariate and multimetric analyses of macroinvertebrate assemblages. *Environmental Pollution* 275: 116655. DOI: 10.1016/j.envpol.2021.116655.

Chapter 11

Assessing the effect of biosolids on soil health

Fiona A. Nicholson, Anne Bhogal, Alison Rollett and John R. Williams, ADAS, UK

1 Introduction

Treated sewage sludge, also known as biosolids, has been recycled to agricultural land for many decades in Europe, the USA and other countries throughout the world. Recent data from the UK, for example, suggest that around 78% of the 3.6 million tonnes (fresh weight) of biosolids produced each year is spread on agricultural land (Black et al., 2016). Currently, the commonly held view is that application to agricultural land is the best practical environmental option for recycling biosolids, the alternatives being incineration, landfill or disposal at sea. In many countries, recycling of biosolids to agricultural land is a well-controlled activity, with a variety of regulations and codes of practice in place to ensure that it is undertaken safely, whilst in others (e.g. Switzerland), the practice is banned altogether.

Biosolids are applied to agricultural land because they are an important source of crop-available nitrogen (N) and phosphorus (P), as well as other major and minor plant nutrients, and with some conditioning processes, they can have value as a liming material. Because they contain valuable quantities of stable organic matter (OM), biosolids can help to replenish or maintain soil OM

http://dx.doi.org/10.19103/AS.2021.0094.15

levels, influencing a range of soil properties that are important for soil health, carbon (C) storage and sustainable crop production.

Despite these benefits, it is widely recognised that biosolids often contain chemicals or other agents which may present a potential risk to the environment or human health. There is good evidence from a substantial body of international research over many years, on the effects of various chemical (e.g. potentially toxic elements (PTEs), persistent organic pollutants) and microbiological agents on the physical, chemical and biological properties of soil. Previous assessments have concluded that the risks associated with these are manageable to an acceptable level. However, more recently, researchers have begun to investigate 'emerging' organic chemicals, as well as other 'new' and potentially harmful contaminants such as microplastics, synthetic fibres, nanoparticles and antimicrobial resistance genes. Here, the evidence and supporting data are less robust, and there are a growing number of scientific papers being published that are raising concerns about the sustainability of the current biosolids recycling practices in terms of the effects these agents may have on the soil and in the wider environment.

In this chapter, we review the evidence for both the beneficial and potentially harmful effects of biosolids application on soil health and highlight areas where further research is still required.

2 Effects on soil nutrient supply and fertility

Biosolids have been demonstrated to be a valuable source of both immediately crop-available and slow-release nutrients (Table 1); indeed this is one of the main reasons why many farmers are keen to apply biosolids to their soils.

2.1 Nitrogen

Most agricultural soils do not contain enough plant-available nitrogen (N) to meet the needs of a crop throughout the growing season, and farmers need to supply additional N in the form of mineral fertilisers. Biosolids contain readily available N (RAN, i.e. N which is present as ammonium-N) and this can be used to replace some of the required mineral fertiliser N (Rouch et al., 2011; Lu et al., 2012). Typically, the total N content of biosolids (digested cake) in the UK is around 11 kg N/t fresh weight (Table 1), with 10–20% of this available to the next crop depending on soil type, application timing and whether or not the biosolids are soil-incorporated (AHDB, 2020).

Biosolids also contain N in organic forms which is not immediately available for crop uptake, but can help to build longer term soil N stocks (Rigby et al., 2009). Potentially mineralisable nitrogen (PMN) is a biological measure of the soil's capacity to supply N through the mineralisation of soil organic N reserves to ammonium-N, which can consequently be converted to nitrate-N

Table 1 Typical nutrient contents of UK biosolids products on a fresh weight basis (AHDB, 2020)

Biosolids type	Dry matter (%)	Total nitrogen (kg/t)	Total phosphate (kg P_2O_5/t)	Total potash (kg K_2O/t)	Total sulphur (kg SO_3/t)	Total magnesium (kg MgO/t)
Digested cake	25	11	11	0.6	8.2	1.6
Thermally dried	95	40	55	2.0	23	6.0
Lime stabilised	40	8.5	7	0.8	7.4	2.4
Composted	60	11	10	3.0	6.1	2.0

(by nitrification processes). PMN provides an indication of the amount of N that will be released in the short to medium term and significantly impacts fertiliser recommendations. There is good evidence that biosolids organic N can be mineralised to plant-available forms, contributing to crop N requirements and reduced mineral N fertiliser costs. For example, Kao et al. (2006) reported that in soil treated with biosolids, the quantity of mineralised N increased by approximately five-fold, and the rate of N mineralisation was increased by about one-fold compared to untreated soil. Notably, the addition of PTEs with biosolids impaired the mineralisation process, particularly lead (see Section 6.1). A review by Rigby et al. (2016) found that the amount of mineralisable N in biosolids depends on the treatment process (i.e. the extent of biological stabilisation) and varies across climatic regions. They concluded that some international fertiliser recommendations may underestimate the amount of mineralisable N in biosolids and hence their fertiliser N value. More work is therefore needed, particularly from field studies, to better ascertain both the quantity and timing of N mineralisation in different biosolids products, climatic regions and soil types.

2.2 Phosphorus

Phosphorus (P) is also an essential plant nutrient but unlike N, the P applied in organic materials, such as biosolids, is generally immobile in soils, which can hold large quantities in forms that are available for crop uptake over several years. As soil fertility falls, organic P forms can be mobilised by soil microbes to provide a source of P for crop growth; biosolids can therefore act as both a short-term and a long-term source of soil P for plant uptake (Lu et al., 2012; Torri et al., 2017). However, care must be taken to avoid excessive levels of soil P (i.e. Olsen P > 60 mg l^{-1}), as this can pose a risk of phosphate losses to surface waters, contributing to eutrophication (Withers et al., 2016). This issue is discussed further in Section 7.2.

2.3 Other nutrients and soil fertility effects

Biosolids only contain small amounts of potash, but can make a significant contribution to crop sulphur (S) requirements (Sagoo et al., 2014), with digested cake typically containing around 8 kg SO_3/t fresh weight (AHDB, 2020). This is important because most soils, especially chalky or sandy soils, store very little S. The continued decline in atmospheric deposition of S due to reduced atmospheric emissions, coupled with our increased reliance on low-S fertilisers (Blair, 2009), means that it is likely that a wider range of crops and soil types will be at risk of S deficiency. As a result, the S in biosolids products may become more valued by farmers. This was demonstrated by Nicholson et al. (2018) who reported that biosolids applications significantly increased S concentrations in breadmaking-quality wheat; enhanced grain S has previously been shown to be related to improved bread quality in terms of dough elasticity and loaf volume (Zhao et al., 1999a,b), thus increasing the commercial value of the grain.

Usually the presence of PTEs such as zinc (Zn), copper (Cu) and nickel (Ni) in biosolids is considered problematic in terms of their consequences for soil health, with many studies reporting the increased uptake of PTEs by various crops following biosolids applications (see Section 6.1). However, some PTEs are also essential micronutrients, and low soil concentrations can lead to deficiency-related problems. In global terms, Zn deficiency is very common in cereal-cultivated soils, particularly those with high pH and high calcite and OM contents (Alloway, 2009). This can result in decreased growth and grain yield, leading to economic losses for farmers and significant decreases in the nutritional quality of grain. There is an argument that the fortification of cereal grain with additional Zn via biosolids application could increase its nutritional value (McGrath et al., 2012). Copper deficiency in crops is associated with sandy soils, shallow soils over chalk and soils high in OM, and so biosolids applications which supply additional Cu may be a potential tool for addressing deficiency in these areas. Nickel can also increase the growth and yield of some crops (e.g. cereals, legumes), although deficiency in animals and humans is unlikely.

Biosolids applications can also increase the soil cation exchange capacity (CEC) because the additional supply of OM increases the electrostatic surface charge available to attract and hold cations (Nicholson et al., 2018; Lu et al., 2012); this translates into improved soil nutrient retention and, hence, fertility. Finally, lime-stabilised biosolids have value as liming materials, with a neutralising value typically in the range of 2–6% per tonne fresh weight (AHDB, 2020). They have been shown to significantly reduce soil acidity (Guo et al., 2012), which can have an important influence on nutrient availability and crop growth.

3 Effects on soil organic matter and carbon sequestration

Soil organic matter (SOM) is a key indicator of soil health; loss of SOM (due to changes in management, land-use and climate) is one of the most important threats facing our soils and can contribute to global warming (Lal, 2016). Biosolids are typically made up of 40–70% OM, with the organic C content ranging from 20% to 50%; the OM has usually undergone some degree of stabilisation through anaerobic or aerobic digestion before being applied to the land (Torri et al., 2014).

Biosolids applications to land have been extensively studied and shown to increase SOM (and hence soil organic carbon (SOC)) across a range of soil types, climates and cropping systems (e.g. Nicholson et al., 2018; Toffey and Brown, 2020). For example, at a site in Texas (USA), Jin et al. (2015)

found that total SOC stocks (0–45 cm) increased in relation to the amount of biosolids applied; in fields receiving 20 years of biosolids applications SOC stocks were 36% higher than those which had received eight years of applications at the same rate (20 t ha^{-1} yr^{-1}). At two sites in Australia, biosolids application of 70 t ha^{-1} increased the surface (0–15 cm) SOC by over 45%; this was attributed to the direct contribution from residual C in the biosolids as well as increased biomass production (Wijesekara et al., 2017). In their UK study, Powlson et al. (2012) reported an average increase in SOC of 180 kg C ha^{-1} yr^{-1} t^{-1} dry solids (ds) from digested biosolids applications (applied at c.8 t ds ha^{-1}), three times higher than the rate for farm manures (60 kg C ha^{-1} yr^{-1} t^{-1} ds).

In a comprehensive meta-analysis of soil C sequestration following biosolids application, Wijesekara et al. (2016) used datasets from 25 countries to examine the effect of time after biosolids application, and the cumulative amount of biosolids applied on the difference in SOC content relative to an untreated control. The short-term (one to three years) application of biosolids resulted in the highest SOC difference (2.5 g kg^{-1}), with a negative correlation observed between increasing time after biosolids application and SOC storage, probably reflecting the mineralisation of OM over time. The highest mean difference in SOC content of 2.8 g kg^{-1} was observed for cumulative biosolids applications of 201–250 t ha^{-1}, although these high rates are only likely to be used for site restoration purposes. Nevertheless, there was a positive correlation between the cumulative biosolids application rate (up to 201–250 t ha^{-1}) and the difference in SOC contents.

These results indicate that biosolids are a good source of stable OM for building up SOM levels, which has profound implications for soil physical properties (see Section 4) and overall soil health. Whether or not this increase in SOM constitutes genuine C sequestration is still a matter for debate, depending as it does on the alternative fate of the biosolids if not applied to land (e.g. incineration or landfill) and whether or not the increase in SOM is finite and reversible.

4 Effects on soil physical properties

Additions of OM are known to improve soil physical properties, such as soil structure and water holding capacity, that are important for maintaining soil health. OM also helps to bind soil mineral particles into aggregates, thus improving the stability of the soil and its resistance to erosive forces. As with other sources of OM (livestock manures, compost, etc.), additions of biosolids have been shown to improve a suite of soil physical properties, and many studies from across the globe have been published on the topic as reviewed by Lu et al. (2012). Some key aspects are outlined below.

Topsoil bulk density has a direct impact on a number of essential soil physical and biological processes including gas exchange, root access, infiltration rates and soil faunal activity and is usually a key measure in the assessment of soil health as an indicator of soil compaction (Merrington et al., 2006; Rickson et al., 2012). Bulk density tends to be inversely related to SOM (Newell-Price et al., 2012), such that Merrington et al. (2006) proposed higher 'trigger values' (i.e. the bulk density above which soil functions may be impaired) for soils with lower SOM content. Bulk density is directly related to total porosity, which is a measure of the space available within the soil for gas and water movement and for root development, and to air capacity which measures the amount of air space in a soil. Many previous studies have demonstrated that the long-term applications of biosolids can decrease soil bulk density (e.g. Cogger et al., 2013; Ozores-Hampton et al., 2011). However, Nicholson et al. (2018) found no statistically significant effect of biosolids treatments on bulk density, porosity or air capacity after 20 years of biosolids additions at 5 sites across the UK, with Jin et al. (2015) also finding that soil bulk density was little affected by long- (20 year) and mid- (eight year) term biosolids applications.

The moisture retention characteristics of soils are controlled by a number of physical properties that influence pore space distribution, particularly particle size distribution (i.e. the percentages of sand, silt and clay in the soil), OM content, stone content and bulk density. One of the most important benefits farmers ascribe to the recycling of biosolids to land is an increase in plant-available water supply. Indeed, drought is a major limiting factor for crop yields in some regions, particularly on sandy textured soils in low rainfall areas. Several studies have found that biosolids additions can increase the soil-available water capacity (AWC) on light/sandy soils (e.g. Ozores-Hampton et al., 2011). This can improve crop yields, particularly on low AWC sandy soils where vegetable crops are grown and reduce the requirement for irrigation, leading to substantial savings in operational costs.

Aggregate stability is a measure of the soil's ability to resist disruptive forces, which is a valuable criterion for distinguishing between soils with different degrees of erosivity. Nicholson et al. (2018) reported small but significant increases in structural stability following biosolids applications to a shallow chalk soil in the UK. Other studies have also reported increased water retention and aggregate stability after biosolids application, although very high application rates of up to 300 t ha^{-1} were sometimes used (Lu et al., 2012; Jin et al., 2015). The magnitude of the observed effects will clearly depend strongly on the type and quantity of the materials applied as well as on the characteristics of the receiving soil and climatic conditions (Jin et al., 2015). Nevertheless, these findings indicate that biosolids additions at normal agronomic rates can help to improve soil stability and play an important role in allowing more water to infiltrate soil more quickly, reducing the need for irrigation and helping to reduce peak flows into watercourses and hence reduce the risk of flooding.

5 Effects on soil biology

5.1 Soil microbial biomass

The soil microbial biomass, consisting of bacteria and fungi, is a measure of the quantity of the living component of the SOM. The microbial community is involved in the decomposition and transformation of OM and nutrients in soils, releasing carbon dioxide (CO_2) and plant-available nutrients. Soil properties (e.g. pH, clay and SOC) and environmental factors, such as temperature and moisture, can influence the size and activity of the microbial biomass. There have been numerous studies showing that biosolids can increase the size and activity of soil microbial biomass and alter the structure and diversity of the soil microbial community (e.g. Barbarick et al., 2004; Pawlett et al., 2015; Sullivan et al., 2006a,b; Zerzghi et al., 2010a,b). The findings from some of the more recent studies investigating the impacts of biosolids on the soil microbial biomass are outlined further.

Four years after application, Schlatter et al. (2019) found that some families of bacteria were more abundant in biosolids-treated soils than in those receiving synthetic fertilisers. Although bacteria associated with biosolids were reported to persist in the soil, potentially pathogenic taxa (*Salmonella*, *Clostridium*) were only present at low levels. The authors concluded that the biosolids had profound effects on soil bacterial communities both by introducing exogenous bacteria and by enriching the indigenous soil populations. Price et al. (2021) studied soils which had received annual applications of alkaline-stabilised biosolids at four rates over ten years, in comparison to a one-time application at ten years previously. The authors reported a rather complicated picture where bacterial community complexity and stability were increased by one-time low rates, but were decreased by both a single high rate and annual moderate rates of biosolids additions.

Early work in the UK (Anderson et al., 2008) reported on the effects of biosolids on the soil fungal community composition. Their data demonstrated that although an increase in SOM content altered fungal diversity and composition, increasing soil concentrations of cadmium, copper and zinc up to the current legislative limits had little additional effect. Schlatter et al. (2017) also reported that biosolids applications substantially altered fungal community composition, but not diversity, compared to synthetic fertilisers, with a small number of taxa highly enriched by biosolids applications. These shifts in fungal communities were correlated with differences in soil characteristics, especially C, N, and P concentrations, and persisted for at least three years after the initial biosolids application. The results suggest that biosolids can have strong and lasting impacts on soil fungal communities, probably due to their effects on soil nutrient supply, and selective impacts on a small number of fungi capable of utilising biosolids as a food source. Interestingly, Price et al. (2021) found that fungi and eukaryotes were less influenced by annual biosolids applications than was the bacterial community structure.

These studies illustrate how biosolids can alter the soil microbial community structure and diversity, with the effects depending on the timing and rate of the applications. However, the picture is complicated by the presence of antimicrobials and other contaminants that may be present in biosolids (e.g. PTEs, pharmaceuticals and nanomaterials) and which may affect the soil microbiome; this issue is discussed further in Section 6.

5.2 Soil macrofauna

Earthworms have a major influence on soil health due to their role in breaking down OM, improving soil structure and allowing water/oxygen to move through the soil profile. Also of importance are collembola. These abundant and ubiquitous soil microarthropods can affect the size and activity of the microbial community by selectively feeding on fungi and bacteria, breaking down OM and altering nutrient availability (Coulibaly et al., 2019).

In general, farming systems that return greater amounts of OM to the soil have higher earthworm populations (Scullion et al., 2002). However, there have sometimes been inconsistent results from field studies with respect to the effects of organic manures on earthworm populations. Kinney et al. (2012) suggested that this might be a result of variations between the applied materials in terms of both their nutrient and potentially toxic compound concentrations, as well as differences in soil characteristics.

There have been a number of studies which have looked at the effects of biosolids on soil macrofauna populations. For example in laboratory tests, Artuso et al. (2011) used five biosolids from different sewage works in Ireland, together with a sixth biosolid high in selenium (Se) obtained from the USA, to investigate the effects of different types and rates of biosolids applications on earthworms (*Eisenia fetida*) and collembola (*Folsomia candida*). They found that low biosolids application rates (2–5 t ha^{-1}) did not cause adult earthworm mortality, although higher rates (10–20 t ha^{-1}) did. At application rates up to 20 t ha^{-1}, all biosolids-treated soils had significantly fewer juvenile worms relative to the control. Increasing the rate from 2 t ha^{-1} to 10 t ha^{-1} had no effect on adult collembola numbers, but at 20 t ha^{-1} there were significantly fewer adults. There were significantly fewer juvenile collembola for biosolids applied at 2 t ha^{-1} compared with controls, and also when biosolids were applied at 5–20 t ha^{-1} relative to 2 t ha^{-1}. Some significant differences between biosolids types were observed, but in general, the negative effects were not related to biosolids PTE concentrations. The authors concluded that biosolids applied at low rates (about 2 t ha^{-1}) are unlikely to be detrimental to earthworms or adult collembola but could have a detrimental effect on collembola reproduction. Similarly, Coors et al. (2016) found only weak evidence for the negative long-term effects on soil fauna (nematodes, enchytraeids and earthworms) of biosolids applied at commercial rates at two sites in Canada. This finding is also supported by data from the UK (Nicholson et al., 2018), where the number and weight of earthworms was found to have approximately doubled where biosolids with low PTE content were applied (Fig. 1). Importantly, applications of biosolids with elevated concentrations of Zn and cadmium (Cd) did not decrease earthworm numbers relative to the untreated control.

Somewhat contradictory results were reported in a recent study from China (Li et al., 2020). Here, increases in soil PTE concentrations following nine years of biosolids applications (fresh domestic, dried domestic and fresh industrial sludges) in field conditions led to significant decreases in the total numbers of soil macrofauna including epigeic and endogenic collembola, nematodes and enchytraeid worms. Their analysis indicated that the soil total Zn concentration was the main factor affecting collembola abundance after biosolids amendment, with soil nutrient concentrations of less importance.

The evidence from these studies is rather mixed, but overall the results seem to suggest that biosolids, providing they have low PTE concentrations,

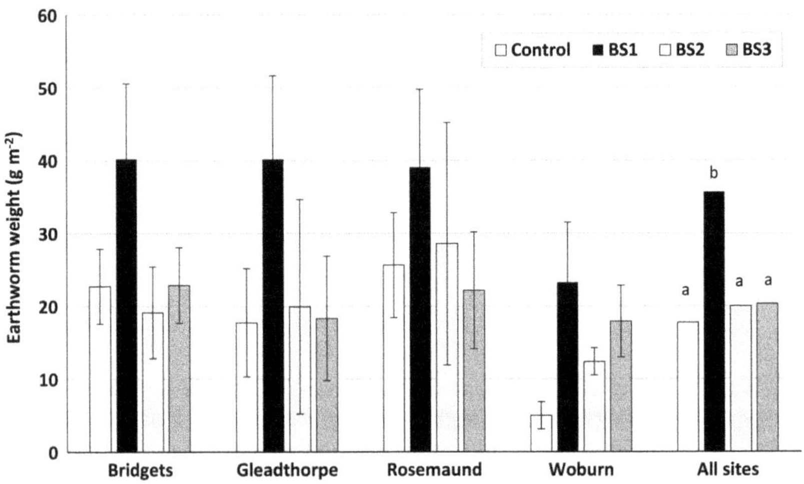

Figure 1 Effect of biosolids applications on earthworm weight (error bars show standard errors; columns labelled with different letters are significantly ($P < 0.05$) different from each other). BS1: low-metal concentration biosolids; BS2: biosolids with elevated Zn concentration; BS3: biosolids with elevated Cd and Cu concentrations. From Nicholson et al. (2018).

have few negative impacts on the soil macrofauna and could have beneficial effects as a result of the OM inputs providing an additional food source.

6 Potentially harmful effects on soil health

Much research has been undertaken on the potentially adverse effects that long-term biosolids recycling to agricultural land may have on soil health. There have been numerous reviews which summarise, integrate and critically assess the evidence from individual investigations on PTEs (also known as heavy metals) and organic chemical contaminants (OCCs) (e.g. Chambers et al., 2010a,b; Sutherland and Taylor, 2012; Rigby et al., 2015), pathogens (e.g. Gale and Stanfield, 2001; Gale, 2005) and pharmaceuticals and personal care products (PPCPs) (e.g. Wu et al., 2015) in biosolids. In this section, we have summarised and reappraised these previously published reviews to identify any contaminants that continue to pose concern and to highlight any knowledge gaps. The fate and behaviour of 'novel' contaminants in biosolids such as microplastics, fibres and nanoparticles have received less attention and we have attempted to draw a picture of the emerging evidence base.

6.1 PTEs

In Europe and North America, regulations have been in place for many years to limit the content of certain PTEs (i.e. cadmium, copper, lead, mercury, nickel, zinc and chromium) in biosolids intended for land application and/or their concentrations in the receiving soils. Despite the comprehensive legislative framework for regulating PTE concentrations of biosolids and the improvements in biosolids quality, there remain understandable concerns regarding the risks of PTE accumulation in the soil (Mamindy-Pajany et al., 2014) and food crops (Latare et al., 2014), of transport to groundwater and the food chain (Kidd et al., 2007), and their effects on the soil microbial biomass.

Many studies have investigated PTE uptake by food crops grown on land where biosolids have been applied (e.g. Latare et al., 2014; McBride et al., 1999; McBride et al., 2004). These have confirmed that PTEs are taken up by crops, and therefore the authors concluded that there are positive risk factors for PTEs through biosolids land application. However, many of these studies were laboratory-based or used PTE-enriched biosolids and are therefore not representative of standard agricultural practices (Prosser and Sibley, 2015). Field and greenhouse trials are more realistic. However a combination of soil, crop and environmental factors means that measured PTE uptakes can vary greatly (Harrison et al., 1999). Nicholson et al. (2015) reported results from field studies at four UK sites where 20 annual applications of high Zn and Cd-rich biosolids significantly increased grain Zn, Cu and Ni concentrations compared with the low-metal biosolids and control treatments (although as discussed in Section 2.3, this is not necessarily a cause for concern because these are essential micronutrients). These authors also found that wheat grown at one site exceeded the soil Cd limit value of 0.2 mg kg^{-1} (EC, 2006) on all treatments, probably because the chalky soil contained relatively high background levels (c.1 mg Cd kg^{-1}); none of the grain grown at the other sites exceeded the limit values. The Cd-rich biosolids treatment significantly increased grain Cd concentrations compared with the low-metal and control treatments at all the sites; similarly, grain Pb was significantly increased on the Zn-rich biosolids treatments when all the sites were considered together. There were no statistically significant differences in grain Cr or Hg concentrations between the different treatments. Deeks et al. (2013) and Antille et al. (2017) also undertook field scale trials in the UK where an organo-mineral fertiliser (based on nutrient-balanced dried digested sludge cake) and biosolids granules were applied to combinable crops. Whilst crop PTE concentrations were not measured, soil PTE (Zn, Cu, Cd and Pb) concentrations and addition rates were well below legally permitted levels, suggesting that soil application of these products does not pose an increased risk of transfer to the food chain via crop uptake.

There have also been many studies looking at the effects of PTEs of biosolids on the soil microbial biomass, with sometimes contradictory outcomes. For example, incubation experiments by Kao et al. (2006) indicated that large nutrient, microorganism and C sources from biosolids were beneficial to microbial respiration. However, biosolids supplemented with Cu, Pb and Zn reduced the amount of mineralised C by roughly 36%, probably because a portion of the PTEs were complexed with OM, preventing organic C decomposition by microorganisms. In the UK, a long-term experiment was established in 1994 where metal-rich sludge cakes each with high Zn, Cu or Cd concentrations were applied annually for four years to nine sites in increasing quantities to establish PTE dose-response curves. The remaining 'uncontaminated' treatments contained PTE concentrations typical for biosolids produced in the UK at the time and were applied at a single rate to control for the effect of applying organic C to soil; an untreated soil was also included. The initial data from the experiment provided no evidence that PTEs applied with biosolids damaged soil microbial activity in the short term after the cessation of biosolids addition (Gibbs et al., 2006). However, a more recent meta-analysis of the data (Charlton et al., 2016a) found that there had been significant decreases in biomass C even where the soil total concentrations of Zn and Cu were below the current UK statutory limits. In a parallel study, Charlton et al. (2016b) reported a decrease in rhizobium MPN (most probable number) in treatments where Zn-rich biosolids had been applied, whilst no significant effect was noted following the application of Cu-rich biosolids. In contrast, the application of biosolids predominantly contaminated with Cd had no effect on biomass C or rhizobium MPN at soil concentrations below the current UK statutory limit.

6.2 Emerging elements of concern

Current legislative frameworks only require monitoring for regulated substances, and thus the routine testing of biosolids does not include all the numerous potential contaminants that might be present. In a European study, unregulated metal elements including silver (Ag), barium (Ba), antimony (Sb), selenium (Se), titanium (Ti) and vanadium (V) were analysed and found to be present in biosolids samples from 15 countries. This finding prompted the Joint Research Council (JRC) to recommend that emerging contaminants should be monitored because the available database is poor (EC, 2012).

Recently, Healy et al. (2016) measured concentrations of a number of metals (Cd, Cr, Cu, Fe, Hg, Mo, Ni, Pb, Sb, Se, Sn and Zn) in biosolids collected in Ireland. Regulated metals were below the maximum allowable concentrations for biosolids use in agriculture in the European Union. However, some priority metals, such as Sb and Sn, which are potentially harmful to human health, were identified in some of the samples analysed, and the authors recommended that the regulations governing metal limits in biosolids for use in agriculture should be extended to include these metals. Marguí et al. (2016) reported that the mobility of Sb, Ag and Se in soils amended with biosolids was low ($< 10\%$), suggesting low environmental risks. Similarly, in a study investigating depth profiles of metal concentrations in biosolids-amended soil, Yang et al., (2014) reported that unregulated elements including Sb, hafnium (Hf), Mo, niobium (Nb), gold (Au), Ag, tantalum (Ta), Sn, tungsten (W) and zirconium (Zr) have low vertical mobility. Therefore, these elements, even if present in relatively high concentrations in biosolids, may not have a significant impact on soil health in the short term. However, accumulation in the soil following repeated applications of biosolids, may be problematic (Healy et al., 2016) and soil pH along with other soil properties (e.g. soil clay and OM content) can influence mobility in soil (Marguí et al., 2016; Mosquera-Losada et al., 2011).

6.3 Organic chemical contaminants

Organic chemical contaminants (OCCs) persist in the environment, build up through the food chain, and pose a potential risk to human health and the environment. They include manufactured industrial chemicals (e.g. flame retardants) and chemicals produced as accidental by-products of industrial processes or combustion (e.g. dioxins). OCCs can enter wastewater treatment plants from urban, agricultural, domestic and industrial sources and also via atmospheric deposition. OCCs tend to combine with or dissolve in fats and are unable to mix with water; attachment to sewage sludge solids is the primary pathway for their removal from wastewater. Thus, the application of biosolids to land is likely to expose the soil and any associated crops to OCCs.

Research has typically focussed on those OCCs which are regulated in some countries, such as polycyclic aromatic hydrocarbons (PAHs), polychlorinated biphenyls (PCBs), polychlorinated dioxins and furans (PCDD/Fs) and di(2-ethyhexyl)phthalate (DEHP), and as a result, these OCCS have been well characterised. Soil is an extremely effective filter and buffer, and most of these OCCs are retained in the soil matrix by adsorption onto mineral particles or by forming stable complexes with SOM (e.g. Earl et al., 2003). In many countries, there are no specified 'safe' limits for OCC concentrations in agricultural soils (or in soil amendments such as biosolids), although a set of preliminary, human health-related numerical limits for OCCs in soil was developed by Chang et al. (2002) for land application of wastewater and sewage sludge. Suominen et al. (2014) quantified the potential loading of OCCs from the application of a range of organic materials (including biosolids) to arable soils in Finland. They found that annual loadings of dioxins, furans, phthalates and PCBs were similar to or lower than those from atmospheric deposition in Scandinavia and were therefore of low risk to food safety. This is an important finding for the sustainable use of these materials on agricultural land used for food production, increasing confidence that agricultural soil health will not be impaired by a build-up of potentially harmful OCCs.

However, there are also a large number of 'emerging' OCCs which have not been as well studied and which may pose risks to soil health. Many of these emerging OCCs have been measured in biosolids, and some research has been undertaken on their fate and behaviour in soils, as outlined in the following paragraphs.

Perfluorochemicals (PFCs) have been used since the late 1950s to make products resistant to heat, oil, stains, grease and water. Since they are more water soluble than many other OCCs, the barriers that normally prevent human and ecological exposures to OCCs in biosolids-amended soil may not prevent the movement of PFCs. Indeed, a recent review has shown that perfluorooctane sulphonate (PFOS) and perfluorooctanoate (PFOA) accumulate in both plants and worms (Kwak et al., 2017) and concluded that further research on toxicity, bioaccumulation and biomagnification in plants and soil organisms is required. This review did not focus on PFCs from biosolids because atmospheric deposition, agro-chemicals and other organic materials are also important sources, with recent research in China suggesting that < 1% of total PFC additions to soils were from biosolids (Liu et al., 2017).

Polychlorinated alkanes (PCAs) are highly lipophilic and have low water solubility. Commercial products are usually mixtures of different C-chain lengths and varying degrees of chlorination, and consist of a complex mixture of isomers and congeners. Technical mixtures of PCAs, often referred to as chlorinated paraffins (CPs), have been produced since the 1930s for use as extreme pressure lubricant additives, plasticisers, flame retardants and paint additives (IPCS, 1996). Because of their relatively high assimilation and accumulation potential, the short-chain chlorinated paraffins (SCCPs) have been the most widely studied (e.g. Van Mourik et al., 2016). Consequently, restrictions on production and use have been enforced in several countries/ regions. Although CPs tend to be preferentially partitioned to sewage sludge during wastewater treatment (Zeng et al., 2011), reports of CP concentrations in biosolids are very limited, mainly as a result of the complexity of CP mixtures; their analysis, toxicity and environmental fate assessments are, therefore, extremely challenging. Published data particularly on medium and long-chain CPs in the environment is scarce, suggesting that more research is needed to further understand their potential impact on soil health.

Mixtures of polychlorinated naphthalenes (PCNs) have been used as dielectric fluids, engine oil additives, electroplating masking compounds, wood preservatives, lubricants and for dye production (Falandysz, 1998), but were voluntarily phased out in the 1960s and 1970s. There is very limited data on PCNs in biosolids, although there is an indication that concentrations are decreasing (e.g. Guo et al., 2009; Zhang et al., 2014). Clarke and Smith (2011) highlighted the importance of further research on PCNs due to their persistence in soil, potential to enter the human food chain, possible bioaccumulation and/ or soil ecotoxicity and lack of empirical data. However, they also concluded that PCN concentrations measured in biosolids internationally were 'unlikely to pose a threat to human health or the environment'.

Halogenated flame retardants (HFRs) are either brominated (e.g. polybrominated diphenyl ethers - PBDEs) or chlorinated. PBDEs are bioaccumulative endocrine disruptors and toxic; their half-life in soil has been reported to be 4–20 years (Sellstrom et al., 2005; Eljarrat et al., 2008). PBDEs are lipophilic and hence are sorbed to biosolids (e.g. De la Torre, 2011). A recent review noted a downward trend in biosolids concentrations, reflecting the on-going efforts to phase-out PBDEs (Kim et al., 2017). Whilst they can accumulate in both plants and worms (Kwak et al., 2017; Navarro et al., 2016, 2017), both authors recommended that further research on toxicity, bioaccumulation and biomagnification in plants and soil organisms is required. As the use of PBDEs becomes increasingly restricted worldwide, the use of alternative and new flame retardants is expected to increase (Davis et al., 2012). Although alternative flame retardants have been detected in biosolids, there is very limited data available on their potential uptake and thus it is not possible to be certain of their potential for bioaccumulation or toxicity to soil organisms.

Brominated dioxins and furans (PBDD/Fs) are similar to PCDD/Fs but contain bromines instead of chlorines and are therefore more lipophilic and less volatile than the corresponding chlorinated substances. Their stability in soil is reported to be similar to chlorinated dioxins (Remberger et al., 2014). Benzothiazoles (BTHs) and their derivatives are industrial chemicals and are also used in a variety of drugs. BTHs in wastewater mainly originate from urban runoff and household sources, but accumulation in biosolids is not thought to be important (Asimakopoulos et al., 2013).

6.4 Pharmaceuticals and personal care products (PPCPs)

Many chemical compounds are approved as safe for use in soaps, shampoos, deodorants, cosmetics and other personal care products. These compounds, along with pharmaceuticals (e.g. painkillers such as diclofenac, antibiotics such as clarithromycin) and their breakdown products end up in wastewater, where they are transported to treatment plants. What happens to those compounds during treatment will vary from one compound to another. Some, such as the antimicrobial compounds triclosan (TCS) and triclocarban (TCC) are hydrophobic and attach to solid particles suspended in the wastewater. Many compounds are broken down during wastewater treatment, but some pharmaceuticals are particularly resistant to biological breakdown and are therefore transferred to the biosolids.

There is an increasing body of evidence showing that certain PPCP compounds migrate into soils amended with biosolids and are taken up by crop plants (e.g. Wu et al., 2010; Holling et al., 2012. Shargil et al., 2015). Other impacts are also evident. For example, Carter et al. (2016) showed that earthworms could accumulate a number of pharmaceutical compounds (fluoxetine, carbamazepine, diclofenac and orlistat), while Konradi and Vogel (2013) studied ciprofloxacin, sulphamethoxazole, ofloxacin and clarithromycin, and recommended that the presence of these and other antibiotics in biosolids be monitored, to mitigate risks to soil microbial populations. Bioaccumulation in earthworms has been demonstrated for both TCS and TCC, anti-microbial agents widely used in soaps, toothpastes, deodorants and shaving gels (Higgins et al., 2009; Kinney et al., 2008). Laboratory experiments have shown that both compounds degraded in aerobic soil, with half-lives of 108 days and 18 days, respectively, and no negative effects on soil microbial activity were observed during the experiments (Ying et al., 2007). However, preliminary risk assessment (based on limited toxicity data) suggests that TCC and TCS might pose high risks to soil organisms (Chen et al., 2014), with a study by Waller and Kookana (2009) demonstrating that TCS at concentrations below 10 mg/kg can disturb the N cycle in some soils. Carrot root length and hyphal length of the arbuscular mycorrhizal fungus *Glomus intraradices* have been reported to be impacted in experiments where soil 17α-ethynylestradiol (EE2) concentrations increased from 10 μg/l to 1000 μg/l (Hillis et al., 2008); based on this data, and the known recalcitrance of the compound, further studies are required to assess its impact on soil biota.

Toxicity and fate data are still absent (or insufficient) for many emerging PPCPs (Higgins et al., 2010), although risk assessments in some countries have already been undertaken for a number of compounds (Eriksen et al., 2009; Smith, 2009; Clarke and Smith, 2011; Jensen et al., 2012; WEAO, 2001, 2010; Healy et al., 2017). These have concluded that the very low concentrations of most PPCPs in biosolids mean that there is minimal risk to human health. However, for a small number of compounds (in particular TCS, TTC and EE2) there may be risks to soil health, particularly with respect to soil fauna, including bacteria and earthworms.

6.5 Microplastics and fibres

Microplastics are a heterogeneous mixture of differently shaped materials referred to as fragments, fibres, spheroids, granules, pellets, flakes or beads; these can enter sewage treatment works from diverse sources, the most important of which are thought to be personal care products and cosmetics ('microbeads'); tyres could also be a significant source via surface run-off from roads. When synthetic textile materials are washed, they shed microfibres into the sewage system. Given the volume of synthetic fibres manufactured and used throughout the world, it is suspected that microfibre contamination of the environment is higher than that of particulate plastics (Carr, 2017).

Wastewater treatment plants are effective in removing the majority of microplastics and fibres from wastewater during treatment (Iyare et al., 2020); they are however retained in the biosolids. Therefore, it is likely that land-applied biosolids could be a significant source of microplastics in soils (Guo et al., 2020). Indeed there is evidence from the USA that the levels of plastic fibres in soil up to 15 years post-biosolids application exceeded those at sites with no history of biosolids application (Zubris and Richards, 2005).

A recent review by Guo et al. (2020) listed a range of potential impacts of microplastics and fibres on soil health and function, including effects on soil structure; the nature of such effects will depend on the quantity, composition and size of the microplastic particles as well as on the properties of the receiving soil. For example, de Souza Machado et al. (2018) found that polyester fibres could integrate into soil aggregates, significantly increasing soil water holding capacity and significantly decreasing bulk density and water-stable aggregation. In contrast, Zhang et al. (2019) observed no change in soil bulk density and a negative effect on the water holding capacity as a result of polyester microfibre treatments, compared with the control treatment. Other potential impacts summarised by Guo et al. (2020) included effects on soil enzyme activity and hence nutrient cycling, soil microbial activity, the transport and behaviour of organic contaminants and PTEs, and transfer to higher trophic levels with subsequent consequences for soil macrofauna and plants.

It is likely that biosolids will continue to be a source of microplastics and fibres to soils for many years to come, and research is needed to better understand, control and mitigate the risks to soil health.

6.6 Nanoparticles

Engineered nanoparticles (NPs) are used in a wide range of consumer products. However, there is lack of information on the types and concentrations of NPs in biosolids and soil, partly because measurement techniques are expensive and time consuming. Sources of NPs to soils include atmospheric deposition, fertilisers and organic materials; there is no evidence that biosolids are the largest contributor, although they are likely to be an important source (Durenkamp et al., 2016; Jesmer et al., 2017). Concerns about engineered NPs include soil runoff, uptake of nano-and micron-scale metals and metal oxides by plants and risks to soil microbial and invertebrate communities (Yang et al., 2014). In particular, the antimicrobial/toxic properties of silver (Ag) and titanium (Ti) NPs could inhibit a wide range of soil microorganisms and affect soil health.

Kiser et al. (2012) reported that processes such as anaerobic digestion and biosolids decomposition in soils may transform NP surfaces and significantly alter their fate in the environment, whilst agricultural practices (such as biosolids and fertiliser applications) can affect AgNP bioavailability (Doolette et al., 2015). Plant uptake was demonstrated by Priester et al. (2012) in a study using soy beans; here, ZnNPs were taken up and distributed throughout the edible tissues, and cerium (Ce) NPs were shown to reduce plant growth, yield and N fixation.

Several studies have investigated the toxicity of AgNPs to plants and soil microbes after biosolids application (Judy et al., 2016; Wang et al., 2016), with the results indicating that AgNPs entering soils through this route pose low risk to plants and soil microbial communities. This was ascribed to the transformation of AgNPs into recalcitrant Ag_2S during wastewater treatment, resulting in decreased Ag bioavailability and toxicity. Colman et al. (2013) showed that biosolids artificially amended with AgNPs increased nitrous oxide fluxes, changed the microbial community composition and reduced plant biomass; the effect of the AgNP treatment was as large, or larger, than the treatment with $AgNO_3$ added at four times the concentration. Similarly, Jesmer et al. (2017) reported greater toxicity in plants and invertebrates (earthworm and collembola) exposed to biosolids dosed with AgNPs compared to $AgNO_3$. Judy et al. (2015) also reported that the plant tissue Zn concentration was greater in biosolids treated with NPs than in the bulk/dissolved metal treatment, although this difference was not noted for Ti or Ag; large reductions in plant growth and significant shifts in soil microbial community composition were found for the NP treatment compared to the bulk/dissolved metal treatment. In contrast, Durenkamp et al. (2016) reported similar leachate composition and soil microbial community responses in soils treated with biosolids enriched with either engineered NPs or metal salts.

Kunhikrishnan et al. (2015) reviewed various studies demonstrating the uptake of NPs by soil macrofauna. Subsequently, Carbone et al. (2016) found that CeO_2 and SnO_2 NPs did not bioaccumulate in earthworms, whereas other researchers have found that Ag and Au NPs did accumulate in earthworms resulting in reduced reproductive activity (Kwak and An, 2015) and transfer to higher trophic levels (e.g. Kwak and An, 2016). The potential for bioaccumulation of gold (Au) NPs in terrestrial ecosystems was demonstrated by Judy et al. (2011, 2012). However, these studies were undertaken under controlled laboratory conditions using spiked AuNPs.

It is important to be aware that many of the studies outlined above were undertaken in conditions which were not representative of the real world, in that the soil or biosolids were often dosed with pristine NPs at higher concentrations than environmentally relevant, or the studies were conducted in artificial soils or in hydroponic settings. Harmful levels of NPs have not been established because environmental and biological exposure pathways are still largely unknown, and the potential risks to soil health have not been fully quantified.

6.7 Pathogens and antimicrobial resistance

In most developed countries, land application of biosolids is strictly regulated, with sludge treatment processes being required to minimise or eliminate the presence of pathogens. If the resulting biosolids are applied to land, any pathogenic microorganisms still present will be retained in the topsoil to a degree depending on the soil type and rainfall. On tilled soils, incorporating biosolids provides an initial immediate dilution, which may be increased if the pathogens are subsequently transported to depth via percolation. The types of pathogens in biosolids generally reflect the prevalence in the local human population, and some of these are better adapted to survival in the soil environment than others. Depending on the pathogen, the soil physical conditions (i.e. soil type, temperature and moisture content) and soil biotic processes (e.g. predation), there will be varying degrees of pathogen decay in the soil (Pritchard et al., 2010; van Elsas et al., 2012; McLaughlin et al., 2011).

Recently, a concern has been expressed that PTEs are associated with a trend towards increased antimicrobial resistance and that PTEs may indirectly enhance the spread of antibiotic-resistance genes (ARGs). Correlations between the presence of PTEs and antibiotics in biosolids have been demonstrated, especially for Zn, lead (Pb) and tetracyclines (Cheng et al., 2014). Some PTEs may cause an increase in the adsorption of antibiotics into high OM biosolids. For example, Cu may increase the formation of complexes between OM and tetracycline (MacKay and Canterbury, 2005; Zhao et al., 2013) and can enhance the adsorption of tetracycline onto clays (Wang et al., 2008). It is well established that PTEs and other stressors in the environment are responsible for increasing ARG numbers (Gullberg et al., 2014). Indeed, some findings suggest that the number of intrinsic ARGs positively correlates with the content of Cu, Cr, Ni, Pb and iron (Fe) even in uncontaminated soils (Knapp et al., 2011).

Copper can be persistent in agricultural soils with an estimated half-life of hundreds to thousands of years (Berg et al., 2005) and thus may provide a long-standing selection pressure for antibiotic resistance. In China, Hu et al. (2016) provided field-based evidence that long-term Cu contamination significantly changed the diversity, abundance and mobility potential of environmental antibiotic resistance and cautioned against the risk of ARG dissemination in PTE-polluted environments. Another study from China (Hu et al., 2017) showed that increased Ni bioavailability also increased antibiotic resistance. Bondarczuk et al. (2016) concluded that information on the effect of PTEs on antimicrobial resistance is insufficient and that there is an urgent need to broaden our knowledge of this topic in general, and not just in relation to biosolids.

Research into the fate and behaviour of micro-organisms in biosolids used in agriculture has understandably focussed on human pathogens and the risks to human health. However, questions remain over the relative importance of biosolids recycling as a contributor to the environmental reservoir of ARGs; there has to date been no formal risk assessment and the impact on soil health is unknown.

6.8 Radionuclides

In most countries, some liquid discharges are authorised into the sewer system (e.g. from hospitals and research facilities). Thus, land application of biosolids is a potential pathway for radionuclides to enter the soil. Studies have indicated that the effects of biosolids applications on soil radionuclide concentrations are likely to be undetectable, and hence the implications for soil and human health are minimal (Punt et al., 2007; McGowan et al., 2014). However, based on current clinical pressures, there could in future be an increase in medical use, including as yet clinically unexploited radionuclides (Sundell-Bergman et al., 2008). Thus regular monitoring of radionuclide concentrations in biosolids and receiving soils is recommended.

7 Other environmental effects

There are a range of other potential issues that can arise following biosolids applications to land, including pollutant losses to water and air via nitrate leaching, surface runoff (phosphorus) and ammonia and greenhouse gas (GHG) (nitrous oxide, methane and CO_2) emissions.

7.1 Nitrate leaching

There have been few studies measuring nitrate leaching losses from biosolids additions, although two major research programs were completed in the UK in the 1990s which studied N losses from biosolids-amended soil (Smith et al., 1994; Misselbrook et al., 1996; Shepherd, 1996).

Smith et al. (1994) found that N recovery by a winter wheat crop following liquid mesophilic anaerobic-digested biosolids applied and incorporated into the soil increased from 25% in October to 55% in December. Although it was not directly measured, they attributed this to greater nitrate leaching losses following the October application. Shepherd (1996) measured nitrate leaching losses following applications of raw liquid, digested liquid and dewatered digested cake, applied in September, November or January ahead of either winter or spring barley. The earlier the application the greater the leaching risk, with losses highest following the injected liquid digested sludge (which had the highest RAN content) and the lowest losses following broadcast dewatered cake applications. Misselbrook et al. (1996) conducted a similar study, but the biosolids applications were made to two freely draining grassland soils. Autumn applications gave rise to greater losses than winter or spring applications. Soil injection of liquid digested sludge exacerbated nitrate leaching losses compared with surface broadcasting, because preventing N loss via ammonia volatilisation meant that more N was subsequently available to be lost through nitrate leaching.

These studies confirmed that the basic principles established from research with livestock manures also apply to biosolids, that is, the amount of N lost via leaching is dependent on biosolids type - in particular on its RAN content (see Section 2.1) - and on the timing and method of application.

7.2 Phosphorus losses to water

To be able to predict P solubility, availability to plants and leaching losses to water, it is important to understand the forms of P present in biosolids. Biosolids differ markedly from other organic manures due to the treatment processes that they have undergone (e.g. enhanced anaerobic digestion, dosing with Ca, Fe or Al, thermal drying or composting) prior to land application. To meet statutory obligations for the P concentration in discharges from wastewater treatment plants, P is often removed from wastewater using chemical or biological treatment processes; chemical dosing is the most commonly used option (Bunce et al., 2018). Phosphate forms stable complexes with an extensive range of cations, and chemical precipitation with iron (Fe), aluminium (Al) and calcium (Ca) salts removes P from solution into the solid fraction. As a result, biosolids produced from wastewater treatment using these enhanced P removal methods have higher total P, Fe, Al and Ca contents.

The P buffering capacity of soils depends on the quantities and forms of Fe, Al and Ca in the soil (Withers, 2011). The immobilization of P in soil by these cations increases the P sorption capacity and reduces P availability to crops and the risk of losses to water. Flynn and Withers (2007) found that soil P sorption capacity and binding energy have a large influence on the release of P into the soil solution (leachate) and to runoff water. Soils release P much more easily when they become P saturated (e.g. from overuse of P fertilisers) and the soil P buffering capacity is exceeded. Where soils receive P additions from fertiliser and livestock manures and become more P saturated, the binding energy of P to soil surfaces falls, releasing P into the soil solution (Holford et al., 1997; Hooda et al., 2001). However, the addition of biosolids containing large amounts of P-binding elements (e.g. Fe and Al) reverses this trend (Abdala et al., 2015). For example, Maguire et al. (2000a) found that the application of higher quantities of biosolids amended with Fe and Al did not increase P in runoff because the degree of P saturation in soil was not increased; this was because the biosolids increased the P sorption capacity (Lu and O'Connor, 2001). A number of experiments have shown that Fe-treated biosolids do not increase and in some cases may reduce the risk of P losses to water (Maguire et al., 2000a,b; White et al., 2010; Flynn and Withers, 2007). Withers et al. (2001) measured P losses in surface water runoff from field plots that had previously received P from fertiliser, livestock manures and biosolids. The authors concluded that biosolids posed the lowest risk of P loss compared with other agricultural amendments applied at similar P rates. However, the risk of P loss from biosolids-amended soil depended on both the type of biosolids applied and the nature and degree of soil P saturation.

Overall, the literature indicates that P applied to soils in biosolids is less prone to runoff and subsurface drainage loss than the same amount of P applied in livestock manures and chemical fertilisers.

7.3 Ammonia emissions

Ammonia (NH_3) emissions to air contribute to acid deposition and can cause eutrophication of sensitive ecosystems, with re-deposition of emitted NH_3 also contributing to indirect nitrous oxide emissions from soils. In addition, NH_3 reacts with acids in the atmosphere forming particulate matter which may pose a threat to human health (Webb et al., 2014).

Early laboratory-based work in the USA by Donovan and Logan (1983) found greater NH_3 losses from lime-stabilised sludge than from anaerobic and aerobically digested sludges and compost, with the majority of losses occurring within 12 hours of application. A later laboratory incubation study found that more NH_3 was volatilised from biosolids than from compost, which was probably due to the higher total N concentration and lower C/N ratio of the biosolids (He et al., 2003). There have been very few field measurements of NH_3 losses from biosolids. However, an Australian study (Robinson and Polglase, 2000) calculated volatilisation from dewatered biosolids using a mass balance approach and found that most N loss occurred within 1 week of application, with 71-81% of the RAN applied lost after three weeks. In a field study from Sweden, Willen et al. (2016) showed that 55-65% of NH_3 losses occurred in the first three hours after biosolids application, confirming the efficacy of rapid soil incorporation in reducing NH_3 losses.

There is some early laboratory-based research which investigated various other soil and environmental factors that might affect NH_3 losses following biosolids spreading to land. For example, Donovan and Logan (1983) studied the effect of a single variable on NH_3 emissions from biosolids. They concluded that NH_3 emissions were higher from moist soils than from air-dried soils, that NH_3 volatilisation increased with increasing soil pH and temperature, and that when the biosolids contained large particles, NH_3 loss increased with vegetative cover (the cover had no effect when the sludge was well homogenised). There is little evidence from field studies to support these findings. However, Beauchamp et al. (1978) applied anaerobically digested sewage sludge in the field in May and October and measured NH_3 losses using the horizontal flux method. The fluxes generally followed a diurnal pattern, with maximums occurring at about midday, and decreased with time in an exponential manner. Air temperature appeared to be most closely related to flux rate especially in the two to three days following application.

Clearly the significance of NH_3 losses following biosolids application depends on the quantity applied to agricultural land and emissions from other sources. In the UK, for example, Misselbrook and Gilhespy (2019) estimated that 4.2 kt of NH_3 were emitted in 2017, equivalent to just 2% of the total emission from UK agriculture.

7.4 Nitrous oxide emissions

Nitrous oxide (N_2O) is a GHG with a global warming potential about 300-fold greater than CO_2. GHG inventory estimates indicate that agricultural soils are a major contributor to national N_2O emissions (Brown et al., 2019), although these losses are generally small in agronomic terms.

Studies have found that the C/N ratio in biosolids is the most important factor driving N_2O emissions, so that biosolids products with lower C/N ratios have higher cumulative N_2O emissions (e.g. Charles et al., 2017; Singh, 2020). Thermal drying has also been found to reduce the mineralisation of organic N in anaerobically digested biosolids and thus decrease N_2O emissions (e.g. Case et al., 2016). There is some evidence that soil incorporation following land application can increase N_2O losses relative to surface broadcasting (Willen et al., 2016), although this is not always the case (Singh, 2020). Where seasonal differences in N_2O emissions have been reported (e.g. Kamal, 2019), these are thought to be a result of complex interactions between temperature, rainfall (soil moisture), biosolids composition or management and other soil and environmental factors, as well as crop N uptake. The effect of these factors on N_2O losses from biosolids applications, together with the influence of biosolids contaminants (see Section 6.6), are issues which still need to be fully explored.

7.5 Methane emissions

Because biosolids products applied to land have been anaerobically digested, there will usually be little C remaining in a form which is readily lost as methane (CH_4). Nevertheless, there have been a small number of research studies where CH_4 losses following biosolids application to land have been measured. Ball et al. (2004) reported that CH_4 emissions generally only lasted for two to three days after the injection of liquid digested sludge into grassland, with negligible emissions measured from composted sewage sludge and thermally dried sludge pellets. Later at the same site, Jones et al. (2005) reported that CH_4 emissions from thermally dried sludge pellets did not differ from the untreated control.

A review by Andrews et al. (2009) concluded that the available UK field evidence supported the Intergovernmental Panel on Climate Change (IPCC) view that CH_4 emissions following the application of organic materials to land are minimal, a position reinforced by a Swedish study where Willen et al. (2016) found that CH_4 emissions from digested and dewatered sewage sludge applications were negative or negligible.

7.6 Carbon dioxide emissions

As organic compounds in the soil are decomposed by soil microorganisms, CO_2 is generated via respiration (enzymatic oxidation processes). Many factors can influence the rate at which SOM is broken down, including temperature, soil structure (which influences how much oxygen is available in the soil), soil moisture, nutrient availability and the quantity and nature of the OM present.

Torri et al. (2014) explained how biosolids application modifies soil nutrient stoichiometry, which regulates the microbial processes governing OM decomposition. Biosolids provide labile C which produces a large initial CO_2 flush due to the rapid decomposition of labile OM and microbial growth. The shape of the curve is generally related to N availability, the quality of the biosolids' OM, and can be affected by the presence of PTEs or organic pollutants in the biosolids (Renella et al., 2007). In contrast, soil application of composted or thermally dried sludge may reduce soil microorganism decomposition rates due to the higher proportion of recalcitrant C added with these products (Domene et al., 2008).

Scott et al. (2000) applied high rates of digested sludge cake to grassland in the UK and incorporated it prior to reseeding. CO_2 fluxes from the soil were monitored two to four times per day, for six months after sludge incorporation, and reflected diurnal temperature variations. Although emissions decreased in the winter, temperatures below 4°C stimulated short, sharp fluxes as temperature increased. In Spain, De Andres et al. (2012) showed that a large proportion of the OM added to the soil with biosolids was easily biodegradable in the three years following application. Whilst biosolids application generally increased soil CO_2 emissions, this was not proportional to the quantity applied. Differences in soil humidity were associated with differences in soil CO_2 flux over the year, with soil water content the major factor affecting soil respiration. In Brazil, Carmo et al. (2014) reported the highest CO_2 fluxes from composted sludge than from uncomposted sludge, mineral fertiliser and a control treatment, although on all treatments fluxes were correlated with soil temperature. Over the 172-day evaluation period, 2.7% and 0.7% of the total C added to the soil was lost as CO_2 from the sewage sludge and sewage sludge compost treatments, respectively. At two sites in Australia, Wijesekara et al. (2017) found that temperature was a more important factor for controlling microbial degradation of C in biosolids than soil moisture. They showed that whilst biosolids increased the surface (0 to 15 cm) soil TOC by > 45%, enhanced CO_2 fluxes indicated limited C sequestration potential (see Section 3).

8 Conclusion and future trends

The beneficial effects of biosolids recycling on soil health are well understood. This is a mature research area and there have been many papers published on the topic by research groups across the globe. In this chapter we have summarised some of this literature showing how biosolids applied to agricultural land are a source of major and minor plant nutrients, in particular nitrogen, phosphorus, sulphur, copper and zinc, and can be used to replace mineral fertiliser applications. There is extensive evidence that biosolids additions provide valuable amounts of OM, which is vital for building and maintaining healthy soils, and potentially contributing to soil carbon sequestration.

However, biosolids may also contain a range of contaminants including PTEs, organic chemicals, PCPs, microplastics and nanoparticles, all of which may have an adverse impact on soil health. Research is continuing on these new and emerging threats, and whilst the results from these studies are sometimes inconclusive, the water industry in many countries is managing increasing concerns over the acceptability of recycling biosolids to agricultural land, and legislation in parts of Europe is tightening.

We are in a position where society as whole needs to take more responsibility for what we often unthinkingly dispose of into the sewage system; so we should ensure that biosolids are as 'clean' as possible. If not, we run the risk of causing more environmental damage as a result of having to find alternative outlets for the biosolids. For example, if all biosolids had to be incinerated rather than applied to land, this would generate increased CO_2 and OCC emissions; OM and nutrients such as N and S would be lost, and the ash (which could contain high concentrations of potentially toxic substances) might have to be disposed of to a landfill. If the land recycling route for biosolids is no longer available, there would be financial costs to farmers of replacing some or all the nutrients with manufactured fertilisers, as well as additional C emissions associated with the production of these additional fertilisers. It is difficult to assess the overall impact of the loss of the OM inputs on soil health, but whilst biosolids application to land is by no means a perfect solution, it does complete natural nutrient and OM cycles and minimises the loss of these valuable resources.

9 Where to look for further information

9.1 Useful books

- E. Epstein. *Land Application of Sewage Sludge and Biosolids*. 1st Edition. CRC Press. 2019.
- I. K. Kalavrouziotis (Ed.) *Wastewater and Biosolids Management*. IWA Publishing. 2017.

9.2 Key societies, professional organisations and other websites

- The UK Biosolids Assurance Scheme website is a useful source of information: https://assuredbiosolids.co.uk/about-biosolids/.
- The Water Environment Federation (WEF) provides a number of resources relevant to biosolids: http://biosolidsresources.org/OE/.
- A summary of EU sewage sludge/biosolids legislation can be found at https://ec.europa.eu/environment/waste/sludge/.

9.3 Key journals and conferences

The European Biosolids & Bioresources Conference is held annually in the UK and is attended by water companies, researchers and other stakeholders.

In the USA, the Water Environment Federation (WEF) organises the Residuals and Biosolids Conference.

RAMIRAN (Research Network on Recycling of Agricultural and Industrial Residues in Agriculture) holds conferences in Europe every two years.

10 References

Abdala, D. B., da Silva, I. R., Vergütz, L. and Sparks, D. L. (2015). Long-term manure application effects on phosphorus speciation, kinetics and distribution in highly weathered agricultural soils. *Chemosphere* 119: 504–514.

AHDB (2020). Nutrient management guide (RB209). Section 2: organic materials. Available at: https://projectblue.blob.core.windows.net/media/Default/Imported%20Publication%20Docs/RB209/RB209_Section2_2020_200116_WEB.pdf [accessed 29/12/20].

Alloway, B. J. (2009). Soil factors associated with zinc deficiency in crops and humans. *Environ. Geochem. Health* 31(5): 537–548.

Anderson, I. C., Parkin, P. I. and Campbell, C. D. (2008). DNA- and RNA-derived assessments of fungal community composition in soil amended with sewage sludge rich in cadmium, copper and zinc. *Soil Biol. Biochem.* 40(9): 2358–2365.

Andrews, J., Chambers, B., Davey, A., Galletti, S., Hobson, J., Hunt, D., Thorman, R. and Walker, I. (2009). Carbon accounting in the water industry: non-CO2 emissions. UKWIR Report Reference: 09/CL/01/10.

Antille, D. L., Godwin, R. J., Sakrabani, R., Seneweera, S., Tyrrel, S. F. and Johnston, A. E. (2017). Field-scale evaluation of biosolids-derived organomineral fertilizers applied to winter wheat in England. *Agron. J.* 109(2): 654–674.

Artuso, N., Kennedy, T. F., Connery, J., Grant, J. and Schmidt, O. (2011). Effects of biosolids at varying rates on earthworms (*Eisenia fetida*) and springtails (*Folsomia candida*). *Appl. Environ. Soil Sci.* 2011: 1–10. doi: 10.1155/2011/519485.

Asimakopoulos, A. G., Ajibola, A., Kannan, K. and Thomaidis, N. S. (2013). Occurrence and removal efficiencies of benzotriazoles and benzothiazoles in a wastewater treatment plant in Greece. *Sci. Total Environ.* 452–453: 163–171.

Ball, B. C., McTaggart, I. P. and Scott, A. (2004). Mitigation of greenhouse gas emissions from soil under silage production by use of organic manures or slow-release fertilizer. *Soil Use Manag.* 20(3): 287–295.

Barbarick, K. A., Doxtader, K. G., Redente, E. F. and Brobst, R. B. (2004). Biosolids effects on microbial activity in shrubland and grassland soils. *Soil Sci.* 169(3): 176–187.

Beauchamp, E. G., Kidd, G. E. and Thurtell, G. (1978). Ammonia volatilization from sewage sludge applied in the field. *J. Environ. Qual.* 7(1): 141–146.

Berg, J., Tom-Petersen, A. and Nybroe, O. (2005). Copper amendment of agricultural soil selects for bacterial antibiotic resistance in the field. *Lett. Appl. Microbiol.* 40(2): 146–151.

Black, S., Grieve, D. and Llewellyn, A. (2016). Biosolids assurance scheme. 21st European Biosolids and Organic Resour. Conference and Exhibition. Edinburgh, November 2016.

Blair, G. J. (2009). Sulphur Enhanced Fertilizer (SEF). A new generation of fertilizers. UC Davis: Department of Plant Sciences. Available at: https://escholarship.org/uc/item/16h5b2dm.

Bondarczuk, K., Markowicz, A. and Piotrowska-Seget, Z. (2016). The urgent need for risk assessment on the antibiotic resistance spread via sewage sludge land application. *Environ. Int.* 87: 49–55.

Brown, P., Broomfield, M., Cardenas, L., Choudrie, S., Jones, S., Karagianni, E., et al. (2019). UK Greenhouse Gas Inventory, 1990–2017: annual report for submission under the framework convention on climate change. Ricardo-AEA, Didcot, UK, April 2019.

Bunce, J. T., Ndam, E., Ofiteru, I. D., Moore, A. and Graham, D. W. (2018). A review of phosphorus removal technologies and their applicability to small-scale domestic wastewater treatment systems. *Front. Environ. Sci.* 6(8). doi: 10.3389/fenvs.2018.00008.

Carbone, S., Hertel-Aas, T., Joner, E. J. and Oughton, D. H. (2016). Bioavailability of CeO_2 and SnO_2 nanoparticles evaluated by dietary uptake in the earthworm *Eisenia fetida* and sequential extraction of soil and feed. *Chemosphere* 162: 16–22.

Carmo, J. B., Urzedo, D. Id, Ferreira Filho, P. J., Pereira, E. A. and Pitombo, L. M. (2014). CO_2 emission from soil after reforestation and application of sewage sludge. *Bragantia* 73(3): 312-318.

Carr, S. A. (2017). Sources and dispersive modes of micro-fibers in the environment: environmental microfiber sources. *Integr. Environ. Assess. Manag.* 13(3): 466–469.

Carter, L. J., Ryan, J. J. and Boxall, A. B. A. (2016). Effects of soil properties on the uptake of pharmaceuticals into earthworms. *Environ. Pollut.* 213: 922-931.

Case, S. D., Gómez-Muñoz, B., Magid, J. and Jensen, L. S. (2016). Increasing thermal drying temperature of biosolids reduced nitrogen mineralisation and soil N_2O emissions. *Environ. Sci. Pollut. Res. Int.* 23(14): 14383–14392.

Chambers, B. J., Comber, S., Gardner, M., Smith, S., Sutherland, J. and Taylor, M. (2010a). Investigation of organic chemicals in sludge. UKWIR Report No. 10/RG/07/19.

Chambers, B. J., Clarke, B., Comber, S., Gardner, M., Smith, S., Sutherland, J. and Taylor, M. (2010b). A review of the scientific basis for proposed EU limit values for organic chemicals in sludge. UKWIR Report No. 11/RG/07/23.

Chang, A. C., Pan, G., Page, A. L. and Takashi Asano, T. (2002). Developing human health-related chemical guidelines for reclaimed water and sewage sludge applications in agriculture. Prepared for World Health Organization. Available at: http://www.who.int/water_sanitation_health/wastewater/gwwuchemicals.pdf.

Charles, A., Rochette, P., Whalen, J. K., Angers, D. A., Chantigny, M. H. and Bertrand, N. (2017). Global nitrous oxide emission factors from agricultural soils after addition of organic amendments: a meta-analysis. *Agric. Ecosyst. Environ.* 236: 88–98.

Charlton, A., Sakrabani, R., Tyrrel, S., Casado, M. R., McGrath, S. P., Crooks, B., Cooper, P. and Campbell, C. D. (2016a). Long-term impact of sewage sludge application on soil microbial biomass: an evaluation using meta-analysis. *Environ. Pollut.* 219: 1021–1035.

Charlton, A., Sakrabani, R., McGrath, S. P. and Campbell, C. D. (2016b). Long-term impact of sewage sludge application on. *Rhizobium Leguminisorum*. *J. Environ. Qual.* 45(5): 1572–1587.

Chen, F., Ying, G. G., Ma, Y. B., Chen, Z. F., Lai, H. J. and Peng, F. J. (2014). Field dissipation and risk assessment of typical personal care products TCC, TCS, AHTN and HHCB in biosolid-amended soils. *Sci. Total Environ.* 470–471: 1078–1086.

Cheng, M., Wu, L., Huang, Y., Luo, Y. and Christie, P. (2014). Total concentrations of heavy metals and occurrence of antibiotics in sewage sludges from cities throughout China. *J. Soils Sediments* 14(6): 1123–1135.

Clarke, B. O. and Smith, S. R. (2011). Review of 'emerging' organic contaminants in biosolids and assessment of international research priorities for the agricultural use of biosolids. *Environ. Int.* 37(1): 226–247.

Cogger, C. G., Bary, A. I., Kennedy, A. C. and Fortuna, A. M. (2013). Long-term crop and soil response to biosolids applications in dryland wheat. *J. Environ. Qual.* 42(6): 1872–1880.

Colman, B. P., Arnaout, C. L., Anciaux, S., Gunsch, C. K., Hochella Jr., M. F., Kim, B., Lowry, G. V., McGill, B. M., Reinsch, B. C., Richardson, C. J., Unrine, J. M., Wright, J. P., Yin, L. and Bernhardt, E. S. (2013). Low concentrations of silver nanoparticles in biosolids cause adverse ecosystem responses under realistic field scenario. *PLoS ONE* 8(2): e57189.

Coors, A., Edwards, M., Lorenz, P., Rombke, J., Schmelz, R. M., Topp, E., Waszak, K., Wilkes, G. and Lapen, D. R. (2016). Biosolids applied to agricultural land: influence on structural and functional endpoints of soil fauna on a short- and long-term scale. *Sci. Total Environ.* 562: 312-326.

Coulibaly, S. F. M., Winck, B. R., Akpa-Vinceslas, M., Mignot, L., Legras, M., Forey, E. and Chauvat, M. (2019). Functional assemblages of Collembola determine soil microbial communities and associated functions. *Front. Environ. Sci.* 7. Available at: https://www.frontiersin.org/article/10.3389/fenvs.2019.00052.

Davis, E. F., Klosterhaus, S. L. and Stapleton, H. M. (2012). Measurement of flame retardants and triclosan in municipal sewage sludge and biosolids. *Environ. Int.* 40: 1-7.

De Andrés, E. F., Tenorio, J. L., del Mar Albarran, M. and Walter, I. (2012). Carbon dioxide flux in a soil treated with biosolids under semiarid conditions. *Compost Sci. Util.* 20(1): 43-48.

Deeks, L. K., Chaney, K., Murray, C., Sakrabani, R., Gedara, S., Le, M. S., Tyrrel, S., Pawlett, M., Read, R. and Smith, G. H. (2013). A new sludge-derived organo-mineral fertilizer gives similar crop yields as conventional fertilizers. *Agron. Sustain. Dev.* 33(3): 539-549.

De la Torre, A., Sverko, E., Alaee, M. and Ángeles-Martínez, M. Á (2011). Concentrations and sources of dechlorane plus in sewage sludge. *Chemosphere* 82(5): 692-697.

de Souza Machado, A. A., Lau, C. W., Till, J., Kloas, W., Lehmann, A., Becker, R. and Rillig,M. C. (2018). Impacts of microplastics on the soil biophysical environment. *Glob. Change Biol.* 24: 1405-1416.

Domene, X., Ramírez, W., Mattana, S., Alcañiz, J. M. and Andrés, P. (2008). Ecological risk assessment of organic waste amendments using the species sensitivity distribution from a soil organisms test battery. *Environ. Pollut.* 155(2): 227-236.

Donovan, W. C. and Logan, T. J. (1983). Factors affecting ammonia volatilisation from sewage sludge applied to soil in a laboratory study. *J. Environ. Qual.* 12(4): 584-590.

Doolette, C. L., McLaughlin, M. J., Kirby, J. K. and Navarro, D. A. (2015). Bioavailability of silver and silver sulphide nanoparticles to lettuce (*Lactuca sativa*): effect of agricultural amendments on plant uptake. *J. Hazard. Mater.* 300: 788-795.

Durenkamp, M., Pawlett, M., Ritz, K., Harris, J. A., Neal, A. L. and McGrath, S. P. (2016). Nanoparticles within WWTP sludges have minimal impact on leachate quality and soil microbial community structure and function. *Environ. Pollut.* 211: 399-405.

Earl, N., Cartwright, C. D., Horrocks, S. J., Worboys, M., Swift, S., Kirton, J. A., Askan, A. U., Kelleher, H. and Nancarrow, D. J. (2003). *Review of the Fate and Transport of Selected Contaminants in the Soil Environment*. Environment Agency, Bristol, UK.

EC (2006). Commission Regulation 1881/2006 setting maximum levels for certain contaminants in foodstuffs. *Official Journal of the European Union*. L 364/5.

EC (2012). Occurrence and levels of selected compounds in Europe sewage sludge samples. Results of a Pan-European Screening Experience (FATE SEES).

Eljarrat, E., Marsh, G., Labandeira, A. and Barceló, D. (2008). Effect of sewage sludges contaminated with polybrominated diphenylethers on agricultural soils. *Chemosphere* 71(6): 1079-1086.

Eriksen, G. S., Amundsen, C. E., Bernhoft, A., Eggen, T., Grave, K., Halling-Sørensen, B., Källqvist, T., Sogn, T. and Sverdrup, L. (2009). Risk assessment of contaminants in sewage sludge applied on Norwegian soils. Opinion of the panel on contaminants in the Norwegian Scientific Committee for Food Safety. Norwegian Scientific Committee for Food Safety (VKM). 05/511-22-final.

Falandysz, J. (1998). Polychlorinated naphthalenes: an environmental update. *Environ. Pollut.* 101(1): 77–90.

Flynn, N. J. and Withers, P. J. A. (2007). Application of phosphorus in biosolids to agricultural soils. UKWIR Report Ref. No. 07/SL/02/7.

Gale, P. (2005). Land application of treated sewage sludge: quantifying pathogen risks from consumption of crops. *J. Appl. Microbiol.* 98(2): 380–396.

Gale, P. and Stanfield, G. (2001). Towards a quantitative risk assessment for BSE in sewage sludge. *J. Appl. Microbiol.* 91(3): 563–569.

Gibbs, P. A., Chambers, B. J., Chaudri, A. M., McGrath, S. P., Carlton-Smith, C. H., Bacon, J. R., Campbell, C. D. and Aitken, M. N. (2006). Initial results from a long-term, multi-site field study of the effects on soil fertility and microbial activity of sludge cakes containing heavy metals. *Soil Use Manag.* 22(1): 11–21.

Gullberg, E., Albrecht, L. M., Karlsson, C., Sandegren, L. and Andersson, D. I. (2014). Selection of a multidrug resistance plasmid by sublethal levels of antibiotics and heavy metals. *mBio* 5(5): e01918–e01914.

Guo, J. J., Huang, X. P., Xiang, L., Wang, Y. Z., Li, Y. W., Li, H., Cai, Q. Y., Mo, C. H. and Wong, M. H. (2020). Source, migration and toxicology of microplastics in soil. *Environ. Int.* 137: 105263.

Guo, L., Zhang, B., Xiao, K., Zhang, Q. and Zheng, M. (2009). Levels and distributions of polychlorinated biphenyls in sewage sludge of urban wastewater treatment plants. *J. Environ. Sci. (China)* 21(4): 468–473.

Guo, M., Song, W. and Kazda, R. (2012). Fertilizer value of lime-stabilized biosolids as a soil amendment. *Agron. J.* 104(6): 1679–1686.

Harrison, E. Z., McBride, M. B. and Bouldin, D. R. (1999). Land application of sewage sludges: an appraisal of the US regulations. *Int. J. Environ. Pollut.* 11(1): 1–26.

He, Z. L., Calvert, D. V., Alva, A. K., Li, Y. C., Stoffella, P. J. and Banks, D. J. (2003). Nitrogen transformation and ammonia volatilization from biosolids and compost applied to calcareous soil. *Compost Sci. Util.* 11(1): 81–88.

Healy, M. G., Fenton, O., Cummins, E., Clarke, R., Peyton, D., Fleming, G., Wall, D., Morrison, L. and Cormican, M. (2017). Health and water quality impacts arising from land spreading of biosolids. EPA Research Report 200.

Healy, M. G., Fenton, O., Forrestal, P. J., Danaher, M., Brennan, R. B. and Morrison, L. (2016). Metal concentrations in lime stabilised, thermally dried and anaerobically digested sewage sludges. *Waste Manag.* 48: 404–408.

Higgins, C. P., Paesani, Z. J., Chalew, T. E. A. and Halden, R. U. (2009). Bioaccumulation of triclocarban in Lumbriculus variegates. *Environ. Toxicol. Chem.* 65: 141–148.

Higgins, C. P., Sharp, J. O., Sepulvado,J. G., Littrell, B., O'Connor, G., Snyder, E. and McAvoy, D. (2010). Trace organic chemicals in biosolids-amended soils. State-of-the-Science Review. Water Environment Research Foundation, IWA Publishing. SRSK5T09.

Hillis, D. G., Antunes, P., Sibley, P. K., Klironomos, J. N. and Solomon, K. R. (2008). Structural responses of *Daucus carota* root-organ cultures and the arbuscular mycorrhizal fungus, *Glomus intraradices*, to 12 pharmaceuticals. *Chemosphere* 73(3): 344–352.

Holford, I. C. R., Hird, C. and Lawrie, R. (1997). Effects of animal effluents on the phosphorus sorption characteristics of soils. *Soil Res.* 35(2): 365-373.

Holling, C. S., Bailey, J. L., Heuvel, B. V. and Kinney, C. A. (2012). Uptake of human pharmaceuticals and personal care products by cabbage (*Brassica campestris*) from fortified and biosolids-amended soils. *J. Environ. Monit.* 11: 3029-3036.

Hooda, P. S., Truesdale, V. W., Edwards, A. C., Withers, P. J. A., Aitken, M. N., Miller, A. and Rendell, A. R. (2001). Manuring and fertilization effects on phosphorus accumulation in soils and potential environmental implications. *Adv. Environ. Res.* 5(1): 13-21.

Hu, H. W., Wang, J. T., Li, J., Li, J. J., Ma, Y. B., Chen, D. and He, J. Z. (2016). Field-based evidence for copper contamination induced changes of antibiotic resistance in agricultural soils. *Environ. Microbiol.* 18(11): 3896-3909.

Hu, H. W., Wang, J. T., Li, J., Shi, X. Z., Ma, Y. B., Chen, D. and He, J. Z. (2017). Long-term nickel contamination increases the occurrence of antibiotic resistance genes in agricultural soils. *Environ. Sci. Technol.* 51(2): 790-800.

IPCS (1996). Environmental health criteria 181. Chlorinated paraffins. World Health Organization, Geneva.

Iyare, P. U., Oukia, S. K. and Bond, T. (2020). Microplastics removal in wastewater treatment plants: a critical review. *Environ. Sci.: Water Res. Technol.* 6(10): 2664-2675.

Jensen, J., Ingvertsen, S. T. and Magid, J. (2012). Risk evaluation of five groups of persistent organic contaminants in sewage sludge. Danish Ministry of the Environment. Environmental Protection Agency. Environmental Project No. 1406.

Jesmer, A. H., Velicogna, J. R., Schwertfeger, D. M., Scroggins, R. P. and Princz, J. I. (2017). The toxicity of silver to soil organisms exposed to silver nanoparticles and silver nitrate in biosolids-amended field soil. *Environ. Toxicol. Chem.* 36(10): 2756-2765.

Jin, V. L., Potter, K. N., Johnson, M.-V. V., Harmel, R. D. and Arnold, J. G. (2015). Surface-applied biosolids enhance soil organic carbon and nitrogen stocks but have contrasting effects on soil physical quality. *Appl. Environ. Soil Sci.* 2015: 1-10.

Jones, S. K., Rees, R. M., Skiba, U. M. and Ball, B. C. (2005). Greenhouse gas emissions from a managed grassland. *Glob. Planet. Change* 47(2-4): 201-211.

Judy, J. D., McNear Jr., D. H., Chen, C., Lewis, R. W., Tsyusko, O. V., Bertsch, P. M., Rao, W., Stegemeier, J., Lowry, G. V., McGrath, S. P., Durenkamp, M. and Unrine, J. M. (2015). Nanomaterials in biosolids inhibit nodulation, shift microbial community composition, and result in increased metal uptake relative to bulk/dissolved metals. *Environ. Sci. Technol.* 49(14): 8751-8758.

Judy, J. D., Unrine, J. M. and Bertsch, P. M. (2011). Evidence for biomagnification of gold nanoparticles within a terrestrial food chain. *Environ. Sci. Technol.* 45(2): 776-781.

Judy, J. D., Unrine, J. M., Rao, W. and Bertsch, P. M. (2012). Bioaccumulation of gold nanomaterials by *Manduca sexta* through dietary uptake of surface contaminated plant tissue. *Environ. Sci. Technol.* 46(22): 12672-12678.

Judy, J. D., Kirby, J. K., McLaughlin, M. J., McNear, D. and Bertsch, P. M. (2016). Symbiosis between nitrogen-fixing bacteria and *Medicago truncatula* is not significantly affected by silver and silver sulfide nanomaterials. *Environ. Pollut.* 214: 731-736.

Kamal, A. (2019). Reduction of nitrous oxide emissions from biosolids-amended corn agroecosystems. MSc Thesis. McGill University, Montreal, Canada.

Kao, P. H., Huang, C. C. and Hseu, Z. Y. (2006). Response of microbial activities to heavy metals in a neutral loamy soil treated with biosolid. *Chemosphere* 64(1): 63-70.

Kidd, P. S., Domínguez-Rodríguez, M. J., Díez, J. and Monterroso, C. (2007). Bioavailability and plant accumulation of heavy metals and phosphorus in agricultural soils amended by long-term application of sewage sludge. *Chemosphere* 66(8): 1458–1467.

Kim, M., Li, L. Y., Gorgy, T. and Grace, J. R. (2017). Review of contamination of sewage sludge and amended soils by polybrominated diphenyl ethers based on meta-analysis. *Environ. Pollut.* 220(B): 753–765.

Kinney, C. A., Campbell, B. R., Thompson, R., Furlong, E. T., Kolpin, D. W., Burkhardt, M. R., Zaugg, S. D., Werner, S. L. and Hay, A. G. (2012). Earthworm bioassays and seedling emergence for monitoring toxicity, aging and bioaccumulation of anthropogenic waste indicator compounds in biosolids-amended soil. *Sci. Total Environ.* 433: 507–515.

Kinney, C. A., Furlong, E. T., Kolpin, D. W., Burkhardt, M. R., Zaugg, S. D., Werner, S. L., Bossio, J. P. and Benotti, M. J. (2008). Bioaccumulation of pharmaceuticals and other anthropogenic waste indicators in earthworms from agricultural soil amended with biosolid or swine manure. *Environ. Sci. Technol.* 42(6): 1863–1870.

Kiser, M. A., Ladner, D. A., Hristovski, K. D. and Westerhoff, P. K. (2012). Nanomaterial transformation and association with fresh and freeze-dried wastewater activated sludge: implications for testing protocol and environmental fate. *Environ. Sci. Technol.* 46(13): 7046–7053.

Knapp, C. W., McCluskey, S. M., Singh, B. K., Campbell, C. D., Hudson, G. and Graham, D. W. (2011). Antibiotic resistance gene abundances correlate with metal and geochemical conditions in archived Scottish soils. *PLoS ONE* 6(11): e27300.

Konradi, S. and Vogel, I. (2013). Fate of pharmaceuticals (PhCs) in sewage sludge and proposal of indicator substances for monitoring. Available at: https://www.umweltbundesamt.de/sites/default/files/medien/376/dokumente/konradi-presentation.pdf [accessed 22/12/20].

Kunhikrishnan, A., Shon, H. K., Bolan, N. S., El Saliby, I. and Vigneswaran, S. (2015). Sources, distribution, environmental fate, and ecological effects of nanomaterials in wastewater streams. *Crit. Rev. Env. Sci. Tec.* 45(4): 277–318.

Kwak, J. I. and An, Y.-J. (2015). Ecotoxicological effects of nanomaterials on earthworms: a review. *Hum. Ecol. Risk Assess.* 21(6): 1566–1575.

Kwak, J. I. and An, Y.-J. (2016). Trophic transfer of silver nanoparticles from earthworms disrupts the locomotion of springtails (Collembola). *J. Hazard. Mater.* 315: 110–116.

Kwak, J. I., Moon, J., Kim, D. and An, Y.-J. (2017). Soil ecotoxicity of seven endocrine-disrupting chemicals: a review. *Eur. J. Soil Sci.* 68(5): 621–649.

Lal, R. (2016). Soil health and carbon management. *Food Energy Secur.* 5(4): 212–222.

Latare, A. M., Kumar, O., Singh, S. K. and Gupta, A. (2014). Direct and residual effect of sewage sludge on yield, heavy metals content and soil fertility under rice-wheat system. *Ecol. Eng.* 69: 17–24.

Li, S., Zhu, L., Li, J., Ke, X., Wu, L., Luo, Y. and Christie, P. (2020). Influence of long-term biosolid applications on communities of soil fauna and their metal accumulation: a field study. *Environ. Pollut.* 260: 114017.

Liu, Z., Lu, Y., Wang, P., Wang, T., Liu, S., Johnson, A. C., Sweetman, A. J. and Baninla, Y. (2017). Pollution pathways and release estimation of perfluorooctane sulfonate (PFOS) and perfluorooctanoic acid (PFOA) in central and eastern China. *Sci. Total Environ.* 580: 1247–1256.

Lu, P. and O'Connor, G. A. (2001). Biosolid effects on phosphorus retention and release in some sandy Florida soils. *J. Environ. Qual.* 30(3): 1059–1063.

Lu, Q., He, Z. L. and Stoffella, P. J. (2012). Land application of biosolids in the USA: a review. *Appl. Environ. Soil Sci.* 2012: 1-11.

MacKay, A. A. and Canterbury, B. (2005). Oxytetracycline sorption to organic matter by metalbridging. *J. Environ. Qual.* 34(6): 1964-1971.

Maguire, R. O., Sims, J. T. and Coale, F. J. (2000a). Phosphorus solubility in biosolids-amended farm soils in the Mid-Atlantic region of the USA. *J. Environ. Qual.* 29(4): 1225-1233.

Maguire, R. O., Sims, J. T. and Coale, F. J. (2000b). Phosphorus fractionation in biosolids-amended soils: relationships to soluble and desorbable phosphorus. *Soil Sci. Soc. Am. J.* 64(6): 2018-2024.

Mamindy-Pajany, Y., Sayen, S., Mosselmans, J. F. W. and Guillon, E. (2014). Copper, nickel and zinc speciation in a biosolid-amended soil: pH adsorption edge, μ-XRF and μ-XANES investigations. *Environ. Sci. Technol.* 48(13): 7237-7244.

Marguí, E., Iglesias, M., Camps, F., Sala, L. and Hidalgo, M. (2016). Long-term use of biosolids as organic fertilizers in agricultural soils: potentially toxic elements occurrence and mobility. *Environ. Sci. Pollut. Res. Int.* 23(5): 4454-4464.

McBride, M. B., Richards, B. K. and Steenhuis, T. (2004). Bioavailability and crop uptake of trace elements in soil columns amended with sewage sludge products. *Plant Soil* 262(1/2): 71-84.

McBride, M. B., Richards, B. K., Steenhuis, T. and Spiers, G. (1999). Long-term leaching of trace elements in a heavily sludge-amended silty clay loam soil. *Soil Sci.* 164(9): 613-623.

McGowan, D. R., Pratt, B. E., Hinton, P. J., Peet, D. J. and Crawley, M. T. (2014). Iodine-131 monitoring in sewage plant outflow. *J. Radiol. Prot.* 34(1): 1-14.

McGrath, S. P., Chambers, B. J., Taylor, M. J. and Carlton-Smith, C. H. (2012). Biofortification of zinc in wheat grain by the application of sewage sludge. *Plant Soil* 361(1-2): 97-108.

McLaughlin, H. P., Casey, P. G., Cotter, J., Gahan, C. G. M. and Hill, C. (2011). Factors affecting survival of *Listeria monocytogenes* and *Listeria innocua* in soil samples. *Arch. Microbiol.* 193(11): 775-785.

Merrington, G. (2006). The development and use of soil quality indicators for assessing the role of soil in environmental interactions. The Environment Agency. Science Report SC030265.

Misselbrook, T. H. and Gilhespy, S. L. (2019). Inventory of ammonia emissions from UK agriculture 2017. DEFRA contract SCF0107. Inventory Submission Report. Available at: https://uk-air.defra.gov.uk/assets/documents/reports/cat09/1903141332_UK_Agriculture_Ammonia_Emission_Report_1990-2017.pdf [accessed 17/03/21].

Misselbrook, T. H., Shepherd, M. A. and Pain, B. F. (1996). Sewage sludge applications to grassland: influence of sludge type, time and method of application on nitrate leaching and herbage yield. *J. Agric. Sci.* 126(3): 343-352.

Mosquera-Losada, M. R., Morán-Zuloaga, D. and Rigueiro-Rodriguez, A. (2011). Effects of lime and sewage sludge on soil, pasture production, and tree growth in a six-year-old *Populus canadensis* Moench silvopastoral system. *J. Plant Nut. Soil Sci.* 174(1): 145-153.

Navarro, I., de la Torre, A., Sanz, P., Porcel, M. Á, Pro, J., Carbonell, G. and de los Ángeles Martínez, M. L. (2017). Uptake of perfluoroalkyl substances and halogenated flame retardants by crop plants grown in biosolids-amended soils. *Environ. Res.* 152: 199-206.

Navarro, I., de la Torre, A., Sanz, P., Pro, J., Carbonell, G. and de los Ángeles Martínez, M. L. Á (2016). Bioaccumulation of emerging organic compounds (perfluoroalky substances and halogenated flame retardants) by earthworm in biosolid amended soils. *Environ. Res.* 149: 32–39.

Newell Price, P., Chambers, B. and Whittingham, M. (2012). WP1 characterisation of soil compaction under grassland. Grassland soil compaction assessment–stages 1 and 2. Report to Defra–BD5001: characterisation of soil structural degradation under grassland and development of measures to ameliorate its impact on biodiversity and other soil functions. 96 pp.

Nicholson, F., Bhogal, A., Taylor, M., McGrath, S. and Withers, P. (2018). Long-term effects of biosolids on soil quality and fertility. *Soil Sci.* 183(3): 89–98.

Nicholson, F., Taylor, M., Bhogal, A., McGrath, S. and Withers, P. (2015). Biosolids: soil quality and fertility benefits. UKWIR Report Reference: 15/SL/01/8.

Ozores-Hampton, M., Stansly, P. A. and Salame, T. P. (2011). Soil chemical, physical, and biological properties of a sandy soil subjected to long-term organic amendments. *J. Sustain. Agric.* 35(3): 243–259.

Pawlett, M., Deeks, L. K. and Sakrabani, R. (2015). Nutrient potential of biosolids and urea derived or-gano-mineral fertilisers in a field scale experiment using ryegrass (*Lolium perenne* L.). *Field Crops Res.* 175: 56–63.

Powlson, D. S., Bhogal, A., Chambers, B. J., Coleman, K., Macdonald, A. J., Goulding, K. W. T. and Whitmore, A. P. (2012). The potential to increase soil carbon stocks through reduced tillage or organic material additions in England and Wales: a case study. *Agric. Ecosyst. Environ.* 146(1): 23–33.

Price, G. W., Langille, M. G. I. and Yurgel, S. N. (2021) Microbial co-occurrence network analysis of soils receiving short- and long-term applications of alkaline treated biosolids. *Sci. Total Environ.* 751: 141687.

Priester, J. H., Ge, Y., Mielke, R. E., Horst, A. M., Moritz, S. C., Espinosa, K., Gelb, J., Walker, S. L., Nisbet, R. M., An, Y. J., Schimel, J. P., Palmer, R. G., Hernandez-Viezcas, J. A., Zhao, L., Gardea-Torresdey, J. L. and Holden, P. A. (2012). Soybean susceptibility to manufactured nanomaterials with evidence for food quality and soil fertility interruption. *Proc. Natl. Acad. Sci. U. S. A.* 109(37): E2451–E2456.

Prosser, R. S. and Sibley, P. K. (2015). Human health risk assessment of pharmaceuticals and personal care products in plant tissue due to biosolids and manure amendments, and wastewater irrigation. *Environ. Int.* 75: 223–233.

Pritchard, D. L., Penney, N., McLaughlin, M. J., Rigby, H. and Schwarz, K. (2010). Land application of sewage sludge (biosolids) in Australia: risks to the environment and food crops. *Water Sci. Technol.* 62(1): 48–57.

Punt, A., Woods, M. and Rose, D. (2007). Radionuclide discharges to sewer–a field investigation. EA Science Report SC020150/SR2.

Remberger, M., Kaj, L., Hansson, K., Bibi, M., Brorström-Lundén, E., Haglund, P., Liljelind, P., Bergek, S., Andersson, R. and Kitti-Sjöström, A. (2014). *Screening of Emerging Brominated Flame Retardants (BFRs) and Polybrominated Dibenzofurans (PBDFs).* IVL Swedish Environmental Research Institute Ltd.

Renella, G., Chaudri, A. M., Falloon, C. M., Landi, L., Nannipieri, P. and Brookes, P. C. (2007). Effects of Cd, Zn, or both on soil microbial biomass and activity in a clay loam soil. *Biol. Fertil. Soils* 43(6): 751–758.

Rickson, J., Deeks, L. K., Corstanje, R., Newell-Price, P., Kibblewhite, M. G., Chambers, B., Bellamy, P., Holman, I., James, I. T., Jones, R., Kechavarsi, C., Mouazen, A. M., Ritz, K.

and Waine, T. (2012). Indicators of soil quality–physical properties (SP1611). Final Report to Defra.

Rigby, H., Clarke, B. O., Pritchard, D. L., Meehan, B., Beshah, F., Smith, S. R. and Porter, N. A. (2016). A critical review of nitrogen mineralization in biosolids-amended soil, the associated fertilizer value for crop production and potential for emissions to the environment. *Sci. Total Environ.* 541: 1310–1338.

Rigby, H., Dowding, A., Fernandes, A., Humphries, D., Petch, R. G., Reynolds, C. K., Rose, M. and Smith, S. R. (2015). Organic contaminant content and physico-chemical characteristics of waste materials recycled in agriculture. *Agriculture* 5(4): 1289–1328.

Rigby, H., Perez-Viana, F., Cass, J., Rogers, M. and Smith, S. R. (2009). The influence of soil and biosolids type, and microbial immobilisation on nitrogen availability in biosolids-amended agricultural soils–implications for fertiliser recommendations. *Soil Use Manag.* 25(4): 395–408.

Robinson, M. B. and Polglase, P. J. (2000). Volatilization of nitrogen from dewatered biosolids. *J. Environ. Qual.* 29(4): 1351–1355.

Rouch, D. A., Fleming, V. A., Pai, S., Deighton, M., Blackbeard, J. and Smith, S. R. (2011). Nitrogen release from air-dried biosolids for fertilizer valueNitrogen release from air-dried biosolids for fertilizer value. *Soil Use Manag.* 27: 294–304.

Sagoo, E., Smith, K. A., Nicholson, F. A. and McGrath, S. P. (2014). Sulphur availability from organic materials applied to winter wheat. *Int. Fert. Soc. Proc.* 742: 1–26.

Schlatter, D. C., Paul, N. C., Shah, D. H., Schillinger, W. F., Bary, A. I., Sharratt, B. and Paulitz, T. C. (2019). Biosolids and tillage practices influence soil bacterial communities in dryland wheat. *Microb. Ecol.* 78(3): 737–752.

Schlatter, D. C., Schillinger, W. F., Bary, A. I., Sharratt, B. and Paulitz, T. C. (2017). Biosolids and conservation tillage: impacts on soil fungal communities in dryland wheat-fallow cropping systems. *Soil Biol. Biochem.* 115: 556–567.

Scott, A., Ball, B. C., Crichton, I. J. and Aitken, M. N. (2000). Nitrous oxide and carbon dioxide emissions from grassland amended with sewage sludge. *Soil Use Manag.* 16(1): 36–41.

Scullion, J., Neale, S. and Philipps, L. (2002). Comparisons of earthworm populations and cast properties in conventional and organic arable rotations. *Soil Use Manag.* 22: 181–190.

Sellström, U., de Wit, C. A., Lundgren, N. and Tysklind, M. (2005). Effect of sewage-sludge application on concentrations of higher-brominated diphenyl ethers in soils and earthworms. *Environ. Sci. Technol.* 39(23): 9064–9070.

Shargil, D., Gerstl, Z., Fine, P., Nitsan, I. and Kurtzman, D. (2015). Impact of biosolids and wastewater effluent application to agricultural land on steroidal hormone content in lettuce plants. *Sci. Total Environ.* 505: 357–366.

Shepherd, M. A. (1996). Factors affecting nitrate leaching from sewage sludges applied to a sandy soil in arable agriculture. *Agric. Ecosys. Environmentalist* 58: 171–185.

Singh, G. (2020). Effect of the land application of biosolids on greenhouse gas emissions Under Atlantic Canadian conditions. MSC Thesis. Dalhousie University, Halifax, Nova Scotia.

Smith, S. R. (2009). Organic contaminants in sewage sludge (biosolids) and their significance for agricultural recycling. *Philos. Trans. A Math. Phys. Eng. Sci.* 367(1904): 4005–4041.

Smith, S. R., Hallett, J. E., Reynold, S. E., Brookman, S. J., Carlton-Smith, C. H., Woods, V. and Sweet, N. (1994). Nitrate leaching losses from sewage sludge treated agricultural

land. Report number UM 1448. Common Interest Research Programme Reference U-0915, WRc, Medmenham.
Suominen, K., Verta, M. and Marttinene, S. (2014). Hazardous organic compounds in biogas plant end products–soil burden and risk to food safety. *Sci. Total Environ.* 491-492: 192-199.
Sullivan, T. S., Stromberger, M. E., Paschke, M. W. and Ippolito, J. A. (2006a). Long-term impacts of infrequent biosolids applications on chemical and microbial properties of a semi-arid rangeland soil. *Biol. Fertil. Soils* 42(3): 258-266.
Sullivan, T. S., Stromberger, M. E. and Paschke, M. W. (2006b). Parallel shifts in plant and soil microbial communities in response to biosolids in a semi-arid grassland. *Soil Biol. Biochem.* 38(3): 449-459.
Sundell-Bergman, S., de la Cruz, I., Avila, R. S. and Hasselblad, S. (2008). A new approach to assessment and management of the impact from medical liquid radioactive waste. *J. Environ. Radioact.* 99(10): 1572-1577.
Sutherland, J. and Taylor, M. (2012). Investigation of organic compounds in sludge–persistent organic pollutants; steroid oestrogens; and selected pharmaceuticals. United Kingdom Water Industry Research, London, UK.
Toffey, W. and Brown, S. (2020). Biosolids and ecosystem services: making the connection explicit. *Environ. Sci. Health* 14: 51-55.
Torri, S. I., Corrêa, R. S. and Renella, G. (2014). Soil carbon sequestration resulting from biosolids application. *Appl. Environ. Soil Sci.* 2014: 1-9.
Torri, S. I., Corrêa, R. S. and Renella, G. (2017). Biosolid application to agricultural land–a contribution to global phosphorus recycle: a review. *Pedosphere* 27(1): 1-16.
van Elsas, J. D., Chiurazzi, M., Mallon, C. A., Elhottovā, D., Krištůfek, V. and Salles, J. F. (2012). Microbial diversity determines the invasion of soil by a bacterial pathogen. *Proc. Natl. Acad. Sci. U. S. A.* 109(4): 1159-1164.
van Mourik, L. M., Gaus, C., Leonards, P. E. G. and de Boer, J. (2016). Chlorinated paraffins in the environment: a review on their production, fate, levels and trends between 2010 and 2015. *Chemosphere* 155: 415-428.
Waller, N. J. and Kookana, R. S. (2009). Effect of triclosan on microbial activity in Australian soils. *Environ. Toxicol. Chem.* 28(1): 65-70.
Wang, P., Menzies, N. W., Dennis, P. G., Guo, J., Forstner, C., Sekine, Lombi, E., Kappen, P., Bertsch, P. M. and Kopittke, P. M. (2016). Silver nanoparticles entering soils via the wastewater-sludge-soil pathway pose low risk to plants but elevated Cl concentrations increase Ag bioavailability. *Environ. Sci. Technol.* 50(15): 8274 -8281.
Wang, Y. J., Jia, D. A., Sun, R. J., Zhu, H. W. and Zhou, D. M. (2008). Adsorption and cosorption of tetracycline and copper(II) on montmorillonite as affected by solution pH. *Environ. Sci. Technol.* 42(9): 3254-3259.
WEAO (2001). Fate and significance of selected contaminants in sewage biosolids applied to agricultural land through literature review and consultation with stakeholder groups. Water Environment Association of Ontario.
Webb, J., Thorman, R. E., Fernanda-Aller, M. and Jackson, D. R. (2014). Emission factors for ammonia and nitrous oxide emissions following immediate manure incorporation on two contrasting soil types. *Atmos. Environ.* 82: 280-287.
White, J. W., Coale, F. J., Sims, J. T. and Shober, A. L. (2010). Phosphorus runoff from waste water treatment biosolids and poultry litter applied to agricultural soils. *J. Environ. Qual.* 39(1): 314- 323.

Wijesekara, H., Bolan, N. S., Thangavel, R., Seshadri, B., Surapaneni, A., Saint, C., Hetherington, C., Matthews, P. and Vithanage, M. (2017). The impact of biosolids application on organic carbon and carbon dioxide fluxes in soil. *Chemosphere* 189: 565-573.

Wijesekara, H., Colyvas, K., Seshadri, B., Ok, Y., Awad, Y., Surapaneni, A., Saint, C., Kumarathilaka, P.. and Vithanage, M. (2016). Biosolids application and soil organic carbon dynamics: a meta-analysis. Proceedings of 5th Annual International Research Conference on "Value Addition Evidence Based Innovation from Management Sciences to Society", 405-409.

Willén, A., Jönsson, H., Pell, M. and Rodhe, L. (2016). Emissions of nitrous oxide, methane and ammonia after field application of digested and dewatered sewage sludge with or without addition of urea. *Waste Biomass Valor.* 7(2): 281-292.

Withers, P. J. A. (2011). The agronomic and environmental impacts of phosphorus in biosolids applied to agricultural land: a review of UK research. UKWIR Report 11/SL/02/10.

Withers, P. J. A., Clay, S. D. and Breeze, V. G. (2001). Phosphorus transfer in runoff following application of fertilizer, manure, and sewage sludge. *J. Environ. Qual.* 30(1): 180-188.

Withers, P. J. A., Flynn, N. J., Warren, G. P., Taylor, M. and Chambers, B. J. (2016). Sustainable management of biosolid phosphorus: a field study. *Soil Use Manag.* 32 (Suppl. 1): 54-63.

Wu, C., Spongberg, A. L., Witter, J. D., Fang, M. and Czajkowski, K. P. (2010). Uptake of pharmaceutical and personal care products by soybean plants from soils applied with biosolids and irrigated with contaminated water. *Environ. Sci. Technol.* 44(16): 6157-6161.

Wu, X., Dodgen, L. K., Conkle, J. L. and Gan, J. (2015). Plant uptake of pharmaceutical and personal care products from recycled water and biosolids: a review. *Sci. Total Environ.* 536: 655-666.

Yang, Y., Wang, Y., Westerhoff, P., Hristovski, K., Jin, V. L., Johnson, M. V. and Arnold, J. G. (2014). Metal and nanoparticle occurrence in biosolid-amended soils. *Sci. Total Environ.* 485-486: 441-449.

Ying, G. G., Yu, X. Y. and Kookana, R. S. (2007). Biological degradation of triclocarban and triclosan in a soil under aerobic and anaerobic conditions and comparison with environmental fate modelling. *Environ. Pollut.* 150(3): 300-305.

Zeng, L. X., Wang, T., Wang, P., Liu, Q., Han, S. L., Yuan, B., Zhu, N. I., Wang, Y. W. and Jiang, G. (2011). Distribution and trophic transfer of short-chain chlorinated paraffins in an aquatic ecosystem receiving effluents from a sewage treatment plant. *Environ. Sci. Technol.* 45(13): 5529-5535.

Zerzghi, H., Gerba, C. P., Brooks, J. P. and Pepper, I. L. (2010a). Long-term effects of land application of Class B biosolids on the soil microbial populations, pathogens, and activity. *J. Environ. Qual.* 39(1): 402-408.

Zerzghi, H., Brooks, J. P., Gerba, C. P. and Pepper, I. L. (2010b). Influence of long-term land application of Class B biosolids on soil bacterial diversity. *J. Appl. Microbiol.* 109(2): 698-706.

Zhang, G. S., Zhang, F. X. and Li, X. T. (2019). Effects of polyester microfibers on soil physical properties: perception from a field and a pot experiment. *Sci. Total Environ.* 670: 1-7.

Zhang, H., Xiao, K., Liu, J., Wang, T., Liu, G., Wang, Y. and Jiang, G. (2014). Polychlorinated naphthalenes in sewage sludge from wastewater treatment plants in China. *Sci. Total Environ.* 490: 555-560.

Zhao, F. J., Salmon, S. E., Withers, P. J. A., Evans, E. J., Monaghan, J., Shewry, P. R. and McGrath, S. P. (1999a). Responses of breadmaking quality to sulphur in three wheat varieties. J. Sci. Food Agric. 7: 1865–1874. https://doi.org/10.1016/j.scitotenv.2019.03.149.

Zhao, F. J., Salmon, S. E., Withers, P. J. A., Monaghan, J. M., Evans, E. J., Shewry, P. R. and McGrath, S. P. (1999b). Variation in the breadmaking quality and rheological properties of wheat in relation to sulphur nutrition under field conditions. *J. Cereal Sci.* 30(1): 19–31.

Zhao, Y., Tan, Y., Guo, Y., Gu, X., Wang, X. and Zhang, Y. (2013). Interactions of tetracycline with Cd (II), Cu (II) and Pb (II) and their cosorption behavior in soils. *Environ. Pollut.* 180: 206–213.

Zubris, K. A. V. and Richards, B. K. (2005). Synthetic fibres as an indicator of land application of sludge. *Environ. Pollut.* 138(2): 201–211.

Chapter 12

Biofertilizers: assessing the effects of arbuscular mycorrhizal fungi on soil health

M. J. Salomon, The Waite Research Institute and The School of Agriculture, Food and Wine, The University of Adelaide, Australia; S. F. Bender, Agroscope, Switzerland; T. R. Cavagnaro, The Waite Research Institute and The School of Agriculture, Food and Wine, The University of Adelaide, Australia; and M. G. A. van der Heijden, Agroscope and University of Zurich, Switzerland

1 Introduction

Soils provide a variety of important ecosystem services and are the foundation of global biogeochemical cycles such as carbon, water and plant nutrients. They host an abundance of microorganisms, ranging from the microscopic to the macroscopic level (Adhikari and Hartemink, 2016). Healthy soils have the capacity to provide those ecosystem functions that are appropriate to its surrounding and to do so in a sustainable way. As such, healthy soils are the foundation of most food production systems, ecosystems and urban settlements (Keesstra et al., 2016, Salomon et al., 2020). Many developments of the past and present have led to land loss and land degradation. Anthropogenic activities that heavily impact soils on a global scale include intensive agriculture, deforestation and land disturbances, such as for urban settlements or mining. Those impacts can be complex and, as many systems are interwoven with one another, can result in a cascade of events with broader

http://dx.doi.org/10.19103/AS.2021.0094.17

implications on its surroundings. Estimates suggest that up to 45% of global land areas are degraded, which may undermine the well-being of 1.5 billion people (Gibbs and Salmon, 2015). Major drivers of land degradation are soil erosion, acidification, land clearance, salination or pollution through heavy metals and petrochemicals (Olsson et al., 2019).

The previous decades saw a variety of concepts and names that were used for soil assessment. Among those are soil fertility, soil quality, soil capability and, recently, soil health. Similarly, the objectives and methods of each concept progressed throughout the years. The earliest forms of soil assessment were emphasizing the suitability for crop growth, whereas newer concepts consider the multifunctionality of whole ecosystem services. Soil assessment can be done on soil biology (e.g. microbial communities), chemistry (e.g. pH and nutrients) and physics (e.g. bulk density and texture). Current developments in soil analysis allowed a shift towards the inclusion of soil biology as commonly used indicators, which were previously more focused on soil chemistry and physics. Following, we adhere to the term 'soil health' to describe sustainable and resilient ecosystem services (Bünemann et al., 2018).

The sustainability of many current agricultural practices has been questioned. For example, one consequence of suboptimal soil management is the release of soil carbon into the atmosphere which is about the same magnitude as carbon emissions caused by current deforestation events (Sanderman et al., 2017). The concentration of carbon in the soil is tightly linked with important soil characteristics, including soil aggregation and soil microbial biomass (Wilson et al., 2009b). As a consequence, the loss of soil carbon leads to deteriorated soil health and contributes to elevated atmospheric carbon dioxide (CO_2) concentrations (Rumpel et al., 2020). Intense farming systems put further pressure on soil fertility and can result in nutrient depletion. Such soils lack the capacity of replenishing essential plant nutrients, and crop yields are only sustained through the application of fertilizers (Tan et al., 2005). As a further consequence, soils might see a loss of biodiversity which deprives them of their essential ecosystem functions (van der Heijden et al., 2008).

Soils are home to a wide range of biota that have beneficial effects on plant growth and soil functioning. One of those groups is the mycorrhizal fungi. Mycorrhizal fungi follow a cosmopolitan distribution and are one of the main drivers of soil microbial interactions. Their hyphae can build up to 40 m g^{-1} of soil (Smith et al., 2004) and reach a biomass of 700–900 kg ha^{-1} (Wallander et al., 2001). Mycorrhizal fungi can be broadly categorized into ectomycorrhizas, ericoid mycorrhizas, orchid mycorrhizas and arbuscular mycorrhizas (Smith and Read, 2008). These mycorrhizas differ in their physiology, symbiotic strategies and taxonomic classification. However, they all form mycorrhizal associations with host plants that can be described as a mutualistic symbiosis. Only in particular cases or circumstances are these associations parasitic (Smith and Read, 2008).

Arbuscular mycorrhizas are plant-fungal associations in which the fungus enters the plant root and forms specialized structures, the arbuscules, which are used for nutrient exchange with the host plant (see Fig. 1). Arbuscular mycorrhizal fungi (AMF) are obligate symbionts for 80% of terrestrial plant species and most crop plants. This mutualistic relationship is improving plant nutrition by aiding in the uptake of phosphorus and zinc. In return, the plant delivers photosynthates and lipids to the fungal symbiont (Smith and Read, 2008). Besides improved plant nutrition, AMF play a crucial role in soil health and are key indicators for describing soil quality. AMF have been found to improve many aspects of soil health and counteract the negative impacts caused by inappropriate soil management (Jeffries et al., 2003). At the same time, common agricultural practices led to a diminished abundance of AMF in the soil. Such practices include simplified crop rotations, application of mineral fertilizer or soil disturbances, such as cultivation (Verbruggen et al., 2010).

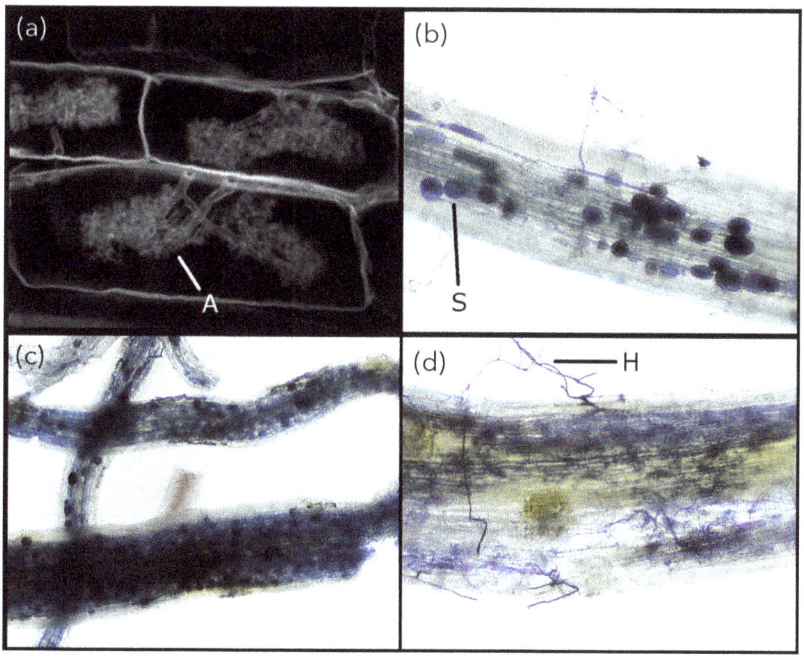

Figure 1 (a) Mycorrhizal arbuscules inside Pisum sativum root cells; (b-d) colonized maize roots with many intraradical spores and extraradical hyphae. A = arbuscule, S = spore, H = hyphae. Figure 1 (a): Photo courtesy of Ryan Geil, published with kind permission from Peterson and Massicotte (2004) and NRC Press, © Canadian Science Publishing or its licensors.

One proposed way to increase soil health in a sustainable way is the application of AMF as a biofertilizer. AMF biofertilizers are designed to bolster natural mycorrhizal communities when those have been impaired or to introduce new mycorrhizal isolates with improved functional traits. The AMF inoculum is embedded in a carrier material which might include further additives, such as organic additives (e.g. humic acids) or other plant growth-promoting microorganisms such as *Trichoderma spp.*, *Bacillus spp.* or other microorganisms. Such biofertilizers have been proposed as an alternative to mineral fertilizers and pesticides and have been known to increase yield resilience by supporting plants against abiotic stress (Berruti et al., 2016). While AMF biofertilizers hold enormous potential to improve sustainability, their integration into broad-scale applications has been challenging. The mass production of AMF propagules is defined by its symbiotic lifecycle in which mycorrhizal fungi require a host plant to propagate. This production method is linked to phytosanitary issues and requires much care to exclude plant pathogens. Although AMF can be cultured axenically, it is currently not economical to do so on a broad scale. Furthermore, most root culture systems require genetically modified hairy roots, which are produced using tumour-inducing (Ti) plasmids (Berruti et al., 2016, Adholeya et al., 2005). Recent studies demonstrated that AMF cannot produce fatty acids themselves and rely on their host plants for providing them. Thus, it is possible that specific growth media might be designed in the future, in which it is possible to grow AMF without host plants (Kameoka et al., 2019). Further issues arise when applying AMF inoculum to field soil, which already contains an established microbial community. Under those circumstances, survival and establishment of the introduced AMF needs to be evaluated (Rodriguez and Sanders, 2015, Bender et al., 2019). Furthermore, AMF show a certain host selectivity and might not perform consistently in a crop rotation (Hoeksema et al., 2010).

Although the production and application of AMF biofertilizer come with certain challenges, an increasing number of companies are attracted by its potential economic value (Vosátka et al., 2012). The global market for microbial inoculants is expected to reach US$3.622 billion by 2024, of which the mycorrhizal fungi are a major sector (Knowledge Sourcing Intelligence LLP, 2019). In most countries, the term 'biofertilizer' is not legally defined and is therefore lacking regulations and minimum standard requirements. This situation led to an unregulated market in which a high percentage of commercial AMF biofertilizers fail to induce mycorrhizal colonization. To this date, a number of scientific studies are available where the majority of tested inoculants showed unsatisfying results (Tarbell and Koske, 2007, Corkidi et al., 2017) (Salomon et al., in press). This situation is undermining the potential of AMF biofertilizer with bigger implications for sustainability in food production and consumer protection. Scientific research might be sabotaged when researchers rely on commercial mycorrhizal inoculants.

AMF biofertilizers hold the potential to significantly reduce our carbon footprint on this earth. The management of AMF can be a sustainable option to reduce agrochemicals and to increase soil health at the same time. As outlined above, the production and application of AMF biofertilizers can be challenging and needs further research before they can be adapted on a broader scale. However, those challenges should not be a reason to undermine the potential of AMF, nor should the interest in AMF applications be belittled as a 'recurring evolution' (Hart et al., 2018). Instead, the management and understanding of the soil microbiome should be treated as a valuable tool for humanity to stay within its planetary boundaries (Rockström et al., 2009). Following, we will address the key issues of how AMF can help to improve soil health and outline promising developments towards the use of AMF biofertilizers (see Table 1).

2 Arbuscular mycorrhizal fungi and soil health: addressing the key issues

2.1 *Improved soil structure and stability*

Soil structure is defined as the spatial arrangement of soil particles to form a three-dimensional matrix consisting of mineral and organic particles (aggregates) and porous spaces in between. Soil stability describes the disintegration forces necessary to disrupt this matrix. Soil structure is a key indicator of soil health as it influences a variety of important soil characteristics. Well-structured soils facilitate root growth, aeration and water infiltration into deeper soil layers, while providing water retention at the same time. High soil stability prevents soil from wind and water erosion. Soil structure and stability are generally influenced by soil physical, chemical and microbial influences. Within the microbial influences, AMF are one of the main contributors towards soil structure and stability.

The effects of AMF on soil structure are of direct and indirect nature. The direct effects include processes in which fungal mycelium enmeshes soil particles into bigger units, including soil microaggregates (<0.25 mm) and macroaggregates (>0.25 mm) (Miller and Jastrow, 1990). The indirect effects are describing how AMF can influence plants and microbial communities which then influence the soil structure (Tisdall and Oades, 1982). At the macroaggregate level, soil structure through AMF is mainly improved due to the physical force provided by the hyphal entanglement of soil particles (Miller and Jastrow, 2000). As mentioned before, AMF hyphae can build up to 40 m g^{-1} of soil (Smith et al., 2004). At the same time, the hyphal diameter is about ten times smaller than that of fine roots, allowing hyphae to penetrate even micropores (<30 μm) (Smith and Read, 2008). The well-known biochemical compound Glomalin and the group of Glomalin-related soil proteins are a biochemical pathway for soil aggregation. This group of proteins is thought to act like a glue for soil particles (Driver et al.,

Table 1 Impact of AMF on soil health: overview of soil ecosystem functions provided by AMF and their underlying mechanism

Soil ecosystem function	Mechanism	Reference
Uptake of plant nutrients and stimulation of plant growth	Improved uptake of phosphorus, nitrogen and micronutrients	Watts-Williams and Cavagnaro (2012)
	Increased plant biomass	van der Heijden et al. (1998)
	Improved drought resistance	Sanchez-Diaz and Honrubia (1994)
	Protective effects against root diseases	Hol and Cook (2005)
	Induction of systemic pathogen resistance	Pozo and Azcón-Aguilar (2007)
Improved soil structure and stability	Hyphal entanglement of soil particles	Miller and Jastrow (1990)
	Carbon deposition	Rillig and Mummey (2006)
	Changes in surface polarity of soil particles	Rillig (2005)
	Alignment of soil particles	Tisdall (1991)
	Eliminating spatial constraints	Six et al. (2004)
	Indirect effects through changes in plant and microbial communities	Rillig and Mummey (2006)
Alleviation of soil contamination	Immobilization of metals	French (2017)
	Upregulation of plant detoxification genes	Jiang et al. (2016)
	Removal of contaminants from plants through fungal structures	Göhre and Paszkowski (2006)
	Increased plant vigour through improved plant nutrition	Vogel-Mikuš and Regvar (2006)
	Degradation of organic pollutants	Wang et al. (2020)
Carbon sequestration	Significant carbon sink into the soil	Douds et al. (2000)
	Increased plant community productivity	Zhang et al. (2012)
	Improved soil structure	Wilson et al. (2009)
Nutrient retention	Improved nutrient uptake of plants	Smith and Read (2008)
	Immobilization in fungal structures	Watts-Williams and Cavagnaro (2012)
	Improved soil structure	Cavagnaro et al. (2015)

2005). However, there are open questions regarding their quantification and release into the soil and if they are specific to AMF (Rosier et al., 2006). Fungal growth is further linked to a variety of other extracellular organic compounds that have been shown to improve the soil structure, for example, through changes in the surface polarity (Gebbink et al., 2005) or carbon deposition. Mycorrhizal effects on the formation of microaggregates are less researched but hypothesized to work through physical forces on primary soil particles. The turgor pressure of hyphae during their growth could eliminate spatial constraints that would otherwise prevent the formation of microaggregates. Similarly, this physical force could align particles and bind them with organic matter (Rillig and Mummey, 2006).

Indirect effects of AMF on soil structure describe its effect on plants and soil communities, which, in return, can influence soil structure and stability. The influence of plant communities on the soil structure has been repeatedly shown, for example, in the context of agricultural (Munkholm et al., 2013) or natural ecosystems (Pérès et al., 2013). The host-selectivity of AMF thereby promotes certain plant species over others, which ultimately leads to changes in soil structure (van der Heijden et al., 2006). The effects of AMF on single plants are mainly evolving around an increase in the ratio of root to shoot biomass as a result of the mycorrhizal symbiosis. Following is a cascade of events with potential positive outcomes on the soil structure, such as increased rhizodeposition, soil entanglement by fine roots, increased root decomposition and changes to the soil water regime (Piotrowski et al., 2004). AMF can have further effects on the composition and quantity of soil microorganisms, which in return lead to changes to the soil structure (van der Heijden et al., 2008, Rillig and Mummey, 2006).

2.2 Soil contamination

Soils are considered contaminated once they contain substances (e.g. organic pollutants), elements (e.g. toxic metals) or microorganisms (e.g. pathogens) above normal concentrations or above national guideline levels (FAO, 2015). Soil contamination is a diverse topic which makes it difficult to quantify it on a global scale. The extent of soil contamination is nevertheless alarming and considered one of the major threats to soil functioning. Although soil contamination can be of natural origin, anthropogenic activities are considered as the main causes (FAO, 2015).

One of the main issues of soil contamination is the accumulation of toxic metals, for example, through the application of agrochemicals, biosolids, wastewater or mining (Wuana and Okieimen, 2011). The earliest studies about the protective effects of mycorrhizal fungi against toxic metals involved ericoid mycorrhizal fungi (Bradley et al., 1981). Since then it has been repeatedly demonstrated that also AMF can alleviate stress caused by increased metal concentrations in soils (Hildebrandt et al., 2007). It is almost a confusing phenomenon, since some metals are necessary micronutrients for plant life and their uptake is promoted by AMF (Watts-Williams and Cavagnaro, 2012). This uptake of micronutrients and the protective effect against toxic levels demonstrate the complex interaction between soils, the soil microbiome and plants. Studies at contaminated sites revealed the presence of AMF communities (Vogel-Mikuš et al., 2005), albeit their diversity is often reduced compared to non-contaminated sites. However, those remaining AMF species might be better adapted to high concentrations of toxic metals (Del Val et al., 1999, Galli et al., 1994). The underlying mechanisms are either changing the fate of the metals in the soil or change the plant's response towards them. One established mechanism is the immobilization of metals in intra- and extraradical fungal structures through metallothioneins or other chelating agents (Lanfranco, 2007, French, 2017). Interactions between metals and Glomalin-related soil proteins have been reported as well, leading to similar forms of immobilization (Yang et al., 2017). Other proposed mechanisms are related to changes in the plant physiology which are caused by the AM symbiosis. These changes involve the upregulation of plant genes involved in detoxification (Jiang et al., 2016) or the removal of contaminants from plant roots through fungal structures such as arbuscules (Göhre and Paszkowski, 2006). More recent studies also investigated the effects of AMF on organic pollutants in soil. The mechanisms involved are similar to those of potentially toxic metals but also include the degradation of pollutants through enzymes. Again, the overall protective effect is composed of multiple mechanisms and interactions between soil microorganisms and plants (Wang et al., 2020). The effects of many other common pollutants such as pesticides and micro-plastics are still poorly investigated. Especially the effects of mixed pesticides in combination with other soil stress factors on soil health are still poorly understood (Rillig et al., 2019b).

2.3 Carbon sequestration

Soils are the largest terrestrial carbon storage. The amount of carbon that is stored in soils exceeds that of the atmosphere and the global plant biomass combined (Scharlemann et al., 2014). The carbon cycle is one of the most important processes on Earth to which almost every organism is linked in one way or another. Plants are one of the main drivers of this process and as such also their symbionts at the root-soil interface. Plants forming associations with AMF allocate up to 20% of carbon to AMF and plants forming ectomycorrhizas can allocate up to 50% of their assimilates, demonstrating that mycorrhizal associations are substantially involved in the carbon cycle (Soudzilovskaia et al., 2015). The prehistoric events that saw a decline in atmospheric carbon are linked to the emergence of deeply rooted trees around 450 million years ago and angiosperms around 130 million years ago. Mycorrhizal associations have been confirmed in root fossils dating back more than 400 million years ago (Remy et al., 1994) and are, therefore, considered crucial for the evolution of terrestrial plants, indicating that their role in carbon cycling might be of equal importance (Taylor et al., 2009). Soil carbon concentration is an important factor for soil health and is the foundation of almost all other soil components. Having high concentrations of soil carbon is therefore considered valuable for 'healthy soils'. However, due to unsustainable management, soils are increasingly releasing carbon into the atmosphere in the form of CO_2. This development is aggravating climate change when soils could actually be used as carbon storage and bind atmospheric CO_2 into the ground. One initiative predicts that an annual growth rate of 0.4% soil carbon could have significant effects on soil health and contribute to limit global warming at the same time (Kon Kam King et al., 2018).

Studying the impact of AMF on the soil carbon cycle is a difficult undertaking on an ecosystem level. Pot studies under controlled conditions allow to separate the carbon inputs between roots and AMF. However, on an ecosystem level, this becomes increasingly complicated, as an almost unmanageable amount of processes interact with AMF and vice versa. On the level of a single plant, mycorrhizal associations are translocating photosynthates from the plant directly into the soil. Hexose is the preferred metabolite for AMF which is sourced from plants and then used for growth and reproduction. Estimates suggest that between 5% and 20% of plant-derived carbon is translocated to AMF associations (Douds and Shachar-Hill, 2000). The overall carbon sink through AMF mycelium thereby accumulates to 50–900 kg carbon ha^{-1} (Zhu and Michael Miller, 2003). Although this carbon drain might suggest negative impacts on plant growth, most mycorrhizal plants react to mycorrhizal colonization by producing more biomass. AMF provide many advantages such as improved nutrient uptake which then leads to higher photosynthetic rates. Another mechanism on how AMF improve carbon sequestration is their positive effect on soil stability and soil aggregation. Mycorrhizas are critical components of the terrestrial carbon cycle and shape plants and soils alike. Conversely, depriving soils of their mycorrhizal potential leads to implications that go beyond the effects on a single plant (Wilson et al., 2009a).

2.4 Nutrient retention

High concentrations of plant nutrients can be challenging when those are not fully integrated into the ecosystem. Nutrient availability that is exceeding its demand leads to nutrient loss, either through leaching, gaseous emissions or erosion. This situation is mainly observed after the application of fertilizers, for example, in the agricultural context. Nutrients that are mobile within the soil matrix can leach out and make them inaccessible to the root system. Various forms of nitrogen are at risk of being lost through gaseous emissions, such as in the form of nitrous oxide (N_2O). Immobile nutrients such as some forms of phosphorus are more susceptible to soil erosion than leaching. Although phosphorus can also bind to mobile particles and then leach through the soil. Such excess nutrients can have negative impacts on their surroundings. Leached or wind-dispersed nutrients contaminate water ways and groundwater, thereby damaging the aquatic biodiversity. N_2O is a potent greenhouse gas with significant global warming potential. The overall fertilizer efficiency is sobering, especially when considering the global scale of fertilizer application. It is estimated that 30% of nitrogen and 15–30% of phosphorus are lost in the same year of application (Quan et al., 2020, Syers et al., 2008).

One method by which AMF can improve nutrient retention is obviously due to its role in plant nutrition, especially in the case of phosphorus. Without mycorrhizal associations, plant roots are inside a depletion zone once all immobile nutrients (like phosphorus) around the roots are taken up, making them dependent on nutrients moving to roots via mass flow, usually a slow process. Plants with mycorrhizal colonization can escape this depletion zone through the hyphae which can penetrate a bigger soil volume than roots by itself. This way, phosphorus gets bound into fungal and plant biomass which immobilizes it and protects it from erosion and leaching. Furthermore, mycorrhiza show enhanced mineralization of complex-bound or organic forms of phosphorus which would otherwise be unavailable to plant roots and susceptible of leaching. Another essential plant nutrient that is impacted by AMF is nitrogen. To this date, the exact interactions between AMF and the various forms of nitrogen are still to be investigated (Hodge and Fitter, 2010). However, it is known that AMF can take up nitrogen in the form of ammonium (NH_4^+) and amino acids. Evidence suggests that even other organic forms of nitrogen might be involved, especially since AMF are equipped with nitrogen reductase genes. The net balance of AMF and its role in nitrogen uptake on plants differ between studies. Whereas some studies show almost no contribution, other studies come to the opposite result. Although the effect of AMF on the nitrogen uptake of the plant is variable, its impact on reducing nitrogen losses has been proven repeatedly (Cavagnaro et al., 2015).

Another important mechanism on how AMF help in nutrient retention is through effects on soil and plant water relations. There is some evidence that AMF affect plant water transport and can support plant water uptake (Augé et al., 2015, Bowles et al., 2016). Enhanced plant water uptake will consequently reduce the amount of leachate and the amount of nutrients being transported down the soil profile, which can also reduce N_2O emissions (Lazcano et al., 2014). In addition, the above-mentioned effects of AMF on enhanced soil aggregation can lead to significant effects on water retention. Both mechanisms together directly impact the amount of leaching in soils (Cavagnaro et al., 2015). Altogether, the impact of nutrient retention due to AMF is compelling. Nutrient retention of up to 80% has been reported for nitrogen and up to 60% for phosphorus (van der Heijden, 2010, Corkidi et al., 2011). Gaseous emissions of N_2O were reduced between about 30% and 50% after the application of nitrate fertilizer (Bender et al., 2014, 2015).

3 Arbuscular mycorrhizal fungi biofertilizer production

As outlined in the previous sections, mycorrhizal fungi have been linked to increased soil health and increased plant vigour. They are key players in the soil microbiome and at the soil-root interface. These characteristics make them promising alternatives to agrochemicals which can be used in sustainable agriculture and ecosystem restoration. The idea of using AMF as biofertilizers is almost as old as the systematic research of AMF itself. However, most work to transfer the scientific research into efficient applications has been rather futile up to this date. AMF biofertilizers are still considered a niche product and are mostly limited to the hobby market or some horticulture systems (Rouphael et al., 2015). The main reasons for the narrow application range being high costs (Berruti et al., 2016), unreliable product quality (Salomon et al., in press) and questions regarding the establishment under field conditions (Rodriguez and Sanders, 2015). Regardless of those issues, their huge environmental and economic potential is fuelling continuous research and major investments from agrochemical companies (Vosátka et al., 2008).

The earliest systems for the propagation of AMF used host plants in pot cultures and sterilized soil, which were then inoculated with the desired species. Preferred host plants are maize or sorghum, as their fast-growing root system allows substantial sporulation (Ijdo et al., 2011). To this date, it is still the most common method for mass production of mycorrhizal fungi, as it is relatively cheap and can be easily upscaled (Ijdo et al., 2011). This *in vivo* system underwent multiple modifications, such as the closed bag system which would reduce the costs for maintenance and helped to exclude potential phytopathogens (Walker and Vestberg, 1994). Multiple publications describe the use of soil-less substrate, such as sand (Jentschke et al., 1999), which could further help with phytosanitary issues and eliminate the need for soil sterilization. It is even possible to produce on-site inoculum on agricultural waste products, yielding 3600 spores in 100 mL^{-1} of soil (Chaiyasen et al., 2017). The simplicity of *in vivo* systems allows easy propagation and maintenance of mycorrhizal cultures. Major bottlenecks for the use in commercial biofertilizer production are phytosanitary issues and the difficulty to extract spores from the substrate, for example, to be further concentrated or to be used with different carrier materials (see Fig. 2a).

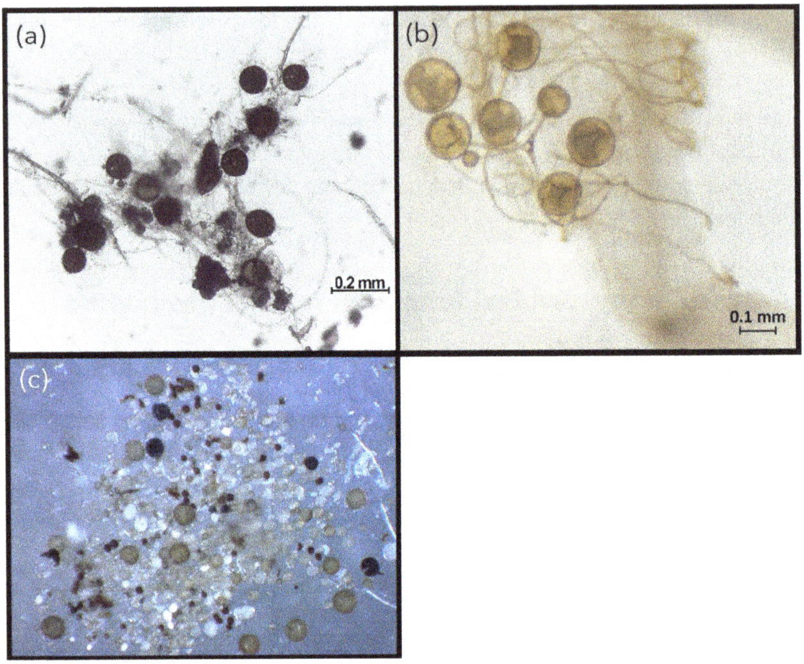

Figure 2 (a) Extracted AMF spores from a commercial product; (b) spores of an in vitro culture of Rhizophagus clarus growing on tomato roots; (c) spores of various sizes and colours as can be found in natural soils. Photo courtesy of Luise Köhl.

The need for axenic AMF systems led to the development of *in vitro* cultures on Ti-transformed hairy roots (Declerck et al., 1996) (see Fig. 2b). The motivation behind this development was not only its potential use for the production of biofertilizers but also providing a research tool for the study of AMF. The original method made it possible to produce up to 10 000 spores of *Glomus versiforme* within 4 months. The spore production could be further increased by using a split-plate design (65 000 spores) (Douds et al., 2000) or whole plants instead of hairy roots (Voets et al., 2005). The latest research even demonstrated the asymbiotic sporulation of *Rhizophagus irregularis* using culture medium with fatty acids (Kameoka et al., 2019). AMF propagules in axenic systems can be easily extracted and are free of plant pathogens by default. They can be added to a variety of carrier materials which makes them applicable for most production systems. However, axenic cultures require trained personal and high material costs which make it too expensive for most biofertilizer applications. Also, not all AMF species can be cultured *in vitro* and continuous subculturing has been shown to lead to changes in functional traits (Kokkoris and Hart, 2019). Some producers question the ability of axenic cultures to establish in the field (e.g. because the AMF did not adapt and grow under real soil conditions), and this is an issue that needs further testing.

Other approaches for the production of AMF inoculum include hydroponic or aeroponic systems. To this date, there have been a variety of adaptations, each with customized designs and slightly different nutrient solutions. One of the earliest works used the nutrient flow technique (NFT) to produce *Glomus mosseae* propagules on maize, yielding up to 50% mycorrhizal root infection (Elmes and Mosse, 1984). This system provided good growth conditions with relatively little root disturbance to protect the fungal structures. But also deep water culture systems with reduced aeration times have been successfully used (Hawkins and George, 1997). Aeroponic systems have certain advantages over hydroponics, such as increased root aeration which supports plant growth and reduces the likelihood of certain diseases. Nutrients and moisture are applied as aerosols around the roots. Common systems convert the nutrient solution into micro-droplets (spray nozzles), mist (atomizers) or ultrasonic fog (piezo elements) (Jarstfer and Sylvia, 1995). With regard to the production of mycorrhizal inoculum, aeroponics put less physical stress onto the root systems than hydroponics. Production rates of up to 50 spores cm^{-1} of colonized roots have been reported (Hung and Sylvia, 1988). Aeroponics and hydroponics allow easy spore extraction through sieving the roots and moist roots can be further processed to sheared-root inoculum using a common food blender. Sheared-root inoculum can only be made from wet roots and then stored for up to 1 month, whereas unprocessed and dried roots retain their inoculum potential for up to 2 years, but are not suitable for the production of sheared-root inoculum (Sylvia and Jarstfer, 1992). Those examples provide proof of the flexibility of AMF production systems and explain how they can be advanced to fulfil specific requirements. Most companies do not disclose their production methods due to issues about intellectual property. However, we can assume that successful companies developed efficient production systems which might be based on the mentioned hydroponic or aeroponic systems. Systems that have been made public involve an airlift bioreactor (Jolicoeur et al., 1999) or a semi-hydroponic set-up (Declerck et al., 2009).

4 Managing arbuscular mycorrhizal fungi for soil health

4.1 Agriculture

The management of AMF in agriculture has been subject to recent discussions which were mainly focused on insufficient correlations between the yield of agricultural crops and their mycorrhizal colonization (Ryan and Graham, 2018). It is obvious that many plants can grow without AMF assuming that sufficient amounts of fertilizers are added. However, it is important to consider the whole spectrum of ecosystem services which are provided by AMF, especially those related to soil health (Chen et al., 2018). Those additional services are often not as visible as crop yields but nevertheless of high economic and ecological importance (Rillig et al., 2019a). Especially in times of extreme weather events, good soil health and functioning mycorrhizal symbiosis can be considered as a 'health insurance' for crops to achieve yield resilience (Rivero et al., 2018). However, such things are often difficult to measure, scientifically and monetarily. Moreover, AMF can help to reduce the reliance on fertilizers and make agriculture more sustainable. Future research should focus on the role that AMF can play for soil ecological engineering, producing the same amount of food with less inputs.

Mycorrhizal colonization of crop plants through indigenous AMF can be increased through changes in agricultural practices. Key variables which have been identified are the use of cover crops and the type of tillage system (Bowles et al., 2017). Those results indicate that mycorrhizal colonization of summer crops increases by 30% when working with minimal soil disturbance and cover crops instead of barrow fallows. Furthermore, the AMF species richness increased by 11%, highlighting the sensitivity of certain species to soil disturbance. Further research focused on the comparison between organic and conventional agriculture, whereas organic systems are usually associated with higher AMF abundance and richness. Again, tillage and cover crops have been considered as major influences, as well as the impact of fertilization (Borriello et al., 2012). Mineral fertilizers are generally associated with reduced AMF abundance and species richness, caused by their high availability and effects on soil chemistry (Oehl et al., 2004). Most organic farms also incorporate more diverse crop rotations and take advantage of the many positive effects of legumes and mixed crops on soil health (Verbruggen et al., 2010).

Research shows that soils can be actively managed to support AMF and increase soil health. Such measures include organic farming, permanent crop cover, crop rotation and the inclusion of temporary pastures in rotations. However, these methods come with limitations. Organic farming is relying on ploughing and tilling as means to control weeds and phytopathogens. Arid farmlands in many parts of the world do not support multiple crops within one season. It is especially those harsh environments that could benefit from the many advantages of AMF. Canola is an important cash crop in organic and conventional farming systems alike. However, being a non-mycorrhizal plant, it has negative impacts on following crops in terms of mycorrhizal colonization (Valetti et al., 2016). Where options to manage indigenous AMF are limited or weakened by non-mycorrhizal crops, soils can be actively inoculated with AMF biofertilizer. This inoculum could be adapted to the local conditions and provide optimized plant and ecosystem functions (Sanders, 2010).

Although AMF provide multiple ecosystem services, they need to provide clear economic benefits to find their way into most broad-scale applications.

Assuming that most farmers do not get direct fiscal support to protect their soils, the costs of biofertilizer need to be weighed against the cost of potential fertilizer-savings or yield and quality gains. Major factors for this cost-benefit analysis are the amount of necessary inoculum and the specific mycorrhizal growth response of the crop. Due to its relatively low seed density, potato is a crop with strong prospects for profitable AMF biofertilization. Hijri (2016) demonstrated in field trials over 4 years an average yield increase of about 4 t ha^{-1}. The profitability threshold was reached in almost 80% of all trials while an average of 71 spores was applied per seed potato at the time of planting. Tawaraya et al. (2012) achieved significant savings in phosphorus fertilization for the Welsh onion (*Allium fistulosum*). Typical for this crop, the seeds were pre-planted under greenhouse conditions where seedlings were either treated with *Glomus etunicatum* 'R10' or left uninoculated. Plants were then transplanted into the field under various fertilization treatments. The yield of the inoculated plants at 300 mg P_2O_5 kg^{-1} soil was similar to those in 1000 mg P_2O_5 kg^{-1} soil and not inoculated, whereas the costs of the difference in superphosphate application were double the costs of the inoculum. Not part of this equation is the savings in environmental costs due to potential fertilizer run-off and/or leaching. For future research, it is important that on-farm experiments are being performed under real-world agricultural conditions and that inoculation tools that are practically feasible for farmers are provided.

Other common agricultural crops are more difficult to inoculate at reasonable costs, mostly due to higher seeding rates or smaller yield gains after inoculation. Most studies in the field of AMF do not include an economic analysis and use laboratory cultures of AMF rather than commercial inoculants. Economic analysis on self-propagated cultures would be unreasonable as they are not produced on a commercial scale and therefore produced at much higher costs. Ignoring the economic factors of biofertilization, myriads of studies show that field inoculation with AMF is producing promising outcomes for plant growth (Thompson et al., 2013, Cely et al., 2016, Al-Karaki et al., 2004, Pellegrino and Bedini, 2014) and soil health (see Tables 2 and 3). In terms of efficient application of AMF for large seed quantities, seed coating is a promising solution to deliver not only AMF but also other beneficial microorganisms directly to the plant (Rocha et al., 2019b). Successful implementation of this technique, in combination with AMF and agricultural crops, has been reported (Rocha et al., 2019a, Oliveira et al., 2016), even in combination with fungicidal seed coating (Cameron et al., 2017).

4.2 Ecological restoration

The science of ecological restoration is focused on trajectories of change, that is, on the succession, assembly and state transition of natural communities. The

Table 2 Various meta-analysis on AMF involving their effect on plant growth and nutrient uptake. No additional analysis is presented here

Analyzed interaction	Outcome	Reference
Photosynthesis under salt stress	Alleviation of salinity stress and increased photosynthetic rate	Wang et al. (2019)
Wetland plant performance	Significant benefits, even under flood conditions	Ramírez-Viga et al. (2018)
Nutritional and non-nutritional factors in crops	Positive effects on nutrient uptake, soil aggregation, water flow and disease resistance	Delavaux et al. (2017)
Potato yield	Increased yields due to AMF inoculation, with economic benefits	Hijri (2016)
Response of wheat to AMF	AMF field inoculation increased nutrient uptake and dry weight	Pellegrino et al. (2015)
Copper, manganese and iron concentration in crops	Significant role for copper and iron and limited role for manganese	Lehmann and Rillig (2015)
Zinc nutrition in crop plants	Positive impacts on zinc concentration in shoot, root and fruit	Lehmann et al. (2014)
Nutrient uptake of tomato	Particularly beneficial for phosphorus and zinc	Watts-Williams and Cavagnaro (2014)
Plant growth under water stress	Improved drought resistance of plants	Jayne and Quigley (2014)
Mycorrhizal responsiveness in crop plants and wild relatives	No evidence of decreased mycorrhizal responsiveness	Lehmann et al. (2012)
Context-dependant plant response to AMF	Plant response most positive when plants are phosphorus limited	Hoeksema et al. (2010)

specific aims range from re-introduction of single species to population and community restorations. Broadly, ecological restoration can be defined as the assisted recovery of ecosystems that have been degraded or destroyed. Due to the various positive effects of AMF on the ecosystem and plants, they are considered a promising tool to drive this restoration process. In that way, they can increase the survivability of plant species in contaminated soils or favour certain plants over others, such as unwanted neophytes. Ultimately, restoration efforts in combination with AMF lead to the enhanced establishment of plant communities with positive effects on soil health (Asmelash et al., 2016).

In most cases, natural ecosystems do not allow active soil management such as in the case of agriculture. Most restoration efforts are focusing on transplanting and seeding of plants. Various studies demonstrated that AMF inoculation can be successfully merged with those methods. Zhang et al. (2012)

Table 3 Studies involving AMF field inoculation (biofertilization) and their effects on improving soil health

Host plant	Primary AMF ecosystem function	Outcome	Reference
Ricinus communis	Soil contamination	Increased plant survival and amelioration of lead soil pollution	González-Chávez et al. (2019)
Elymus trachycaulus, *Lotus corniculatus*	Soil contamination	Increased plant growth in coarse tailings sands	Boldt-Burisch et al. (2018)
Eight grassland plant species	Soil contamination	Higher plant cover in post-mine sandpits	Ohsowski et al. (2018)
Prunus discadenia, *Prunus dictyneura*, *Xanthoceras sorbifolium*, *Armeniaca sibirica*	Carbon sequestration, soil contamination	Increased carbon sequestration in coalfield soil	Wang et al. (2016)
Oryza glaberrima	Nutrient retention	Decreased nitrogen run-off at various fertilization levels	Zhang et al. (2016)
Viburnum tinus	Soil structure, soil contamination	Improved soil structure and growth of plants with wastewater treatment	Gómez-Bellot et al. (2015)
Oryza glaberrima	Nutrient retention	Decreased N_2O	Zhang et al. (2015)
Miscanthus × giganteus	Soil contamination	Alleviation of plant stress in metal contaminated sites	Firmin et al. (2015)
Agrostis capillaris	Soil contamination	Protective effect against arsenic toxicity in plants	Neagoe et al. (2014)
Meta-analysis	Soil structure	Overall positive effect of AMF inoculation on soil aggregation	Leifheit et al. (2014)
Soil macrocosm	Soil structure	Increased soil aggregate stability	Martin et al. (2012)
Erodium oxyrrhynchum, *Hyalea pulchella*, *Trigonella arcuata*, *Schismus arabicus*	Soil contamination	Increased plant cover and community productivity	Zhang et al. (2012)
Chrysopogon zizanioides	Soil contamination	Protective effects in lead and zinc mine tailings	Wu et al. (2010)
Medicago sativa	Soil contamination	Enhanced phytoremediation in polychlorinated biphenyl (PCB) contaminated soils	Teng et al. (2010)

Timeframe: 2020–2010.
Source: Scopus search: 'AMF mycorrhiza field inoculation'.

applied a mix of three lab-cultured AMF species to grassland by drilling holes and refilling them with the inoculum. Over 3 years, the number of established seedlings and their community productivity was significantly higher than in the uninoculated control. White et al. (2008) drilled seeds with inoculum into the soil, as well as broadcasting both onto the surface, achieving similar colonization results. However, in this case, there were no positive mycorrhizal effects due to the inoculation, which might be caused by high levels of phosphorus in the soil and a high abundance of native AMF communities. The efficiency between native and introduced AMF is still to be debated. Williams et al. (2012) pre-inoculated tree cuttings, whereas inoculum made of native AMFs and a natural forest performed better than a commercial one. The positive effects of seedling inoculation with natural AMF communities had also been demonstrated in other studies (Emam, 2016). It becomes evident that field inoculation of natural ecosystems with AMF is a feasible option for ecological restorations, and, as shown in some studies, can have long-lasting effects over multiple years. To this date, the species effectiveness between native and introduced AMF is not fully understood. However, we do have a good understanding of the importance that soil microorganisms have on above-ground ecological complexes. Managing below-ground mycorrhizas seems to be one important step in order to harbour the full potential of restoration efforts. Khan (2006) defined the term 'Mycorrhizoremediation' as an enhanced form of phytoremediation.

5 Conclusion

In this chapter we highlighted how AMF can enhance soil health and how they could be used as biofertilizers, either in agriculture or ecological restoration. The main mechanisms by which AMF improve soil health include improved soil structure and the immobilization of potentially toxic metals. The latest research even showed that AMF can enhance the degradation of organic pollutants. Due to the nature of their symbiosis, they act as a significant carbon sink for the soil, thereby increasing soil carbon. Carbon sequestration for soils is not only improving soil health but can also be used to trap atmospheric CO_2 back into the soil, thereby contributing to mitigate climate change. In the context of climate change, we also addressed the potential of AMF to reduce gaseous nutrient losses from soils, such as in the form of N_2O. This gas is mineralized from other forms of nitrogen in the soil and is a potent climate gas. AMF further reduce the leaching of nitrogen into the ground water and phosphorus-loss through soil erosion. Nutrient loss from soils is not only impacting soil fertility but is also critical with regard to the surrounding ecosystem, especially the aquatic biodiversity.

The highlighted effects of AMF on soil health can be used to increase sustainability in food production and to restore degraded soils. It is obvious that AMF could be applied as biofertilizers in various scenarios. Their economic and ecological potential was also commercialized by a variety of companies all around the world. The production and application of AMF is challenging and requires expertise. For people working in the field of mycorrhizal research, it is not surprising that many of those products fail to induce mycorrhizal colonization. This was also confirmed by some studies where only a fraction of the tested products were considered viable. During the past centuries, a variety of AMF production systems have been proposed which uses soil, hydroponics, aeroponics, or axenic cultures. Each system has its advantages and disadvantages, and to this date, the mass production of high-quality AMF propagules is an expensive endeavour. Nevertheless, investments in mycorrhizal products are fuelled by an endless stream of mycorrhizal research that highlights again and again the positive effects of AMF and how it could be used to reduce our footprint on this earth. The value of AMF field inoculation has been proven for a variety of scenarios, such as agriculture systems or to enhance phytoremediation efforts. In some cases where the targeted host plant is exceptionally responsive to mycorrhiza and where only relatively small amounts of inoculum are necessary, the use of AMF biofertilizer can be economically viable already today.

6 Future trends in research

In recent years, particularly with the advent of high throughput sequencing, our knowledge of soil microbial communities has increased greatly. It has long been known that soil microorganisms, including AMF, provide essential ecosystem services which we are now able to better quantify. The next great step in soil ecology involves the assignment of soil functions to microbial communities. In doing so, we will be able to quantify, and then predict, the impacts that management and other causes of environmental change (e.g. climate change) have on soils and the ecosystem services they provide. Armed with this knowledge we will then be able to better understand how to best manage systems to support the benefits of the soil microbiome. Such knowledge will also assist in the development of reliable biofertilizers. Those biofertilizers are not limited to AMF but the whole spectrum of beneficial soil microorganisms. One utopian vision of the future would involve a replacement of most agrochemicals with biofertilizers.

We have observed and also predict a continued interest in more sustainable farming systems in the future. Such systems seek to produce more food, on less land and with fewer inputs. At the same time, they are equipped against adverse conditions, such as the increasing occurrence of extreme weather events. This is a tall order but one we cannot afford to not meet in the context of an increasing global population and increased food requirements in the coming decades.

Moreover, we have observed a growing interest in agricultural paradigms that is shifting towards biologically regulated nutrient supply, rather than the importation of externally sourced synthetic inputs. This development would help to 'close the loop' and minimize the movement of resources on and off farm.

This is an exciting time to be studying soil ecology. There are many challenges in the here and now, and in the future to come. We contend that the soil microbiome, including mycorrhizal fungi, may hold many of the answers to meet those challenges and to do so in a sustainable way.

7 Where to look for further information

The following articles and books provide a good overview of the subject:

- Smith, S. E. and Read, D. (2008): *Mycorrhizal Symbiosis,* 3rd Edition, Academic Press, London.
- Brundrett, M., Bougher, N., Dell, B., Grove, T. and Malajczuk, N. (1995): *Working with Mycorrhizas in Forestry and Agriculture*, Australian Centre for International Agricultural Research, Canberra.
- Declerck, S., Strullu, D.-G. and Fortin, A. (2005): *In Vitro Culture of Mycorrhizas*, Springer, Berlin Heidelberg.
- Fisseha, A., Bekele, T. and Birhane, E. (2016): The Potential Role of Arbuscular Mycorrhizal Fungi in the Restoration of Degraded Lands, *Front Microbiol.* 7: 1095.

Research associations:

- International Mycorrhiza Society (http://mycorrhizas.org).

AMF collections:

- Banque Européenne des Glomeromycota (BEG) (www.kent.ac.uk/bio/beg).
- Glomeromycota In Vitro Collection (GINCO) (http://res2.agr.gc.ca/ecorc/ginco-can/index_e.htm).
- International Culture Collection of Vesicular-Arbuscular Mycorrhizal Fungi (INVAM) (http://invam.caf.wvu.edu).

8 References

Adhikari, K. and Hartemink, A. E. 2016. Linking soils to ecosystem services–a global review. *Geoderma* 262, 101-111.

Adholeya, A., Tiwari, P. and Singh, R. 2005. Large-scale inoculum production of arbuscular mycorrhizal fungi on root organs and inoculation strategies. In: Declerck, S., Fortin,

J. A. and Strullu, D.-G. (Eds) *In Vitro Culture of Mycorrhizas* (pp. 315-338). Berlin, Heidelberg: Springer.

Al-Karaki, G., McMichael, B. and Zak, J. 2004. Field response of wheat to arbuscular mycorrhizal fungi and drought stress. *Mycorrhiza* 14(4), 263-269.

Asmelash, F., Bekele, T. and Birhane, E. 2016. The potential role of arbuscular mycorrhizal fungi in the restoration of degraded lands. *Frontiers in Microbiology* 7, 1095.

Augé, R. M., Toler, H. D. and Saxton, A. M. 2015. Arbuscular mycorrhizal symbiosis alters stomatal conductance of host plants more under drought than under amply watered conditions: a meta-analysis. *Mycorrhiza* 25(1), 13-24.

Bender, S. F., Conen, F. and van der Heijden, M. G. A. 2015. Mycorrhizal effects on nutrient cycling, nutrient leaching and N2O production in experimental grassland. *Soil Biology and Biochemistry* 80, 283-292.

Bender, S. F., Plantenga, F., Neftel, A., Jocher, M., Oberholzer, H. R., Koehl, L., Giles, M., Daniell, T. J. and van der Heijden, M. G. A. 2014. Symbiotic relationships between soil fungi and plants reduce N_2O emissions from soil. *ISME Journal* 8(6), 1336-1345.

Bender, S. F., Schlaeppi, K., Held, A. and van der Heijden, M. G. A. 2019. Establishment success and crop growth effects of an arbuscular mycorrhizal fungus inoculated into Swiss corn fields. *Agriculture, Ecosystems and Environment* 273, 13-24.

Berruti, A., Lumini, E., Balestrini, R. and Bianciotto, V. 2016. Arbuscular mycorrhizal fungi as natural biofertilizers: let's benefit from past successes. *Frontiers in Microbiology* 6, 1559.

Boldt-Burisch, K., Naeth, M. A., Schneider, U., Schneider, B. and Hüttl, R. F. 2018. Plant growth and arbuscular mycorrhizae development in oil sands processing by-products. *Science of the Total Environment* 621, 30-39.

Borriello, R., Lumini, E., Girlanda, M., Bonfante, P. and Bianciotto, V. 2012. Effects of different management practices on arbuscular mycorrhizal fungal diversity in maize fields by a molecular approach. *Biology and Fertility of Soils* 48(8), 911-922.

Bowles, T. M., Barrios-Masias, F. H., Carlisle, E. A., Cavagnaro, T. R. and Jackson, L. E. 2016. Effects of arbuscular mycorrhizae on tomato yield, nutrient uptake, water relations, and soil carbon dynamics under deficit irrigation in field conditions. *Science of the Total Environment* 566-567, 1223-1234.

Bowles, T. M., Jackson, L. E., Loeher, M., Cavagnaro, T. R. and Nuñez, M. 2017. Ecological intensification and arbuscular mycorrhizas: a meta-analysis of tillage and cover crop effects. *Journal of Applied Ecology* 54(6), 1785-1793.

Bradley, R., Burt, A. J. and Read, D. J. 1981. Mycorrhizal infection and resistance to heavy metal toxicity in Calluna vulgaris. *Nature* 292(5821), 335-337.

Bünemann, E. K., Bongiorno, G., Bai, Z., Creamer, R. E., De Deyn, G., De Goede, R., Fleskens, L., Geissen, V., Kuyper, T. W., Mäder, P., Pulleman, M., Sukkel, W., Van Groenigen, J. W. and Brussaard, L. 2018. Soil quality–a critical review. *Soil Biology and Biochemistry* 120, 105-125.

Cameron, J. C., Lehman, R. M., Sexton, P., Osborne, S. L. and Taheri, W. I. 2017. Fungicidal seed coatings exert minor effects on arbuscular mycorrhizal fungi and plant nutrient content. *Agronomy Journal* 109(3), 1005-1012.

Cavagnaro, T. R., Bender, S. F., Asghari, H. R. and van der Heijden, M. G. A. 2015. The role of arbuscular mycorrhizas in reducing soil nutrient loss. *Trends in Plant Science* 20(5), 283-290.

Cely, M. V. T., De Oliveira, A. G., De Freitas, V. F., De Luca, M. B., Barazetti, A. R., Dos Santos, I. M. O., Gionco, B., Garcia, G. V., Prete, C. E. C. and Andrade, G. 2016. Inoculant of arbuscular mycorrhizal fungi (*Rhizophagus clarus*) increase yield of soybean and cotton under field conditions. *Frontiers in Microbiology* 7, 720.

Chaiyasen, A., Chaiya, L., Douds, D. D. and Lumyong, S. 2017. Influence of host plants and soil diluents on arbuscular mycorrhizal fungus propagation for on-farm inoculum production using leaf litter compost and agrowastes. *Biological Agriculture and Horticulture* 33(1), 52–62.

Chen, M., Arato, M., Borghi, L., Nouri, E. and Reinhardt, D. 2018. Beneficial services of arbuscular mycorrhizal fungi–from ecology to application. *Frontiers in Plant Science* 9, 1270.

Corkidi, L., Allen, E. B., Merhaut, D. J., Allen, M. F., Downer, J. A., Bohn, J. and Evans, M. 2017. Assessing the infectivity of commercial mycorrhizal inoculants in plant nursery conditions. *Journal of Environmental Horticulture* 22(3), 149–154.

Corkidi, L., Merhaut, D. J., Allen, E. B., Downer, J., Bohn, J. and Evans, M. 2011. Effects of mycorrhizal colonization on nitrogen and phosphorus leaching from nursery containers. *HortScience* 46(11), 1472–1479.

Declerck, S., Ijdo, M., Suarez, K. F., Voets, L. and De La Providencia, I. 2009. Method and system for in vitro mass production of arbuscular mycorrhizal fungi. Worldwide Patent Application.

Declerck, S., Strullu, D. G. and Plenchette, C. 1996. In vitro mass-production of the arbuscular mycorrhizal fungus, Glomus versiforme, associated with Ri T-DNA transformed carrot roots. *Mycological Research* 100(10), 1237–1242.

Del Val, C., Barea, J. M. and Azcón-Aguilar, C. 1999. Diversity of arbuscular mycorrhizal fungus populations in heavy-metal-contaminated soils. *Applied and Environmental Microbiology* 65(2), 718–723.

Delavaux, C. S., Smith-Ramesh, L. M. and Kuebbing, S. E. 2017. Beyond nutrients: a meta-analysis of the diverse effects of arbuscular mycorrhizal fungi on plants and soils. *Ecology* 98(8), 2111–2119.

Douds, D. and Shachar-Hill, Y. 2000. Carbon partitioning, cost, and metabolism of arbuscular mycorrhizas. In: Arbuscular Mycorrhizas: Physiology and Function (pp. 107–129). Springer, Dordrecht.

Douds, D. D., Pfeffer, P. E. and Shachar-Hill, Y. 2000. Application of in vitro methods to study carbon uptake and transport by AM fungi. *Plant and Soil* 226(2), 255–261.

Driver, J. D., Holben, W. E. and Rillig, M. C. 2005. Characterization of glomalin as a hyphal wall component of arbuscular mycorrhizal fungi. *Soil Biology and Biochemistry* 37(1), 101–106.

Elmes, R. P. and Mosse, B. 1984. Vesicular-arbuscular endomycorrhizal inoculum production. II. Experiments with maize (*Zea mays*) and other hosts in nutrient flow culture. *Canadian Journal of Botany* 62(7), 1531–1536.

Emam, T. 2016. Local soil, but not commercial AMF inoculum, increases native and non-native grass growth at a mine restoration site. *Restoration Ecology* 24(1), 35–44.

FAO 2015. Status of the world's soil resources (SWSR)-main report. *Food and Agriculture Organization of the United Nations and intergovernmental technical panel on soils*, Rome, Italy, 650. Available at: http://www.fao.org/documents/card/en/c/c6814873-efc3-41db-b7d3-2081a10ede50/.

Firmin, S., Labidi, S., Fontaine, J., Laruelle, F., Tisserant, B., Nsanganwimana, F., Pourrut, B., Dalpé, Y., Grandmougin, A., Douay, F., Shirali, P., Verdin, A. and Lounès-Hadj Sahraoui, A. 2015. Arbuscular mycorrhizal fungal inoculation protects Miscanthus ×

giganteus against trace element toxicity in a highly metal-contaminated site. *Science of the Total Environment* 527, 91-99.

French, K. E. 2017. Engineering mycorrhizal symbioses to alter plant metabolism and improve crop health. *Frontiers in Microbiology* 8, 1403.

Galli, U., Schüepp, H. and Brunold, C. 1994. Heavy metal binding by mycorrhizal fungi. *Physiologia Plantarum* 92, 364-368.

Gebbink, M. F., Claessen, D., Bouma, B., Dijkhuizen, L. and Wösten, H. A. 2005. Amyloids—a functional coat for microorganisms. *Nature Reviews. Microbiology* 3(4), 333-341.

Gibbs, H. K. and Salmon, J. M. 2015. Mapping the world's degraded lands. *Applied Geography* 57, 12-21.

Göhre, V. and Paszkowski, U. 2006. Contribution of the arbuscular mycorrhizal symbiosis to heavy metal phytoremediation. *Planta* 223(6), 1115-1122.

Gómez-Bellot, M. J., Ortuño, M. F., Nortes, P. A., Vicente-Sánchez, J., Martín, F. F., Bañón, S. and Sánchez-Blanco, M. J. 2015. Protective effects of *Glomus iranicum var. tenuihypharum* on soil and *Viburnum tinus* plants irrigated with treated wastewater under field conditions. *Mycorrhiza* 25(5), 399-409.

González-Chávez, M. D. C. A., Carrillo-González, R., Cuellar-Sánchez, A., Delgado-Alvarado, A., Suárez-Espinosa, J., Ríos-Leal, E., Solís-Domínguez, F. A. and Maldonado-Mendoza, I. E. 2019. Phytoremediation assisted by mycorrhizal fungi of a Mexican defunct lead-acid battery recycling site. *Science of the Total Environment* 650(2), 3134-3144.

Hart, M. M., Antunes, P. M., Chaudhary, V. B. and Abbott, L. K. 2018. Fungal inoculants in the field: is the reward greater than the risk? *Functional Ecology* 32(1), 126-135.

Hawkins, H.-J. and George, E. 1997. Hydroponic culture of the mycorrhizal fungus *Glomus mosseae* with *Linum usitatissimum* L., Sorghum bicolor L. and *Triticum aestivum* L. *Plant and Soil* 196(1), 143-149.

Hijri, M. 2016. Analysis of a large dataset of mycorrhiza inoculation field trials on potato shows highly significant increases in yield. *Mycorrhiza* 26(3), 209-214.

Hildebrandt, U., Regvar, M. and Bothe, H. 2007. Arbuscular mycorrhiza and heavy metal tolerance. *Phytochemistry* 68(1), 139-146.

Hodge, A. and Fitter, A. H. 2010. Substantial nitrogen acquisition by arbuscular mycorrhizal fungi from organic material has implications for N cycling. *Proceedings of the National Academy of Sciences of the United States of America* 107(31), 13754-13759.

Hoeksema, J. D., Chaudhary, V. B., Gehring, C. A., Johnson, N. C., Karst, J., Koide, R. T., Pringle, A., Zabinski, C., Bever, J. D., Moore, J. C., Wilson, G. W. T., Klironomos, J. N. and Umbanhowar, J. 2010. A meta-analysis of context-dependency in plant response to inoculation with mycorrhizal fungi. *Ecology Letters* 13(3), 394-407.

Hol, W. H. G. and Cook, R. 2005. An overview of arbuscular mycorrhizal fungi-nematode interactions. *Basic and Applied Ecology* 6(6), 489-503.

Hung, L. L. and Sylvia, D. M. 1988. Production of vesicular-arbuscular mycorrhizal fungus inoculum in aeroponic culture. *Applied and Environmental Microbiology* 54(2), 353-357.

Ijdo, M., Cranenbrouck, S. and Declerck, S. 2011. Methods for large-scale production of AM fungi: past, present, and future. *Mycorrhiza* 21(1), 1-16.

Jarstfer, A. G. and Sylvia, D. M. 1995. Aeroponic culture of VAM fungi. In: Varma, A. and Hock, B. (Eds) *Mycorrhiza: Structure, Function, Molecular Biology and Biotechnology* (pp. 427-441). Berlin, Heidelberg: Springer.

Jayne, B. and Quigley, M. 2014. Influence of arbuscular mycorrhiza on growth and reproductive response of plants under water deficit: a meta-analysis. *Mycorrhiza* 24(2), 109–119.

Jeffries, P., Gianinazzi, S., Perotto, S., Turnau, K. and Barea, J.-M. 2003. The contribution of arbuscular mycorrhizal fungi in sustainable maintenance of plant health and soil fertility. *Biology and Fertility of Soils* 37(1), 1–16.

Jentschke, G., Brandes, B., Heinzemann, J., Marschner, P. and Godbold, D. L. 1999. Sand culture of mycorrhizal plants. *Journal of Plant Nutrition and Soil Science* 162(1), 107–112.

Jiang, Q. Y., Zhuo, F., Long, S. H., Zhao, H. D., Yang, D. J., Ye, Z. H., Li, S. S. and Jing, Y. X. 2016. Can arbuscular mycorrhizal fungi reduce Cd uptake and alleviate Cd toxicity of *Lonicera japonica* grown in Cd-added soils? *Scientific Reports* 6, 21805.

Jolicoeur, M., Williams, R. D., Chavarie, C., Fortin, J. A. and Archambault, J. 1999. Production of *Glomus intraradices* propagules, an arbuscular mycorrhizal fungus, in an airlift bioreactor. *Biotechnology and Bioengineering* 63(2), 224–232.

Kameoka, H., Tsutsui, I., Saito, K., Kikuchi, Y., Handa, Y., Ezawa, T., Hayashi, H., Kawaguchi, M. and Akiyama, K. 2019. Stimulation of asymbiotic sporulation in arbuscular mycorrhizal fungi by fatty acids. *Nature Microbiology* 4(10), 1654 –1660.

Keesstra, S. D., Bouma, J., Wallinga, J., Tittonell, P., Smith, P., Cerdà, A., Montanarella, L., Quinton, J. N., Pachepsky, Y., Van Der Putten, W. H., Bardgett, R. D., Moolenaar, S., Mol, G., Jansen, B. and Fresco, L. O. 2016. The significance of soils and soil science towards realization of the United Nations sustainable development goals. *SOIL* 2(2), 111–128.

Khan, A. G. 2006. Mycorrhizoremediation–an enhanced form of phytoremediation. *Journal of Zhejiang University. Science. B* 7(7), 503–514.

Knowledge Sourcing Intelligence LLP 2019. Agricultural inoculants market–forecasts from 2019 to 2024. Available at: https://www.researchandmarkets.com/reports/4895565/agricultural-inoculants-market-forecasts-from.

Kokkoris, V. and Hart, M. 2019. In vitro propagation of arbuscular mycorrhizal fungi may drive fungal evolution. *Frontiers in Microbiology* 10, 2420.

Kon Kam King, J., Granjou, C., Fournil, J. and Cecillon, L. 2018. Soil sciences and the French 4 per 1000 initiative–the promises of underground carbon. *Energy Research and Social Science* 45, 144–152.

Lanfranco, L. 2007. The fine-tuning of heavy metals in mycorrhizal fungi. *New Phytologist* 174(1), 3–6.

Lazcano, C., Barrios-Masias, F. H. and Jackson, L. E. 2014. Arbuscular mycorrhizal effects on plant water relations and soil greenhouse gas emissions under changing moisture regimes. *Soil Biology and Biochemistry* 74, 184–192.

Lehmann, A., Barto, E. K., Powell, J. R. and Rillig, M. C. 2012. Mycorrhizal responsiveness trends in annual crop plants and their wild relatives–a meta-analysis on studies from 1981 to 2010. *Plant and Soil* 355(1–2), 231–250.

Lehmann, A. and Rillig, M. C. 2015. Arbuscular mycorrhizal contribution to copper, manganese and iron nutrient concentrations in crops–a meta-analysis. *Soil Biology and Biochemistry* 81, 147–158.

Lehmann, A., Veresoglou, S. D., Leifheit, E. F. and Rillig, M. C. 2014. Arbuscular mycorrhizal influence on zinc nutrition in crop plants–a meta-analysis. *Soil Biology and Biochemistry* 69, 123–131.

Leifheit, E. F., Veresoglou, S. D., Lehmann, A., Morris, E. K. and Rillig, M. C. 2014. Multiple factors influence the role of arbuscular mycorrhizal fungi in soil aggregation–a meta-analysis. *Plant and Soil* 374(1-2), 523-537.

Martin, S. L., Mooney, S. J., Dickinson, M. J. and West, H. M. 2012. Soil structural responses to alterations in soil microbiota induced by the dilution method and mycorrhizal fungal inoculation. *Pedobiologia* 55(5), 271-281.

Miller, R. M. and Jastrow, J. D. 1990. Hierarchy of root and mycorrhizal fungal interactions with soil aggregation. *Soil Biology and Biochemistry* 22(5), 579-584.

Miller, R. M. and Jastrow, J. D. 2000. Mycorrhizal fungi influence soil structure. In: Kapulnik, Y. and Douds, D. D. (Eds) *Arbuscular Mycorrhizas: Physiology and Function* (pp. 3-18). Dordrecht, Netherlands: Springer.

Munkholm, L. J., Heck, R. J. and Deen, B. 2013. Long-term rotation and tillage effects on soil structure and crop yield. *Soil and Tillage Research* 127, 85-91.

Neagoe, A., Stancu, P., Nicoară, A., Onete, M., Bodescu, F., Gheorghe, R. and Iordache, V. 2014. Effects of arbuscular mycorrhizal fungi on *Agrostis capillaris* grown on amended mine tailing substrate at pot, lysimeter, and field plot scales. *Environmental Science and Pollution Research International* 21(11), 6859-6876.

Oehl, F., Sieverding, E., Mader, P., Dubois, D., Ineichen, K., Boller, T. and Wiemken, A. 2004. Impact of long-term conventional and organic farming on the diversity of arbuscular mycorrhizal fungi. *Oecologia* 138(4), 574-583.

Oliveira, R. S., Rocha, I., Ma, Y., Vosátka, M. and Freitas, H. 2016. Seed coating with arbuscular mycorrhizal fungi as an ecotechnological approach for sustainable agricultural production of common wheat (Triticum aestivum L.). *Journal of Toxicology and Environmental Health. Part A* 79(7), 329-337.

Olsson, L., Barbosa, H., Bhadwal, S., Cowie, A., Delusca, K., Flores-Renteria, D., Hermans, K., Jobbagy, E., Kurz, W., Li, D., Sonwa, D. J. and Stringer, L. 2019. Land Degradation: IPCC Special Report on Climate Change, Desertification, Land 5 Degradation, Sustainable Land Management, Food Security, and 6 Greenhouse gas fluxes in Terrestrial Ecosystems. Available at: https://www.ipcc.ch/srccl/chapter/chapter-4/.

Ohsowski, B. M., Dunfield, K., Klironomos, J. N. and Hart, M. M. 2018. Plant response to biochar, compost, and mycorrhizal fungal amendments in post-mine sandpits. *Restoration Ecology* 26(1), 63-72.

Pellegrino, E. and Bedini, S. 2014. Enhancing ecosystem services in sustainable agriculture: biofertilization and biofortification of chickpea (*Cicer arietinum* L.) by arbuscular mycorrhizal fungi. *Soil Biology and Biochemistry* 68, 429-439.

Pellegrino, E., Öpik, M., Bonari, E. and Ercoli, L. 2015. Responses of wheat to arbuscular mycorrhizal fungi: a meta-analysis of field studies from 1975 to 2013. *Soil Biology and Biochemistry* 84, 210-217.

Pérès, G., Cluzeau, D., Menasseri, S., Soussana, J. F., Bessler, H., Engels, C., Habekost, M., Gleixner, G., Weigelt, A., Weisser, W. W., Scheu, S. and Eisenhauer, N. 2013. Mechanisms linking plant community properties to soil aggregate stability in an experimental grassland plant diversity gradient. *Plant and Soil* 373(1-2), 285-299.

Peterson, R. L. and Massicotte, H. B. 2004. Exploring structural definitions of mycorrhizas, with emphasis on nutrient-exchange interfaces. Canadian Journal of Botany 82(8), 1074-1088.

Piotrowski, J. S., Denich, T., Klironomos, J. N., Graham, J. M. and Rillig, M. C. 2004. The effects of arbuscular mycorrhizas on soil aggregation depend on the interaction between plant and fungal species. *New Phytologist* 164(2), 365-373.

Pozo, M. J. and Azcón-Aguilar, C. 2007. Unraveling mycorrhiza-induced resistance. *Current Opinion in Plant Biology* 10(4), 393–398.

Quan, Z., Li, S., Zhang, X., Zhu, F., Li, P., Sheng, R., Chen, X., Zhang, L.-M., He, J.-Z., Wei, W. and Fang, Y. 2020. Fertilizer nitrogen use efficiency and fates in maize cropping systems across China: field 15N tracer studies. *Soil and Tillage Research* 197, 104498.

Ramírez-Viga, T. K., Aguilar, R., Castillo-Argüero, S., Chiappa-Carrara, X., Guadarrama, P. and Ramos-Zapata, J. 2018. Wetland plant species improve performance when inoculated with arbuscular mycorrhizal fungi: a meta-analysis of experimental pot studies. *Mycorrhiza* 28(5–6), 477–493.

Remy, W., Taylor, T. N., Hass, H. and Kerp, H. 1994. Four hundred-million-year-old vesicular arbuscular mycorrhizae. *Proceedings of the National Academy of Sciences of the United States of America* 91(25), 11841–11843.

Rillig, M. C. 2005. A connection between fungal hydrophobins and soil water repellency? *Pedobiologia* 49(5), 395–399.

Rillig, M. C., Aguilar-Trigueros, C. A., Camenzind, T., Cavagnaro, T. R., Degrune, F., Hohmann, P., Lammel, D. R., Mansour, I., Roy, J., van der Heijden, M. G. A. and Yang, G. 2019a. Why farmers should manage the arbuscular mycorrhizal symbiosis. *New Phytologist* 222(3), 1171–1175.

Rillig, M. C., Ryo, M., Lehmann, A., Aguilar-Trigueros, C. A., Buchert, S., Wulf, A., Iwasaki, A., Roy, J. and Yang, G. 2019b. The role of multiple global change factors in driving soil functions and microbial biodiversity. *Science* 366(6467), 886–890.

Rillig, M. C. and Mummey, D. L. 2006. Mycorrhizas and soil structure. *New Phytologist* 171(1), 41–53.

Rivero, J., Álvarez, D., Flors, V., Azcón-Aguilar, C. and Pozo, M. J. 2018. Root metabolic plasticity underlies functional diversity in mycorrhiza-enhanced stress tolerance in tomato. *New Phytologist* 220(4), 1322–1336.

Rocha, I., Ma, Y., Carvalho, M. F., Magalhães, C., Janoušková, M., Vosátka, M., Freitas, H. and Oliveira, R. S. 2019a. Seed coating with inocula of arbuscular mycorrhizal fungi and plant growth promoting rhizobacteria for nutritional enhancement of maize under different fertilisation regimes. *Archives of Agronomy and Soil Science* 65(1), 31–43.

Rocha, I., Ma, Y., Souza-Alonso, P., Vosátka, M., Freitas, H. and Oliveira, R. S. 2019b. Seed coating: a tool for delivering beneficial microbes to agricultural crops. *Frontiers in Plant Science* 10, 1357.

Rockström, J., Steffen, W., Noone, K., Persson, A., Chapin, F. S., Lambin, E. F., Lenton, T. M., Scheffer, M., Folke, C., Schellnhuber, H. J., Nykvist, B., De Wit, C. A., Hughes, T., Van Der Leeuw, S., Rodhe, H., Sörlin, S., Snyder, P. K., Costanza, R., Svedin, U., Falkenmark, M., Karlberg, L., Corell, R. W., Fabry, V. J., Hansen, J., Walker, B., Liverman, D., Richardson, K., Crutzen, P. and Foley, J. A. 2009. A safe operating space for humanity. *Nature* 461(7263), 472–475.

Rodriguez, A. and Sanders, I. R. 2015. The role of community and population ecology in applying mycorrhizal fungi for improved food security. *The ISME Journal* 9(5), 1053–1061.

Rosier, C. L., Hoye, A. T. and Rillig, M. C. 2006. Glomalin-related soil protein: assessment of current detection and quantification tools. *Soil Biology and Biochemistry* 38(8), 2205–2211.

Rouphael, Y., Franken, P., Schneider, C., Schwarz, D., Giovannetti, M., Agnolucci, M., Pascale, S. D., Bonini, P. and Colla, G. 2015. Arbuscular mycorrhizal fungi act as biostimulants in horticultural crops. *Scientia Horticulturae* 196, 91–108.

Rumpel, C., Amiraslani, F., Chenu, C., Garcia Cardenas, M., Kaonga, M., Koutika, L. S., Ladha, J., Madari, B., Shirato, Y., Smith, P., Soudi, B., Soussana, J. F., Whitehead, D. and Wollenberg, E. 2020. The 4p1000 initiative: opportunities, limitations and challenges for implementing soil organic carbon sequestration as a sustainable development strategy. *Ambio* 49(1), 350-360.

Ryan, M. H. and Graham, J. H. 2018. Little evidence that farmers should consider abundance or diversity of arbuscular mycorrhizal fungi when managing crops. *New Phytologist* 220(4), 1092-1107.

Salomon, M. J., Watts-Williams, S. J., McLaughlin, M. J. and Cavagnaro, T. R. 2020. Urban soil health: a city-wide survey of chemical and biological properties of urban agriculture soils. *Journal of Cleaner Production* 275, 122900.

Sanchez-Diaz, M. and Honrubia, M. 1994. Water relations and alleviation of drought stress in mycorrhizal plants. In: *Impact of Arbuscular Mycorrhizas on Sustainable Agriculture and Natural Ecosystems* (pp. 167-178). Basel: Birkhäuser.

Sanderman, J., Hengl, T. and Fiske, G. J. 2017. Soil carbon debt of 12,000 years of human land use. *Proceedings of the National Academy of Sciences* 114(36), 9575 -9580.

Sanders, I. R. 2010. 'Designer' mycorrhizas?: using natural genetic variation in AM fungi to increase plant growth. *The ISME Journal* 4(9), 1081-1083.

Scharlemann, J. P. W., Tanner, E. V. J., Hiederer, R. and Kapos, V. 2014. Global soil carbon: understanding and managing the largest terrestrial carbon pool. *Carbon Management* 5(1), 81-91.

Six, J., Bossuyt, H., Degryze, S. and Denef, K. 2004. A history of research on the link between (micro) aggregates, soil biota, and soil organic matter dynamics. *Soil and Tillage Research* 79(1), 7-31.

Smith, S. E. and Read, D. 2008. Mycorrhizal symbiosis. In: Smith, S. E. and Read, D. (Eds) *Mycorrhizal Symbiosis* (3rd edn., pp. 11-145). London: Academic Press.

Smith, S. E., Smith, F. A. and Jakobsen, I. 2004. Functional diversity in arbuscular mycorrhizal (AM) symbioses: the contribution of the mycorrhizal P uptake pathway is not correlated with mycorrhizal responses in growth or total P uptake. *New Phytologist* 162(2), 511-524.

Soudzilovskaia, N. A., van der Heijden, M. G. A., Cornelissen, J. H. C., Makarov, M. I., Onipchenko, V. G., Maslov, M. N., Akhmetzhanova, A. A. and Van Bodegom, P. M. 2015. Quantitative assessment of the differential impacts of arbuscular and ectomycorrhiza on soil carbon cycling. *New Phytologist* 208(1), 280-293.

Syers, J., Johnston, A. and Curtin, D. 2008. Efficiency of soil and fertilizer phosphorus use. *FAO Fertilizer and Plant Nutrition Bulletin* 18. Rome, Italy: Food and Agriculture Organization of the United Nations.

Sylvia, D. M. and Jarstfer, A. G. 1992. Sheared-root inocula of vesicular-arbuscular mycorrhizal fungi. *Applied and Environmental Microbiology* 58(1), 229-232.

Tan, Z. X., Lal, R. and Wiebe, K. D. 2005. Global soil nutrient depletion and yield reduction. *Journal of Sustainable Agriculture* 26(1), 123-146.

Tarbell, T. J. and Koske, R. E. 2007. Evaluation of commercial arbuscular mycorrhizal inocula in a sand/peat medium. *Mycorrhiza* 18(1), 51-56.

Tawaraya, K., Hirose, R. and Wagatsuma, T. 2012. Inoculation of arbuscular mycorrhizal fungi can substantially reduce phosphate fertilizer application to Allium fistulosum L. and achieve marketable yield under field condition. *Biology and Fertility of Soils* 48(7), 839-843.

Taylor, L. L., Leake, J. R., Quirk, J., Hardy, K., Banwart, S. A. and Beerling, D. J. 2009. Biological weathering and the long-term carbon cycle: integrating mycorrhizal evolution and function into the current paradigm. *Geobiology* 7(2), 171-191.

Teng, Y., Luo, Y., Sun, X., Tu, C., Xu, L., Liu, W., Li, Z. and Christie, P. 2010. Influence of arbuscular mycorrhiza and rhizobium on phytoremediation by alfalfa of an agricultural soil contaminated with weathered PCbs: a field study. *International Journal of Phytoremediation* 12(5), 516-533.

Thompson, J. P., Clewett, T. G. and Fiske, M. L. 2013. Field inoculation with arbuscular-mycorrhizal fungi overcomes phosphorus and zinc deficiencies of linseed (Linum usitatissimum) in a vertisol subject to long-fallow disorder. *Plant and Soil* 371(1-2), 117-137.

Tisdall, J. 1991. Fungal hyphae and structural stability of soil. *Soil Research* 29(6), 729-743.

Tisdall, J. M. and Oades, J. M. 1982. Organic matter and water-stable aggregates in soils. *Journal of Soil Science* 33(2), 141-163.

Valetti, L., Iriarte, L. and Fabra, A. 2016. Effect of previous cropping of rapeseed (Brassica napus L.) on soybean (Glycine max) root mycorrhization, nodulation, and plant growth. *European Journal of Soil Biology* 76, 103-106.

van der Heijden, M. G. 2010. Mycorrhizal fungi reduce nutrient loss from model grassland ecosystems. *Ecology* 91(4), 1163-1171.

van der Heijden, M. G. A., Bardgett, R. D. and Van Straalen, N. M. 2008. The unseen majority: soil microbes as drivers of plant diversity and productivity in terrestrial ecosystems. *Ecology Letters* 11(3), 296-310.

van der Heijden, M. G. A., Klironomos, J. N., Ursic, M., Moutoglis, P., Streitwolf-Engel, R., Boller, T., Wiemken, A. and Sanders, I. R. 1998. Mycorrhizal fungal diversity determines plant biodiversity, ecosystem variability and productivity. *Nature* 396(6706), 69-72.

van der Heijden, M. G. A., Streitwolf-Engel, R., Riedl, R., Siegrist, S., Neudecker, A., Ineichen, K., Boller, T., Wiemken, A. and Sanders, I. R. 2006. The mycorrhizal contribution to plant productivity, plant nutrition and soil structure in experimental grassland. *New Phytologist* 172(4), 739-752.

Verbruggen, E., Röling, W. F. M., Gamper, H. A., Kowalchuk, G. A., Verhoef, H. A. and van der Heijden, M. G. A. 2010. Positive effects of organic farming on below-ground mutualists: large-scale comparison of mycorrhizal fungal communities in agricultural soils. *New Phytologist* 186(4), 968-979.

Voets, L., Dupré De Boulois, H., Renard, L., Strullu, D. G. and Declerck, S. 2005. Development of an autotrophic culture system for the in vitro mycorrhization of potato plantlets. *FEMS Microbiology Letters* 248(1), 111-118.

Vogel-Mikuš, K., Drobne, D. and Regvar, M. 2005. Zn, Cd and Pb accumulation and arbuscular mycorrhizal colonisation of pennycress *Thlaspi praecox* Wulf. (Brassicaceae) from the vicinity of a lead mine and smelter in Slovenia. *Environmental Pollution* 133(2), 233-242.

Vogel-Mikuš, K. and Regvar, M. 2006. Arbuscular mycorrhiza as a tolerance strategy in metal contaminated soils: prospects in phytoremediation. *New Topics in Environmental Research*, 37-56.

Vosátka, M., Albrechtová, J. and Patten, R. 2008. The international market development for mycorrhizal technology. In: Varma, A. (Ed) *Mycorrhiza: State of the Art, Genetics and Molecular Biology, Eco-Function, Biotechnology, Eco-Physiology, Structure and Systematics* (pp. 419-438). Berlin, Heidelberg: Springer.

Vosátka, M., Látr, A., Gianinazzi, S. and Albrechtová, J. 2012. Development of arbuscular mycorrhizal biotechnology and industry: current achievements and bottlenecks. *Symbiosis* 58(1-3), 29-37.

Walker, C. and Vestberg, M. 1994. A simple and inexpensive method for producing and maintaining closed pot cultures of arbuscular mycorrhizal fungi. *Agricultural and Food Science* 3(3), 233-240.

Wallander, H., Nilsson, L. O., Hagerberg, D. and Bååth, E. 2001. Estimation of the biomass and seasonal growth of external mycelium of ectomycorrhizal fungi in the field. *New Phytologist* 151(3), 753-760.

Wang, F., Adams, C. A., Yang, W., Sun, Y. and Shi, Z. 2020. Benefits of arbuscular mycorrhizal fungi in reducing organic contaminant residues in crops: implications for cleaner agricultural production. *Critical Reviews in Environmental Science and Technology* 50(15), 1580-1612.

Wang, Y., Wang, J., Yan, X., Sun, S. and Lin, J. 2019. The effect of arbuscular mycorrhizal fungi on photosystem II of the host plant Under salt stress: a meta-analysis. *Agronomy* 9(12), 806.

Wang, Z. G., Bi, Y. L., Jiang, B., Zhakypbek, Y., Peng, S. P., Liu, W. W. and Liu, H. 2016. Arbuscular mycorrhizal fungi enhance soil carbon sequestration in the coalfields, northwest China. *Scientific Reports* 6, 34336.

Watts-Williams, S. J. and Cavagnaro, T. R. 2012. Arbuscular mycorrhizas modify tomato responses to soil zinc and phosphorus addition. *Biology and Fertility of Soils* 48(3), 285-294.

Watts-Williams, S. J. and Cavagnaro, T. R. 2014. Nutrient interactions and arbuscular mycorrhizas: a meta-analysis of a mycorrhiza-defective mutant and wild-type tomato genotype pair. *Plant and Soil* 384(1-2), 79-92.

White, J. A., Tallaksen, J. and Charvat, I. 2008. The effects of arbuscular mycorrhizal fungal inoculation at a roadside prairie restoration site. *Mycologia* 100(1), 6-11.

Williams, A., Norton, D. A. and Ridgway, H. J. 2012. Different arbuscular mycorrhizal inoculants affect the growth and survival of *Podocarpus cunninghamii* restoration plantings in the Mackenzie Basin, New Zealand. *New Zealand Journal of Botany* 50(4), 473-479.

Wilson, G. W., Rice, C. W., Rillig, M. C., Springer, A. and Hartnett, D. C. 2009. Soil aggregation and carbon sequestration are tightly correlated with the abundance of arbuscular mycorrhizal fungi: results from long-term field experiments. *Ecology Letters* 12(5), 452-461.

Wu, S. C., Wong, C. C., Shu, W. S., Khan, A. G. and Wong, M. H. 2010. Mycorrhizo-remediation of lead/zinc mine tailings using vetiver: a field study. *International Journal of Phytoremediation* 13(1), 61-74.

Wuana, R. A. and Okieimen, F. E. 2011. Heavy metals in contaminated soils: a review of sources, chemistry, risks and best available strategies for remediation. *ISRN Ecology* 2011, 1-20, 402647.

Yang, Y., He, C., Huang, L., Ban, Y. and Tang, M. 2017. The effects of arbuscular mycorrhizal fungi on glomalin-related soil protein distribution, aggregate stability and their relationships with soil properties at different soil depths in lead-zinc contaminated area. *PLoS ONE* 12(8), e0182264.

Zhang, S., Wang, L., Ma, F., Zhang, X. and Fu, D. 2016. Reducing nitrogen runoff from paddy fields with arbuscular mycorrhizal fungi under different fertilizer regimes. *Journal of Environmental Sciences* 46, 92-100.

Zhang, T., Sun, Y., Shi, Z. and Feng, G. 2012. Arbuscular mycorrhizal fungi can accelerate the restoration of degraded spring grassland in Central Asia. *Rangeland Ecology and Management* 65(4), 426–432.

Zhang, X., Wang, L., Ma, F. and Shan, D. 2015. Effects of arbuscular mycorrhizal fungi on N20 emissions from rice paddies. *Water, Air, and Soil Pollution* 226(7), 222.

Zhu, Y. G. and Michael Miller, R. M. 2003. Carbon cycling by arbuscular mycorrhizal fungi in soil-plant systems. *Trends in Plant Science* 8(9), 407–409.

Chapter 13

Biofertilizers: assessing the effects of plant growth-promoting bacteria (PGPB) or rhizobacteria (PGPR) on soil and plant health

Elisa Zampieri, Institute for Sustainable Plant Protection, Italy; Iakovos S. Pantelides, Cyprus University of Technology, Cyprus; and Raffaella Balestrini, Institute for Sustainable Plant Protection, Italy

1 Introduction

The green revolution was driven by agricultural intensification resulting in increased productivity and incomes, but also a dependence on chemical substances in many developing countries of the world (Aeron et al., 2020). The uncontrolled application of synthetic agrochemicals imposes serious negative impacts on the environment and human and animal health, leading to a reduction in soil fertility and microbial diversity, soil pollution and environmental degradation and the development of resistance in phytopathogens and pests (Aeron et al., 2020). However, global demand for agricultural crops is increasing with yields still insufficient to face the ever-growing food demand (Timmusk et al., 2017). In this context, unsustainable agricultural intensification has often led to pollution, overexploitation of natural areas and resources, loss of soil fertility, soil erosion, salinization, runoff and in some cases desertification (IPCC, 2019, special

http://dx.doi.org/10.19103/AS.2021.0094.18

report; https://www.ipcc.ch/srccl/). Drought and land degradation following the salinization of soil are considerably increasing worldwide, and the ongoing climate change could amplify the negative effects of this scenario (Corwin, 2021). For these reasons, together with the increasing awareness of consumers about healthy food, sustainable agricultural practices are encouraged as alternatives to mineral fertilizers and synthetic pesticides (Brodt et al., 2011).

Sustainable agricultural management practices include the use of beneficial microorganisms, such as mycorrhizal fungi, rhizobia and other plant growth-promoting bacteria (PGPB) or rhizobacteria (PGPR), to support plant protection and nutrition and assist water conservation. Today, these beneficial microorganisms (i.e. arbuscular mycorrhizal fungi (AMF), PGPB or PGPR) are considered a key factor for managing crop productions (Schlaeppi and Bulgarelli, 2015). However, their application in agriculture is still a challenge due to inconsistent and unpredictable results, which often are context-dependent (dos Santos et al., 2020; Compant et al., 2019). There are many aspects that need to be considered for a successful implementation of AMF and PGPB/PGPR as microbial inoculants with desired outputs in different crop genotypes and upon different (and combined) stress conditions (Pascale et al., 2020). These microbial inoculants are living microorganisms that colonize the rhizosphere (i.e. the zone surrounding the roots that is directly influenced by plant root secretions) and/or the inner regions of plant tissues and promote plant growth or act as biological control agents (BCAs) against soilborne and seedborne plant pathogens (Aeron et al., 2020; Khatoon et al., 2020; Raj et al., 2020; Tsolakidou et al., 2019; Orozco-Mosqueda et al., 2018; Bhattacharyya and Jha, 2012). Additionally, understanding the effect of cropland management on soil microorganism dynamics is fundamental for designing better management practices to restore soil function in intensively managed agricultural systems (Baritz et al., 2018; http://www.fao.org/3/a-bl813e.pdf).

2 Mechanisms mediated by plant growth-promoting bacteria/rhizobacteria

Generally, about 2–5% of the total rhizospheric bacteria are PGPB/PGPR (Antoun and Prévost, 2006). Features that allow bacterial survival in the rhizosphere and plant tissue colonization are motility, chemotaxis, attachment, growth and stress resistance (Bulgarelli et al., 2013). Some PGPB/PGPR are considered biofertilizers that augment the availability of nutrients in a form that can be easily assimilated by plants, while others act as biocontrol agents or biopesticides that suppress or control plant disease (Timmusk et al., 2017). Many PGPB/PGPR can solubilize insoluble soil phosphate to release soluble phosphorus and making it available to plants. This trait is very interesting since the phosphorus content of soil is about 0.05% (w/w) but only 0.1% of this fraction is available to plants. Phosphorus is an essential element involved in many important metabolic pathways, such as photosynthesis, respiration, electron transport chain, biosynthesis of macromolecules and signal transduction (Khan et al., 2010). It also influences root growth, seed development and normal crop maturity (Heydari et al., 2019). Many bacterial and fungal strains, such as *Bacillus*, *Pseudomonas* or *Penicillium*, that release organic acids or phosphatases are capable to solubilize phosphorus and are, therefore, promising as PGPR (Khatoon et al., 2020; Bulgarelli et al., 2013). Apart from phosphate solubilisation (Figure 1), many other mechanisms mediated by PGPB/PGPR can lead to plant-growth promotion and improve plant tolerance/resistance to abiotic and biotic stresses (Glick, 2012), such as synthesis of hormones (abscisic acid (ABA), gibberellic acid, cytokinins and auxins) (Pérez-Flores et al., 2017; Bhattacharyya et al., 2015), nitrogen fixation (Ashraf et al., 2011), solubilization of other nutrients (Zn, K) (Vikram and Hamzehzarghani, 2008; Etesami et al., 2017; Zaheer et al., 2019), production of siderophores (Sinha and Parli, 2020; Calvo et al., 2017), ethylene (ET) control in emerging plants under stress conditions through the production of aminocyclopropane-1-carboxylate (ACC) deaminase (Ravanbakhsh et al., 2018; Glick, 2012), secretion of several molecules, including antibiotics, hydrolytic enzymes and volatile organic compounds, alleviating biotic stress effects and contributing to systemic resistance (Meena et al., 2020; Kour et al., 2019; Orozco-Mosqueda et al., 2018), production of exopolysaccharides (EPS) and biofilm formation (Dimkpa et al., 2009), heavy metal detoxification (Sharma and Archana, 2016; Tak et al., 2013; Ma et al., 2011) and pest management (Subbanna et al., 2018).

An increasing number of plants have been reported to benefit from PGPB/PGPR presence (Santos et al., 2019), including the model plant *Arabidopsis thaliana* (Lee et al., 2020) and several crops, such as winter wheat (*Triticum aestivum*) (Awan et al., 2020; Turan et al., 2012) and wheat (*Triticum durum*) (Bechtaoui et al., 2019), rice (*Oryza sativa*) (Xiao et al., 2020; Andreozzi et al., 2019;), sunflower (*Helianthus annuus*) (Ambrosini et al., 2012), rape (*Brassica napus*) (Farina et al., 2012), runner bean (*Phaseolus coccineus*) (Stefan et al., 2013) and faba bean (*Vicia faba*) (Bechtaoui et al., 2019), corn (*Zea mays*) (Batista et al., 2018; Tchuisseu Tchakounté et al., 2018; Arruda et al., 2013), soybean (*Glycine max*) (Batista et al., 2018), chickpea (*Cicer arietinum*) (Bisht et al., 2019), tomato (*Solanum lycopersicum*) (Kalam et al., 2020; Pellegrini et al., 2020), potato (*Solanum tuberosum*) (Pellegrini et al., 2020), flax (*Linum usitatissimum*) (Planchon et al., 2021), coriander (*Coriandrum sativum*) (Jiménez-Gómez, et al., 2020) and spinach (*Spinacia oleracea*) (Zafar-Ul-Hye et al., 2020). Apart from increased plant biomass, PGPR have demonstrated positive effects on total oil content and lipid composition in *G. max*, *B. napus* and *Buglossoides arvensis* that are important sources of oleic, linoleic, α-linolenic and omega-3 stearidonic acids (Jiménez et al., 2020). In addition, PGPR have been reported to improve carotenoids, tocopherols, and folate content in horseradish (*Moringa oleifera*) (Sonbarse et al, 2020).

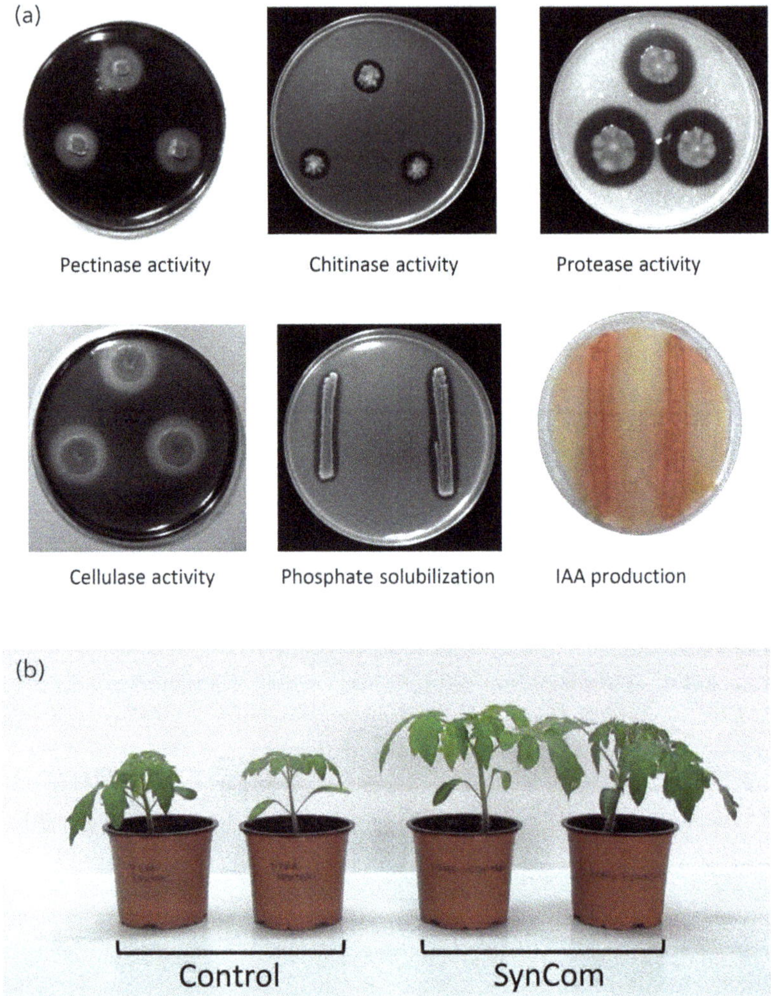

Figure 1 (a) *In vitro* screening of plant growth promotion traits; (b) Tomato plants not inoculated (left) and inoculated (right) with a microbial synthetic community (SynCom). [Photographs by I. S. Pantelides, Cyprus University of Technology].

PGPR have also been known to mediate biotic stress tolerance in plants through the production of antimicrobial compounds and the induction of plant defence responses. Ali et al. (2020) recently isolated bacteria from maize, rice, wheat, potato, sunflower and soybean rhizosphere and verified the antifungal activity against *Fusarium oxysporum*, *Fusarium moniliforme*, *Rhizoctonia solani*, *Colletotrichum gloeosporioides*, *Colletotrichum falcatum*, *Aspergillus niger* and *Aspergillus flavus*. The PGPR showing the highest antagonistic activity belonged to *Pseudomonas* and *Bacillus* species (Ali et al., 2020). The same genera have also suppressed *Phytophthora capsic*i infections in chilli pepper (Hyder et al., 2020). Multiple strains of *Bacillus* spp. together with a strain of *Stenotrophomonas rhizophila* were also effective in reducing *Meloidogyne incognita* population density and improving turfgrass root growth (Groover et al., 2020). In addition, *Bacillus amyloliquefaciens* strain S1 exhibited high *in vitro* antagonistic activity against *Clavibacter michiganensis* ssp. *michiganensis,* suggesting its possible employment in controlling bacterial canker in tomato plants (Gautam et al., 2019).

Following a transcriptomics approach, Gamez et al. (2019) highlighted that PGPR inoculation in banana (*Musa acuminata* Colla) cv. Williams resulted in differential expression of genes related to growth promotion and regulation of specific functions (flowering, photosynthesis, glucose catabolism and root growth) as well as genes involved in plant defence. Jiang et al. (2019) also demonstrated that the watermelon gene expression profile was altered in the presence of a *Bacillus* strain in combination with *F. oxysporum* f. sp. *niveum.* The *Bacillus* strain enhanced plant disease resistance against the pathogen through activation of defence-related genes and phytohormone signal factors (Jiang et al., 2019). Recently, Bertani et al. (2021) showed the expression of rice genes involved in ET and auxin pathways together with genes coding for a metallothionein-like protein and a multiple stress-responsive zinc-finger protein when the plants are inoculated with *Pseudomonas chlororaphis* ST9.

3 Tolerance to abiotic stresses

Over the past years, several studies indicated that PGPB/PGPR inoculation can induce plant tolerance against different abiotic stresses (Alagna et al., 2020; Gamalero et al., 2020; Sangiorgio et al., 2020; Meena et al., 2017). Nevertheless, the level of tolerance depends on the microbial capability to induce the expression of stress-responsive transcription factors in plants as well as the production of enzymes involved in the detoxification of reactive oxygen species (ROS), synthesizing proline and EPS and biomass stabilization (Aeron et al., 2020).

Salt stress is one of the major threats to agriculture, negatively affecting crop yield and growth (Shrivastava and Kumar, 2015). It induces osmotic and ionic stress in plants, causing nutritional imbalance, morphological damages, less photosynthetic capacity and death (Ahmad et al., 2013). Unfortunately, high salinity areas are increasing every year, and agriculture has therefore to manage salt stress maintaining a sufficient crop production to satisfy food demand (Panwar et al., 2016). PGPR can alleviate the negative effects of salt by incrementing seed germination rate and leaf area, improving chlorophyll and protein content, increasing plant growth, productivity, and nutrient availability, delaying leaf senescence and enhancing tolerance to stresses

(Saghafi et al., 2019; Habib et al., 2016). PGPR ameliorate salt stress tolerance through several mechanisms, for example, accumulation of osmolytes operating in ion homeostasis, improvement of nutrient uptake (N, P, K, Zn and Si), production of ACC deaminase, indole acetic acid (IAA), sideropheres and EPS, and alteration of the antioxidant defence system (Saghafi et al., 2019 and reference therein). Upadhyay and Singh (2015) have demonstrated that salt-tolerant PGPR improved both growth and dry mass of wheat grown in pots, as well as root dry weight and shoot biomass in field conditions. Palaniyandi et al. (2014) demonstrated that inoculation of tomato plants with *Streptomyces* sp. strain PGPA39A under salt stress increased plant biomass and chlorophyll content, while leaf proline content decreased. In another study, it was shown that strains belonging to *Streptomyces* ameliorated salt stress tolerance in *Stevia* crops (Tolba et al., 2019). Panwar et al. (2016) suggested that using a combination of two PGPR (bacterial strains belonging to genus *Pantoea* and *Enterococcus*) on mung bean (*Vigna radiate*) plants resulted in enhanced growth and yield, a reduced Na^+ concentration, less membrane damage and more antioxidants, such as ascorbic acid and glutathione, under salt stress. In the study of Khan et al. (2019), isolation and application of halotolerant PGPR on soybean plants grown under salt stress resulted in an increase in the antioxidant level, K^+ uptake, plant growth attributes and chlorophyll content and a reduction of the Na^+ ion concentration and the ABA level. Recently, Galicia-Campos et al. (2020) showed that the use of PGPR strains improved stress tolerance and water use efficiency in olive plants grown under saline stress.

Drought can also have a negative impact on crops causing significant yield reductions (Zhang et al., 2009; Breitkreuz et al., 2019). Many crops, including rice and winter wheat, need irrigation with big quantities of water in order to grow and produce acceptable yields. The research carried out by Zhang et al. (2020) showed that the association of rice roots with *Enterobacter aerogenes* is involved in rhizosheath (i.e. the layer of soil around the root containing a mixture of exudates, mucilage and exopolymers, which increases the wettability and water use efficiency of the root system) formation under moderate soil drying. It has been proposed that root-bacteria associations substantially contribute to this process by mechanisms that involve the ET response, considering that an ACC deaminase-deficient mutant of *E. aerogenes* failed to enhance rice rhizosheath formation. Breitkreuz et al. (2019) showed the positive role of *Phyllobacterium* in phosphate solubilization in rhizosphere under drought conditions. Brilli et al. (2019) demonstrated that tomato plants inoculated with *Pseudomonas chlororaphis* subsp. *aureofaciens* strain M71 have more proline and an improved antioxidant activity under mild water stress, thus reducing ROS presence and enhancing stress tolerance. The presence of the M71 strain also had an impact on stomatal closure, increasing ABA level in leaves and improving water use efficiency and biomass in water-stressed plants (Brilli et al., 2019). Rolli et al. (2015) demonstrated that PGPB have the ability to increase grapevine root biomass in field conditions under drought stress, while Saleem et al. (2018) showed that two PGPR strains improve velvet bean growth under drought conditions, by reducing ET production through ACC deaminase activity, which acts on the ET precursor ACC. Rubin et al. (2017) using a meta-analysis reported that PGPR can contribute to drought amelioration and water conservation, increasing shoot biomass and yield, especially under drought conditions.

Application of PGPR in combination with salicylic acid (SA) on maize plants (Khan et al., 2020) resulted in significant increases in nutrients content (Ca, K, Mg, Zn and Fe) in the shoots and the rhizosphere of plants and alleviated the adverse effects of low moisture stress of soil. Previously, Khan and Bano (2019) showed that the combination of PGPR and SA on wheat under drought stress led to a significant increase in leaf protein and sugar contents and higher chlorophyll content, chlorophyll fluorescence and performance index (Khan and Bano, 2019), suggesting the adoption of a mixed approach including both biological and chemical priming (Alagna et al., 2020).

In natural conditions, abiotic stresses can occur simultaneously, for example, salinity and phosphorous deficiency. It has been demonstrated that, under phosphate (Pi) limitation and salt stress, PGPR can support plant growth in plant genotype- and bacterial strain-dependent way (Tchuisseu Tchakounté et al., 2018). Osmotic stress and limitation of resources can also affect ornamental plants. It was shown that inoculation of petunia with *Pseudomonas* strains increased plant biomass and flowers number (Nordstedt et al., 2019). A study by Liu et al. (2019) focused on the physiological features and growth of North China red elder (*Sambucus williamsii*) under drought stress and in the presence of PGPR. *Acinetobacter calcoaceticus* X128 significantly increased stomatal conductance (Liu et al., 2019). The bacterium was able to increase cytokinins levels in the leaves that promote the stomatal opening, mitigating the inhibition of the photosynthetic rate in arid locations (Liu et al., 2019). In addition, the application of the PGPR strain might increase the permeability of roots to water or improve the transport of ions into the xylem, with an intensification of ABA transport, resulting in a decrease or complete closure in the stomatal opening (Liu et al., 2019). Generally, PGPR inoculation under drought conditions improved the adaptability of red elder plants to the arid environment by affecting phytohormones content in plants (Liu et al., 2019).

4 Beneficial effects against biotic stresses

The use of PGPB/PGPR is an eco-friendly tool that can be used for biocontrol of plant pathogens either by suppressing pathogenic microorganisms directly or by improving plant defence against pathogens (Lugtenberg and Kamilova, 2009). Controlling plant diseases by microorganisms is a complex process involving the biocontrol agent, the pathogen and the host, but also the indigenous microorganisms of the rhizosphere, other native macrobiota and the plant growth substrate. To act efficiently, the biocontrol microbe should remain active under varying conditions, such as temperature, moisture, pH and other soil properties.

Various mechanisms have been reported to be involved in biocontrol. The production of antibiotics and other antimicrobial metabolites is considered as a primary mechanism of biocontrol by PGPB and PGPR and is the most effective antagonistic activity to suppress phytopathogens. Diffusible antibiotics produced like phenazines, rhamnolipids, cyclic lipopeptides, zwittermycin A, kanosamine, oomycin A, ecomycins, butyrolactones and volatiles, such as hydrogen cyanide, ammonia, 2,3-butanediol and other blends of aldehydes, alcohols, ketones and sulphides, are known to possess antimicrobial and growth-promoting activities (Kai et al., 2009; Fernando et al, 2005). These compounds are toxic towards phytopathogens at concentrations depending on the compound and the target. Modes of action are not fully understood for many antimicrobial metabolites yet. In fungal pathogens, they may affect the cell membrane and zoospores (biosurfactants), inhibit the respiratory electron transport (phenazines, pyrrolnitrin) or cytochrome c oxidases and other metalloenzymes (hydrogen cyanide) (Raaijmakers, et al., 2006; Haas and Défago, 2005).

Another important mechanism in biocontrol is the production of hydrolytic enzymes by PGPB/PGPR directed against plant pathogens. Many biocontrol agents synthesize and secrete catabolic enzymes that can contribute to the suppression of phytopathogens through the hydrolysis of fungal cell wall components, such as cellulose, chitin, β-glucans and proteins (Abdullah et al., 2008; Dunne et al., 1997; Chernin et al., 1995). Production of β-1,3-glucanase by *Streptomyces* and *Paenibacillus* strains was shown to have an inhibitory effect on *F. oxysporum*, while *Bacillus cepacia* with glucanase activity showed inhibitory effect on many soilborne pathogens, including *Rhizoctonia solani*, *Sclerotium rolfsii* and *Pythium ultimum* (Compant et al., 2005). Several microorganisms were reported to show chitinolytic activity, including many *Bacillus, Streptomyces*, *Serratia* and *Pseudomonas* strains (Tsolakidou et al., 2019; Felse and Panda, 2000;). Co-cultivation of *Rhizoctonia solani* with the chitinolytic *Serratia marcescens* B2 strain led to several abnormalities of the mycelia (e.g. swelling, curling or bursting), suggesting degradation of the hyphal cell wall or hyphal cell death. Moreover, the application of *Serratia marcescens* B2 strain on cyclamen plants suppressed the diseases caused by *Rhizoctonia solani* and *F. oxysporum* f. sp. *cyclaminis* (Someya et al., 2000). Chitinases and cellulases are also involved in predation and parasitism, the major biocontrol mechanism used by *Trichoderma* and *Gliocladium* species (Harman et al., 2004a). This form of antagonism affects various fungal pathogens, such as *Sclerotinia*, *Rhizoctonia*, *Verticillium* and *Gaeumannomyces* (Harman et al., 2004b), and involves tropic growth of the BCA towards the target organism, coiling and dissolution of the pathogen's cell wall or membrane through enzymatic activity (Djonović et al., 2006; Woo et al., 2006; Zeilinger et al., 1999).

Apart from the mechanisms where a BCA produces substances with direct inhibitory effect for phytopathogens, it is possible for some PGPB/PGPR to outcompete the phytopathogens, either for space at the root surface or for nutrients, especially those secreted by the roots. This competition excludes pathogens by the physical occupation of binding sites on the root or by the depletion of food. Competition can take place for organic compounds necessary for pathogen proliferation and subsequent root colonization and for micronutrients that are essential for the growth and activity of the pathogen (Raaijmakers et al., 2009). Biocontrol based on competition for micronutrients has long been recognized, especially for nutrients that are not readily available for plants and microorganisms. Iron is a characteristic example of a micronutrient that is extremely limited in soils, and its availability depends on soil pH. In oxidized soils, iron is in the ferric form that is insoluble in water (Lindsay, 1979), and its concentration is too low to support the growth of microorganisms. To survive, microorganisms produce and secrete high-affinity chelators called siderophores (Neilands, 1995). Siderophore-producing PGPB/PGPR show increased efficiency in iron uptake making iron unavailable to pathogens and thus preventing their proliferation around the root, especially in soils with high pH (Kumar et al., 2015). Competition for iron as well as competition for carbon is an important mode of action of many biocontrol agents (Alabouvette et al., 2006; Lemanceau et al., 1992).

Besides functioning as BCAs, several PGPB and PGPR can induce a systemic response in the plant, leading to the activation of plant defence mechanisms against a wide range of phytopathogens (Pieterse et al., 2014). This form of resistance is referred to as induced systemic resistance (ISR) and is described as an enhanced defence capability of the plant against multiple pathogens (Conrath et al., 2015). ISR is induced by non-pathogenic PGPR, PGPB and fungi and can reduce the activity of pathogenic microorganisms via a complex system mediated by jasmonic acid (JA) and ET signalling (Pieterse et al., 2014; Van Loon, 1997). In contrast to classical biological control, in which the BCA is active against one or a few pathogens, ISR is effective against a broad spectrum of pathogens (Hariprasad et al., 2014). Several cell surface components and compounds produced by PGPR/PGPB can trigger ISR, including lipopolysaccharides and flagella, (Pieterse et al., 2003; Haas and Défago; 2005), siderophores (Meziane et al., 2005), volatiles (Ryu et al., 2004), hydrogen cyanide (HCN) (Defago et al., 1990), diacetylphloroglucinol (DAPG) (Weller et al., 2007) and cyclic lipopeptide surfactants (Ongena et al., 2007). The first reports of ISR were published back in 1991 and provided evidence that certain PGPR strains can stimulate the plant immune system and promote plant health (Alström, 1991; Van Peer et al., 1991; Wei et al., 1991). Since then, many studies have reported the ability of non-pathogenic microorganisms to trigger ISR including bacteria (e.g. *Pseudomonas*, *Serratia*, *Bacillus*), fungi (*F. oxysporum*, *Trichoderma*, *Piriformospora indica*) and symbiotic AMF (Pieterse et al., 2014). Enhancement in the plant's defence capability by ISR involves the activation of many biochemical pathways leading to fortification of structural barriers, such as thickened cell walls, suberization and deposition of lignin and callose (Raj et al., 2012; Benhamou et al., 1998). The phenomenon of ISR is also associated with increased expression of defence-related enzymes, such as phenylalanine ammonia lyases, peroxidases, lipoxygenases, polyphenol oxidases and synthesis of antimicrobial compounds, such as pathogenesis-related proteins, phytoalexins, phenolic compounds and cell wall peroxidases (Stringlis et al., 2018; van Loon et al., 1998; Zdor and Anderson, 1992; van Peer et al., 1991; Mauch et al., 1988).

5 Interaction between plant growth-promoting bacteria/rhizobacteria and arbuscular mycorrhizal fungi

The rhizosphere harbours a diverse community of microorganisms, such as bacteria and fungi that can interact with the plant, influencing plant growth, nutrition and health and protecting them from biotic and abiotic stresses in agro-ecosystems and in natural ecosystems (Philippot et al., 2013). AMF are one among the soilborne fungi that form symbiotic interactions with the majority of terrestrial plants. AMF are actively involved in the uptake of water and nutrients (such as phosphorus, nitrogen, zinc, copper, etc.) and increase resistance or tolerance of plants to biotic and abiotic stresses (Balestrini and Lumini, 2018).

AMF may interact synergistically with PGPR, leading to enhanced plant growth compared to single inoculation with either of them (Nanjundappa et al., 2019).

The review of Nanjundappa et al. (2019) focusing on the interaction between AMF and *Bacillus* concluded that combined inoculation leads to enhanced growth of plants, such as *Medicago sativa* (Medina et al., 2003), *Lactuca sativa* (Adriana et al., 2007), *Calendula officinalis* (Flores et al., 2007), *Artemisia annua* (Awasthi et al., 2011), *Pelargonium graveolens* (Alam et al., 2011), and *Cucumis sativus* (Rabab, 2014) as compared to single inoculation with either of them. Cely et al. (2016) also demonstrated that AMF and PGPR increased wood yield of *Schizolobium parahyba* var. *amazonicum* with respect to a fertilizer addition. Recently, Rocha et al. (2020) confirmed by field trials the positive role of co-presence of *Pseudomonas libanensis* and multiple AM fungal isolates of *Rhizophagus irregularis* in cowpea (*Vigna unguiculata*).

The plants showed increased shoot dry weight, pods and seeds per plant and grain yield (Rocha et al., 2020). In another study, a consortium of PGPR-rhizobia-AMF affected positively fava bean and wheat, improving shoot and root dry weight, leaf number, productivity and sugar, protein N, P, Ca, K and Na content (Raklami et al., 2019). Bona et al. (2015, 2017) demonstrated that the AMF-bacterium combined application can also affect fruit crop quality and nutritional value of strawberry and tomato (increased sugar content, fruit size, quantity and flowers) in conditions of reduced chemical inputs. The interaction between fungi and bacteria can also protect plants, by inducing systemic resistance to soilborne pathogens (Nanjundappa et al., 2019). For example, Jaizme-Vega et al. (2006) demonstrated a reduction of *Meloidogyne* infestation in AMF-PGPR-inoculated papaya plants, while Phirke et al. (2008) showed reduced *Fusarium* wilt in addition to improved yield in mycorrhized banana inoculated with rhizobacteria. The co-presence of AMF and PGPB/PGPR also improved tolerance to drought and salt stress in *Lactuca sativa* (Vivas et al., 2003), *Retama sphaerocarpa* (Marulanda et al., 46), *Z. mays* (Armada et al., 2015), *Trifolium repens* (Ortiz et al., 2015), *Lavandula dentate* (Armada et al., 2016) and *Acacia gerrardii* (Hashem et al., 2016). Recently, Inculet et al. (2019) demonstrated that inoculation of an irrigated tomato cultivar with AMF, PGPR and *Trichoderma*-based products increased plant length, fruit number, yield and quality traits based on lycopene and polyphenol content. Mannino et al. (2020), using different microbial inocula based on AM fungi or PGPR tolerant to salt, demonstrated that the tomato responses to water limitation depended on the inoculum composition. Balestrini et al. (2017) showed that the response of grapevine changed in the presence of a mixed inoculum composed by bacterial and fungal consortium compared to that with an inoculum based on *Funneliformis mosseae* only. Thus, the strategy of using a combination of AMF and PGPR in agricultural practice may improve soil health management, aiding nutrient solubilization and uptake and reduce the necessary fertilizer quantity. Nevertheless, more field studies are needed in order to verify the successful performance of the combined inoculations under real conditions (Nanjundappa et al., 2019).

The study of Todeschini et al. (2018) highlighted the importance of selecting the optimal combination of AMF and PGPR to positively influence physiological parameters, yield and quality in strawberry. The results of this study showed that application of the AMF affected the parameters associated with the plant vegetative portion, while application of the bacterium affected the fruit yield and quality. Interestingly, the volatile profile and elemental composition of the strawberry fruit were affected by the presence of a specific fungal-bacterial combination (Todeschini et al., 2018). This study showed for the first time that different soil microorganisms are able to influence the fruit concentration of some elements and/or volatiles (Todeschini et al., 2018).

Previous studies tested the ability of *Pseudomonas fluorescens* PGPR strains to form biofilm on mycorrhized and non- mycorrhized roots and on extraradical mycelium of an AM fungus (Bianciotto et al., 2001a, b). The nonmucoid wild-type strain *Pseudomonas fluorescens* CHA0 adhered very little on all surfaces, whereas two mucoid strains with increased production of acidic extracellular polysaccharides formed a dense and patchy bacterial layer on the roots and fungal structures (Bianciotto et al., 2001a). The results of this study suggest that increased adhesive properties of PGPR may lead to more stable interactions in mixed inocula and the rhizosphere. In another study, the bacterial components possibly involved in the attachment of two other PGPR (*Azospirillum* and *Rhizobium*) to AM roots and AM fungal structures were evaluated; mutants affected in EPS were tested in *in vitro* adhesion assays and shown to be strongly impaired in the attachment to both types of surfaces as well as to quartz fibres (Bianciotto et al., 2001b). Anchoring of PGPR to AMF seems to be a significant trait for a stable fungus–bacteria association that would improve the development of mixed inocula.

6 Conclusion and future trends in research

PGPB/PGPR can be promising economical and healthy alternatives to chemical fertilizers, antibiotics, herbicides, pesticides, with their abilities to improve agro-ecological sustainability. However, it is important to realize that PGPR showing a positive effect on a plant species may not have the same effect on others (Raj et al., 2020; Zeller et al., 2007). As explained by Timm et al. (2016), not all the microbes present in the soil have positive functions, so it is important to understand which microbial species should be employed to maximize plant growth, development and health (Xiao et al., 2020; Yuan et al., 2016; Mueller and Sachs, 2015). Recently, Finkel et al. (2020) demonstrated that a single bacterial genus in a complex microbiome modulates root growth. Interestingly, Guerrieri et al. (2020) suggested that using a consortium of native PGPR strains may represent a suitable solution in sustainable agriculture, to guarantee crop yield and quality, reducing the chemical input application. Apart from the studies on the efficacy of microbial inoculants on plants, their potential risks to other plants, animals, and humans must also be evaluated (Martínez-Hidalgo et al., 2019). Also, isolation, purification and characterization of microorganisms from saline habitats and inoculation of agricultural plants with them could be a successful strategy to increase tolerance and productivity of the plants grown under stress conditions (Saghafi et al., 2019). Escudero-Martinez and Bulgarelli (2019) highlighted that the genetic diversity of the crop microbiota is reduced compared to that of wild plants and that in combination with the application of human inputs, the agroecosystem resilience and sustainability to various stressors (e.g. climate change) is undermined. It is, therefore, desirable to carry out genetic mapping analyses, crossing interfertile wild and modern varieties, to discover host traits putatively influencing the recruitment and maintenance of the microbiota (Perez-Jaramillo et al., 2016; Schlaeppi and Bulgarelli, 2015). A concept named 'breeding for the plant microbiota' based on the development of plant varieties able to recruit specific microbial taxa may result in future crops that are less dependent on external inputs to produce acceptable yields (Escudero-Martinez and Bulgarelli, 2019; Bulgarelli et al., 2013; Wissuwa et al., 2009). Moreover, the prospect of using microbial mixtures as inoculants that can positively affect plant performance is gaining research interest. A substantial number of studies suggests that complex microbial consortia provide plants with increased growth and health as compared to single strains. However, our understanding of how members of microbial consortia interact with one another and with their hosts in nature is critical for the successful implementation of microbial synthetic communities (SynComs) with desired host outputs (Pascale et al., 2020; Tsolakidou et al., 2019). On the basis of these approaches, it will be possible to deal with challenges that agriculture shall meet in the coming years.

7 Acknowledgement

RB was supported by the CNR project FOE-2019 DBA.AD003.139.

8 Where to look for further information

8.1 Special issues on plant growth promoting rhizobacteria

- https://www.mdpi.com/journal/plants/special_issues/PGPB.
- https://www.sciencedirect.com/journal/microbiological-research/special-issue/10P22CLD85N.

8.2 Key research organizations

The Asian PGPR Society for Sustainable Agriculture (http://asianpgpr.com/index.php).

9 References

Abdullah, M. T., Ali, N. Y. and Suleman, P. (2008). Biological control of *Sclerotinia sclerotiorum* (Lib.) de Bary with *Trichoderma harzianum* and *Bacillus amyloliquefaciens*, *Crop Prot.* 27(10), 1354–1359.

Adriana, M. A., Rosario, A., Juan, R. L. and Ricardo, A. (2007). Differential effects of a *Bacillus megaterium* strain on *Lactuca sativa* plant growth depending on the origin of the arbuscular mycorrhizal fungus coinoculated: physiologic and biochemical traits, *J. Plant Growth Regul.* 27(10), 10–18.

Aeron, A., Khare, E., Jha, C. K., Meena, V. S., Aziz, S. M. A., Islam, M. T., Kim, K., Meena, S. K., Pattanayak, A., Rajashekara, H., Dubey, R. C., Maurya, B. R., Maheshwari, D. K., Saraf, M., Choudhary, M., Verma, R., Meena, H. N., Subbanna, A. R. N. S., Parihar, M., Shukla, S., Muthusamy, G., Bana, R. S., Bajpai, V. K., Han, Y. K., Rahman, M., Kumar, D., Singh, N. P. and Meena, R. K. (2020). Revisiting the plant growth-promoting rhizobacteria: lessons from the past and objectives for the future, *Arch. Microbiol.* 202(4), 665–676.

Ahmad, M., Zahir, Z. A., Khalid, M., Nazli, F. and Arshad, M. (2013). Efficacy of Rhizobium and *Pseudomonas* strains to improve physiology, ionic balance and quality of mung bean under salt-affected conditions on farmer's fields, *Plant Physiol. Biochem.* 63, 170–176.

Alabouvette, C., Olivain, C. and Steinberg, C. (2006). Biological control of plant diseases: the European situation, *Eur. J. Plant Pathol.* 114(3), 329–341.

Alagna, F., Balestrini, R., Chitarra, W., Marsico, A. D. and Nerva, L. (2020). Chapter 3–Getting ready with the priming: innovative weapons against biotic and abiotic crop enemies in a global changing scenario. In: Hossain, M. A., Liu, F., Burritt, D. J., Fujita, M. and Huang, B. (Eds), *Priming-Mediated Stress and Cross-Stress Tolerance in Crop Plants*, Academic Press, pp. 35–56. https://www.elsevier.com/books/priming-mediated-stress-and-cross-stress-tolerance-in-crop-plants/hossain/978-0-12-817892-8

Alam, M., Khaliq, A., Sattar, A., Shukla, R. S., Anwar, M. and Seema Dharni, S. (2011). Synergistic effect of arbuscular mycorrhizal fungi and *Bacillus subtilis* on the biomass and essential oil yield of rose-scented geranium (*Pelargonium graveolens*), *Arch. Agron. Soil Sci.* 57(8), 889–898.

Ali, S., Hameed, S., Shahid, M., Iqbal, M., Lazarovits, G. and Imran, A. (2020). Functional characterization of potential PGPR exhibiting broad-spectrum antifungal activity, *Microbiol. Res.* 232, 126389.

Alström, S. (1991). Induction of disease resistance in common bean susceptible to halo blight bacterial pathogen after seed bacterization with rhizosphere pseudomonads, *J. Gen. Appl. Microbiol.* 37(6), 495-501.

Ambrosini, A., Beneduzi, A., Stefanski, T., Pinheiro, F. G., Vargas, L. K. and Passaglia, L. M. P. (2012). Screening of plant growth promoting rhizobacteria isolated from sunflower (*Helianthus annuus* L.), *Plant Soil* 356(1-2), 245–264.

Andreozzi, A., Prieto, P., Mercado-Blanco, J., Monaco, S., Zampieri, E., Romano, S., Valè, G., Defez, R. and Bianco, C. (2019). Efficient colonization of the endophytes *Herbaspirillum huttiense* RCA24 and *Enterobacter cloacae* RCA25 influences the physiological parameters of *Oryza sativa* L. cv. Baldo rice, *Environ. Microbiol.* doi: 10.1111/1462-2920.14688.

Antoun, H. and Prévost, D. (2006). Ecology of plant growth promoting rhizobacteria. In: Siddiqui, Z. A. (Ed), *PGPR Biocontrol Biofertilization*, Springer, New York, pp. 1–38.

Armada, E., Azcón, R., López-Castillo, O. M., Calvo-Polanco, M. and Ruiz-Lozano, J. M. (2015). Autochthonous arbuscular mycorrhizal fungi and *Bacillus thuringiensis* from a degraded Mediterranean area can be used to improve physiological traits and performance of a plant of agronomic interest under drought conditions, *Plant Physiol. Biochem.* 90, 64–74.

Armada, E., Probanza, A., Roldán, A. and Azcón, R. (2016). Native plant growth promoting bacteria *Bacillus thuringiensis* and mixed or individual mycorrhizal species improved drought tolerance and oxidative metabolism in *Lavandula dentata* plants, *J. Plant Physiol.* 192, 1–12.

Arruda, L., Beneduzi, A., Martins, A., Lisboa, B., Lopes, C., Bertolo, F., Passaglia, L. M. P. and Vargas, L. K. (2013). Screening of rhizobacteria isolated from maize (*Zea mays* L.) in Rio Grande do Sul State (South Brazil) and analysis of their potential to improve plant growth, *Appl. Soil Ecol.* 63, 15–22.

Ashraf, M. A., Rasool, M. and Mirza, M. S. (2011). Nitrogen fixation and indole acetic acid production potential of bacteria isolated from rhizosphere of sugarcane (*Saccharum officinarum* L), *Adv. Biol. Res.* 5(6), 348–355.

Awan, S. A., Ilyas, N., Khan, I., Raza, M. A., Rehman, A. U., Rizwan, M., Rastogi, A., Tariq, R. and Brestic, M. (2020). *Bacillus siamensis* reduces cadmium accumulation and improves growth and antioxidant defense system in two wheat (*Triticum aestivum* L.) varieties, *Plants (Basel)* 9(7), 878.

Awasthi, A., Bharti, N., Nair, P., Singh, R., Shukla, A. K., Gupta, M. M., Darokar, M. P. and Kalra, A. (2011). Synergistic effect of *Glomus mosseae* and nitrogen fixing *Bacillus subtilis* strain Daz26 on artemisin content in *Artemisia annua* L., *Appl. Soil Ecol.* 49, 125–130.

Balestrini, R. and Lumini, E. (2018). Focus on mycorrhizal symbioses, *Appl. Soil Ecol.* 123, 299–304.

Balestrini, R., Salvioli, A., Dal Molin, A., Novero, M., Gabelli, G., Paparelli, E., Marroni, F. and Bonfante, P. (2017). Impact of an arbuscular mycorrhizal fungus versus a mixed microbial inoculum on the transcriptome reprogramming of grapevine roots, *Mycorrhiza* 27(5), 417–430.

Baritz, R., Wiese, L., Verbeke, I. and Vargas, R. (2018). Voluntary guidelines for sustainable soil management: global action for healthy soils. In: Ginzky, H., Dooley, E., Heuser, I., Kasimbazi, E., Markus, T. and Qin, T. (Eds), *International Yearbook of Soil Law and Policy* (vol. 2017), Springer, Cham, pp. 17–36.

Batista, B. D., Lacava, P. T., Ferrari, A., Teixeira-Silva, N. S., Bonatelli, M. L., Tsui, S., Mondin, M., Kitajima, E. W., Pereira, J. O., Azevedo, J. L. and Quecine, M. C. (2018). Screening of tropically derived, multi-trait plant growth- promoting rhizobacteria and evaluation of corn and soybean colonization ability, *Microbiol. Res.* 206, 33–42.

Bechtaoui, N., Raklami, A., Tahiri, A. I., Benidire, L., El Alaoui, A., Meddich, A., Göttfert, M. and Oufdou, K. (2019). Characterization of plant growth promoting rhizobacteria and their benefits on growth and phosphate nutrition of faba bean and wheat, *Biol. Open* 8(7), bio043968.

Benhamou, N., Kloepper, J. W. and Tuzun, S. (1998). Induction of resistance against *Fusarium* wilt of tomato by combination of chitosan with an endophytic bacterial strain: ultrastructure and cytochemistry of the host response, *Planta* 204(2), 153–168.

Bertani, I., Zampieri, E., Bez, C., Volante, A., Venturi, V. and Monaco, S. (2021). Isolation and Characterization of *Pseudomonas chlororaphis* Strain ST9; Rhizomicrobiota and in Planta Studies, *Plants* 10(7), 1466.

Bhattacharyya, D., Garladinne, M. and Lee, Y. H. (2015). Volatile indole produced by Rhizobacterium *Proteus vulgaris* JBLS202 stimulates growth of *Arabidopsis thaliana* through auxin, cytokinin, and brassinosteroid pathways, *J. Plant Growth Regul.* 34(1), 158–168.

Bhattacharyya, P. N. and Jha, D. K. (2012). Plant growth-promoting rhizobacteria (PGPR): emergence in agriculture, *World J. Microbiol. Biotechnol.* 28(4), 1327–1350.

Bianciotto, V., Andreotti, S., Balestrini, R., Bonfante, P. and Perotto, S. (2001a). Mucoid mutants of the biocontrol strain pseudomonas fluorescens CHA0 show increased ability in biofilm formation on mycorrhizal and nonmycorrhizal carrot roots, *Mol. Plant Microbe Interact.* 14(2), 255–260.

Bianciotto, V., Andreotti, S., Balestrini, R., Bonfante, P. and Perotto, S. (2001b). Extracellular polysaccharides are involved in the attachment of *Azospirillum brasilense* and *Rhizobium leguminosarum* to arbuscular mycorrhizal structures, *Eur. J. Histochem.* 45(1), 39–49.

Bisht, N., Tiwari, S., Singh, P. C., Niranjan, A. and Singh Chauhan, P. (2019). A multifaceted rhizobacterium *Paenibacillus lentimorbus* alleviates nutrient deficiency-induced stress in *Cicer arietinum* L., *Microbiol. Res.* 223–225, 110–119.

Bona, E., Cantamessa, S., Massa, N., Manassero, P., Marsano, F., Copetta, A., Lingua, G., D'Agostino, G., Gamalero, E. and Berta, G. (2017). Arbuscular mycorrhizal fungi and plant growth-promoting pseudomonads improve yield, quality and nutritional value of tomato: a field study, *Mycorrhiza* 27(1), 1–11.

Bona, E., Lingua, G., Manassero, P., Cantamessa, S., Marsano, F., Todeschini, V., Copetta, A., D'Agostino, G., Massa, N., Avidano, L., Gamalero, E. and Berta, G. (2015). AM fungi and PGP pseudomonads increase flowering, fruit production, and vitamin content in strawberry grown at low nitrogen and phosphorus levels, *Mycorrhiza* 25(3), 181–193.

Breitkreuz, C., Buscot, F., Tarkka, M. and Reitz, T. (2019). Shifts Between and Among populations of wheat rhizosphere *Pseudomonas*, *Streptomyces* and *Phyllobacterium* suggest consistent phosphate mobilization at different wheat growth stages Under abiotic stress, *Front. Microbiol.* 10, 3109.

Brilli, F., Pollastri, S., Raio, A., Baraldi, R., Neri, L., Bartolini, P., Podda, A., Loreto, F., Maserti, B. E. and Balestrini, R. (2019). Root colonization by *Pseudomonas chlororaphis* primes tomato (*Lycopersicum esculentum*) plants for enhanced tolerance to water stress, *J. Plant Physiol.* 232, 82–93.

Brodt, S., Six, J., Feenstra, G., Ingels, C. and Campbell, D. (2011). Sustainable agriculture, *Nat. Educ. Knowl.* 3(10), 1.

Bulgarelli, D., Schlaeppi, K., Spaepen, S., Ver Loren van Themaat, E. and Schulze-Lefert, P. (2013). Structure and functions of the bacterial microbiota of plants, *Annu. Rev. Plant Biol.* 64, 807–838.

Calvo, P., Watts, D. B., Kloepper, J. W. and Torbert, H. A. (2017). Effect of microbial-based inoculants on nutrient concentrations and early root morphology of corn (*Zea mays*), *J. Plant Nutr. Soil Sci.* 180(1), 56–70.

Cely, M. V. T., Siviero, M. A., Emiliano, J., Spago, F. R., Freitas, V. F., Barazetti, A. R., Goya, E. T., Lamberti, G. S., Dos Santos, I. M., De Oliveira, A. G. and Andrade, G. (2016). Inoculation of *Schizolobium parahyba* with mycorrhizal fungi and plant growth-promoting rhizobacteria increases wood yield under field conditions, *Front. Plant Sci.* 7, 1708.

Chernin, L., Ismailov, Z., Haran, S. and Chet, I. (1995). Chitinolytic *Enterobacter agglomerans* antagonistic to fungal plant pathogens, *Appl. Environ. Microbiol.* 61(5), 1720–1726.

Compant, S., Duffy, B., Nowak, J., Clement, C. and Barka, E. A. (2005). Use of plant growth-promoting bacteria for biocontrol of plant diseases: principles, mechanisms of action, and future prospects, *Appl. Environ. Microbiol.* 71(9), 4951 –4959.

Compant, S., Samad, A., Faist, H. and Sessitsch, A. (2019). A review on the plant microbiome: ecology, functions, and emerging trends in microbial application, *J. Adv. Res.* 19, 29–37.

Conrath, U., Beckers, G. J., Langenbach, C. J. and Jaskiewicz, M. R. (2015). Priming for enhanced defense, *Annu. Rev. Phytopathol.* 53, 97–119.

Corwin, D. L. (2021). Climate change impacts on soil salinity in agricultural areas, *Eur. J. Soil Sci.* 72(2), 842–862. doi: 10.1111/ejss.13010.

Defago, G., Berling, C. H., Burger, U., Hass, D., Kahr, G., Keel, C., Voisard, C., Wirthner, P. and Wuthrich, B. (1990). Suppression of black root rot of tobacco and other root disease by strains of *Pseudomonas fluorescens*: potential applications and mechanisms. In: Schippers, D., Cook, R. J., Henis, Y., Ko, W. H., Rovira, A. D., Schippers, B. and Scott, P. R. (Eds), *Biological Control of Soil Borne Plant Pathogens*, CABI Publishing, Oxfordshire, UK, pp. 93-108.

Dimkpa, C., Weinand, T. and Asch, F. (2009). Plant-rhizobacteria interactions alleviate abiotic stress conditions, *Plant Cell Environ.* 32(12), 1682–1694.

Djonović, S., Pozo, M. J. and Kenerley, C. M. (2006). Tvbgn3 a b-16-glucanase from the biocontrol fungus *Trichoderma virens* is involved in mycoparasitism and control of *Pythium ultimum*, *Appl. Environ. Microbiol.* 72(12), 7661-7670.

dos Santos, R. M., Diaz, P. A. E., Lobo, L. L. B. and Rigobelo, E. C. (2020). Use of plant growth-promoting rhizobacteria in maize and sugarcane: characteristics and applications, *Front. Sustain. Food Syst.* 4, 136.

Dunne, C., Crowley, J. J., Moënne-Loccoz, Y., Dowling, D. N., Bruijn, S. and O'Gara, F. (1997). Biological control of *Pythium ultimum* by *Stenotrophomonas maltophilia* W81 is mediated by an extracellular proteolytic activity, *Microbiology (Reading)* 143(12), 3921–3931.

Escudero-Martinez, C. and Bulgarelli, D. (2019). Tracing the evolutionary routes of plant-microbiota interactions, *Curr. Opin. Microbiol.* 49, 34–40.

Etesami, H., Emami, S. and Alikhani, H. A. (2017). Potassium solubilizing bacteria (KSB): mechanisms, promotion of plant growth, and future prospects: a review, *J. Soil Sci. Plant Nutr.* 17(4), 897–911.

Farina, R., Beneduzi, A., Ambrosini, A., Campos, S. B., Lisboa, B. B., Wendisch, V., Vargas, L. K. and Passaglia, L. M. P. (2012). Diversity of plant growth-promoting rhizobacteria communities associated with the stages of canola growth, *Appl. Soil Ecol.* 55, 44–52.

Felse, P. A. and Panda, T. (2000). Production of microbial chitinases–a revisit, *Bioprocess Eng.* 23(2), 127–134.

Fernando, W. G. D., Nakkeeran, S. and Zhang, Y. (2005). Biosynthesis of antibiotics by PGPR and its relation in biocontrol of plant diseases. In: Siddiqui, Z. A. (Ed), *PGPR: Biocontrol and Biofertilization*, Springer, Dordrecht, pp. 67–109.

Finkel, O. M., Salas-González, I., Castrillo, G., Conway, J. M., Law, T. F., Teixeira, P. J. P. L., Wilson, E. D., Fitzpatrick, C. R., Jones, C. D. and Dangl, J. L. (2020). A single bacterial genus maintains root growth in a complex microbiome, *Nature* 587(7832), 103–108. doi: 10.1038/s41586-020-2778-7.

Flores, A. C., Luna, A. A. E. and Portugal, V. O. (2007). Yield and quality enhancement of marigold flowers by inoculation with *Bacillus subtilis* and *Glomus fasciculatum*, *J. Sustain. Agr.* 31(1), 21–31.

Galicia-Campos, E., Ramos-Solano, B., Montero-Palmero, M. B., Gutierrez-Mañero, F. J. and García-Villaraco, A. (2020). Management of plant physiology with beneficial bacteria to improve leaf bioactive profiles and plant adaptation under saline stress in *Olea europea* L, *Foods* 9(1), 57.

Gamalero, E., Bona, E., Todeschini, V. and Lingua, G. (2020). Saline and arid soils: impact on bacteria, plants, and their interaction, *Biology (Basel)* 9(6), 116.
Gamez, R. M., Rodríguez, F., Vidal, N. M., Ramirez, S., Vera Alvarez, R., Landsman, D. and Mariño-Ramírez, L. (2019). Banana (*Musa acuminata*) transcriptome profiling in response to rhizobacteria: *Bacillus amyloliquefaciens* Bs006 and *Pseudomonas fluorescens* Ps006, *BMC Genomics* 20(1), 378.
Gautam, S., Chauhan, A., Sharma, R., Sehgal, R. and Shirkot, C. K. (2019). Potential of *Bacillus amyloliquefaciens* for biocontrol of bacterial canker of tomato incited by *Clavibacter michiganensis* ssp. michiganensis, *Microb. Pathog.* 130, 196-203.
Glick, B. R. (2012). Plant growth-promoting bacteria: mechanisms and applications, *Scientifica (Cairo)* 2012, 963401.
Groover, W., Held, D., Lawrence, K. and Carson, K. (2020). Plant growth-promoting rhizobacteria: a novel management strategy for *Meloidogyne incognita* on turfgrass, *Pest Manag. Sci.* 76(9), 3127-3138. doi: 10.1002/ps.5867.
Guerrieri, M. C., Fanfoni, E., Fiorini, A., Trevisan, M. and Puglisi, E. (2020). Isolation and screening of extracellular PGPR from the rhizosphere of tomato plants after long-term reduced tillage and cover crops, *Plants (Basel)* 9(5), 668.
Haas, D. and Défago, G. (2005). Biological control of soil-borne pathogens by fluorescent pseudomonads, *Nat. Rev. Microbiol.* 3(4), 307-319.
Habib, S. H., Kausar, H. and Saud, H. M. (2016). Plant growth-promoting rhizobacteria enhance salinity stress tolerance in okra through ROS-scavenging enzymes, *BioMed Res. Int.* 2016, 6284547.
Hariprasad, P., Chandrashekar, S., Singh, S. B. and Niranjana, S. R. (2014). Mechanism of plant growth promotion and disease suppression by *Pseudomonas aeruginosa* strain 2apa, *J. Basic Microbiol.* 54(8), 792-801.
Harman, G. E., Howel, C. R., Viterbo, A., Chet, I. and Lorito, M. (2004a). *Trichoderma* species-opportunistic, avirulent plant symbionts, *Nat. Rev. Microbiol.* 2(1), 43-56.
Harman, G. E., Petzoldt, R., Comis, A. and Chen, J. (2004b). Interactions between *Trichoderma harzianum* strain T22 and maize inbred line Mo17 and effects of these interactions on diseases caused by *Pythium ultimum* and *Colletotrichum graminicola*, *Phytopathology* 94(2), 147-153.
Hashem, A., Abd_Allah, E. F., Alqarawi, A. A., Al-Huqail, A. A. and Shah, M. A. (2016). Induction of osmoregulation and modulation of salt stress in acacia gerrardii benth. by Arbuscular Mycorrhizal Fungi and Bacillus subtilis (BERA 71), *BioMed Res. Int.* 2016, 1-11, 6294098.
Heydari, M. M., Brook, R. M. and Jones, D. L. (2019). The role of phosphorus sources on root diameter, root length and root dry matter of barley (*Hordeum vulgare* L.), *J. Plant Nutr.* 42(1), 1-15.
Hyder, S., Gondal, A. S., Rizvi, Z. F., Ahmad, R., Alam, M. M., Hannan, A., Ahmed, W., Fatima, N. and Inam-Ul-Haq, M. (2020). Characterization of native plant growth promoting rhizobacteria and their anti-oomycete potential against *Phytophthora capsici* affecting chilli pepper (*Capsicum annum* L.), *Sci. Rep.* 10(1), 13859.
Inculet, C. S., Mihalache, G., Sellitto, V. M., Hlihor, R. M. and Stoleru, V. (2019). The effects of a microorganisms-based commercial product on the morphological, biochemical and yield of tomato plants under two different water regimes, *Microorganisms* 7(12), 706.

Jaizme-Vega, MdC., Rodriguez-Romero, A. S. and Barroso Nunez, L. A. (2006). Effect of the combined inoculation of arbuscular mycorrhizal fungi and plant growth promoting rhizobacteria on papaya (*Carica papaya* L.) infected with the root-knot nematode *Meloidogyne incognita*, *Fruits* 61(3), 151–162.

Jiang, C. H., Yao, X. F., Mi, D. D., Li, Z. J., Yang, B. Y., Zheng, Y., Qi, Y. J. and Guo, J. H. (2019). Comparative transcriptome analysis reveals the biocontrol mechanism of *Bacillus velezensis* F21 Against *Fusarium* Wilt on watermelon, *Front. Microbiol.* 10, 652.

Jiménez, J. A., Novinscak, A. and Filion, M. (2020). *Pseudomonas fluorescens* LBUM677 differentially increases plant biomass, total oil content and lipid composition in three oilseed crops, *J. Appl. Microbiol.* 128(4), 1119–1127.

Jiménez-Gómez, A., García-Estévez, I., García-Fraile, P., Escribano-Bailón, M. T. and Rivas, R. (2020). Increase in phenolic compounds of *Coriandrum sativum* L. after the application of a *Bacillus* halotolerans biofertilizer, *J. Sci. Food Agric.* 100(6), 2742–2749.

Kai, M., Haustein, M., Molina, F., Petri, A., Scholz, B. and Piechulla, B. (2009). Bacterial volatiles and their action potential, *Appl. Microbiol. Biotechnol.* 81(6), 1001–1012.

Kalam, S., Basu, A. and Podile, A. R. (2020). Functional and molecular characterization of plant growth promoting *Bacillus* isolates from tomato rhizosphere, *Heliyon* 6(8), e04734.

Khan, M. A., Asaf, S., Khan, A. L., Adhikari, A., Jan, R., Ali, S., Imran, M., Kim, K. M. and Lee, I. J. (2019). Halotolerant rhizobacterial strains mitigate the adverse effects of NaCl stress in soybean seedlings, *BioMed Res. Int.* 2019, 9530963.

Khan, N. and Bano, A. (2019). Exopolysaccharide producing rhizobacteria and their impact on growth and drought tolerance of wheat grown under rainfed conditions, *PLoS ONE* 14(9), e0222302.

Khan, N., Bano, A. and Curá, J. A. (2020). Role of beneficial microorganisms and salicylic acid in improving rainfed agriculture and future food safety, *Microorganisms* 8(7), 1018.

Khan, M. S., Zaidi, A., Ahemad, M., Oves, M. and Wani, P. A. (2010). Plant growth promotion by phosphate solubilizing fungi–current perspective, *Arch. Agron. Soil Sci.* 56(1), 73–98.

Khatoon, Z., Huang, S., Rafique, M., Fakhar, A., Kamran, M. A. and Santoyo, G. (2020). Unlocking the potential of plant growth-promoting rhizobacteria on soil health and the sustainability of agricultural systems, *J. Environ. Manage.* 273, 111118.

Kour, D., Rana, K. L., Yadav, N., Yadav, A. N., Kumar, A., Meena, V. S. and Saxena, A. K. (2019). Rhizospheric microbiomes: biodiversity, mechanisms of plant growth promotion, and biotechnological applications for sustainable agriculture. In: Kumar A., Meena V. (Eds), *Plant Growth Promoting Rhizobacteria for Agricultural Sustainability*, Springer, Singapore, pp. 19–65.

Kumar, A., Bahadur, I., Maurya, B. R., Raghuwanshi, R., Meena, V. S., Singh, D. K. and Dixit, J. (2015). Does a plant growth promoting rhizobacteria enhance agricultural sustainability?, *J. Pur. Appl. Microbiol.* 9, 715–724.

Lee, S., Trịnh, C. S., Lee, W. J., Jeong, C. Y., Truong, H. A., Chung, N., Kang, C. S. and Lee, H. (2020). *Bacillus subtilis* strain L1 promotes nitrate reductase activity in *Arabidopsis* and elicits enhanced growth performance in *Arabidopsis*, lettuce, and wheat, *J. Plant Res.* 133(2), 231–244.

Lemanceau, P., Bakker, P. A. H. M., de Kogel, W. J., Alabouvette, C. and Schippers, B. (1992). Effect of pseudobactin 358 production by *Pseudomonas putida* WCS358 on suppression of Fusarium wilt of carnation by nonpathogenic *Fusarium oxysporum* Fo47, *Appl. Environ. Microbiol.* 58(9), 2978–2982.

Lindsay, W. L. (1979). *Chemical Equilibria in Soils*, John Wiley & Sons, Inc., New York.

Liu, F., Ma, H., Du, Z., Ma, B., Liu, X. H., Peng, L. and Zhang, W. X. (2019). Physiological response of North China red elder container seedlings to inoculation with plant growth-promoting rhizobacteria under drought stress, *PLoS ONE* 14(12), e022 6624.

Lugtenberg, B. and Kamilova, F. (2009). Plant-growth-promoting rhizobacteria, *Annu. Rev. Microbiol.* 63, 541–556.

Ma, Y., Prasad, M. N., Rajkumar, M. and Freitas, H. (2011). Plant growth promoting rhizobacteria and endophytes accelerate phytoremediation of metalliferous soils, *Biotechnol. Adv.* 29(2), 248–258.

Mannino, G., Nerva, L., Gritli, T., Novero, M., Fiorilli, V., Bacem, M., Bertea, C. M., Lumini, E., Chitarra, W. and Balestrini, R. (2020). Effects of different microbial inocula on tomato tolerance to water deficit, *Agronomy* 10(2), 170.

Martínez-Hidalgo, P., Maymon, M., Pule-Meulenberg, F. and Hirsch, A. M. (2019). Engineering root microbiomes for healthier crops and soils using beneficial, environmentally safe bacteria, *Can. J. Microbiol.* 65(2), 91–104.

Mauch, F., Mauch-Mani, B. and Boller, T. (1988). Antifungal hydrolasses in pea tissue: II. Inhibition of fungal growth by combinations of chitinases and two β-1,3-glucanases, *Plant Physiol.* 87, 936–942.

Medina, A., Probanza, A., Gutierrez Mañero, F. J. and Azcón, R. (2003). Interactions of arbuscular-mycorrhizal fungi and *Bacillus* strains and their effects on plant growth, microbial rhizosphere activity (thymidine and leucine incorporation) and fungal biomass (ergosterol and chitin), *Appl. Soil Ecol.* 22(1), 15–28.

Meena, K. K., Sorty, A. M., Bitla, U. M., Choudhary, K., Gupta, P., Pareek, A., Singh, D. P., Prabha, R., Sahu, P. K., Gupta, V. K., Singh, H. B., Krishanani, K. K. and Minhas, P. S. (2017). Abiotic stress responses and microbe-mediated mitigation in plants: the omics strategies, *Front. Plant Sci.* 8, 172.

Meena, M., Swapnil, P., Divyanshu, K., Kumar, S., Harish, T., Tripathi, Y. N., Zehra, A., Marwal, A. and Upadhyay, R. S. (2020). PGPR-mediated induction of systemic resistance and physiochemical alterations in plants against the pathogens, current perspectives, *J. Basic Microbiol.* 60(10), 828–861.

Meziane, H., Van der Sluis, I., Van Loon, L. C., Hofte, M. and Bakker, P. A. (2005). Determinants of *Pseudomonas putida* WCS358 involved in inducing systemic resistance in plants, *Mol. Plant Pathol.* 6(2), 177–185.

Mueller, U. G. and Sachs, J. L. (2015). Engineering microbiomes to improve plant and animal health, *Trends Microbiol.* 23(10), 606–617.

Nanjundappa, A., Bagyaraj, D. J., Saxena, A. K., Kumar, M. and Chakdar, H. (2019). Interaction between arbuscular mycorrhizal fungi and *Bacillus* spp. in soil enhancing growth of crop plants, *Fungal Biol. Biotechnol.* 6, 23.

Neilands, J. B. (1995). Siderophores: structure and function of microbial iron transport compounds, *J. Biol. Chem.* 270(45), 26723–26726.

Niranjan Raj, S., Lavanya, S. N., Amruthesh, K. N., Niranjana, S. R., Reddy, M. S. and Shetty, H. S. (2012). Histo-chemical changes induced by PGPR during induction of resistance in pearl millet against downy mildew disease, *Biol. Control* 60(2), 90–102.

Nordstedt, N. P., Chapin, L. J., Taylor, C. G. and Jones, M. L. (2019). Identification of *Pseudomonas* spp. that increase ornamental crop quality during abiotic stress, *Front. Plant Sci.* 10, 1754.

Ongena, M., Jourdan, E., Adam, A., Paquot, M., Brans, A., Joris, B., Arpigny, J. L. and Thonart, P. (2007). Surfactin and fengycin lipopeptides of *Bacillus subtilis* as elicitors of induced systemic resistance in plants, *Environ. Microbiol.* 9(4), 1084–1090.

Orozco-Mosqueda, M. D. C., Rocha-Granados, M. D. C., Glick, B. R. and Santoyo, G. (2018). Microbiome engineering to improve biocontrol and plant growth-promoting mechanisms, *Microbiol. Res.* 208, 25–31.

Ortiz, N., Armada, E., Duque, E., Roldán, A. and Azcón, R. (2015). Contribution of arbuscular mycorrhizal fungi and/or bacteria to enhancing plant drought tolerance under natural soil conditions: effectiveness of autochthonous or allochthonous strains, *J. Plant Physiol.* 174, 87–96.

Palaniyandi, S. A., Damodharan, K., Yang, S. H. and Suh, J. W. (2014). *Streptomyces* sp. strain PGPA39 alleviates salt stress and promotes growth of "Micro Tom" tomato plants, *J. Appl. Microbiol.* 117(3), 766–773.

Panwar, M., Tewari, R. and Nayyar, H. (2016). Native halo-tolerant plant growth promoting rhizobacteria *Enterococcus* and *Pantoea* sp. improve seed yield of Mungbean (*Vigna radiata* L.) under soil salinity by reducing sodium uptake and stress injury, *Physiol. Mol. Biol. Plants* 22(4), 445–459.

Pascale, A., Proietti, S., Pantelides, I. S. and Stringlis, I. A. (2020). Modulation of the root microbiome by plant molecules: the basis for targeted disease suppression and plant growth promotion, *Front. Plant Sci.* 10, 1741.

Pellegrini, M., Ercole, C., Di Zio, C., Matteucci, F., Pace, L. and Del Gallo, M. (2020). *In vitro* and *in planta* antagonistic effects of plant growth-promoting rhizobacteria consortium against soilborne plant pathogens of *Solanum tuberosum* and *Solanum lycopersicum*, *FEMS Microbiol. Lett.* 367(13), fnaa099.

Pérez-Flores, P., Valencia-Cantero, E., Altamirano-Hernández, J., Pelagio-Flores, R., López-Bucio, J., García-Juárez, P. and Macías-Rodríguez, L. (2017). *Bacillus methylotrophicus* M4-96 isolated from maize (*Zea mays*) rhizoplane increases growth and auxin content in *Arabidopsis thaliana* via emission of volatiles, *Protoplasma* 254(6), 2201–2213.

Perez-Jaramillo, J. E., Mendes, R. and Raaijmakers, J. M. (2016). Impact of plant domestication on rhizosphere microbiome assembly and functions, *Plant Mol. Biol.* 90(6), 635–644.

Philippot, L., Raaijmakers, J. M., Lemanceau, P. and van der Putten, W. H. (2013). Going back to the roots: the microbial ecology of the rhizosphere, *Nat. Rev. Microbiol.* 11(11), 789–799.

Phirke, N. V., Kothari, R. M. and Chincholkar, S. B. (2008). Rhizobacteria in mycorrhizosphere improved plant health and yield of banana by offering proper nourishment and protection against diseases, *Appl. Biochem. Biotechnol.* 151(2–3), 441–451.

Pieterse, C. M. J., van Pelt, J. A., Verhagen, B. W. M., Ton, J., van Wees, S. C. M., Leon-Kloosterziel, K. M. and van Loon, L. C. (2003). Induced systemic resistance by plant growth-promoting rhizobacteria, *Symbiosis* 35, 39–54.

Pieterse, C. M. J., Zamioudis, C., Berendsen, R. L., Weller, D. M., Van Wees, S. C. M. and Bakker, P. A. H. M. (2014). Induced systemic resistance by beneficial microbes, *Annu. Rev. Phytopathol.* 52(1), 347–375.

Planchon, A., Durambur, G., Besnier, J. B., Plasson, C., Gügi, B., Bernard, S., Mérieau, A., Trouvé, J. P., Dubois, C., Laval, K., Driouich, A., Mollet, J. C. and Gattin, R. (2021). Effect of a *Bacillus subtilis* strain on flax protection against *Fusarium oxysporum* and its impact on the root and stem cell walls, *Plant Cell Environ.* 44(1), 304-322. doi: 10.1111/pce.13882.

Raaijmakers, J. M., de Bruijn, I. and de Kock, M. J. D. (2006). Cyclic lipopeptide production by plant-associated *Pseudomonas* species: diversity, activity, biosynthesis and regulation, *Mol. Plant-Microb. Interact.* 19, 699-710.

Raaijmakers, J. M., Paulitz, T. C., Steinberg, C., Alabouvette, C. and Moënne-Loccoz, Y. (2009). The rhizosphere: a playground and battlefield for soilborne pathogens and beneficial microorganisms, *Plant Soil* 321(1-2), 341-361.

Rabab, M. (2014). Interaction of *Bacillus subtilis* and *Trichoderma harzianum* with mycorrhiza on growth and yield of cucumber (*Cucumis sativus* L.), *Int. J. Curr. Res.* 6(8), 7754-7758.

Raj, M., Kumar, R., Lal, K., Sirisha, L., Chaudhary, R. and Kumar Patel, S. (2020). Dynamic role of plant growth promoting rhizobacteria (PGPR) in agriculture, *Int. J. Chemic, Stud* 8(5), 105-110.

Raklami, A., Bechtaoui, N., Tahiri, A. I., Anli, M., Meddich, A. and Oufdou, K. (2019). Use of rhizobacteria and mycorrhizae consortium in the open field as a strategy for improving crop nutrition, productivity and soil fertility, *Front. Microbiol.* 10, 1106.

Ravanbakhsh, M., Sasidharan, R., Voesenek, L. A. C. J., Kowalchuk, G. A. and Jousset, A. (2018). Microbial modulation of plant ethylene signaling: ecological and evolutionary consequences, *Microbiome* 6(1), 52.

Rocha, I., Souza-Alonso, P., Pereira, G., Ma, Y., Vosátka, M., Freitas, H.. and Oliveira, R. S. (2020). Using microbial seed coating for improving cowpea productivity under a low-input agricultural system, *J. Sci. Food Agric.* 100(3), 1092-1098.

Rolli, E., Marasco, R., Vigani, G., Ettoumi, B., Mapelli, F., Deangelis, M. L., Gandolfi, C., Casati, E., Previtali, F., Gerbino, R., Pierotti Cei, F., Borin, S., Sorlini, C., Zocchi, G. and Daffonchio, D. (2015). Improved plant resistance to drought is promoted by the root-associated microbiome as a water stress-dependent trait, *Environ. Microbiol.* 17(2), 316-331.

Rubin, R. L., van Groenigen, K. J. and Hungate, B. A. (2017). Plant growth promoting rhizobacteria are more effective under drought: a meta-analysis, *Plant Soil* 416(1-2), 309-323.

Ryu, C. M., Farag, M. A., Hu, C. H., Reddy, M. S., Kloepper, J. W. and Pare, P. W. (2004). Bacterial volatiles induce systemic resistance in *Arabidopsis*, *Plant Physiol.* 134(3), 1017-1026.

Saghafi, D., Delangiz, N., Lajayer, B. A. and Ghorbanpour, M. (2019). An overview on improvement of crop productivity in saline soils by halotolerant and halophilic PGPRs, *3 Biotech* 9(7), 261.

Saleem, A. R., Brunetti, C., Khalid, A., Della Rocca, G., Raio, A., Emiliani, G., De Carlo, A., Mahmood, T. and Centritto, M. (2018). Drought response of *Mucuna pruriens* (L.) DC. inoculated with ACC deaminase and IAA producing rhizobacteria, *PLoS ONE* 13(2), e0191218.

Sangiorgio, D., Cellini, A., Donati, I., Pastore, C., Onofrietti, C. and Spinelli, F. (2020). Facing climate change: application of microbial biostimulants to mitigate stress in horticultural crops, *Agronomy* 10(6), 794.

Santos, M. S., Nogueira, M. A. and Hungria, M. (2019). Microbial inoculants: reviewing the past, discussing the present and previewing an outstanding future for the use of beneficial bacteria in agriculture, *AMB Express* 9(1), 205.

Schlaeppi, K. and Bulgarelli, D. (2015). The plant microbiome at work, *Mol. Plant Microbe Interact.* 28(3), 212–217.

Sharma, R. K. and Archana, G. (2016). Cadmium minimization in food crops by cadmium resistant plant growth promoting rhizobacteria, *Appl. Soil Ecol.* 107, 66–78.

Shrivastava, P. and Kumar, R. (2015). Soil salinity: a serious environmental issue and plant growth promoting bacteria as one of the tools for its alleviation, *Saudi J. Biol. Sci.* 22(2), 123–131.

Sinha, A. K. and Parli, B. V. (2020). Siderophore production by bacteria isolated from mangrove sediments: a microcosm study, J. Exp. Mar. Ecol. 524, 151290.

Someya, N., Kataoka, N., Komagata, T., Hirayae, K., Hibi, T. and Akutsu, K. (2000). Biological control of cyclamen soilborne diseases by *Serratia marcescens* strain B2, *Plant Dis.* 84(3), 334–340.

Sonbarse, P. P., Kiran, K., Sharma, P. and Parvatam, G. (2020). Biochemical and molecular insights of PGPR application for the augmentation of carotenoids, tocopherols, and folate in the foliage of *Moringa oleifera*, *Phytochemistry* 179, 112506.

Stefan, M., Munteanu, N., Stoleru, V., Mihasan, M. and Hritcu, L. (2013). Seed inoculation with plant growth promoting rhizobacteria enhances photosynthesis and yield of runner bean (*Phaseolus coccineus* L.), *Sci. Hortic.* 151, 22–29.

Stringlis, I. A., Yu, K., Feussnerb, K., de Jonge, R., Van Bentum, S., Van Verk, M. C., Berendsen, R. L., Bakker, P. A. H. M., Feussner, I. and Pieterse, C. M. J. (2018). MYB72-dependent coumarin exudation shapes root microbiome assembly to promote plant health, *Proc. Natl. Acad. Sci. U. S. A.* 115(22), E5213–E5222.

Subbanna, A. R. N. S., Rajasekhara, H., Stanley, J., Mishra, K. K. and Pattanayak, A. (2018). Pesticidal prospectives of chitinolytic bacteria in agricultural pest management, *Soil Biol. Biochem.* 116, 52–66.

Tak, H. I., Ahmad, F. and Babalola, O. O. (2013). Advances in the application of plant growth-promoting rhizobacteria in phytoremediation of heavy metals, *Rev. Environ. Contam. Toxicol.* 223, 33–52.

Tchuisseu Tchakounté, G. V., Berger, B., Patz, S., Fankem, H. and Ruppel, S. (2018). Community structure and plant growth-promoting potential of cultivable bacteria isolated from Cameroon soil, *Microbiol. Res.* 214, 47–59.

Timm, C. M., Pelletier, D. A., Jawdy, S. S., Gunter, L. E., Henning, J. A., Engle, N., Aufrecht, J., Gee, E., Nookaew, I., Yang, Z., Lu, T. Y., Tschaplinski, T. J., Doktycz, M. J., Tuskan, G. A. and Weston, D. J. (2016). Two poplar-associated bacterial isolates induce additive favorable responses in a constructed plant-microbiome system, *Front. Plant Sci.* 7, 497.

Timmusk, S., Behers, L., Muthoni, J., Muraya, A. and Aronsson, A. C. (2017). Perspectives and challenges of microbial application for crop improvement, *Front. Plant Sci.* 8, 49.

Todeschini, V., AitLahmidi, N., Mazzucco, E., Marsano, F., Gosetti, F., Robotti, E., Bona, E., Massa, N., Bonneau, L., Marengo, E., Wipf, D., Berta, G. and Lingua, G. (2018). Impact of beneficial microorganisms on strawberry growth, fruit production, nutritional quality, and Volatilome, *Front. Plant Sci.* 9, 1611.

Tolba, S. T. M., Ibrahim, M., Amer, E. A. M. and Ahmed, D. A. M. (2019). First insights into salt tolerance improvement of *Stevia* by plant growth-promoting *Streptomyces* species, *Arch. Microbiol.* 201(9), 1295–1306.

Tsolakidou, M. D., Stringlis, I. A., Fanega-Sleziak, N., Papageorgiou, S., Tsalakou, A. and Pantelides, I. S. (2019). Rhizosphere-enriched microbes as a pool to design synthetic communities for reproducible beneficial outputs, *FEMS Microbiol. Ecol.* 95(10), fiz138.

Turan, M., Gulluce, M., von Wirén, N. and Sahin, F. (2012). Yield promotion and phosphorus solubilization by plant growth-promoting rhizobacteria in extensive wheat production in Turkey, *J. Plant Nutr. Soil Sci.* 175(6), 818–826.

Upadhyay, S. K. and Singh, D. P. (2015). Effect of salt-tolerant plant growth-promoting rhizobacteria on wheat plants and soil health in a saline environment, *Plant Biol. (Stuttg.)* 17(1), 288–293.

Van Loon, L. C. (1997). Induced resistance in plants and the role of pathogenesis-related proteins, *Eur. J. Plant Pathol.* 103(9), 753–765.

Van Loon, L. C., Bakker, P. A. H. M. and Pieterse, C. M. J. (1998). Systemic resistance induced by rhizosphere bacteria, *Annu. Rev. Phytopathol.* 36, 453–483.

Van Peer, R., Niemann, G. and Schippers, B. (1991). Induced resistance and phytoalexin accumulation in biological control of *Fusarium* wilt of carnation by *Pseudomonas* sp. strain WCS 417 r, *Phytopathology* 81(7), 728–734.

Vikram, A. and Hamzehzarghani, H. (2008). Effect of phosphate solubilizing bacteria on nodulation and growth parameters of greengram (*Vigna radiata* L. Wilczek), *Res. J. Microbiol.* 3(2), 62–72.

Vivas, A., Marulanda, A., Ruiz-Lozano, J. M., Barea, J. M. and Azcon, R. (2003). Influence of a *Bacillus* sp. on physiological activities of two arbuscular mycorrhizal fungi and on plant responses to PEG-induced drought stress, *Mycorrhiza* 13(5), 249-256.

Wei, G., Kloepper, J. W. and Tuzun, S. (1991). Induction of systemic resistance of cucumber to *Colletotrichum orbiculare* by select strains of plant growth-promoting rhizobacteria, *Phytopathology* 81(12), 1508–1512.

Weller, D. M., Landa, B. B., Mavrodi, O. V., Schroeder, K. L., De La Fuente, L., Bankhead, S. B., Molar, R. A., Bonsall, R. F., Mavrodi, D. M. and Thomashow, L. S. (2007). Role of 2,4-diacetylphloroglucinol producing fluorescent *Pseudomonas* spp. in plant defense, *Plant Biol.* 9, 4-20.

Wissuwa, M., Mazzola, M. and Picard, C. (2009). Novel approaches in plant breeding for rhizosphere-related traits, *Plant Soil* 321(1–2), 409–430.

Woo, S. L., Scala, F., Ruocco, M. and Lorito, M. (2006). The molecular biology of the interactions between *Trichoderma* spp. phytopathogenic fungi and plants, *Phytopathology* 96(2), 181–185.

Xiao, A. W., Li, Z., Li, W. C. and Ye, Z. H. (2020). The effect of plant growth-promoting rhizobacteria (PGPR) on arsenic accumulation and the growth of rice plants (*Oryza sativa* L.), *Chemosphere* 242, 125136.

Yuan, Z., Druzhinina, I. S., Labbé, J., Redman, R., Qin, Y., Rodriguez, R., Zhang, C., Tuskan, G. A. and Lin, F. (2016). Specialized microbiome of a halophyte and its role in helping non-host plants to withstand salinity, *Sci. Rep.* 6, 32467.

Zafar-Ul-Hye, M., Tahzeeb-Ul-Hassan, M., Abid, M., Fahad, S., Brtnicky, M., Dokulilova, T., Datta, R. and Danish, S. (2020). Potential role of compost mixed biochar with rhizobacteria in mitigating lead toxicity in spinach, *Sci. Rep.* 10(1), 12159.

Zaheer, A., Malik, A., Sher, A., Mansoor Qaisrani, M., Mehmood, A., Ullah Khan, S., Ashraf, M., Mirza, Z., Karim, S. and Rasool, M. (2019). Isolation, characterization, and effect of phosphate-zinc-solubilizing bacterial strains on chickpea (*Cicer arietinum* L.) growth, *Saudi J. Biol. Sci.* 26(5), 1061–1067.

Zdor, R. E. and Anderson, A. J. (1992). Influence of root colonizing bacteria on the defense responses of bean, *Plant Soil* 140(1), 99–107.

Zeilinger, S., Galhaup, C., Payer, K., Woo, S. L., Mach, R. L., Fekete, C., Lorito, M. and Kubicek, C. P. (1999). Chitinase gene expression during mycoparasitic interaction of *Trichoderma harzianum* with its host, *Fungal Genet. Biol.* 26(2), 131–140.

Zeller, S. L., Brandl, H. and Schmid, B. (2007). Host-plant selectivity of rhizobacteria in a crop/weed model system, *PLoS ONE* 2(9), e846.

Zhang, H., Xue, Y., Wang, Z., Yang, J. and Zhang, J. (2009). An alternate wetting and moderate soil drying regime improves root and shoot growth in rice, *Crop Sci.* 49(6), 2246–2260.

Zhang, Y., Du, H., Xu, F., Ding, Y., Gui, Y., Zhang, J. and Xu, W. (2020). Root-bacteria associations boost Rhizosheath Formation in moderately dry soil through ethylene responses, *Plant Physiol.* 183(2), 780–792.

Chapter 14

The role of liming in improving soil health

Richard C. Hayes, Jason R. Condon and Guangdi D. Li, New South Wales Department of Primary Industries, Australia

1 Introduction

Liming is an essential component of sustainable land management in agriculture. All agricultural production systems acidify soils through application of fertilizers such as ammonium-based nitrogen (N) fertilizers or elemental sulfur, and/or export of products (grain, meat, milk, fiber, biomass). A challenge in the sustainable intensification of agriculture is that as productivity increases, so too does the rate of soil acidification. At a minimum, liming is required to counter the acidification caused by the agricultural production enterprise. On more naturally acidic soils or degraded landscapes where soil acidity has extended to subsurface soil layers, additional inputs of alkalinity are required to address toxicities of ions such as aluminum (Al) and/or manganese (Mn) that are often associated with low soil pH and inhibit plant growth. Agricultural production on such soils is more vulnerable to extreme climatic events or incursion from pests and diseases because those plants are less robust and root systems less well developed. Liming is, therefore, a priority on soils that are already acidic and where agricultural production is at risk.

http://dx.doi.org/10.19103/AS.2021.0094.19

The term 'liming' comes from the age-old practice of applying finely ground limestone ($CaCO_3$) to soil. However, many other inorganic products such as dolomite [$CaMg(CO_3)_2$], or organic products such as manures, mulches, and composts contain alkali that can be used as alternatives, or in combination with limestone to ameliorate soil acidity. The term 'liming' in this chapter is used in a general sense inferring the application of any alkaline product for the amendment of soil acidity. The term 'lime' refers specifically to $CaCO_3$.

Lime is commonly mistaken as a fertilizer. In fact, it is better described as a soil amendment. The difference is that a fertilizer is used to supply a nutrient that is at insufficient levels for optimal plant growth. By contrast, the application of lime seeks to change the chemistry of the soil volume by raising soil pH, which has many indirect effects on chemical, biological and, sometimes, physical attributes of the soil. Notwithstanding, lime can have a direct fertilizing effect with the high concentration of calcium (Ca) in $CaCO_3$. Similarly, many organic manures and fertilizers can have liming properties in addition to their fertilizer attributes. In this review, we draw a distinction between soil amendments and fertilizers but acknowledge that many products share both properties.

The aim of this chapter is to examine the circumstances in which liming is an essential management strategy in maintaining or improving soil health. To understand the liming requirements of a soil in a particular production enterprise, it is first necessary to understand some fundamental processes driving soil acidification. Research into acid soils gained momentum in the 1960s and 1970s around the world and there now exists a plethora of literature on the topic. This chapter provides a brief overview of methods of assessing soil acidity, as pH can be measured using different protocols that are not directly comparable. The key processes contributing to soil acidification are also described, as the sustainable management of acidic soils can only be achieved with an understanding of the causal agents. Attention is then devoted to liming to ameliorate soil acidity, including products that are commonly applied to soil, rates required, and incorporation of lime with the soil volume. We then present a case study of the Australian experience with soil acidity and liming and explore some of the practical considerations, such as the financial implications, tolerant species, and new recommendations based on more recent research.

Australia has some of the poorest and most degraded soils on the planet and so, perhaps not surprisingly, has been a key hub for acid soil research over many decades. The topsoil of much of Australia's agriculturally productive landscape is seldom deeper than 15 cm and is commonly stratified (Paul et al., 2003), having relatively higher pH in the surface few centimeters overlying a more acidic subsurface zone between 5 cm and 15 cm depth (Fig. 1). The stratification of soil pH impacts acid soil management and biological responses to acidity. Some of the lessons from Australia can be applied to other regions of the world, especially where soil acidity has been less of a focus for land

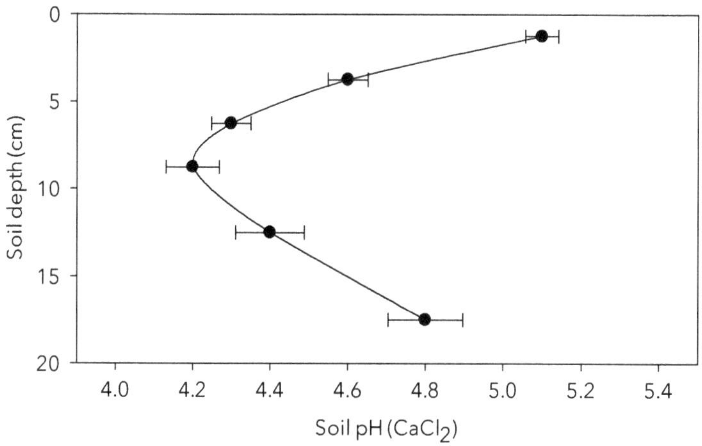

Figure 1 Soil pH is commonly stratified within the topsoil layer. Survey data from ten agriculturally managed fields, horizontal bar is standard error. Source: Adapted from Burns and Norton, 2018.

managers leading to more soil acidity now compared to three or four decades ago. However, it is first necessary to consider soil acidity and liming as they relate to soil health.

2 Soil health, soil acidity and liming

We accept the definition proposed by Lehmann et al. (2020) that *soil health is the continued capacity of soil to function as a vital living ecosystem that sustains plants, animals and humans, and connects agricultural and soil science to policy, stakeholder needs and sustainable supply chain management*. Whilst the terms 'soil quality' and 'soil health' are commonly used interchangeably in contemporary public discourse (Creamer et al., 2022), we acknowledge the distinction drawn by Lehmann et al. (2020) that the two terms differ and that soil health focuses on broader sustainability goals such as planetary health in contrast to soil quality, which concerns ecosystem services and soil functions from a human perspective. A key element of this definition, as it relates to the topic at hand, is the *continued capacity to function*. Soil acidification is a natural process and if it leads to reduced soil function, then liming to redress soil acidification on acidic soil types must be a critical plank of management for improved, or even sustained, soil health. It is therefore prudent to consider how soil function is impacted by low soil pH.

Soil pH is a fundamental characteristic of the chemistry of a soil and can therefore impact a broad diversity of soil characteristics, both directly and indirectly, as well as the broader soil ecosystem. The most obvious effect is on the soil chemistry, as pH is a measure of the concentration of hydrogen ions (H^+;

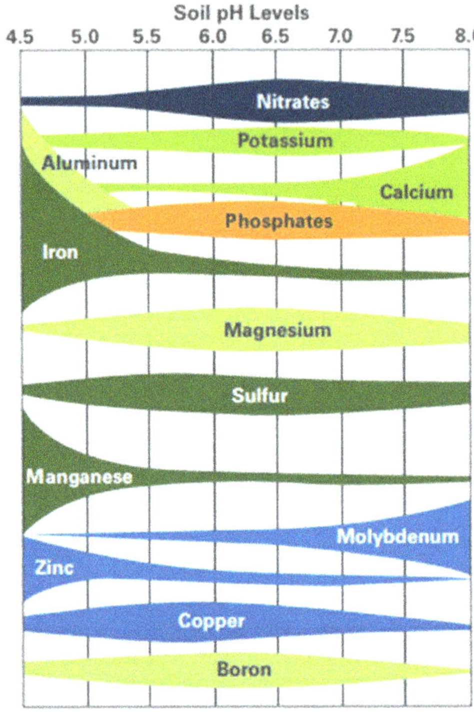

Figure 2 Relative change in availability of nutrients due to soil pH. Source: https://www.lawnstarter.com/blog/lawn-care-2/ph-soil-test-lawn-grass/ accessed April 2022.

see next section). Moreover, pH affects the availability of a range of essential nutrients (Fig. 2), either through direct chemical reaction (e.g. phosphorus (P), Al, biological transformations [e.g. nitrogen (N)] or indirect effects due to competition on exchange sites [e.g. Ca or magnesium (Mg) being displaced by Al]. Soil acidification can therefore lead to toxicities of some nutrients, especially Al and Mn, and deficiencies of other nutrients, such as molybdenum (Mo). Nutrient imbalances associated with pH can be exacerbated by negative effects on the organism living in the soil. For example, Al and Mn toxicities both result in stunting of roots of sensitive species such as alfalfa (lucerne; *Medicago sativa* L.; Fig. 3) rendering seedlings, in particular, more vulnerable to mortality due to drought and other stresses (Culvenor et al., 2011). A stunted root system impairs the ability of a plant to explore the soil volume for nutrients and water. Therefore, in acidic soils where the availability of a nutrient is already marginal, Al or Mn toxicity may result in deficiency of that nutrient in acid-sensitive plants that are unable to explore the soil volume effectively. In this example, low soil pH can be both a direct (low availability in soil) and an indirect (impaired ability for plant to forage for nutrients in soil) cause of nutrient deficiencies in plants.

Figure 3 Alfalfa seedlings (on the left) exhibiting depressed root growth attributable to Mn toxicity compared to seedlings (on the right) which were grown in the same soil with lower concentrations of Mn^{2+} (Hayes et al., 2012).

Similarly, low pH can reduce the availability of plant-available N but this can be exacerbated by the negative effects on legume–rhizobia symbiosis, thus N_2 fixation. Many legume species, such as alfalfa, are sensitive to soil acidity (Bouton, 1996). In part this is due to the sensitivity of alfalfa to toxicities of Al or Mn associated with acidity in many soils (Hayes et al., 2011, 2012) and in part it is due to the sensitivity of root nodule bacteria essential for N_2 fixation (Charman et al., 2008). Root nodule bacteria also require adequate nutrition for growth which has been demonstrated to be constrained on acidic soils deficient in essential nutrients, such as Mo (Anderson and Moye, 1952; O'Hara, 2001). In their Fig. 1, Coventry and Evans (1989) describe the potential breadth of the impact of soil acidity in undermining the effectiveness of the association between legumes and rhizobia. Interruption to the symbiosis of these organisms not only limits their growth and development but also reduces the N inputs into the soil for other species relying on N derived from biological N_2 fixation (Peoples et al., 2012).

Effects of soil acidity on soil biology extend well beyond root nodule bacteria, although this group of organisms has received substantial research attention over many decades due to their pivotal role in agriculture. In general, pH has been shown to change the composition of microbial communities, but differences in abundance and species richness are more variable and associated with local site factors (Lee et al., 2022; Narendrula-Kotha and Nkongolo, 2017; Pang et al., 2019; Yin et al., 2021). For example, in their assessment of soil fungal communities, Yin et al. (2021) demonstrated that the *Lasiosphaeriaceae*, *Piskurozymaceae*, and *Sordariaceae* families were positively correlated with Al and soil organic matter and negatively correlated with pH. Again, it might be expected that soil pH would have both direct and indirect effects on the microbiome. In their assessment of soil bacterial and fungal communities, Pang et al. (2019) demonstrated that soil fertility parameters including soil pH, available N, available P, available potassium (K), Ca^{2+}, Mg^{2+}, soil organic carbon (SOC), total N, and total P explained >80% of the shift in bacterial and fungal communities. In that study, only Ca^{2+}, Mg^{2+}, and pH of the soil were significantly different due to lime. Similarly, Kaveney et al. (2020) reported differences in abundance and function on nitrifying organisms within surface soils due to stratification of soil pH. They reported that liming acid soils enabled bacterial nitrifiers to be more effective than the archaeal nitrifiers present in acidic layers. These results infer that shifts in microbial populations can occur due to relatively subtle changes in soil pH and that multiple soil fertility factors can influence microbial populations, helping to explain why different studies report different responses to lime.

Lime can also impact soil physical characteristics, both directly and indirectly. Chan et al. (2007) demonstrated that lime improved the structure of an acidic hardsetting soil, associated with a reduction in electrical conductivity and sodium absorption ratio as well as increase in SOC, observed 4 years after lime application. On a different soil type, Chan and Heenan (1998) observed a temporary reduction in aggregate stability due to lime on a soil that was not cultivated. This may have been due to an increase in the negative charge of clay colloids immediately following liming, increasing the repulsive forces between particles (Haynes and Naidu, 1998). However, where the soil was cultivated, lime led to an improvement in soil structure compared to the unlimed soil (Chan and Heenan, 1998). Hayes et al. (2018) observed a small but significant difference in gravimetric water content attributable to the application of low rates (up to 5 t/ha) of dairy compost applied to soil, which they ascribed to be likely a result of increased infiltration associated with compost application. Organic soil amendments convey liming properties to soil, but the addition of organic C may also be expected to enhance the physical characteristics of the soil, in addition to any liming benefit (Haynes and Naidu, 1998).

The direct effects of lime on soil physical properties are likely to be soil-specific, driven by the conditions and mineralogy of the particular soil, whereas indirect effects of liming on soil physical properties are likely to be less soil-type specific and associated more with plant growth, or the lack thereof. Haynes and Naidu (1998) suggest that where liming increases crop yields, soil organic matter content increases as does soil aggregation. Conversely, the lack of liming a highly acidic soil has been shown to lead to significant erosion of topsoil. For example, the siltation of the Pejar Dam near Goulburn, New South Wales (NSW), was shown to be associated with extensive soil acidity in the 14 200-ha catchment (Cumming, 1982). In this example, it was a reduction in plant growth and ground cover associated with soil acidity that led to erosion of topsoil. This was of major concern due to premature siltation, over just 50 years, previously observed at the nearby Sooley Dam (Cumming and Garrard, 1981).

The role of liming to improve the C sequestration potential of soils has received some research attention. In their assessment of soil chemical properties at the conclusion of three long-term field experiments, Conyers et al. (2012) asserted that the build-up of C in soil requires accumulation of N and that the build-up of C and N fertility comes at the cost of increased soil acidity. The consequence of increased soil acidification requires liming to maintain agricultural production. The emission of CO_2 when limestone reacts with soil acids, plus the C cost of limestone mining, transport, and application, will negate a proportion of the gains from C sequestration as organic matter in soil. In theory, from just the reaction of lime alone, 1 tonne of $CaCO_3$ has potential to produce ≈ 0.44 tonnes of CO_2, which is equivalent to 0.12 tonnes of C (Rayment and Lyons, 2011).

Liming has also been implicated in the reduction of emissions of other greenhouse gases, notably nitrous oxide (N_2O; Goulding, 2016; Hénault et al., 2019; Holland et al., 2018). However, the many direct and indirect effects of soil acidity on soil fertility, especially as it relates to the C and N cycles, and impacts on soil microorganisms and their activity, result in changes of fluxes of CO_2, N_2O, and methane (CH_4). A comprehensive review of this topic by Kunhikrishnan et al. (2016) highlights the range of complex interactions that can lead to both increases and reductions in greenhouse gas emissions due to liming, making simple generalizations difficult. For example, interventions that may decrease nitrification-induced N_2O emissions (Hénault et al., 2019) can be somewhat countered by an increase in nitrate (NO_3^-) concentrations, a substrate for N_2O emission (Kunhikrishnan et al., 2016), due to increased nitrification as a result of liming (Kaveney et al., 2020). The net impact of liming on greenhouse gas emissions is therefore highly site- and context-specific, as it is with soil physical condition or C sequestration potential, highlighting the importance of a mechanistic understanding of key soil processes for land managers to achieve meaningful improvements in soil health.

3 Soil acidity

Soil acidity refers to the activity of hydrogen ions (H^+) in the soil solution. It is measured on the pH scale; a scale ranging from 0 to 14, with pH 7 being neutral, less than pH 7 being acidic, and greater than pH 7 alkaline. The pH is given as the negative log base 10 of the H^+ concentration ($pH = -\log_{10} [H^+]$). As the H^+ concentration increases, the pH decreases; 1 unit pH decrease represents a 10-fold increase in the H^+ concentration compared to the previous pH unit. Therefore, pH 4 has 100 times the H^+ concentration as pH 6.

3.1 Measuring soil acidity

Soil pH can be measured using pH indicators, either as a liquid or paper test strips, on soil slurries in the field or by using electrodes, often in solutions created from a mass of soil in a stated ratio with an extract solution (soil:solution). Unfortunately, countries/regions use different ratios of soil:solution and different extractant solutions [water, calcium chloride ($CaCl_2$), or potassium chloride (KCl)] at different concentrations (0.01, 1.0 M). There is no one correct method but the method needs to be noted as it influences the numeric result of pH measurement (Table 1). In general, as the volume of solution increases relative to soil mass analyzed, the H^+ concentration is diluted and the pH value will be higher. For a given soil:solution ratio, as the concentration of the extractant salt increases, more H^+ will be displaced from exchange sites in the soil resulting in a higher H^+ concentration in solution analyzed and a lower pH value. In general, the pH in water (1:5) is 0.6–1.2 pH units higher than that in $CaCl_2$ (0.01 M) and 0.9–1.4 pH units higher than that in KCl (1.0 M). The pH field kit reported in Table 1 is calibrated to equivalence of 1:5 soil:water (Raupach and Tucker, 1959). A detailed analysis of 286 soils analyzed at the Wagga Wagga Agricultural Institute showed a strong relationship ($P < 0.001$) between pH measured in 0.01 M $CaCl_2$ and in 1 M KCl (Fig. 4), although the relationship was not linear. At low (<4.0) pH, the difference in values was only 0.1 pH units, increasing to a difference of 0.7 units at pH values approaching 7.0. Understanding the relationship between different extractants is important. For example, extraction in 1 M KCl is a standard procedure for quantifying concentrations of mineral N in soils (Rayment and Lyons, 2011). Analyzing pH using the same extractant can be more cost-effective than preparing new extractants, particularly in research concerning large sample numbers.

Table 1 The influence of method of soil pH measurement on recorded pH for 4 agricultural soils (J. R. Condon, unpublished). Note: All extractions were conducted using 10 g soil with the volume of extractant adjusted to achieve ratios shown.

		pH in H_2O			pH in $CaCl_2$	pH in KCl	Soil pH
		Soil:solution ratio					Field kit
Soil type	Soil texture	1:1	1:5	1:10	1:5 (0.01 M)	1:5 (1.0 M)	slurry
Vertisol	Heavy clay	8.4	9.0	9.2	7.8	7.6	9.0
Ultisol	Light clay	6.1	6.7	7.0	5.9	5.5	6.5
Alfisol	Sandy loam	4.8	5.3	5.6	4.7	4.4	5.5
Alfisol	Sandy loam	5.8	6.3	6.4	5.5	5.0	6.0

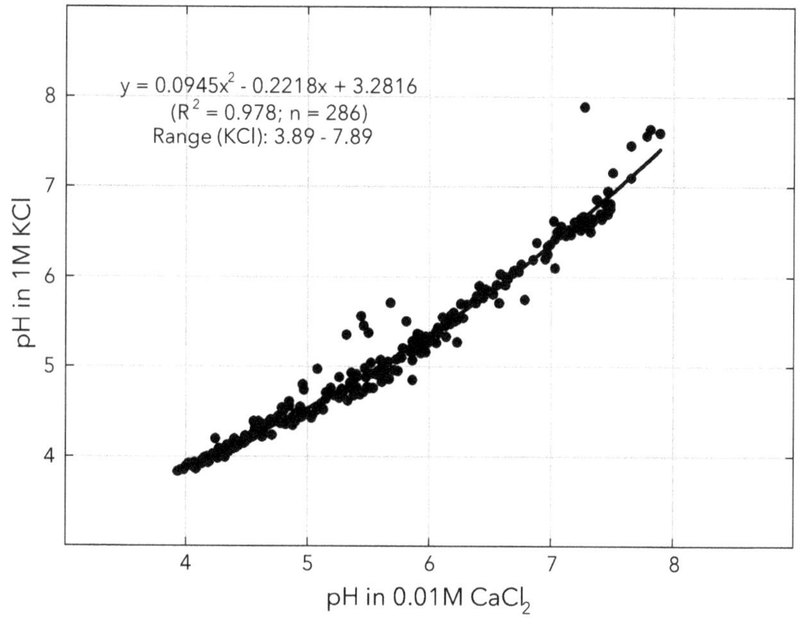

Figure 4 The relationship of soil pH measured in 1 M KCl and in 0.01 M $CaCl_2$ on soil samples (n=286) analyzed at the Wagga Wagga Agricultural Institute soils laboratory.

3.2 Measuring toxicities associated with acidic soils

Toxicities of Al and Mn are common in acidic soils and are the primary drivers of depressed yields but are not captured by measuring pH. Both require specific laboratory tests to estimate relative concentrations which may then be related to plant growth.

Aluminum toxicity is driven by the concentration of toxic Al ions present in the soil solution. Extraction of Al ions, such as in 0.01 M $CaCl_2$, is possible and was found to be highly correlated with exchangeable Al and with percentage Al saturation of the effective cation exchange capacity (Bromfield et al., 1983). Quantifying the effective cation exchange capacity (ECEC) of a soil is a relatively standard soil test (Gillman and Sumpter, 1986) commonly offered by soils laboratories. The ECEC of a soil is usually the sum of five cations, including Al, K, Ca, Mg, and sodium (Na), each expressed in cmol(+)/kg. The ECEC has more utility in acidic soils than the measure of 'base saturation' or cation exchange capacity (the sum of basic cations Ca, K, Mg, Na), as Al can occupy large proportions of the cation exchange sites in soils. For this reason, calculating the percentage Al saturation of the effective cation exchange capacity (%Al_{ex}), which is the concentration of Al relative to the ECEC and expressed as a percentage, is perhaps the most convenient approach for estimating the relative Al toxicity of a soil. As a rule of thumb, yield reductions might be expected in highly sensitive species when % Al_{ex} approaches 10% and in moderately sensitive species when Al_{ex} approaches 20%. Only highly tolerant species would exist at % Al_{ex} values > 30% (Upjohn et al., 2005).

Measuring Mn toxicity in soil is more problematic. In part, this is due to Mn^{2+} being a transient toxicity that is impacted not only by pH but also by factors that drive oxidation/reduction reactions in soil such as high temperatures and anaerobic conditions (Siman et al., 1974). Non-standard soil tests are required to estimate the potential for Mn toxicity to exist, including extraction by 0.01 M $CaCl_2$ (Bromfield et al., 1983) or tests to quantify easily reducible Mn (e.g. ammonium acetate-hydroquinone-extractable Mn), as easily reducible Mn relates to plant-available Mn (Robertson et al., 1973). There is no simple relationship between exchangeable and easily reducible Mn (Osborne et al., 1988). Unlike Al which is excluded from the plant, Mn is an essential nutrient for growth and is readily taken up. This presents an opportunity to sample and measure Mn concentrations of herbage, such as by inductively coupled plasma-optical emission spectrometry (ICP-OES). In species where thresholds have been established (Reuter and Robinson, 1997), it is possible to diagnose Mn toxicity through tissue analysis. Usually, tissue tests are used to confirm Mn toxicity on plants displaying visual symptoms of toxicity. Such symptoms include yellowing of the leaf and necrosis around the leaf margin (Fig. 5) and are more likely to occur on species such as legumes where the Mn concentration in root and shoot tissue is generally higher compared with grasses (Song et al., 2017).

Figure 5 Symptoms of Mn toxicity in alfalfa from field sites; (a) a plant from an unlimed plot showing obvious symptoms compared to the plant on the right from a limed plot (Hayes et al., 2008) and (b) severe symptoms from a soil with extreme levels of easily reducible Mn concentrations (Hayes et al., 2012).

4 Factors influencing changes in soil pH

Any change in soil pH is the product of processes that either consume or create acidity and the ability of the soil to resist pH change due to those processes, known as the soil pH buffering capacity (pH_{BC}). The pH_{BC} is influenced by the number of sites in soil that are able to hold H^+ out of solution, including sites of H^+ association, cation exchange, and carbonates/bicarbonates. However, carbonates tend not to persist in acidic soils. Therefore, soils with high organic matter content and/or high clay content will have higher pH_{BC}. In terms of soil management, pH_{BC} is altered only by implementing practices that build organic matter, although changes can be made via engineering the soil by importing clay or delving clay from lower soil layers with deep cultivation (Davies et al., 2020).

4.1 Processes leading to soil pH changes

Agriculture is an acidifying industry as it is based on the removal of product from the land on which it was derived. The product removed is alkaline in nature due to the cation-anion balance of nutrients taken up by the plants.

Plants acquire ionic nutrients from the soil solution as cations (positive charged ions, e.g. K^+, Ca^{2+}, ammonium (NH_4^+), Mg^{2+}, Na^+) or anions (negatively charged ions, e.g. phosphate, sulfate, nitrate, chloride). The balance of uptake between cations and anions impacts the internal charge of a plant, which is maintained at neutral by a process called antiport exchange: excreting cations when excess cation uptake occurs and excreting anions when excess anion uptake occurs (Kennedy, 1992). The plant uses processes of the C cycle to balance internal charge with H^+ being an easily excreted cation and alkaline anions excreted as either OH^-, HCO_3^-, or $RCOO^-$ (Paul et al., 2003). In most agricultural soils, net excess cation uptake occurs, resulting in dissociation from organic molecules in the plant and then excretion of H^+ from the plant to the soil. This leaves the rhizosphere more acidic (Marschner et al., 1986) and the plant more alkaline as organic anions are formed in the cells as H^+ is dissociated and excreted. This alkali can be measured by ashing the plant material and titrating the ash alkalinity. The return of plant material to the soil was thought to neutralize the acidity created by the net excess cation uptake. However, in practice, the plant residue is largely returned to the soil surface, where the alkali will be released to increase soil pH in a few centimeters of the surface only (Paul et al., 2003). As agriculture removes product from the paddock as grain, hay or silage, or as meat or wool if consumed by animals, the alkali is removed from the paddock leaving the soil acidified (Ridley et al., 1990). Alternatively, if plant residues such as compost, manures, and stock feed produced elsewhere are deposited in a paddock, a net gain in alkali will occur, yet the increase in pH will be left at the soil surface unless the organic material contains soluble organic compounds or is incorporated into deeper soil layers (Nguyen et al., 2018). Based on more than 100 years of Rothamsted data, it was estimated that acidification due to cation-anion balance represented less than 10% of system acidification, with processes of the N cycle being dominant (Johnston et al., 1986).

The N cycle contains several processes which can create or consume acid (H^+) or alkali (OH^-) in the soil (Helyar, 1976). These processes, listed below, are represented diagrammatically in Fig. 6.

Use of fertilizer N influences acidity due to the pathway of added N within the N cycle (Fig. 6). The starting point in the cycle being before process ① for urea, immediately after process ① for ammonium-based fertilizers and immediately after process ④ for nitrate fertilizers. It is clear that leaching of nitrate is only acidifying if that nitrate is the product of *in situ* nitrification. Leaching of nitrate from nitrate fertilizer has no direct effect on soil pH.

Whilst other biochemical processes such as oxidation of sulfur from organic matter and oxidation/reduction processes of Mn can influence changes in soil pH, in most agricultural systems their contribution, relative to N cycle processes, are minimal (Conyers et al., 1995b).

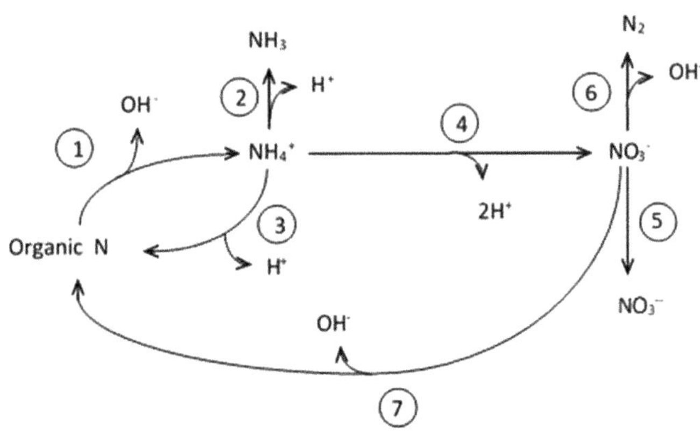

Figure 6 Possible pathways of the nitrogen cycle and the release of pH altering ions (H^+ and OH^-). Processes numbered for reference in text. ① Mineralization: $R\text{-}NH_2 + 2H_2O \rightarrow NH_4^+ + ROH + OH^-$(alkaline); where R is an organic molecule. ② Volatilization: $NH_4^+ \rightarrow NH_{3(gas)} + H^+$ (acidic); ③ Immobilization of ammonium: Biological utilization of ammonium, causing excretion of H^+ by the plant or organism (acidic); ④ Nitrification: $NH_4^+ + 2O_2 \rightarrow NO_3^- + H_2O + 2H^+$ (2 × acidic); ⑤ Leaching: NO_3^- moves downwards with water (neutral); ⑥ Denitrification: $NO_3^- + H_2O \rightarrow N_2 + OH^-$ (alkaline); ⑦ Immobilization of nitrate: Biological utilization of nitrate, causing excretion of OH^- by the plant or organism (alkaline).

4.2 The net effect of pH changing processes

The reactions outlined in Section 4.1 occur at different times of the year in response to rainfall, temperature and plant life cycle fluctuations, different locations within paddocks, and at different soil depths. The net result is that soil pH varies temporally and spatially (horizontally and vertically). Temporal variation in pH of < 0.5 pH units has been reported (Conyers et al., 1997a), which is smaller than the spatial variability that occurred in the same soils (Conyers and Davey, 1990). Horizontally, pH could change > 0.5 pH units within < 1 m (Hayes et al., 2022). Vertical variation in soil pH results from return of organic residues from plants and livestock wastes to the soil surface and subsequent association with soil H^+ prior to biological oxidation (breakdown) of the organic matter, which is then followed by downward movement of ammonium prior to nitrification (Condon et al., 2004; Evans et al., 1998; Paul et al., 1999). Understanding the effects of vertical separation of process changes how management practices are thought to contribute to acidity. For example, it is often stated that complete utilization of urea fertilizer results in no pH change (Goulding, 2016), but this is not the case when N transformation processes are commonly vertically separated (Condon et al., 2004). The vertical separation of processes results in more alkaline reactions occurring in the surface layer (0–2cm) of profiles and net acidifying reactions occurring in the sub-surface (3–8 cm) layers (Condon et al., 2004; Paul et al., 2003). As a consequence, the soil surface has a higher pH than the soil immediately below the surface, thereby creating the formation of acidic subsurface depths (Fig. 1). This pH stratification can occur within a growing season for annual crops (Paul et al., 2001) or within weeks in pastures (Condon et al., 2021). Soil pH stratification is exacerbated by the application of lime without effective incorporation (Conyers et al., 2003). Furthermore, the pH_{BC} and organic matter content tends to be similarly stratified (Paul et al., 2003), causing larger decreases in the pH of the subsurface layers than the alkaline reactions occurring in the soil surface. The stratification of pH_{BC} also has implications for the effectiveness of liming to increase the pH of acidic soil.

4.3 Reducing rates of soil acidification

Product removal, inherent in all agricultural systems, prevents the elimination of acidification. Likewise, organic matter accumulation, although a source of acidification associated with the accumulation of C and N (Conyers et al., 2012), is also desirable for many other positive attributes that it brings to soils, such as increased water-holding capacity, increased friability (Haynes and Naidu, 1998), and increased microbial diversity and abundance (Hayes et al., 2021). Liming is therefore necessary to facilitate organic matter increase whilst avoiding net acidification. However, much of the acidification associated with the N cycle can be minimized through reduced losses of N following nitrification.

Productive perennial vegetation perhaps provides the greatest opportunity to reduce N losses and minimize soil acidification attributable to the N cycle (Ridley et al., 2001; Crews and Peoples, 2004). In contrast to annual crops and pastures, perennial species generally have deeper roots and a longer growing window which enables greater opportunity to explore the soil volume and take up nitrate before it is leached to below the root zone (Dear et al., 2009). Moreover, perennial-based systems reduce the frequency of disturbance events associated with sowing a new crop, such as fallowing with herbicide or cultivation (Crews et al., 2016). Increased rates of mineralization coupled with decreased plant demand resulting from disturbance events leads to an abundance of mineral N in the soil, increasing the risk of N loss (Li et al., 2022). However, the value of perennial vegetation in mitigating soil acidification in agricultural landscapes is limited. First, even if perennial plants increase nitrate uptake deeper in the profile, acidification would have occurred at the site of nitrification, creating acidic subsurface layers within the soil profile (Condon et al., 2021; Paul et al., 2003). Liming is therefore still required to ameliorate acidic soil layers, as well as replace alkali removed in agricultural product. Secondly, minimizing acidification with perennial plants is only feasible in agricultural landscapes where viable perennial options exist. Whilst there are many perennial forage options for grazing, there are few viable broadacre perennial crops. Until the vision for viable perennial crops is realized (Glover et al., 2010), incorporation of perennial plants into cropping systems will remain limited to soils where the use of perennial forages can be incorporated in phased rotations with annual crops (Llewellyn et al., 2014). The effectiveness of phased pasture-crop rotations to reduce soil acidification compared to continuous cropping systems has rarely been demonstrated.

An examination of N sources underpinning the production system is also warranted. Typically, N sources include fertilizer N [products such as urea (46% N)], or organic N from symbiotic N fixation, or from manures and composts. There is little evidence that a particular source of N will necessarily lead to reduced losses through leaching, particularly under dryland production systems, although a greater proportion of N derived from legume residues is generally retained in soil compared to fertilizer N (Crews and Peoples, 2005). Organic fertilizers, such as manures, can be very high in N. The availability of this N will vary with product but often much of the N in manures will not be immediately available to plants as it is bound in organic compounds and will become available over time as those compounds break down. This slow-release process serves to limit N losses from the system attributable to manure application, although it is dependent on the C:N ratio of the product and the rate at which it is applied. Griffiths (2014) suggests that manures with a C:N ratio less than 30:1, such as many poultry manures, are more vulnerable to N losses than other manures such as from cattle. Excessive rates of application also need to be avoided because even slow-release sources of N can lead to losses where the quantity applied exceeds the requirements of the crop. Herein lies the conundrum of managing soil acidification with organic products; on the one hand, for amelioration of soil acidity to be effective, higher rates are likely required on most soils (Hayes et al., 2018), yet high rates of many organic amendments will lead to accelerated acidification through increased N losses. The most effective means to combat soil acidity with organic soil amendments most likely would require relatively high rates of application of products that have a high C:N ratio, but further research is required to develop practical approaches for use on a commercial scale.

5 Liming to ameliorate soil acidity

There are several factors to consider when formulating a liming regime, including determining which product to use, when to apply it, and how often. The amount of lime required is dependent on the current pH and pH_{BC} of the soil, the quality of lime product (see Section 5.2) and the target pH to be achieved. The liming interval may need to be decreased depending on the starting pH, the rate of acidification, the depth of acidic layers, and if lime is incorporated by cultivation (Conyers et al., 2003).

5.1 Calculating lime rates

Lime rate is often calculated with sophisticated decision support systems (Hochman et al., 1989) taking into account important soil attributes, such as organic matter content, texture, and the nature of the clay minerology (Bolan et al., 2005; Condon et al., 2021). The lower the current pH, the greater the quantity of lime required to reach a given pH target at a given pH_{BC}. A soil with higher pH_{BC} is more resistant to pH change compared with soil with lower pH_{BC}, thus more lime is required to reach a similar target pH. Based on pH_{BC} data at 3 soil depths (0–10, 10–20, and 20–30 cm) from 12 sites spanning 6 soil types across NSW and Victoria, Australia, we developed a series of empirical equations to calculate the lime required to increase pH to 6.0 under different pH_{BC} with different initial pH (Fig. 7a), as well as pH increment with different lime rates (Fig. 7b). For example, to lift pH from 4.0 to 6.0, it would only require 2.5 t/ha of lime when pH_{BC} is 15 kmol/ha.10cm/pH but would require 5.0 t/ha of lime when pH_{BC} is 30 kmol/ha.10cm/pH (Fig. 7a). In other words, 2.5 t/ha of lime can increase 2 pH units with pH_{BC} of 15.0 kmol/ha.10cm/pH but only 1 pH unit with pH_{BC} of 30.0 kmol/ha.10cm/pH (Fig. 7b). Due to these relationships, soil

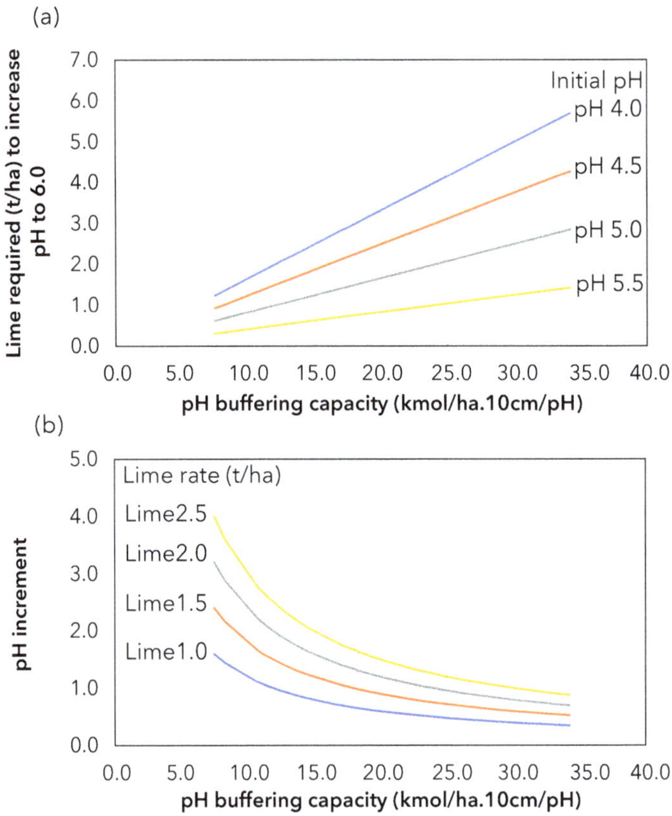

Figure 7 Relationships of pH buffering capacity (kmol/ha.10cm/pH) against (a) lime required (t/ha) with target pH 6.0 in 10 cm of soil at different initial pH_{Ca} and (b) pH increment with different lime rates 12 months after lime is applied, assuming a lime efficiency of 60%.

cation exchange capacity, a property strongly related to both organic matter content and clay content, has been used as a surrogate of pH_{BC} and soil health by farmers and advisors.

Rates of acidification also need to be taken into account and will vary according to the production system and the environment. For example, acidification rates are likely to increase in higher rainfall environments compared to drier environments (Scott et al., 2000). In part this is because of increased levels of production in more favorable environments but also reflects greater opportunities for transformation and loss of N from the system. In a study of the ash alkalinity of different products common in the farming systems of northern Victoria, Slattery et al. (1991) found that hay production was the most acidifying system requiring 270–1260 kg $CaCO_3$/ha.year compared to wool and lamb

(0.8–6.0 kg $CaCO_3$/ha.year) which was the least acidifying. Grain crops in the same environment required 7–20 kg $CaCO_3$/ha.year.

5.2 Lime quality

Lime quality is crucial for the effectiveness and rate of dissolution. There are several key factors that control the lime performance as described by Conyers et al. (1995a). First, particle size determines the surface area of lime exposed to and the distribution of particles throughout a given soil volume. The finer the particle, the greater the surface area of particles for a given mass of lime, the faster the dissolution, and the more particles in each volume of soil after mixing. Therefore, fineness influences not just the rate of dissolution but also the proportion of the soil volume in contact with the particles (Cregan et al., 1989). The residual value of coarser particles might be expected to be greater but may not effectively contribute to soil amelioration within the time frame between lime applications (Conyers et al., 1995a). Secondly, chemical composition, reported as the neutralizing value (% equivalent to 100% $CaCO_3$), influences the quantity of liming agent required to increase soil pH. The higher the neutralizing value of the lime product, the more effective the product will be in increasing soil pH. Thirdly, lime solubility determines the tendency of the lime material to dissolve and hence the rate of reaction of a given particle size, which is closely related to the current status of soil pH as discussed above. In a field experiment in southern NSW, Conyers et al. (2020) demonstrated that fine lime (particle size less than 250 μm) increased soil pH rapidly in year 1, but it was the slow ongoing reaction of lime over 3–6 years that delivered a measurable improvement in subsurface pH and sustained improvements in grain yield for at least eight consecutive years. Conyers et al. (2020) concluded that lime sources with finer particle sizes are the most effective means of achieving these benefits, which did not give way to lesser residual value compared with coarser limestone over the 8-year study period.

In addition, rainfall is another factor that affects lime reaction. The higher the rainfall, the quicker lime will be dissolved and deeper lime will be moved down the soil profile. Tang et al. (2003) reported that soil pH at 10–20 cm did not change 2 years after lime was incorporated at 0–10 cm with an annual rainfall of 362 mm, whereas Caires et al. (2008) demonstrated that soil pH increased significantly down to 20 cm 3 years after lime was applied in a no-till system with an annual rainfall of 1550 mm. Conyers et al. (2020) showed that it took 3 years to dissolve 2.5 t lime ha^{-1} with 1807 mm of rainfall in total, and 6 years to dissolve 5.0 t lime ha^{-1} with 3592 mm of rainfall in total in a field site in southern Australia.

5.3 Lime incorporation

Lime is sparingly soluble (0.013 g/L) and the rate of dissolution slows as the pH increases, particularly above pH 6.0 (Scott et al., 1999). As agricultural production systems form stratified pH layers with higher pH at the soil surface, lime spread on the soil surface without incorporation will react very slowly (Conyers et al., 2020). Furthermore, the application of lime to the soil surface without incorporation will exacerbate pH stratification.

Lime is often incorporated during the cropping phase or prior to the establishment of a pasture as part of standard seed-bed preparation. Scott and Coombes (2006) found that offset discs were more effective to incorporate lime compared with tined implements or harrows due to more thorough mixing of the soil, although effects on crop yields were relatively short-lived. On the vast permanent pasture or grassland areas around the world, there is limited opportunity to incorporate lime due to environmental risks such as wind and water erosion. Furthermore, even on arable land, farmers are reducing the frequency of cultivation of land in minimum tillage systems due to perceived damage to soil structure by cultivation, as well as other benefits such as reduced fuel costs prior to sowing. The consequence for acid soil management has been a reduced opportunity to incorporate lime over an increasing area of agricultural land. On arable land, strategic use of tillage within an otherwise minimum-till system only generated minor and short-term impacts on soil conditions on three contrasting soils in a semi-arid environment (Conyers et al., 2019). However, in many instances, particularly in extensive grasslands, even occasional tillage will not be a viable option. There are three key 'compromise' management practices recommended for applying lime in no-till situations where it cannot be incorporated: (a) applying higher rates of lime than calculated for systems incorporating lime, so as to increase the leaching of bicarbonate (Conyers et al., 2003); (b) allowing more time for lime to react with soil where it is surface applied compared to where it is incorporated (Conyers et al., 2003); and (c) anticipating increased pH stratification with surface-applied lime and manage accordingly, such as by sampling soil in shallower depth increments (Condon et al., 2021) and concurrent use of moderately acid-tolerant species that are less sensitive to acidic sub-surface soil layers (Scott and Fisher, 1989).

5.4 Liming to depth

Soil acidity below 10 cm is difficult to ameliorate. First, lime moves very slowly down the soil profile, especially in drier environments. Li et al. (2019) reported that pH increased at a rate of 0.04 pH units per year at 15–20 cm, providing an increase of about one pH unit over 25 years, by maintaining pH above 5.5 at 0–10 cm with ongoing lime application. It was estimated that the vertical movement of alkalinity was less than 1 cm per year in the same field experiment (Li et al., 2019). This is broadly consistent with the study of Conyers and Scott (1989) who demonstrated pH change on different soil types to 6 cm below the depth of incorporation after 5 years, where surface soil pH_{Ca} was maintained above 5.5. Second, the lack of appropriate machinery and physical constraints associated with drilling lime to depth is a significant obstacle to directly treating acidity in deeper soil layers.

Deep ripping is one of the alternatives to surface liming to place liming materials at depth. However, the lime effectiveness is dependent on the width of the ripping row and how well the soil and lime materials are mixed within the ripping rows, often related to the engineering of the tine design. Recent research from Australia showed that deep placement (10–30cm) of liming materials (lime, dolomite, reactive phosphate rock (RPR), and magnesium silicate) using a prototype machine with 50 cm tine spacings increased soil pH on the ripping line, but the crop yield responses were variable during the first 3 years (Condon et al., 2018; Li et al., 2018). Clearly, the treatment difference was not large enough to demonstrate the potential of lime materials implemented, given that only a small proportion of the soil volume was treated with lime in this case. Furthermore, there was an obvious negative impact on seedling emergence in the year of application due to the uneven seedbed caused by ripping (Poile et al., 2012).

Deep incorporation of lime to the depth of the acidity in the profile would be the most effective method to maximize the benefit of lime applied by mixing the lime with a greater volume of soil. The need for specialized machinery, the impact of incorporation on soil structure, the risk of wind and water erosion, and fuel costs associated with incorporation would be the major considerations by farmers and advisers, particularly in regions where no-till farming practices are favored. In some instances, the risk of bringing dispersive subsoils to the surface would also be a consideration on sodic soils. The long-term impact of deep incorporation of lime on soil chemical, physical, and biological properties; and responses of pasture, crop, and animal production to soil amelioration remains to be studied.

5.5 Alternative products

There are a range of products that might be used as alternatives to lime ($CaCO_3$). The relative merits of alternative inorganic materials can be assessed by comparing the neutralizing value and particle size (fineness) balanced against the proximity to a local source which will impact freight costs to a farm, and the unit cost of the material. Pure dolomite [$CaMg(CO_3)_2$] has equal molecular ratios of $CaCO_3$ and $MgCO_3$ and has a higher neutralizing value

(108%) than $CaCO_3$ on account of the lower molecular weight of Mg relative to Ca (Barber, 1984).

The liming properties of some more novel products have been less well established. Castro and Crusciol (2013) claimed that magnesium silicate (MgSi) was more efficient than lime in alleviating Al toxicity. Initial field experimentation in Australia showed little benefit of MgSi in raising soil pH or reducing Al toxicity, although further research is warranted (Li et al., 2021b). Calcium nitrate [$Ca(NO_3)_2$] supplies plants with N in the form of nitrate, which would promote anion uptake, leading to the release of OH^- ions into the rhizosphere thus increasing rhizosphere pH (Butterly et al., 2021). There is potential for the effect of root-induced alkalization in the rhizosphere to expand and ameliorate acidity in the bulk soil beyond the rhizosphere as demonstrated in several short-term pot experiments (Weng et al., 2022). However, there was no difference in soil pH and exchangeable Al% between treatments top-dressed with urea or $Ca(NO_3)_2$ at equivalent N rates over 3 years on an acidic soil at Holbrook, NSW (Li et al., 2021b). RPR contains 12.7% P and 35.9% Ca and is used as a P source as well as an ameliorant for soil acidity, including in organic production enterprises. Initial field testing by Li et al. (2021b) in northern Victoria placed RPR at 10–30 cm below the surface at 4 t/ha and 8 t/ha. Both RPR treatments increased soil pH significantly at 10–20 cm and 20–30 cm but were not as effective as the deep lime treatment. However, crop yields responded positively to the RPR treatments, due to the additional P supplied at depth. This effect highlights the potential benefits of using products with both liming and fertilizer properties. In a separate study, a positive yield response was observed with the use of sewage ash, which also contained about 3% P, as an ameliorant for soil acidity (Dann et al., 1989).

Organic materials such as crop residues, animal wastes, and garden composts proved to have potential to increase pH either as an alkali source or by enabling alkaline reactions to occur during the decomposition of organic materials. However, the magnitude of any pH rise varied between type of organic materials used, depending on their ash alkalinity and the concentration of excess base cations, which is reflective of the concentration of stored organic anions (Tang and Yu, 1999). However, based on results from several glasshouse experiments, the acid-neutralizing effect was short-lived and proved less effective than calculations of ash alkalinity suggested (Butterly et al., 2011; Nguyen et al., 2018). In part, this may be attributable to an over-estimation of anions available to ameliorate acidity, where they are tightly bound in organic molecules, as well as the subsequent nitrification which may result, particularly from products with a lower C:N ratio (Helyar, 1976; Li et al., 2021a). For example, in a 3-year field experiment in NSW, results showed that only limed treatments increased soil pH (Fig. 8). For all treatments with organic amendments placed at depth, no significant difference in pH was observed compared with the control

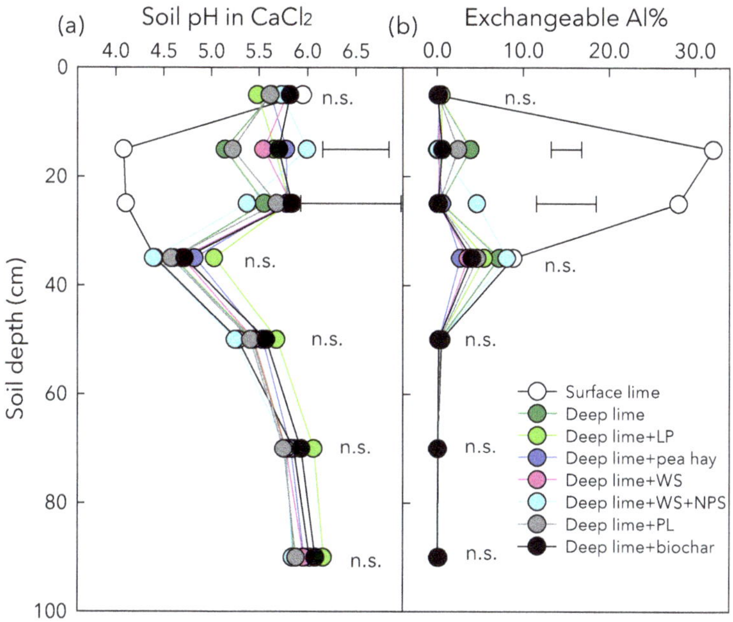

Figure 8 (a) Soil pH in $CaCl_2$ and (b) exchangeable Al% under various organic ameliorants in contrast with lime two years after treatments being implemented at the Holbrook site, NSW, Australia. LP, alfalfa hay pellets; WS, wheat straw pellets; NPS, 5N:2P:1.3S as per (Kirkby et al., 2013); PL, poultry litter pellets. Horizontal bars represent LSD at $P = 0.05$. n.s.: not significant. Source: Adapted from Li et al. (2021b).

treatment at 10–30 cm (Li et al., 2021b). However, organic amendments did reduce exchangeable Al% significantly in the 10–20 cm depth, compared with the surface lime treatment (Fig. 8). This is likely to be the result of the soluble organic molecules from organic materials combining with active Al^{3+} to form insoluble hydroxy-Al compounds (Haynes and Mokolobate, 2001), which would reduce Al toxicity to plant root systems.

It appears that the combination of lime with organic materials could facilitate downward movement of alkalinity from the lime as demonstrated from short-term glasshouse experiments (up to 3 months) (Butterly et al., 2020; Nguyen, 2021). However, results from an intensive soil grid sampling in the field, 4 years after treatments were implemented at a long-term field site, showed that soil pH beyond the ripping depth was lower under alfalfa hay pellet treatments even with the addition of lime (Li et al. unpublished data). This finding indicated that the addition of a large amount of organic material (up to 15 t/ha) could acidify soil further, perhaps doing more damage than good over the longer term. Farmers should take caution when considering applying organic amendments at depth to address subsoil acidity, but further research to examine the potential of surface-applied organic amendments in conjunction with lime to accelerate the movement of alkali material to deeper soil layers is warranted. Furthermore, due to the large volumes required, it is uneconomic and impractical to use organic materials to ameliorate subsurface soil acidity in broadacre agriculture. Even deep placement of lime, which requires reduced volume of product due to its superior efficacy, would take decades to have the capital investment recovered. Therefore, a proactive, preventative management of topsoil pH at 0–10 cm with regular lime addition remains the most cost-effective solution to combat subsoil acidity.

6 Liming alkaline soils

The case for liming alkaline soils is less clear compared to acidic soils. On the one hand, soil acidification is inevitable in agricultural production systems and so liming is necessary to balance the acidification. On the other hand, alkaline soils commonly already have excess lime (carbonates), especially at depth, and so it seems futile to invest in spreading additional lime. Moreover, as detailed in Section 4.2, the solubility of lime decreases as pH increases meaning that on many alkaline soils, lime application would have little consequence other than to increase the volume of free lime at the soil surface. In addition, plant growth on alkaline soils is often constrained by associated factors, such as boron (B) toxicity and salinity (Smith et al., 2013). On such soils, acidification may be beneficial to plant production by alleviating some of these constraints. However, in practice, the high pH_{BC} of such soils would likely limit the pH change observed in most situations in the short-medium term. Farmers on these soils would likely be wise not to invest in lime but monitor pH change in the surface layers over time.

7 Case study: the Australian experience

It is estimated that over 50 million ha of agricultural land is affected by soil acidity (Lockwood et al., 2003), reportedly costing Australian agricultural industries over A$1.6 billion annually in amelioration and lost production (SoE, 2011).

7.1 Causes of widespread acidity

Although many soils are naturally acidic across the agricultural regions of southern Australia, rapid acidification occurred in the post-war era following the 'sub & super' revolution. In the two decades following the end of World War II, there was a large investment in increasing the productivity of agriculture across Australia. A number of factors contributed to this, including the need to provide returned servicemen with employment. The Soldier Settlement Scheme saw the sub-division of smaller (often 400–800 ha) parcels of land from larger stations for which eligible service personnel could 'draw' upon their return (Macintyre, 2015). To be viable, production on these Soldier Settler blocks was substantially intensified compared to larger stations. Production figures from before and after the introduction of the scheme attributable to subdivision are not readily available. However, one anecdotal case from the large Springfield property to the north of Canberra, indicated that a parcel of land which previously ran approximately 4000 wethers prior to subdivision was reported to run >25 000 sheep (including breeding ewes) less than 30 years later (T. Hayes pers. comms). A six-fold increase in production was probably not uncommon in that era.

The uptake of single superphosphate (8.8% P, 11% S), aided by the 'wool boom' of the early 1950s which increased the disposable income of sheep farmers (Massy, 2011) meant that farmers had the means to spread large quantities of fertilizer to address the almost universal deficiency of P in agricultural soils. By addressing the chronic P deficiency, farmers for the first time had the means to grow legumes on a large scale such as the self-regenerating annual, subterranean clover (*Trifolium subterraneum* L.). This species was broadly adapted to mixed (crop/livestock) farming and permanent pasture regions alike due to its high tolerance of uncontrolled grazing and acidic soils (Nichols et al., 2012). Subterranean clover seed was readily broadcast with the fertilizer meaning that its establishment was not reliant upon sophisticated pasture sowing methods or machinery, fostering its rapid spread.

The proliferation of subterranean clover massively increased inputs of N into farming land. More recent research has shown that subterranean clover can fix around 30–35 kg of N per tonne of aboveground legume biomass (Peoples et al., 2012). Therefore, even in a moderate year where there is perhaps just 6 t/ha of cumulative legume biomass, total inputs of N could be in excess of 200 kg N/ha. Many pasture systems at that time were still dominated by relatively unproductive native grass species which often could not fully utilize this bulge of N (Dear and Virgona, 1996; Dear et al., 2009), resulting in rapid soil acidification. In some regions, the rate of decrease in surface soil pH was as much as 1.0 pH unit over just five decades (Cumming and Williams, 1983; Williams, 1980).

7.2 The economics of liming

The productivity decreases associated with the rapid decline in soil pH were, to some extent, masked by the rapid increase in productivity associated with the key causes of widespread acidification; subterranean clover, superphosphate, and subdivision. Reduced productivity was most visible when sensitive plant species started to be grown. For example, soil acidity was implicated as a key impediment to growing alfalfa in some regions (Siman and Flemons, 1970). It was also broadly recognized that canola (*Brassica napus* L.) could not be grown unless soil acidity was addressed (Slattery and Coventry, 1993).

Yet, despite the rapid decrease in soil pH across large regions of Australia's key agricultural zones, the uptake of lime was slow. In part this was due to a physical constraint at the farm level, as lime was harder to spread and required more specialized machinery compared to spreading superphosphate. However, this barrier was not overcome sooner due to ambiguity over whether the farmer could really afford to pay for lime. Few economic assessments were conducted in the early years but the absence of a visible production response to lime in many fields, in contrast to the obvious and almost immediate response following superphosphate application, made farmers question the value of lime. Added to this was the discovery that Mo deficiency associated with many acid soils was a primary constraint to legume growth (Anderson and Spencer, 1950). Subsequent research showed that moderate rates of Mo-enriched superphosphate achieved equivalent legume biomass response as heavy rates of lime (Anderson and Moye, 1952) leading many to conclude that occasional applications of Mo-enriched superphosphate was a substitute for lime application. As a consequence, many researchers and influential agricultural advisors advocated against the use of lime on the basis that it was uneconomic (e.g. Ellington, 1985), with research delivering mixed messages on the topic. For example, using a partial budgeting approach, Khairo et al. (2010) concluded that a merino wether grazing enterprise was more profitable without lime, despite soil pH_{Ca} as low as 4.1 in the surface 10 cm (Norton et al., 2020). In part, their conclusion reflected depressed commodity prices and prolonged drought during the study period. By contrast, Li et al. (2010) demonstrated that liming pastures was profitable over the long term. It was estimated that the pay-back period was 14 years for self-replacing Merino (wool) sheep and 9 years for prime lamb production based on commodity prices of the day. Furthermore, the inclusion of high-value cash crops in the rotation, such as canola or a high protein wheat variety, could significantly shorten the pay-back period for the investment in lime.

In large part, the mixed messaging around the economics of liming remains unresolved and this is thought to stem from the partial budgeting approaches typically used in economic analyses of agricultural practices. It is argued that static measures of financial performance often do not adequately characterize the risk adjusted performance of farming enterprises, resulting in poor recommendations in management practices (Hutchings and Nordblom, 2011). In the case of failing to lime an acid soil, the risk of subsurface soil acidification is imminent (Li et al., 2019) resulting in ongoing depressed yields (Scott et al.,

1997), reduced water utilization (Hayes et al., 2016), reduced groundcover (Norton et al., 2020), and limited species options to support production. Moreover, the cost to ameliorate subsurface soil acidity at a later date increases with the level of acidity, due to the increased time for amelioration, costs of machinery to apply lime at depth (Li, 2020), and higher rates required (Condon et al., 2021). Production risks associated with reduced establishment, biomass accumulation, and persistence of some species also exist and are exacerbated under dry seasonal conditions (Culvenor et al., 2011; Hayes et al., 2008, 2016) but rarely included in economic analyses. A more recent analysis of liming from long-term experiments in the UK suggest that lime should be considered a capital investment due to the long-term nature of responses, meaning that economic analyses need to be conducted over a longer time frame compared to other farm inputs such as fertilizers (Holland and Behrendt, 2021).

7.3 Tolerant plant species

A key reason for the lack of plant response to lime, even on highly acidic soils, relates to the tolerance of many species to acid soil conditions. Many of the earlier reports of improved plant growth due to lime were observed on sensitive species, especially legumes (Cregan et al., 1989). Legumes are at a disadvantage compared to other species because, as previously described, the acid soil tolerance of rhizobia is as important as tolerance of the legume species itself, and the effectiveness of the symbiosis of the two organisms can be interrupted due to conditions associated with acidic soils, such as Mo deficiency (Anderson and Spencer, 1950; Coventry and Evans, 1989). Indeed, a common factor across many of the field studies in Australia which report a nil response to lime of pastures show a low legume content (for example, Peoples et al., 1995), and it is often the legume component that drives the positive response of pastures to lime (Hayes et al., 2008; Norton et al., 2020). Although some important non-legume species such as the perennial grass phalaris (*Phalaris aquatica* L.) respond to lime, many other grasses are less responsive (Li et al., 2006; Ridley and Coventry, 1992).

Plant breeding and selection to improve tolerance to acidic soils has been undertaken on many species important to Australian agriculture, including alfalfa (Hayes et al., 2011; Scott et al., 2008), phalaris (Culvenor et al., 2011), and a range of crop species (Ryan, 2018). In legume species such as alfalfa, this development work has extended to improve nodulation through the selection of tolerant rhizobia strains and the assessment of symbiotic competence under low pH conditions (Wigley et al., 2018). Whilst many of the advances in breeding for acid soil tolerance have resulted in substantial increases in productivity, there is a risk that acid soil tolerance might be viewed as an alternative to liming which has caused some authors to doubt this approach to acid soil management. For example, Fox (1980) likens the use of tolerance to the hazard of flying a low powered aircraft up a narrow canyon. The course starts easily and the scenery is beautiful, but options run out very quickly and to continue is to invite disaster. Scott and Fisher (1989) review this issue in some detail and contend that plant tolerance is an important and complementary management strategy to liming on acid soils, especially in extensive agricultural production systems which are common across Australia. On soils that are acidic to depth, acid soil tolerance helps maximize production during the amelioration phase, which may take several decades with lime (Li et al., 2019). Spatial and temporal variability in pH (Conyers and Davey, 1990; Conyers et al., 1997b), including pH stratification (Condon et al., 2021), and in transient toxins such as Mn (Hayes et al., 2012) can also be better managed with improved tolerance, especially in systems with longer intervals between liming applications. However, plant tolerance could never be considered a sustainable alternative to liming on acid soils because it does nothing to stem the acidification inherent in all production systems.

7.4 New recommendations for acid soil management

There have been two new recommendations released recently by the New South Wales Department of Primary Industries for Australian farmers to better manage soil acidity. The first relates to the appropriate soil sampling depth increment for monitoring pH; the second relates to a revised target pH when setting lime rates.

Standard soil sampling practice often involves the collection of a low volume of soil using narrow diameter (e.g. 2 cm) coring tubes that takes a continuous core of soil to a standard depth, often 0–10 cm (Gourley and Weaver, 2019; Upjohn et al., 2005). In a perfectly homogenized sample, analysis of a subsample of this core provides an average of the soil property in that 0–10 cm layer. This can be problematic for a soil property such as pH, which is highly stratified within the surface layers, with a wide distribution around the mean. An example is given in Fig. 9 where sampling of both limed and unlimed soil has been taken in the surface 20 cm at 5 cm increments. Substantial stratification of pH exists in both profiles regardless of whether lime was applied. Dashed lines have been superimposed to represent the calculated mean pH value in 10 cm increments. In both cases, the average at 10 cm increments substantially underestimates the extent of acidity in the subsurface layers. Sampling at 5 cm rather than 10 cm intervals provides a more accurate reflection of the true extent of soil acidity in the profile without imposing unreasonable cost and complexity of commercial soil sampling (Condon et al., 2021).

As described in Section 5.1 of this chapter, defining lime rates requires an understanding of the target pH. Traditionally, researchers and advisors in Australia have promoted a target pH_{Ca} of around 5.2 (e. g. Upjohn et al., 2005).

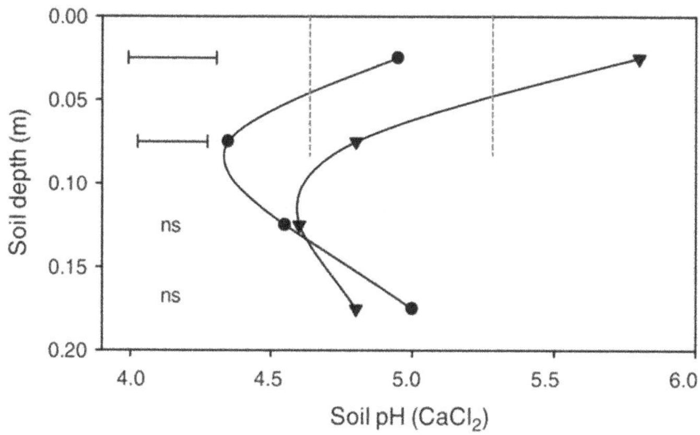

Figure 9 Soil pH_{Ca} stratification resulting from lime application within 5 years (triangle, n = 33) or more than 5 years (circle, n = 15) of sampling from locations between Albury and Cowra, NSW, Australia. Horizontal bars represent l.s.d. at P=0.05, ns: no significant difference, dashed lines represent the approximate mean pH of the 0-10 cm zone of both limed and unlimed profiles. Source: Adapted from (Condon et al., 2021).

Fundamentally, this target reflects the fact that Al toxicity ceases to impact plant growth in most acidic soils at around pH_{Ca} 4.8-5.0. Below pH_{Ca} 4.8, Al toxicity can rapidly increase (Ritchie, 1989) although the extent of the increase will be dependent on the Al buffering capacity of the particular soil. A typical response curve from a low-buffering soil is shown in Fig. 10. A target pH of 5.2 was perceived to achieve the amelioration of Al toxicity for several years following lime application whilst minimizing the cost to producers. However, subsequent research has shown that despite best-practice liming, soils in key

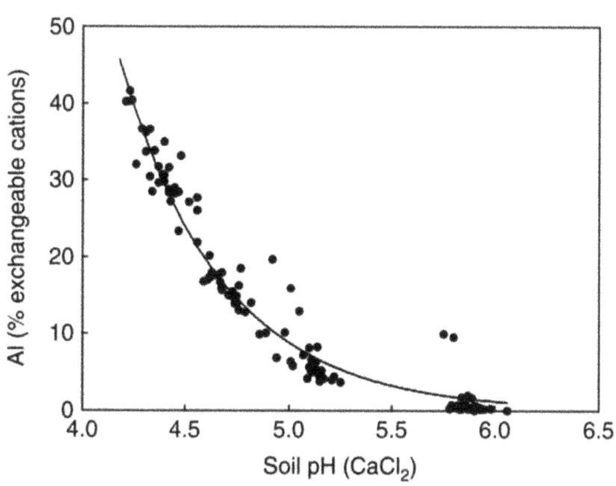

Figure 10 An example of a response curve from an acidic soil with low pH and Al buffering showing the strong negative relationship between exchangeable aluminum and soil pH. Source: Adapted from Guo et al. (2012).

agricultural regions of southern Australia continue to acidify (Condon et al., 2021; Scott et al., 2007), suggesting that current practices are inadequate. This may be due to an underestimation of the extent of acidity through standard soil sampling of stratified soil as described previously (Condon et al., 2021), and skewed variability of pH of composite soil samples where the mean exceeds the mode on some soil types (Conyers and Davey, 1990; Hayes et al., 2022). However, inadequate incorporation of lime (Scott and Coombes, 2006) and underestimation of acidification rates of current agricultural production results in lime being consumed near the surface allowing subsurface soil to continue to acidify. These factors have rendered the previous lime rates, targeted to achieve pH 5.2, inadequate to ameliorate acidity below the depth of incorporation. Conyers and Scott (1989) demonstrated that increasing the pH of the soil surface with higher (8 t/ha) lime rates led to increased pH below the depth of lime incorporation compared with where standard lime rates were used. This was attributed to increased leaching of bicarbonates once pH_{Ca} was maintained above 5.5. The 8 t/ha rate used by Conyers and Scott (1989) was required because the starting soil pH was as low as 4.2. In other soils where the starting pH is higher, a lower lime rate would be required to achieve pH 5.5 (Fig. 7). Several field experiments have demonstrated that where soil pH in the surface 10 cm was increased to $pH_{Ca} \geq 5.5$, subsurface acidity began to be ameliorated over a 10-20 year time frame (Bromfield et al., 1987; Li et al., 2019), prompting a revision of recommendations that liming on acid soils should aim to exceed pH_{Ca} 5.5 in the 0-10 cm zone. The most efficient means of maintaining surface soil $pH_{Ca} \geq 5.5$ remains the focus of ongoing research.

7.5 Key lessons from the Australian experience

The key lesson from the Australian experience is that the prevention of soil acidity is preferable to treating a problem later. To do nothing and invite the inevitable acidification of agricultural soils is to risk less productive and resilient agricultural production systems in the face of increasing biotic and abiotic stresses and to elevate the cost and time required to amend subsurface acidity later. However, liming is a capital cost which can be hard to justify economically, particularly where plant responses to lime are subtle and where static measures and partial budgets are used to analyze the relative financial merits of liming inputs. Liming is an essential component of a sustainable agricultural system on acidic soils. Therefore, we propose that its use should not be determined by the quantum of response in the crop growth but, rather, farmers should seek to maximize the response to lime in order to recoup costs sooner. This might be achieved by sowing higher value, acid-sensitive species to maximize the yield benefit of lime (Li et al., 2001). For example, canola might be included in a cropping rotation (Slattery and Coventry, 1993), or alfalfa might be included in the pasture mixture (Hayes et al., 2016). A farmer might manipulate the timing of lime application to coincide with the establishment of a new pasture where the benefits to newly established seedlings might be more visible than if the same lime was applied to an existing pasture.

8 Conclusion

Soil acidity is a threat to a healthy soil causing both direct and indirect changes in a range of soil chemical, physical, and biological characteristics, resulting in multiple impacts on the organisms growing in that soil as well as negative off-site impacts. Acidity reduces soil function, attributable to reduced plant growth and shifts in the composition of the microbiome. Acidity can also reduce the resilience of agricultural production in the face of biotic and abiotic stresses, through reduced access to resources associated either with reduced availability of nutrients or with reduced capacity of a plant to forage for resources. Acidification of most soils is primarily driven by N-cycle processes, with lower levels of acid-addition contributed by other processes such as the cation–anion balance. Liming is an essential feature of sustainable land management due to the acidifying nature of all agricultural production systems. Only on alkaline soils, which may benefit from acidification due to toxicities associated with alkalinity, would liming not be required.

The application of finely ground limestone remains the most efficient means of ameliorating soil acidity due to its high neutralizing value, but some constraints exist that can act as a disincentive for farmers to invest in lime. The sparingly soluble nature of lime increases the time it takes to react with the soil volume, delaying increases in soil pH. The mixing of lime with the soil volume through cultivation goes some way to overcoming this barrier but is only effective to the depth of incorporation in the short term. Lime movement below the depth of incorporation is very slow and may only occur at a rate of approximately one centimeter per year on acidic soils. Furthermore, highly stratified soil pH profiles are common due to natural processes associated with the uptake of cations and the deposition of organic residues which results in a high pH at the soil surface and an acidic subsurface. The consequence is that acidic subsurface soils are common and lime can have limited effect on the amelioration of subsurface acidity in the short term. Liming, therefore, needs to be part of a long-term program on any farm, aimed to minimize the inevitable acidification of subsurface layers by balancing the acid addition of the production system. Furthermore, lime needs to be considered a capital cost with economic analysis at the farm level conducted over a longer time frame than other farm inputs such as fertilizers. An economic justification for lime may be even more problematic on soils that are not inherently acidic or have a high pH buffering capacity because declines in crop and pasture yields due to acidification are likely to be subtle, especially where production is not limited by associated toxicities of Al or Mn. Nevertheless, soil acidification is inevitable and liming in such situations is required to avoid degradation of the soil resource.

Future research should examine the potential to increase the use of organic amendments, such as manures and crop residues, in the management of soil acidity. Their use is presently impeded by difficulties in assessing their liming value and therefore the quantity required to raise soil pH. The measure of ash alkalinity has limitations when assessing the value of soil inputs because not all the alkaline material becomes available to reduce H^+ concentration in the soil solution. The nature of the product used is also important as some products, especially those with a low C:N ratio, risk further acidification of soil associated with N transformation and losses. The use of organic amendments can therefore exacerbate soil acidity, rather than ameliorate it. Future research should also examine ways to amend subsurface acidity by increasing the rate of movement of bicarbonates into subsurface soil layers. Such endeavors should aim to minimize the use of high soil disturbance, such as deep-ripping, which inevitably comes at a high energy cost and is impractical on many landscapes. The use of lime in conjunction with various products and/or in perennial-based farming systems to increase the speed of movement down the profile would seem to offer a more sustainable solution in the long term, applicable to a much larger proportion of agricultural soils worldwide.

9 Where to look for further information

The book, *Soil Acidity and Plant Growth*, edited by A. D. Robson and published by Academic Press in 1989, is an excellent resource for anyone studying soil acidity. It contains a series of peer-reviewed chapters on the fundamental processes associated with acid soils, written by recognized authorities in their field of research. Notable among these is Dr K. R. Helyar, who in 1976 published his pioneering works on *Nitrogen cycling and soil acidification* (Helyar, 1976). Similarly, the publication, *Soil Acidity and Liming*, edited by Fred Adams and published by the American Society of Agronomy in 1984 provides some useful chapters focused on lime response observed in the North American soils. This publication can be accessed at: https://acsess.onlinelibrary.wiley.com/doi/book/10.2134/agronmonogr12.2ed.

There is relatively scant information on the effects of liming on biodiversity. The recent review out of the UK by Holland et al. (2018) is a useful overview of what is known on those aspects, highlighting that effects of lime can be either positive or negative but very likely to be context-specific.

The *Handbook of Soil Acidity*, edited by Z. Rengal and published by Marcel Dekker Inc. (2003) also has some useful chapters, some of which cover topics not covered in this chapter such as D. L. Godbold's *Managing Acidification and Acidity in Forest Soils* and J. W. Huang and J. Chen's *Role of pH in Phytoremediation of Contaminated Soils*.

For publications more suited to practitioners in the field looking to adapt management strategies to manage or amend acidic soils, we recommend the range of publications available on the NSW Department of Primary Industries website at: https://www.dpi.nsw.gov.au/agriculture/soils/guides.

10 References

Anderson, A. and Moye, D. (1952). Lime and molybdenum in clover development on acid soils, *Australian Journal of Agricultural Research* 3(2), 95-110.

Anderson, A. J. and Spencer, D. (1950). Molybdenum in nitrogen metabolism of legumes and non-legumes, *Australian Journal of Scientific Research (Series B)* 3(4), 414-430.

Barber, S. A. (1984). Liming materials and practices. In: Adams, F. (Ed.), *American Society of Agronomy*, 171-209. https://doi.org/10.2134/agronmonogr12.2ed.c4.

Bolan, N. S., Curtin, D. and Adriano, D. C. (2005). Acidity. In: Hillel, D. (Ed.), *Encyclopedia of Soils in the Environment*, Elsevier, Oxford, pp. 11-17.

Bouton, J. H. (1996). Screening the alfalfa core collection for acid soil tolerance, *Crop Science* 36(1), 198-200.

Bromfield, S. M., Cumming, R. W., David, D. J. and Williams, C. H. (1983). The assessment of available manganese and aluminium status in acid soils under subterranean clover pastures of various ages, *Australian Journal of Experimental Agriculture* 23(121), 192-200.

Bromfield, S. M., Cumming, R. W., David, D. J. and Williams, C. H. (1987). Long-term effects of incorporated lime and topdressed lime on the pH in the surface and subsurface of pasture soils, *Australian Journal of Experimental Agriculture* 27(4), 533-538.

Burns, H. M. and Norton, M. R. (2018). *Legumes in Acidic Soils Maximising Production Potential in South Eastern Australia*, Grains Research and Development Corporation and NSW Department of Industry (Revised June 2021), Canberra.

Butterly, C. R., Costello, B., Lauricella, D., Sale, P., Li, G. and Tang, C. (2020). Alkalinity movement down acid soil columns was faster when lime and plant residues were combined than when either was applied separately, *European Journal of Soil Science* 72(1), 313-325.

Butterly, C. R., Kaudal, B. B., Baldock, J. A. and Tang, C. (2011). Contribution of soluble and insoluble fractions of agricultural residues to short-term pH changes, *European Journal of Soil Science* 62(5), 718-727.

Butterly, C. R., Wang, X., Sale, P., Li, G. and Tang, C. (2021). Liming effect of non-legume residues promotes the biological amelioration of soil acidity via nitrate uptake, *Plant and Soil* 464(1-2), 63-73.

Caires, E. F., Pereira Filho, P. R. S., Zardo Filho, R. and Feldhaus, I. C. (2008). Soil acidity and aluminium toxicity as affected by surface liming and cover oat residues under a no-till system, *Soil Use and Management* 24(3), 302-309.

Castro, G. S. A. and Crusciol, C. A. C. (2013). Effects of superficial liming and silicate application on soil fertility and crop yield under rotation, *Geoderma* 195-196, 234-242.

Chan, K. Y., Conyers, M. K. and Scott, B. J. (2007). Improved structural stability of an acidic hardsetting soil attributable to lime application, *Communications in Soil Science and Plant Analysis* 38(15-16), 2163-2175.

Chan, K. Y. and Heenan, D. P. (1998). Effect of lime ($CaCO_3$) application on soil structural stability of a red earth, *Soil Research* 36(1), 73-86.

Charman, N., Ballard, R. A., Humphries, A. W. and Auricht, G. C. (2008). Improving lucerne nodulation at low pH: contribution of rhizobial and plant genotype to the nodulation of lucerne seedlings growing in solution culture at pH 5, *Australian Journal of Experimental Agriculture* 48(4), 512-517.

Condon, J., Burns, H. and Li, G. (2021). The extent, significance and amelioration of subsurface acidity in southern New South Wales, Australia, *Soil Research* 59(1), 1-11.

Condon, J. R., Black, A. S. and Conyers, M. K. (2004). The role of N transformations in the formation of acidic subsurface layers in stock urine patches, *Soil Research* 42(2), 221-230.

Condon, J. R., Moroni, J. S., Zander, A., Nguyen, H. H. and Li, G. D. (2018). Deep ripping incorporation of novel amendments to ameliorate subsurface acidity. *National Soils Conference 2018*, Canberra, 379-380.

Conyers, M. K., Newton, P., Condon, J., Poile, G., Mele, P. and Ash, G. (2012). Three long-term trials end with a quasi-equilibrium between soil C, N, and pH: an implication for C sequestration, *Soil Research* 50(7), 527-535.

Conyers, M. K. and Scott, B. (1989). The influence of surface incorporated lime on subsurface soil acidity, *Australian Journal of Experimental Agriculture* 29(2), 201-207.

Conyers, M. K., van der Rijt, V., Oates, A., Poile, G., Kirkegaard, J. and Kirkby, C. (2019). The strategic use of minimum tillage within conservation agriculture in southern New South Wales, Australia, *Soil and Tillage Research* 193, 17-26.

Conyers, M. K. and Davey, B. G. (1990). The variability of pH in acid soils of the southern highlands of New South Wales, *Soil Science* 150(4), 695-704.

Conyers, M. K., Heenan, D. P., McGhie, W. J. and Poile, G. P. (2003). Amelioration of acidity with time by limestone under contrasting tillage, *Soil and Tillage Research* 72(1), 85-94.

Conyers, M. K., Scott, B. J., Fisher, R. and Lill, W. (1995a). Predicting the field performance of twelve commercial liming materials from southern Australia, *Fertilizer Research* 44(2), 151-161.

Conyers, M. K., Uren, N. C. and Helyar, K. R. (1995b). Causes of changes in pH in acidic mineral soils, *Soil Biology and Biochemistry* 27(11), 1383-1392.

Conyers, M. K., Scott, B. J. and Whitten, M. G. (2020). The reaction rate and residual value of particle size fractions of limestone in southern New South Wales, *Crop and Pasture Science* 71(4), 368-378.

Conyers, M. K., Uren, N. C., Helyar, K. R., Poile, G. J. and Cullis, B. R. (1997). Temporal variation in soil acidity, *Soil Research* 35(5), 1115-1129.

Coventry, D. R. and Evans, J. (1989). Symbiotic nitrogen fixation and soil acidity. In: Robson, A. D. (Ed.), *Soil Acidity and Plant Growth*, Academic Press, Sydney, pp. 103-138.

Creamer, R. E., Barel, J. M., Bongiorno, G. and Zwetsloot, M. J. (2022). The life of soils: integrating the who and how of multifunctionality, *Soil Biology and Biochemistry* 166, 108561.

Cregan, P. D., Hirth, J. R. and Conyers, M. K. (1989). Amelioration of soil acidity by liming and other amendments. In: Robson, A. D. (Ed.), *Soil Acidity and Plant Growth*, Academic Press, Sydney, pp. 205–264.

Crews, T. E., Blesh, J., Culman, S. W., Hayes, R. C., Jensen, E. S., Mack, M. C., Peoples, M. B. and Schipanski, M. E. (2016). Going where no grains have gone before: from early to mid-succession, *Agriculture, Ecosystems and Environment* 223, 223–238.

Crews, T. E. and Peoples, M. B. (2004). Legume versus fertilizer sources of nitrogen: ecological tradeoffs and human needs, *Agriculture, Ecosystems and Environment* 102(3), 279–297.

Crews, T. E. and Peoples, M. B. (2005). Can the synchrony of nitrogen supply and crop demand be improved in legume and fertiliser-based ecosystems? A review, *Nutrient Cycling in Agroecosystems* 72(2), 101–120.

Culvenor, R. A., McDonald, S. E., Veness, P. E., Watson, D. and Dempsey, W. (2011). The effect of improved aluminium tolerance on establishment of the perennial grass, Phalaris, on strongly acid soils in the field and its relation to seasonal rainfall, *Crop and Pasture Science* 62(5), 413–426.

Cumming, R. W. (1982). pH survey of acid soils in the Pejar Catchment Area, Goulburn, *Journal of the Soil Conservation Service of New South Wales* 38, 13–18.

Cumming, R. W. and Garrard, W. J. (1981). Need for protection of town water supplies - Sooley Dam, Goulburn. *Proceedings of the 3rd Australian Soil Conservation Conference*, pp. 14–21.

Cumming, R. W. and Williams, C. H. (1983). Soil pH in the Pejar catchment area, Goulburn, *Journal of Soil Conservation* 39(1), 106–112.

Dann, P. R., Dear, B. S. and Cunningham, R. B. (1989). Comparison of sewage ash, crushed limestone and cement kiln dust as ameliorants for acid soils, *Australian Journal of Experimental Agriculture* 29(4), 541–549.

Davies, S., Armstrong, R., Macdonald, L., Condon, J. and Petersen, E. (2020). Soil constraints: a role for strategic deep tillage. In: Pratley, J. and Kirkegaard, J. (Eds), *Australian Agriculture in 2020: From Conservation to Automation*, Agronomy Australia and Charles Sturt University, Wagga Wagga, pp. 117–135.

Dear, B. S. and Virgona, J. M. (1996). Legumes in low-input perennial pastures of southern Australia: historical role and future development, *New Zealand Journal of Agricultural Research* 39(4), 579–589.

Dear, B. S., Virgona, J. M., Sandral, G. A., Swan, A. D. and Morris, S. (2009). Changes in soil mineral nitrogen, nitrogen leached, and surface pH under annual and perennial pasture species, *Crop and Pasture Science* 60(10), 975–986.

Ellington, A. (1985). *Soil Acidity - Effects and Remedies: A Review*, Department of Agriculture, Victoria. Technical Report Series No.

Evans, C. M., Conyers, M. K., Black, A. S. and Poile, G. J. (1998). Effect of ammonium, organic amendments, and plant growth on soil pH stratification, *Soil Research* 36(4), 641–653.

Fox, R. L. (1980). Soils with variable charge: agronomic and fertility aspects. In: Theng, B. K. G. (Ed.), *Soils with Variable Charge*, New Zealand Society of Soil Science, Lower Hutt, pp. 195–224.

Gillman, G. and Sumpter, E. (1986). Modification to the compulsive exchange method for measuring exchange characteristics of soils, *Soil Research* 24(1), 61–66.

Glover, J. D., Reganold, J. P., Bell, L. W., Borevitz, J., Brummer, E. C., Buckler, E. S., Cox, C. M., Cox, T. S., Crews, T. E., Culman, S. W., DeHaan, L. R., Eriksson, D., Gill, B. S.,

Holland, J., Hu, F., Hulke, B. S., Ibrahim, A. M. H., Jackson, W., Jones, S. S., Murray, S. C., Paterson, A. H., Ploschuk, E., Sacks, E. J., Snapp, S. S., Tao, D., Van Tassel, D. L., Wade, L. J., Wyse, D. L. and Xu, Y. (2010). Agriculture. Increased food and ecosystem security via perennial grains, *Science* 328(5986), 1638–1639.

Goulding, K. W. T. (2016). Soil acidification and the importance of liming agricultural soils with particular reference to the United Kingdom, *Soil Use and Management* 32(3), 390–399.

Gourley, C. and Weaver, D. (2019). *A Guide for Fit for Purpose Soil Sampling*, Fertilizer Australia, Canberra.

Griffiths, N. (2014). Use of manures, *GRDC Update Papers*, Grains Research and Development Corporation, Available at: https://grdc.com.au/resources-and-publications/grdc-update-papers/tab-content/grdc-update-papers/2014/07/use-of-manures accessed 12 September 2021.

Guo, Y. J., Li, G. D., Hayes, R. C., Dear, B. S. and Price, A. (2012). Tolerance of the annual legumes *Biserrula pelecinus*, *Ornithopus sativa*, *Trifolium spumosum*, *T. vesiculosum* and *T. subterraneum* to soil acidity, *New Zealand Journal of Agricultural Research* 55(1), 1–14.

Hayes, R. C., Conyers, M. K., Li, G. D., Poile, G. J., Price, A., McVittie, B. J., Gardner, M. J., Sandral, G. A. and McCormick, J. I. (2012). Spatial and temporal variation in soil Mn^{2+} concentrations and the impact of manganese toxicity on lucerne and subterranean clover seedlings, *Crop and Pasture Science* 63(9), 875–885.

Hayes, R. C., Dear, B. S., Orchard, B. A., Peoples, M. B. and Eberbach, P. L. (2008). Response of subterranean clover, balansa clover, and gland clover to lime when grown in mixtures on an acid soil, *Australian Journal of Agricultural Research* 59(9), 824–835.

Hayes, R. C., Gupta, V. V. S. R., Li, G. D., Peoples, M. B., Rawnsley, R. P. and Pembleton, K. G. (2021). Contrasting soil microbial abundance and diversity on and between pasture drill rows in the third growing season after sowing, *Renewable Agriculture and Food Systems* 36(2), 163–172.

Hayes, R. C., Li, G. D., Conyers, M. K., Virgona, J. M. and Dear, B. S. (2016). Lime increases productivity and the capacity of lucerne (*Medicago sativa* L.) and Phalaris (*Phalaris aquatica* L.) to utilise stored soil water on an acidic soil in south-eastern Australia, *Plant and Soil* 400(1–2), 29–43.

Hayes, R. C., McCormick, J. I., Oates, A. A., Poile, G. J., Conyers, M. K., Gardner, M. J., Price, A., O'Keeffe, P. and Li, G. D. (2018). Addressing biophysical constraints for Australian farmers applying low rates of composted dairy waste to soil, *Renewable Agriculture and Food Systems* 33(2), 179–193.

Hayes, R. C., Rohan, M., Li, G. D., Orgill, S. E., Poile, G. J., Oates, A. A. and Conyers, M. K. (2022). The nature of spatial variability of four soil chemical properties and the implications for soil sampling, *Journal of Soils and Sediments*. DOI: 10.1007/s11368-022-03285-x.

Hayes, R. C., Scott, B. J., Dear, B. S., Li, G. D. and Auricht, G. C. (2011). Seedling validation of acid soil tolerance of lucerne populations selected in solution culture high in aluminium, *Crop and Pasture Science* 62(9), 803–811.

Haynes, R. J. and Mokolobate, M. S. (2001). Amelioration of Al toxicity and P deficiency in acid soils by additions of organic residues: a critical review of the phenomenon and the mechanisms involved, *Nutrient Cycling in Agroecosystems* 59(1), 47–63.

Haynes, R. J. and Naidu, R. (1998). Influence of lime, fertilizer and manure applications on soil organic matter content and soil physical conditions: a review, *Nutrient Cycling in Agroecosystems* 51(2), 123-137.

Helyar, K. R. (1976). Nitrogen cycling and soil acidification, *Journal of the Australian Institute of Agricultural Science* 42(4), 217-221.

Hénault, C., Bourennane, H., Ayzac, A., Ratié, C., Saby, N. P. A., Cohan, J. P., Eglin, T. and Gall, C. L. (2019). Management of soil pH promotes nitrous oxide reduction and thus mitigates soil emissions of this greenhouse gas, *Scientific Reports* 9(1), 20182.

Hochman, Z., Godyn, D. L. and Scott, B. J. (1989). The integration of data on lime use by modelling. In: Robson, A. D. (Ed.), *Soil Acidity and Plant Growth*, pp. 265-301.

Holland, J. E. and Behrendt, K. (2021). The economics of liming in arable crop rotations: analysis of the 35-year Rothamsted and Woburn liming experiments, *Soil Use and Management* 37(2), 342-352.

Holland, J. E., Bennett, A. E., Newton, A. C., White, P. J., McKenzie, B. M., George, T. S., Pakeman, R. J., Bailey, J. S., Fornara, D. A. and Hayes, R. C. (2018). Liming impacts on soils, crops and biodiversity in the UK: a review, *Science of the Total Environment* 610-611, 316-332.

Hutchings, T. R. and Nordblom, T. L. (2011). A financial analysis of the effect of the mix of crop and sheep enterprises on the risk profile of dryland farms in south-eastern Australia, *Australian Farm Business Management Journal* 8(1), 19-41.

Johnston, A. E., Goulding, K. W. T. and Poulton, P. R. (1986). Soil acidification during more than 100 years under permanent grassland and woodland at Rothamsted, *Soil Use and Management* 2(1), 3-10.

Kaveney, B., Condon, J., Doran, G., Galea, F. and Rigg, J. (2020). The distribution of soil nitrifiers and efficacy of 3,4-dimethylpyrazole phosphate changes with soil depth and calcium carbonate application, *Soil Biology and Biochemistry* 150, 108009.

Kennedy, I. R. (1992). *Acid Soil and Acid Rain*, Research Studies Press, Taunton, Somerset, England.

Khairo, S. A., Norton, M. R., Garden, D., Graham, P., Langford, C., Armstrong, P. and Brassil, T. (2010). The profitability of wool production after surface application of lime and superphosphate on acid soils, *Australian Farm Business Management Journal* 7(1), 1-10.

Kirkby, C. A., Richardson, A. E., Wade, L. J., Batten, G. D., Blanchard, C. and Kirkegaard, J. A. (2013). Carbon-nutrient stoichiometry to increase soil carbon sequestration, *Soil Biology and Biochemistry* 60, 77-86.

Kunhikrishnan, A., Thangarajan, R., Bolan, N. S., Xu, Y., Mandal, S., Gleeson, D. B., Seshadri, B., Zaman, M., Barton, L., Tang, C., Luo, J., Dalal, R., Ding, W., Kirkham, M. B. and Naidu, R. (2016), Chapter one - functional relationships of soil acidification: liming, and greenhouse gas flux. In: *Advances in Agronomy* Sparks, D. L. (Ed.), Academic Press, 1-71.

Lee, S., Sorensen, J. W., Walker, R. L., Emerson, J. B., Nicol, G. W. and Hazard, C. (2022). Soil pH influences the structure of virus communities at local and global scales, *Soil Biology and Biochemistry* 166, 108569.

Lehmann, J., Bossio, D. A., Kögel-Knabner, I. and Rillig, M. C. (2020). The concept and future prospects of soil health, *Nature Reviews. Earth and Environment* 1(10), 544-553.

Li, G. (2020). Acid correction, Ground Cover Supplement - Soil Constraints - Part 2. *Diagnosis and Amendments*, Grains Research and Development (Corporation), Canberra.

Li, G., Condon, J., Hayes, R., Burns, H., Lowrie, R., Lowrie, A., Price, A., Xu, B., Jia, Y. and Poile, G. (2021a). Research update for ameliorating subsoil acidity using organic amendments. In: Slinger, D., Moore, T. and Martin, C. (Eds), *Southern NSW Research Results 2021*, NSW Department of Primary Industries, Orange, NSW, pp. 124–130.

Li, G., Condon, J., Hayes, R., Moroni, S. and Burns, H. (2021b). Innovative approaches to manage subsurface soil acidity in the field. *Soils, Investing in Our Future - 2021 Joint Conference*, Cairns, Australia.

Li, G. D., Condon, J. R., Tavakkoli, E., Moroni, J. S., Hayes, R. C. and Burns, H. (2018). Amelioration of subsurface soil acidity using inorganic amendments. *National Soils Conference 2018*, Canberra, 349–351.

Li, G. D., Conyers, M. K., Helyar, K. R., Lisle, C. J., Poile, G. J. and Cullis, B. R. (2019). Long-term surface application of lime ameliorates subsurface soil acidity in the mixed farming zone of south-eastern Australia, *Geoderma* 338, 236–246.

Li, G. D., Helyar, K. R., Conyers, M. K., Cullis, B. R., Cregan, P. D., Fisher, R. P., Castleman, L. J. C., Poile, G. J., Evans, C. M. and Braysher, B. (2001). Crop responses to lime in long-term pasture-crop rotations in a high rainfall area in south-eastern Australia, *Australian Journal of Agricultural Research* 52(3), 329–341.

Li, G. D., Helyar, K. R., Welham, S. J., Conyers, M. K., Castleman, L. J. C., Fisher, R. P., Evans, C. M., Cullis, B. R. and Cregan, P. D. (2006). Pasture and sheep responses to lime application in a grazing experiment in a high-rainfall area, south-eastern Australia. I. Pasture production, *Australian Journal of Agricultural Research* 57(10), 1045–1055.

Li, G. D., Schwenke, G. D., Hayes, R. C., Lowrie, A. J., Lowrie, R. J. and Price, A. (2022). Perennial pastures reduce nitrous oxide emissions in mixed farming systems in a semi-arid environment, *Science of the Total Environment* 834, 155304.

Li, G. D., Singh, R. P., Brennan, J. P. and Helyar, K. R. (2010). A financial analysis of lime application in a long-term agronomic experiment on the south-western slopes of New South Wales, *Crop and Pasture Science* 61(1), 12–23.

Llewellyn, R. S., Robertson, M. J., Hayes, R. C., Ferris, D., Descheemaeker, K. and Revell, C. (2014). Developing the role of perennial forages for crop-livestock farms: a strategic multi-disciplinary approach, *Crop and Pasture Science* 65(10), 945–955.

Lockwood, P. V., Wilson, B. R., Daniel, H. and Jones, M. J. (2003). Soil acidification and natural resource management. Directions for the future. *NSW Agriculture, Armidale NSW.*

Macintyre, S. (2015). *Australia's Boldest Experiment; War and Reconstruction in the 1940s*, University of New South Wales, New South Publishing, Sydney.

Marschner, H., Romheld, V., Horst, W. J. and Martin, P. (1986). Root-induced changes in the rhizosphere - importance for the mineral-nutrition of plants, *Zeitschrift Fur Pflanzenernahrung Und Bodenkunde* 149(4), 441–456.

Massy, C. (2011). *Breaking the Sheep's Back*, University of Queensland Press, St Lucia.

Narendrula-Kotha, R. and Nkongolo, K. K. (2017). Microbial response to soil liming in damaged ecosystems revealed by Pyrosequencing and phospholipid fatty acid analyses, *PLoS ONE* 12(1), e0168497.

Nguyen, H. H. (2021). *Amelioration of Subsurface Soil Acidity by Surface Incorporation of Organic Amendment*, Charles Sturt University, Wagga Wagga.

Nguyen, H. H., Moroni, J. S., Condon, J. R., Zander, A. and Li, G. (2018). Increasing subsoil pH through addition of lucerne (*Medicago sativa* L.) pellets in the surface layer of an acidic soil, *The 10th International Symposium on Plant-Soil Interactions at Low pH*, Putrajaya, Malaysia, pp. 51-53.

Nichols, P. G. H., Revell, C. K., Humphries, A. W., Howie, J. H., Hall, E. J., Sandral, G. A., Ghamkhar, K. and Harris, C. A. (2012). Temperate pasture legumes in Australia–their history, current use, and future prospects, *Crop and Pasture Science* 63(9), 691-725.

Norton, M. R., Garden, D. L., Orchard, B. A., Armstrong, P. and Brassil, T. (2020). Effects of lime, phosphorus and stocking rate on an extensively managed permanent pasture: botanical composition and groundcover, *Crop and Pasture Science* 71(7), 700-713.

O'Hara, G. W. (2001). Nutritional constraints on root nodule bacteria affecting symbiotic nitrogen fixation: a review, *Australian Journal of Experimental Agriculture* 41(3), 417-433.

Osborne, G. J., Poile, G. J., Hochman, Z. and Fisher, J. (1988). Manganese associations and variations - should we measure "easily reducible" manganese? In: Webb, M. J., Nable, R. O., Graham, R. D. and Hannam, R. J. (Eds), *International Symposium on Manganese in Soils and Plants Manganese Symposium 1988 Inc.*, Adelaide, pp 11-13.

Pang, Z., Tayyab, M., Kong, C., Hu, C., Zhu, Z., Wei, X. and Yuan, Z. (2019). Liming positively modulates microbial community composition and function of sugarcane fields, *Agronomy* 9(12), 808.

Paul, K. I., Black, A. S. and Conyers, M. K. (1999). Influence of moist-dry cycles on pH changes in surface soils, *Australian Journal of Soil Research* 37(6), 1057-1072.

Paul, K. I., Black, A. S. and Conyers, M. K. (2001). Effect of plant residue return on the development of surface soil pH gradients, *Biology and Fertility of Soils* 33(1), 75-82.

Paul, K. I., Black, A. S. and Conyers, M. K. (2003). Development of acidic subsurface layers of soil under various management systems, *Advances in Agronomy* 78(78), 187-214.

Peoples, M. B., Brockwell, J., Hunt, J. R., Swan, A. D., Watson, L., Hayes, R. C., Li, G. D., Hackney, B., Nuttall, J. G., Davies, S. L. and Fillery, I. R. P. (2012). Factors affecting the potential contributions of N_2 fixation by legumes in Australian pasture systems, *Crop and Pasture Science* 63(9), 759-786.

Peoples, M. B., Lilley, D. M., Burnett, V. F., Ridley, A. M. and Garden, D. L. (1995). Effects of surface application of lime and superphosphate to acid soils on growth and N_2 fixation by subterranean clover in mixed pasture swards, *Soil Biology and Biochemistry* 27(4-5), 663-671.

Poile, G., Oates, A., Moroni, S., Lowrie, R., Conyers, M., Swan, T., Peoples, M., Angus, J., Kirkegaard, J., Condon, K., Durham, K., Breust, P., Armstrong, R. and Nuttall, J. (2012). *Canola and Subsoil Constraints - Technical Bulletin* (vol. 12), EH Graham Centre for Agricultural Innovation, Charles Sturt University, Wagga Wagga.

Raupach, M. and Tucker, B. (1959). The field determination of soil reaction, *Journal of the Australian Institute of Agricultural Science* 25, 129-133.

Rayment, G. E. and Lyons, D. J. (2011). *Soil Chemical Methods: Australasia*, Commonwealth Scientific and Industrial Research Organization Publishing, Melbourne.

Reuter, D. J. and Robinson, J. B. (1997). *Plant Analysis an Interpretation Manual Second Edition*, CSIRO Publishing, Collingwood, VIC.

Ridley, A. M. and Coventry, D. R. (1992). Yield responses to lime of Phalaris, cocksfoot, and annual pastures in North-Eastern Victoria, *Australian Journal of Experimental Agriculture* 32(8), 1061–1068.

Ridley, A. M., Slattery, W. J., Helyar, K. R. and Cowling, A. (1990). The importance of the carbon-cycle to acidification of a grazed annual pasture, *Australian Journal of Experimental Agriculture* 30(4), 529–537.

Ridley, A. M., White, R. E., Helyar, K. R., Morrison, G. R., Heng, L. K. and Fisher, R. (2001). Nitrate leaching loss under annual and perennial pastures with and without lime on a duplex (texture contrast) soil in humid southeastern Australia, *European Journal of Soil Science* 52(2), 237–252.

Ritchie, G. S. P. (1989). The chemical behaviour of aluminium, hydrogen and manganese in acid soils. In: Robson, A. D. (Ed.), *Soil Acidity and Plant Growth*, Academic Press, Sydney, pp. 1–60.

Robertson, W. K., Thompson, L. G. and Martin, F. G. (1973). Manganese and copper requirements for soybeans, *Agronony Journal* 65, 641–644.

Ryan, P. R. (2018). Assessing the role of genetics for improving the yield of Australia's major grain crops on acid soils, *Crop and Pasture Science* 69(3), 242–264.

Scott, B. J., Conyers, M. K., Poile, G. J. and Cullis, B. R. (1997). Subsurface acidity and liming affect yield of cereals, *Australian Journal of Agricultural Research* 48(6), 843–854.

Scott, B. J., Conyers, M. K., Poile, G. J. and Cullis, B. R. (1999). Reacidification and reliming effects on soil properties and wheat yield, *Australian Journal of Experimental Agriculture* 39(7), 849–856.

Scott, B. J. and Coombes, N. E. (2006). Poor incorporation of lime limits grain yield response in wheat, *Australian Journal of Experimental Agriculture* 46(11), 1481–1487.

Scott, B. J., Ewing, M. A., Williams, R., Humphries, A. W. and Coombes, N. E. (2008). Tolerance of aluminium toxicity in annual *Medicago* species and lucerne, *Australian Journal of Experimental Agriculture* 48(4), 499–511.

Scott, B. J., Fenton, I. G., Fanning, A. G., Schumann, W. G. and Castleman, L. J. C. (2007). Surface soil acidity and fertility in the eastern Riverina and Western Slopes of southern New South Wales, *Australian Journal of Experimental Agriculture* 47(8), 949–964.

Scott, B. J. and Fisher, J. A. (1989). Selection of genotypes tolerant of aluminium and manganese. In: Robson, A. D. (Ed.), *Soil Acidity and Plant Growth*, Academic Press, Sydney, pp. 167–204.

Scott, B. J., Ridley, A. M. and Conyers, M. K. (2000). Management of soil acidity in long-term pastures of south-eastern Australia: a review, *Australian Journal of Experimental Agriculture* 40(8), 1173–1198.

Siman, A., Cradock, F. W. and Hudson, A. W. (1974). The development of manganese toxicity in pasture legumes under extreme climatic conditions, *Plant and Soil* 41(1), 129–140.

Siman, A. and Flemons, K. (1970). Goulburn lucerne failures linked with induced manganese toxicity, *Agricultural Gazette of New South Wales* 81, 662–663.

Slattery, W. J. and Coventry, D. R. (1993). Response of wheat, triticale, barley, and canola to lime on four soil types in north-eastern Victoria, *Australian Journal of Experimental Agriculture* 33(5), 609–618.

Slattery, W. J., Ridley, A. M. and Windsor, S. M. (1991). Ash alkalinity of animal and plant products, *Australian Journal of Experimental Agriculture* 31(3), 321-324.

Smith, T. E., Grattan, S. R., Grieve, C. M., Poss, J. A., Läuchli, A. E. and Suarez, D. L. (2013). PH dependent salinity-boron interactions impact yield, biomass, evapotranspiration and boron uptake in broccoli (*Brassica oleracea L.*), *Plant and Soil* 370(1-2), 541-554.

SoE (2011). 2.2.5 Soil acidification. *Australia State of the Environment 2011*, State of the Environment 2011, Committee, Canberra, pp. 287-293.

Song, Y., Hayes, R. C., Sandral, G. A., McVittie, B. J., Price, A., Poile, G. J., Zheng, W. and Li, G. D. (2017). Relative tolerance to aluminium and manganese toxicities of Phalaris, cocksfoot, and tall fescue genotypes developed for low rainfall environments, *Journal of Plant Nutrition* 40(11), 1526-1537.

Tang, C., Rengel, Z., Diatloff, E. and Gazey, C. (2003). Responses of wheat and barley to liming on a sandy soil with subsoil acidity, *Field Crops Research* 80(3), 235-244.

Tang, C. and Yu, Q. (1999). Impact of chemical composition of legume residues and initial soil pH on pH change of a soil after residue incorporation, *Plant and Soil* 215(1), 29-38.

Upjohn, B., Fenton, G. and Conyers, M. (2005). Soil acidity and liming, NSW Department of Primary Industries Agfact AC. 19. 3rd edn. Available at: https://www.dpi.nsw.gov.au/__data/assets/pdf_file/0007/167209/soil-acidity-liming.pdf accessed 3 September 2021.

Weng, Z., Li, G., Sale, P. and Tang, C. (2022). Application of calcium nitrate with phosphorus promotes rhizosphere alkalization in acid subsoil, *European Journal of Soil Science* 73(1), e13153. https://doi.org/10.1111/ejss.13153.

Wigley, K., Ridgway, H. J., Humphries, A. W., Ballard, R. A. and Moot, D. J. (2018). Increased lucerne nodulation in acid soils with *Sinorhizobium meliloti* and lucerne tolerant to low pH and high aluminium, *Crop and Pasture Science* 69(10), 1031-1040.

Williams, C. H. (1980). Soil acidification under clover pasture, *Australian Journal of Experimental Agriculture* 20(106), 561-567.

Yin, C., Schlatter, D. C., Kroese, D. R., Paulitz, T. C. and Hagerty, C. H. (2021). Responses of soil fungal communities to lime application in wheat fields in the pacific northwest, *Frontiers in Microbiology* 12, 576763.

Index